Geodiversity

Valuing and Conserving Abiotic Nature

Second Edition

Murray Gray

School of Geography, Queen Mary, University of London, UK

WILEY Blackwell

Registered office: John Wiley & Sons, Ltd, The Atrium, Southern Gate, Chichester, West Sussex, PO19 8SQ, UK

Editorial offices: 9600 Garsington Road, Oxford, OX4 2DQ, UK
The Atrium, Southern Gate, Chichester, West Sussex, PO19 8SQ, UK
111 River Street, Hoboken, NJ 07030-5774, USA

For details of our global editorial offices, for customer services and for information about how to apply for permission to reuse the copyright material in this book please see our website at www.wiley.com/wiley-blackwell.

Library of Congress Cataloging-in-Publication Data

Gray, J. M.
Geodiversity : valuing and conserving abiotic nature / Murray Gray.—2e.
pages cm
Includes bibliographical references and index.
ISBN 978-0-470-74214-3 (cloth)—ISBN 978-0-470-74215-0 (pbk.) 1. Geodiversity. 2. Biodiversity. 3. Conservation of natural resources. I. Title.
QE38.G73 2014
551–dc23

2013014153

A catalogue record for this book is available from the British Library.

Wiley also publishes its books in a variety of electronic formats. Some content that appears in print may not be available in electronic books.

Cover image: Diversity of colour in the sedimentary strata at Paria, Utah, USA.
Cover design by Jeffrey Goh

Set in 10/12pt TimesTen by Laserwords Private Limited, Chennai, India.
Printed and bound in Singapore by Markono Print Media Pte Ltd

1 2013

Contents

Preface to Second Edition

It is almost 10 years since I completed the first edition of this book. This second edition has been extensively revised given the progress that has been made in the subjects of geodiversity and geoconservation over this time. For example, I have created separate chapters for 'World Heritage Sites' and 'Global Geoparks', given the increased interest and success of these site networks. Several of the other chapters have been split to make the subject matter more digestible. Chapter 4 on *Valuing Geodiversity* has been restructured to reflect the 'ecosystem services' approach. Chapter 6 has a new section on the need for conservation of the Global Stratotype Section & Points (GSSP) network. Most of the other chapters have been updated, and I have given greater attention to topical issues such as climate change and marine conservation. And overall, I have tried to make a clearer case for geodiversity being a fundamental basis for geoconservation, geotourism, geoparks, etc.

Since writing the first edition, I have been invited to give presentations in many countries including the Netherlands (twice), Norway, USA (twice), Canada, Portugal, Romania, Poland, Austria, Malaysia, Hong Kong and throughout the UK. I have also attended further conferences in Norway and Portugal as well as in Italy, Croatia, France and Greece. This has given me the opportunity to broaden my experience of geoconservation in these and other countries and this is reflected in new text or illustrations in almost every chapter. In particular, Chapter 13 is a new summary that draws upon this experience of geodiversity and geoconservation methods. I am very grateful for these invitations and opportunities and for the kind comments that colleagues have made about the first edition. I have developed my own ideas on geodiversity significantly over the last few years and incorporated these in this new edition. Inevitably these additions have had to be accompanied by some pruning of material, particularly in reducing the section on geoconservation of geomaterials.

I am very grateful to the many scientists with whom I have had discussions on issues in the book over many years. I am also grateful to Fiona Seymour and Lucy Sayer at Wiley Blackwell for their patience and encouragement when the preparation of this edition overran significantly. Jasmine Chang (Production Editor, Wiley, Singapore) and Sangeetha Parthasarathy (Laserwords, Pvt Ltd, India) skillfully managed the production of the book, assisted by Mitch Fitton (copyeditor) and Gill Whitley (permissions). Ed Oliver, Cartographer in the School of Geography at Queen Mary, University of London kindly drew all the diagrams, and I am grateful to the many publishers and individuals who gave permission to reproduce figures and photographs in this book. As usual I thank Pauline for putting up with other things not being done while I completed this edition. Given my advancing years, I can assure readers that there will not be a third edition. So this is about my final contribution to our subject. I hope you find it interesting and useful and that younger generations can continue to advance a subject that deserves to be much better known and understood.

Murray Gray
Norfolk, January 2013

Preface to First Edition

I began writing this book on 11 September 2001 and completed it as the Iraq War drew to an end in April 2003. The nineteen months spent writing the book were dominated by the 'war on terror' and during that time much was written about globalisation, religious conflicts and cultural diversity. While Fukuyama (2001) sees these events as only blips in the trend towards modernity, Sacks (2002) believes that we should not just be tolerant of difference but should celebrate it. 'Only when we realise the danger of wishing that everyone should be the same—the same faith on the one hand, the same McWorld on the other—will we prevent the clash of civilizations. . . .' (Sacks, 2002, p. 209).

This book is about the value of difference, diversity and distinctiveness in the natural world. A few weeks after 9/11, when Osama bin Laden appeared on television to denounce the aerial bombing of Afghanistan, I predicted that specialists in that country's geology would be able to identify his whereabouts from the bedrock lithology and colour shown in the background. And sure enough, it was not long (*The Times*, 19 October 2001) before newspapers were reporting that Dr John Ford Shroder of the University of Nebraska had identified the rocks as being from the Pliocene Shaigalu or Eocene Siahan Shale Formations of the Katawaz Basin in south-east Afghanistan. This allowed search activities to be focused in these areas. This is a very unusual, but perfectly clear example of one of the applications of the principles of geological diversity. If Afghanistan had been composed of a single rock type, bin Laden's general whereabouts would have been much more difficult to locate.

On the first anniversary of 9/11, I was attending a conference on the geological foundations of landscape in Dublin Castle, Ireland. And as I stood for the minute's silence I was reminded too of Ireland's tragic past and present, but also of the way in which geology and landscape transcend administrative boundaries and can bring people together to value their heritage, conserve its integrity and overcome political barriers and national bureaucracies. In this case, the Royal Irish Academy and Geological Surveys of both the Republic of Ireland and Northern Ireland worked together to organise the conference and field trips.

The mountains and tectonics of Afghanistan, the deserts and rivers of Iraq and the bogs and coastline of Ireland, illustrate landscape geodiversity very clearly, but there is also diversity in their economic geology resources. The oil wealth of Iraq is well known and is attributed by some as the underlining reason for the war.

But Afghanistan also has a huge geological wealth, as yet largely unexploited. This includes an estimated 300 million barrels of oil, about 100 million tonnes of coal and large reserves of copper, gold, iron, chromite and industrial minerals (Stephenson and Penn, 2003). We also know about the Afghan gemstones—the lapis lazuli, ruby and spinel of Badkhshan, the emeralds in the Panjshir valley, the rubies and sapphires of Jegdalek and Gandamak and the pegmatites of Nuriston (Bowersox and Chamberlin, 1995). I hope that the people of Afghanistan will be able to realise the future projected for them by Bowersox and Chamberlin (1995, p. xv) when they said that 'At present, the Afghan gemstone wealth is undetermined, mostly undiscovered, and certainly unexploited. We believe the potential of the country is so great as to promise an acceptable standard of living for every man, woman and child. ....'

This book is aimed at several types of reader. First, I hope it will be of interest to those closely involved in geoconservation whether in universities, nature conservation agencies, geological surveys or other organisations around the world. The book is intended to highlight best practice and I hope there are ideas in it that specialists can use. I apologise if I have omitted brilliant initiatives in geoconservation in one country or another and hope that I will be told about these (j.m.gray@qmul.ac.uk) for a possible second edition. Secondly, the book is intended to stimulate discussion and thought on geoconservation by those whose primary concern has been wildlife conservation. It is therefore at least partly written with biologists and other non-geologists in mind. Thirdly, I hope it will stimulate further university courses in geodiversity and geoconservation and will become an established university textbook for second and third year undergraduates or postgraduate courses.

The book would not have been written without my three decades of experience working in the geosciences and public life, during which time I have benefited from the ideas of countless people. I was born and educated in Edinburgh and must thank my late parents for the educational opportunities they gave me. At the University of Edinburgh I studied both geography and geology and gained greatly from that education. There can be few better places to study the geosciences and conservation than Edinburgh and Scotland given their diverse geology and geomorphology, their association with founders of the subject like James Hutton, John Playfair and Charles Lyell, and the innovative conservation efforts that have been nurtured there from the famous nineteenth century American conservationist, John Muir, who was born in Dunbar, only 30 km east of Edinburgh, to the modern-day geoconservation work of Scottish Natural Heritage.

In public life, I have been a member of South Norfolk District Council and Chairman of its Planning Committee for many years and have served on regional planning panels and Environment Agency committees for the East of England. I am grateful to these organisations for the invaluable opportunities they have given me to understand how geoconservation can be promoted through local and regional government.

The book was largely written during sabbatical leave from my teaching duties at Queen Mary, University of London in 2001–2002 and I wish to thank the College authorities and my Head of Department, Professor Roger Lee, for their generous support and encouragement for this project. My geography colleagues not only covered my teaching, but also made many valuable suggestions for improving the book. Ed Oliver, cartographer in the Geography Department at Queen Mary, skilfully drew all the diagrams for this book to an unfairly short timescale.

During the sabbatical year I benefited from discussions with many people including John Gordon (Scottish Natural Heritage), Stewart Campbell (Countryside Council for Wales), Colin Prosser (English Nature), and by e-mail with Kevin Kiernan, Mike Pemberton and Chris Sharples in Tasmania, Archie Landals and David Welch in Canada and Vince Santucci in the United States. I spent May 2002 in the Canadian and American Rockies learning about geoconservation in those two countries and benefited from discussions with Dave Dalman (Banff National Park), Archie Landals (Alberta government), Dan Fagre (USGS, Glacier National Park), Arvid Aase (Fossil Butte National Monument), Hank Heasler and Lee Whittlesey (Yellowstone National Park), Bill Dolan (Waterton Lakes National Park) and numerous others. Other people, too numerous to mention have helped by providing detailed information or supplying diagrams or photographs and I am grateful to all of them.

The following people kindly read and commented on drafts of all or part of the book—Matthew Bennett, Cynthia Burek, Lars Erikstad, John Gordon, Archie Landals, Andrew McMillan, Mike Pemberton, Colin Prosser, Vince Santucci and Chris Sharples—and I am extremely grateful to all of them for their encouragement and valuable suggestions. This is a better book as a result of their comments. Cynthia Burek kindly allowed me to participate in the workshops aimed at creating a Local Geodiversity Action Plan (LGAP) in Cheshire.

I am grateful to many copyright holders for permission to reproduce figures. The sources are acknowledged in the figure captions, and where I received no reply I have assumed that there is no objection to inclusion of the material in this book.

I must thank John Wiley & Sons for their faith in this project and particularly to Sally Wilkinson and Keily Larkins for their efficient handling of the book material. Finally, my thanks to Pauline for putting up with my obsession with the geosciences in general and this book in particular over a long period of time.

Murray Gray
Norfolk, April 2003

Part I

What is Geodiversity?

1
Defining Geodiversity

If the Lord Almighty had consulted me before embarking upon creation, I should have recommended something simpler.

Alfonso X, King of Castile and Leon (1252–1284), quoted in Mackay (1991)

1.1 A diverse world

Let us begin by imagining the very simplest of planets (see Figure 1.1). A planet composed of a single monomineralic rock such as a pure quartzite. A planet that is a perfect sphere with no topography and where there is no such thing as plate tectonics. Although it has weather, this is very similar everywhere with solid cloud cover, light rain and no winds, so that there is little variation in surface processes or weathering. Consequently the soil is also very uniform. The absence of gradients and surface processes means that there is little erosion, transportation or deposition of sediments. This planet has seen few changes in its 4.6 thousand million year history and there is, in any case, no sedimentary record of these changes. To say the least, this is not a diverse or dynamic planet.

It has to be admitted that our imagined planet has certain attractions. In fact, Alfonso X was a forerunner of many Medieval and Renaissance writers who deplored the rough and disorderly shape of the Earth and 'infestation by mountains which prevented it from being the perfect sphere that God must surely have intended to create' (Midgley, 2001, p.7). Furthermore, there are no natural hazards such as earthquakes or avalanches to cause death and destruction. Civil engineering is very simple given the predictability of the ground conditions. Walking is easy with no gradients to negotiate or rivers to cross. But think of the disadvantages. In a planet made entirely of quartz there are no metals and therefore no metallic products. And in any case, since there is no coal, oil or natural gas, and no geothermal, wave, tidal or wind power, the energy to produce any goods or electricity is lacking. Everywhere looks the same so getting lost is easy and there is no sense of place. Employment and entertainment are limited, given the absence of materials and lack of environmental diversity. The quartzite

Geodiversity: Valuing and Conserving Abiotic Nature, Second Edition. Murray Gray.

Figure 1.1 'Knowledge', the sculpture by Wendy Taylor in Library Square at Queen Mary, University of London, United Kingdom. The steel ball at the centre represents the Earth. Fortunately the real Earth is not like this (see text).

is too hard and massive to quarry in the absence of mechanical equipment or explosives, so the buildings are primitive, being constructed from soil and the simple vegetation types that exist on our planet. For in the absence of physical diversity and habitat variation, little biological evolution of advanced plants and animals has been able to take place. This means that we humans would probably not exist on this planet, but if we did we would certainly find this to be a very primitive and boring place.

Thankfully our world is not like this. It is highly diverse in almost all senses—physical, biological and cultural—and although this produces problems for society and even conflicts and war, would we really want a less diverse and interesting home? The diversity of the physical world is huge and humans have put this diversity to good use even if we often fail to fully appreciate this fact. Diversity also brings with it flexibility of technologies and a greater ability to adapt to change.

Although our medieval ancestors hated the physical chaos of the Earth, our modern aesthetic appreciation of planetary diversity is probably deeply buried in our evolutionary psyche so that we often value it more than uniformity. The broad diversity of places, materials, living things, experiences and peoples not only makes the world a more useful and interesting place, but probably also stimulates creativity and progress in a wide range of ways. Diversity therefore brings a range of values, and it is the thesis of this book that things of value ought to be conserved if they are threatened. And, as we shall see, there are many threats to planetary diversity induced by human actions both directly and indirectly.

The term 'conservation' is used in preference to 'preservation' in this book since the latter implies protection of the *status quo*, whereas nature conservation must allow natural processes to operate and natural change to occur. Unfortunately, human action has often accelerated or tried to stop natural processes, and has thus destroyed much that is valuable in the natural environment. While change through human action is inevitable, we should at least understand the consequences of our actions and hopefully minimise the impacts and losses. Conservation is therefore about the management of change.

In the Preface to the first edition of this book I referred to a growing respect for diversity and a realisation that there is value in difference. Since then, the 'diversity' agenda has taken hold and there has been a blossoming appreciation for the value of local environmental, social and cultural distinctiveness and diversity. This book represents an undervalued aspect of this trend and aims to raise the profile of 'geodiversity'.

1.2 Biodiversity

Nowhere has the trend towards the value of diversity been more evident than in the field of biology. In recent decades the growing concern about species decline and extinction, loss of habitats and landscape change led to a realisation of the multi-functioning nature of the biosphere. For example it acts as a source of fibre, food and medicines, it sustains concentrations of atmospheric gases, it buffers environmental change and it contains millions of species of plants and animals, most of which have unknown value and ecosystem function and deserve respect in their own right. Yet of the 1.5–1.8 million known species, it is estimated that up to a third could be extinct in the next 30 years (Grant, 1995).

Concern for species and habitat loss led to some important international environmental agreements and legislation including the Ramsar Convention on wetland conservation (1971), Convention on International Trade in Endangered Species (CITES) (1973) and the Bonn Convention on Conservation of Migratory Species (1979). More recently the European Union has played an active role in biological conservation, for example through the Habitats Directive and Birds Directive.

An International Convention on Biological Diversity was first proposed in 1974 and during the 1980s the phrase 'biological diversity' started to be shortened to biodiversity. An important meeting of the US National Forum on Biodiversity took place in Washington DC in 1986 under the auspices of the American National Academy of Sciences and Smithsonian Institution. The conference papers (Wilson, 1988) mark an important milestone in the history of nature conservation and caused the issue to be taken seriously by politicians both inside and beyond America.

International recognition of the need for biosphere conservation led to the UN Convention on Biodiversity agreed at the Rio Earth summit in 1992, ratified in 1994 and signed by over 160 countries. The agreement was far reaching and the main features are listed in Table 1.1. Since then great attention has been given at international, national, regional and local levels to protecting and enhancing the biological diversity of the planet. These are usually classified into genetic diversity (conserving the gene pool), species diversity (reducing species loss) and ecosystem diversity (maintaining and enhancing habitats and their biological systems). And biodiversity is not just about numbers of species or ecosystems but about the countless interconnections between them. A wealth of strategies and action plans are being implemented to carry forward the aims of the UN Convention. Every signatory country must prepare a national plan for conserving and sustaining biodiversity, has a responsibility for safeguarding key ecosystems and is responsible for monitoring genetic stock. International designations include Ramsar sites (under the Ramsar Convention) Special Protection Areas (under the European Union Birds Directive) and Special Areas for Conservation (under the European Union Habitats Directive). The International Union for Nature Conservation (IUCN) has helped over 75 countries to prepare and implement national conservation and biodiversity strategies. Jerie, Houshold and Peters (2001, p. 329) have referred to this as 'the torrent of effort being put into the management of biodiversity'. Even some ecologists have spoken of the obsession with loss of species and habitats rather than focusing on the more important issue of functional significance of species in a variety of ecosystems (Dolman, 2000).

In the UK, *Biodiversity: the UK Action Plan* (HMSO, 1994), *Working with the Grain of Nature*, a biodiversity strategy for England (DEFRA, 2002a), *Conserving Biodiversity—the UK approach* (DEFRA, 2007) and *Biodiversity 2020* (DEFRA, 2011a) have been supplemented by many regional and local

Table 1.1 Main features of the Convention on Biological Diversity (after Mather and Chapman, 1995).

- Development of national plans, strategies or programmes for the conservation and sustainable use of biodiversity.
- Inventory and monitoring of biodiversity and of the processes that impact on it.
- Development and strengthening of the current mechanism for conservation of biodiversity both within and outside protected areas, and the development of new mechanisms.
- Restoration of degraded ecosystems and endangered species.
- Preservation and maintenance of indigenous and local systems of biological resource management and equitable sharing of benefits with local communities.
- Assessment of impacts on biodiversity of proposed projects, programmes and policies.
- Recognition of the sovereign right of states over their natural resources.
- Sharing in a fair and equitable way the results of research and development and the benefits arising from commercial and other utilisation of genetic resources.
- Regulation of the release of genetically modified organisms.

initiatives including Local Biodiversity Action Plans (LBAPs), Species Recovery Programmes (SRPs) and Habitat Action Plans (HAPs). These are being implemented by the national conservation bodies (Natural England, Scottish Natural Heritage, Natural Resources Wales, Northern Ireland Environment Agency) in collaboration with local authorities and a wide range of wildlife and conservation organisations (e.g. County Wildlife Trusts, Royal Society for the Protection of Birds, Campaign to Protect Rural England). By 2009 there were 1150 priority species and 65 priority habitats in the UK. In addition, Section 40 of the Natural Environment and Rural Communities Act (2006) requires all public bodies in England and Wales to have regard to biodiversity when carrying out their functions, and is now referred to as the 'biodiversity duty'.

About 100 books have appeared with 'biodiversity' in the title (from Wilson, 1988 to Waterton, Ellis and Wynne, 2012). Wilson (1997, p. 1) refers to 'biodiversity' as 'one of the most commonly used expressions in the biological sciences and has become a household word'. 'Biodiversity science' and 'biodiversity studies' have been born and the origin and maintenance of biodiversity 'pose some of the most fundamental problems of the biological sciences' (Wilson, 1997, p. 2). The Rio+20 conference in 2012 confirmed the importance of biodiversity and the threats to it.

1.3 Geodiversity

Geological and geomorphological conservation (geoconservation) have a long history. In 1668 the Baumannshöle cave in Germany was the subject of a nature conservation decree by Duke Rudolf August (Erikstad, 2008). In the first 20 years of the nineteenth century the quarrying of stone from Salisbury Crags in Edinburgh, Scotland, was having such a serious impact on the city landscape that legal action was taken in 1819 to prevent further deterioration (McMillan, Gillanders and Fairhurst, 1999; Thomas and Warren, 2008). An erratic boulder in Neuchâtel, Switzerland was protected in 1838 (Reynard, 2012). The first geological nature reserve in the world was established at Drachenfels/Siebengebirge in Germany in 1836 and other German hills were protected at Totenstein (1844) and Teufelsmauer (1852). Yosemite was protected by the State of California, United States, in 1864 and Yellowstone was established as the world's first National Park in 1872 largely for its scenic beauty and geological wonders (see Box 6.1). Also in the 1870s, Fritz Muhlberg campaigned to protect giant erratic boulders in Switzerland that were being exploited as kerbstones (Jackli, 1979), and in Scotland the 'Boulder Committee' was established, under the direction of David Milne Home, to identify all remarkable erratics and to recommend measures for their conservation (Milne Home, 1872a, 1872b; Gordon, 1994). Some of the first specific geological sites to be protected were also in Scotland where City Councils acted to enclose Agassiz Rock striations in Edinburgh (1880) and the Fossil Grove Carboniferous lycopod stumps in Glasgow (1887). Other initiatives have followed and many countries now have areas and sites protected at least partially for their geological or landscape interest. But despite many international conferences and books in the past 20 years (e.g. G. Martini, 1994; O'Halloran *et al.*, 1994; Stevens *et al.*, 1994; Wilson, 1994; Barrentino, Vallejo and Gallego, 1999; Gordon and Leys, 2001a; Gray, 2004; Burek and Prosser, 2008; Brocx, 2008; Wimbledon and Smith-Meyer, 2012), in most countries geoconservation is weakly developed and lags severely behind biological conservation.

Geologists and geomorphologists started to use the term ‘geodiversity’ in the 1990s to describe the variety within abiotic nature. The major attention being given to biodiversity and wildlife conservation was simply reinforcing the longstanding imbalance within nature conservation policy and practice between the biotic and abiotic elements of nature. Although geological and geomorphological conservation had been practised for over 100 years, these were usually the ‘Cinderella’ of nature conservation (Gray, 1997a). Many international nature conservation organisations, although using the general term ‘nature conservation’ appeared to see this as synonymous with ‘wildlife conservation’ and focused most or all of their attention on the latter. Milton (2002, p. 115) summarised the situation well in stating that ‘Diversity in nature is usually taken to mean diversity of living nature . . . ’. Pemberton (2001a) believed that nature conservation agencies and governments across Australia, and overseas, ‘tend to emphasize the need for the conservation of biodiversity whilst virtually ignoring the geological foundation on which this is built and has evolved’. He attributed this to the lack of training of earth scientists in geoconservation theory, policy and practice. He made the interesting observation that ‘The majority of earth scientists are trained and employed in the extractive industries. To be involved in conservation could be seen to be contrary to the goals of the profession. . . . ’ and he compares this with the biological sciences where conservation is a major graduate employer. ‘This has generally meant that geoconservation has remained something of an oddity, divorced from mainstream nature conservation, and so it has generally had low priority within land management agencies’ (Pemberton, 2001a). Although geoconservation has not yet been accorded great prominence in Australian nature conservation, Kiernan (1996, p.6) believed that ‘few professional land managers would, having been made aware that a particular landform was, say, the only example in Australia of its type, seriously argue against the validity of safeguarding it, just as they would wish to safeguard the continued existence of a biological species.’

A similar situation has existed in the United Kingdom, where although geoconservation policy and practice have long been actively pursued and developed by several groups and organisations, this has not always been recognised by the wider nature conservation community or public. Therefore, some geologists and geomorphologists saw ‘geodiversity’ not only as a very useful new way of thinking about the abiotic world, but also as a means of promoting geoconservation and putting it on a par with wildlife conservation (Prosser, 2002a). An example of using ‘nature’ and ‘wildlife’ synonymously came in Sir John Lawton’s report (2010) *Making Space for Nature* which was subtitled *A review of England’s wildlife sites and ecological network*. Sadly and amazingly, it was followed by a government White Paper on the natural environment *The Natural Choice: Securing the Value of Nature* (DEFRA, 2011b), that manages, in its 76 pages, to discuss nature in England and Wales without once referring to geology, geomorphology or geodiversity!

In the United States, most nature conservation effort has been directed through the national parks system, and although several units of the system have been established for their geological or geomorphological interest (see Section 9.2.1), this is not always recognised. For example, Sellars' (1997) eloquent history of nature conservation in the US National Parks is almost entirely dominated by wildlife issues, reflecting the major concerns of the parks system over the past 140 years.

It is difficult to trace the first usage of the term 'geodiversity', and indeed it is likely that several earth scientists coined the term independently, as a natural twin to the term 'biodiversity'. Once the Convention on Biodiversity popularised the word and concept of 'biodiversity' in 1992, it became difficult to avoid noting that there is an abiotic equivalent. The early history of geodiversity was reviewed by Gray (2008a) who noted that the first use of the principle of geological diversity and its relevance to geoconservation pre-dates the Convention on Biodiversity. Kevin Kiernan, working for the Tasmanian Forestry Commission in the 1980s was using the terms 'landform diversity' and 'geomorphic diversity' and drawing parallels with biological concepts by using terms such as 'landform species' and 'landform communities' (K. Kiernan, pers. comm.). In one seminar paper in 1991, he pointed out that 'The diversity among landforms is just as valid a target as the diversity of life when developing nature conservation programs. . . . ' (Kiernan, 1991), a remark that was certainly prophetic. The term 'geodiversity' appears to have been first used by F.W. Wiedenbein in a German publication in April 1993 (see Wiedenbein, 1993, 1994), closely followed by Sharples in Tasmania in October of the same year (Sharples, 1993). Subsequently it became widely adopted in studies of geological and geomorphological conservation in Tasmania in particular (Kiernan, 1994, 1996, 1997a; Dixon, 1995, 1996a, 1996b). Sharples (1993) used it to cover 'the diversity of earth features and systems. He also stresses (Sharples, 2002a) the importance of distinguishing the terms 'geodiversity', 'geoconservation' and 'geoheritage'. He defines them as follows:

- 'geodiversity' is the *quality* we are trying to conserve,
- 'geoconservation' is the *endeavour* of trying to conserve it, and
- 'geoheritage' comprises *concrete examples* of it that may be specifically identified as having conservation significance.

The most important landmark for geoconservation in Australia was adoption of the Australian Natural Heritage Charter in 1996 and subsequently updated in 2002 (Australian Heritage Commission, 1996, 2002), which has the term and substance of geodiversity interwoven throughout its Articles (see Section 12.6). As a result, geodiversity is now a widely used and understood term in Australian nature conservation. However, Joyce (1997, p. 39) is critical of the term 'geodiversity' since it '. . . may be attempting to draw too strong a parallel between sites, landscape features and processes in biology and geology'. This issue is discussed in Section 14.1.

In 1995, Ibáñez, De-Alba and Boixadera (1995a), Ibáñez *et al.* (1995b) and McBratney (1995) started to use the word 'pedodiversity' to describe the diversity of soils as an abiotic component of global geodiversity (Ibáñez *et al.*, 1998).

In the United Kingdom, Gray (1997a, p. 323) suggested that 'perhaps one day we will see … a Geodiversity Action Plan for the UK to rank alongside its biological counterpart' and this was launched in September 2011, partly to give a national frameworks for the many Local Geodiversity Action Plans (LGAPs) now published in the country (see Section 12.5). Stanley (2004, p. 48) argued that 'biodiversity is merely part of Geodiversity' and proposed a wide definition (see Table 1.2). Prosser (2002a, 2002b), in useful discussions of terminology, accepted the validity of the term 'geodiversity' and it has now become widely used throughout the United Kingdom.

The Nordic Council of Ministers introduced the term 'geodiversity' into nature conservation in their countries in 1996. Johannson (2000) published an excellent

Table 1.2 Some definitions of geodiversity.

Authors	Definition	Comments
Dixon (1996), Eberhard (1997), Sharples (2002a), Australian Heritage Commission (2002)	'the range or diversity of geological (bedrock), geomorphological (landform) and soil features, assemblages, systems and processes'	A thematic definition
Semeniuk (1997), Brocx (2008)	'the natural variety of geological, geomorphological, pedological, hydrological features of a given area, from the purely static features at one extreme, to the assemblage of products, and at the other, their formative processes'	An area-based definition. Note also the inclusion of hydrological features
Johannson *et al.* (2000)	'the complex variation of bedrock, unconsolidated deposits, landforms and processes that form landscapes… Geodiversity can be described as the diversity of geological and geomorphological phenomena in a defined area'	Note the emphasis on landscapes and defined area.
Nieto (2001)	'the number and variety of structures (sedimentary, tectonics, etc.), geological materials (Minerals, rocks, fossils and soils) that make up the terrain of a region, in which organic activity takes place, including anthropic'	Another area-based definition. Note also the inclusion of the relevance to biodiversity and human activity.
Stanley (2001)	'the link between people, landscapes and culture; it is the variety of geological environments, phenomena and processes that make those landscapes, rocks, minerals, fossils and soils which provide the framework for life on Earth'	A thematic definition strongly linked to human activity and biodiversity. Note also reference to landscapes.

Table 1.2 (*continued*)

Authors	Definition	Comments
Kozlowski (2004)	'the natural variety of the Earth's surface, referring to geological and geomorphological aspects, soils and surface waters, as well as to other systems created as a result of both natural (endogenic and exogenic) processes and human activity'	Note the inclusion of surface hydrology and the link with human activity.
Gray (2004)	'the natural range (diversity) of geological (rocks, minerals, fossils), geomorphological (land form, physical processes) and soil features. It includes their assemblages, relationships, properties, interpretations and systems'	A thematic definition. Note reference to interpretations.
Spanish 'Natural Heritage and Biodiversity' National Law (2007)	'the variety of geological features, including rocks, minerals, fossils, soils, landforms, landscapes, geological formations and units, that are the product and record of the Earth's evolution'	A thematic definition. Note reference to landscapes and Earth history.
Serrano and Ruiz-Flano (2007)	'the variability of abiotic nature, including lithological, tectonic, geomorphological, soil, hydrological, topographical elements and physical processes on the land surface and in the seas and oceans, together with systems generated by natural, endogenous and exogenous, and human processes which cover the diversity of particles, elements and places'	A thematic definition, but note the inclusion of hydrology, seas and oceans and scale factors.
Burek and Prosser (2008)	'the variety of rocks, minerals, fossils, landforms, sediments and soils, together with the natural processes which form and alter them'	A thematic definition.

book on the geodiversity of the Nordic countries, and an English summary was produced in 2003 (Nordic Council of Ministers, 2003). Erikstad and Stabbetorp (2001) used the term in relation to natural areas and environmental impact assessment, and it appears now to be in common usage in the Scandinavian countries. A few authors in Europe (e.g. Panizza, 2009) have begun using the term 'geomorphodiversity' (geomorphological diversity) but this seems an unnecessary addition to the already overly complex geoconservation terminology.

Apart from the publication of a special issue of the *George Wright Society Forum* in 2005 (Santucci, 2005) and an article in *Geoscience Canada* in 2008

(Gray, 2008b), there is little evidence that the term has been adopted in North America.

Table 1.2 shows some of the main definitions of geodiversity that have been developed over the past 15 years, including one by the author that has been widely quoted. The table has a commentary on these definitions that suggests that a modified definition would be appropriate to incorporate hydrological features and landscapes in particular. Consequently, I propose the following new definition:

> Geodiversity: the natural range (diversity) of geological (rocks, minerals, fossils), geomorphological (landforms, topography, physical processes), soil and hydrological features. It includes their assemblages, structures, systems and contributions to landscapes.

1.4 Aims and structure of the book

It is the thesis of this book that 'geodiversity' is a valid and appropriate way in which to look at planet Earth. The specific aims of this book are:

- to raise awareness of the geodiversity of the planet and the value of this diversity;
- to point out the threats to this diversity;
- to examine the ways that this diversity can be conserved, managed and restored;
- to outline the need for a more holistic and integrated approach to nature conservation and land management.

I hope the book will stimulate interest in these topics not just amongst geologists, geomorphologists and soil scientists, many of whom are at least aware of the issues, but also amongst the wider academic and non-academic community—biologists, nature conservationists, landscape architects, planners and politicians—for the world in general has not paid sufficient attention to these issues.

The book is divided into five parts comprising 15 chapters. These follow a natural sequence. The current Chapter 1 begins Part I, which tries to explain what geodiversity is. Chapters 2 and 3 attempt the extremely difficult task of describing the geodiversity of the planet. The aim of these chapters is not to catalogue every variation in rocks, minerals, sediments, processes, landforms, soils, fossils, and so on, but rather to outline the main principles and causes of diversity in the abiotic world. Chapter 2 looks at the early history of the Earth and evolution of global geodiversity. The focus of the book is on the terrestrial systems and the solid Earth (which I shall refer to as the 'geosphere'), rather than conservation or environmental management of all planetary abiotic systems. Consequently I shall say little about conservation of atmosphere and oceans. Chapter 3 examines the local scale. As stated above, biodiversity is often classified into genetic, species and ecosystem diversity, a classification that is based primarily on scale. Such a system is not as appropriate for geodiversity although we can certainly recognise scale differences in the abiotic world. For example, landforms combine to form landscapes which combine to make continents or tectonic plates. Nonetheless, the obvious way to classify abiotic nature is by using a well-tried tripartite subdivision of *material, form* and *process*. The main diversity in earth materials is in lithospheric materials—rocks, minerals, sediments and soils. Form comprises landforms and physical landscapes while numerous processes act on the materials to produce landforms. Anyone who already has a good understanding of the planet's geodiversity can probably skip this chapter.

Part II summarises the values of, and threats to, geodiversity. Chapter 4 discusses the value of the planet's geodiversity in terms of what are called 'abiotic ecosystem services' or 'geosystem services'. A summary diagram at the end of the chapter recognises over 25 types of abiotic value. Chapter 5 outlines the main threats to this valuable geodiversity including mineral extraction, river engineering, fossil collecting, urban expansion, coastal defence, waste disposal, agricultural practices and afforestation. The impact of these activities will be greater on some physical materials and systems than on others, and the issue of sensitivity and vulnerability to human modification emerges from this chapter as very important factors.

The subsequent chapters then follow logically from the previous two using the simple conservation equation:

$$\text{Value} + \text{Threat} = \text{Conservation need.}$$

Part III comprises four chapters describing the 'site-based approach' to geoconservation. Chapter 6 covers the general aspects of international geoconservation, including the important network of Global Stratotype Section and Point (GSSP) sites, while Chapters 7 and 8 describe World Heritage Sites and Global Geoparks respectively. These chapters along with Chapter 9 on National Geoconservation networks and strategies follow traditional methods of conservation, namely by designating protected areas and using legislation and other approaches to conserve and manage them.

Part IV looks at a range of other approaches and new initiatives to extend geoconservation beyond protected areas to the sustainable management of the 'wider landscape', Chapter 10 focusing on the landscape conservation, Chapter 11 on land-use planning and Chapter 12 on policy making.

Finally, in Part V I have attempted to bring ideas together. Chapter 13 tries to draw together some of the main elements of the previous seven Chapters to describe some of the aims and methods of geoconservation and the role of geodiversity as an important basis for geoconservation, geotourism, geoparks, etc. Chapter 14 revisits some important issues for geodiversity in relation to biodiversity and discusses a basic paradox for geodiversity. On the one hand the subject needs to establish itself as a distinctive, independent and essential field of nature conservation, but on the other there is a growing need and trend towards an integrated approach to nature conservation incorporating geological and biological systems and indeed to integrated land management in general. Finally Chapter 15 draws some conclusions and tries to establish a revised vision for the year 2025.

This structure of valuing geodiversity, understanding the threats and conserving and managing the resource and indeed the whole impetus for writing the book follows a logical pattern reflected in the following quote from African conservationist, Baba Dioum (in Rodes and Odell, 1997):

> For in the end we will conserve only what we love.
> We will love only what we understand.
> And we will understand only what we are taught.

I hope this book stimulates a much greater interest in the values of geodiversity and the need to conserve them.

2

Geodiversity: the Global Scale

The wondrous world that we are so familiar with has been shaped by fundamental forces operating on scales that we humans can barely grasp,

Iain Stewart and John Lynch (2007)

The geodiversity of the Earth is hardly less remarkable than its biodiversity. Any attempt to fully describe this geodiversity would fill several books, and so the challenge in the next two chapters is to summarise it in a few thousand words concentrating on the factors producing the diversity rather than an encyclopaedic catalogue of the whole range of minerals, rocks, sediments, fossils, soils, landforms, processes, etc. on the planet. There are several books that describe the geology, geomorphology and pedology of the Earth in more detail, many of them in lavish colour. Readers are referred to those by Summerfield (1991), Grotzinger and Jordan (2010), Holden (2012) and particularly to the superbly illustrated book by Marshak (2012).

Geodiversity applies at various scales from the global scale of continents and oceans to elemental scale of atoms and ions. This is no different to the scale issue in biodiversity where bioscientists have to deal with habitat variations on a global scale, but also with genetic diversity and biotechnology at the microbiological scale. This chapter deals with the global scale—the early history of the Earth and plate tectonics—since they are the keys to understanding much of the planetary geodiversity at many scales.

2.1 Origin the Earth

To understand the current geodiversity of the planet we need to understand how the planet originated and how it has evolved through its 4 600 000 000 year (4.6 Ga) history. Astrophysicists believe that the Earth and other planets of our

Geodiversity: Valuing and Conserving Abiotic Nature, Second Edition. Murray Gray.

solar system developed from a residual disk of gas and dust, called a planetary nebula, left over from the formation of the Sun, perhaps 15 Ga. All 92 naturally occurring elements were already in existence at this time, having been formed by nuclear fusion at very high temperatures from an original mix of hydrogen and helium, the two lightest elements.

There then followed a long series of collisions and combinations in which gaseous atoms coalesced to form molecules, molecules combined to form dust, and dust accreted to become small pieces of rock called planetesimals (Marshak, 2012). Over millions of years further collisions resulted in larger and larger rock lumps that in turn attracted smaller pieces, often as meteorites. Thus 'the Earth was conceived and grew violently, a chaos of impact and fragmentation and annealing. All was instability' (Fortey, 1997, p. 31). Eventually, a series of protoplanets revolving (in the same direction and, with the exception of Pluto, in the same plane) around the Sun evolved into the planetary system we see today, with smaller lumps (asteroids and meteoroids) still hurtling around space and occasionally colliding with the planets. The Earth's Moon is believed to have been formed when a particularly large collision at 4.5 Ga blasted debris into orbit that subsequently coalesced (Canup and Asphaug, 2001). It is still pockmarked by impact craters, whereas the Earth's have been largely eradicated by subsequent geological evolution (see later). The near spherical shape of the planets is produced by gravitational pull aided by geothermal heat as a result of planetesimal collision and radioactive decay of elements.

The Earth's basic composition was established at this early stage with iron (35%), oxygen (30%), silicon (15%) and magnesium (13%) forming most of the Earth's matter (Table 2.1). The solar wind is believed to have blown most of the light elements (hydrogen, helium, etc.) into the outer parts of

Table 2.1 Composition (by weight) of the whole Earth and of the crust. Note that differentiation has created a light crust depleted in iron and rich in oxygen, silicon, aluminium, calcium, potassium and sodium (after Press and Siever, 2000).

Element	Whole Earth	Crust
Iron	35	6
Oxygen	30	46
Silicon	15	28
Magnesium	13	4
Nickel	2.4	< 1
Sulphur	1.9	< 1
Aluminium	1.1	8
Calcium	1.1	2.4
Potassium	< 1	2.3
Sodium	< 1	2.1
Other	< 1	< 1

our solar system where they make up significant parts of Jupiter, Saturn and the other outer planets. But the young Earth would have been 'melted and cauterized by the feverish concantenation of impacts. Elements would have been shuffled and recombined as new minerals in a frantic alchemy of creation' (Fortey, 1997, p. 31).

2.2 Early history of the Earth

Even though at this early stage the Earth had a geodiversity, with meteorites bringing new minerals, which melted on impact and added to the nascent mix, and a surface pockmarked by meteorite craters similar to the Moon, overall the primitive Earth was a fairly homogeneous planet. At this early stage, the Earth's atmosphere also developed from volcanic gases, as the more volatile compounds found their way to the surface in a process known as outgassing. 'You might say that our atmosphere, and the possibility of life itself, was the consequence of a vast, terrestrial flatulence risen from the bowels of the Earth' (Fortey, 1997, p. 35). At this stage the Earth's surface would have been largely molten and the atmosphere would have consisted of an unbreathable cocktail of volcanic gases including nitrogen, ammonia, methane, carbon monoxide, carbon dioxide, sulfur dioxide and water vapour. This atmosphere was held in place by the Earth's gravitational field. The large amount of water vapour subsequently condensed to form clouds and rain, which then filled the oceans.

Box 2.1 Internal structure of the Earth

The Earth's internal layers, comprising crust, mantle and core also formed at an early stage in the life of the planet. Since iron is denser than the other common minerals it quickly migrated in a molten state to the Earth's interior, with the result that the layers increase in density from the surface to the centre (Table 2.1, Figure 2.1). The present-day structure has been deciphered from various types of evidence including research on planetary density and the progress through the Earth of earthquake waves. Three layers are recognised each of which is divided into two.

- *Core* The inner core is believed to be a solid iron alloy with a density of c.13 g/cm^3 and a temperature of over 4300 °C. The outer core is a liquid iron alloy with a slightly lower density and temperature.
- *Mantle* The mantle is also divided into two sections separated by a transition zone. The lower mantle has a density of 4–5.5 g/cm^3 while the density of the upper mantle is closer to 3.5 g/cm^3. Most of the mantle is solid rock, though at a depth of 100–200 km below the ocean floors there is a low-velocity zone where a mixture of molten rock and solid material exists. But even though the mantle is solid, the high temperatures mean that it is able to deform internally at very slow rates (a few centimetres/year) and these movements occur as convection currents (see later).

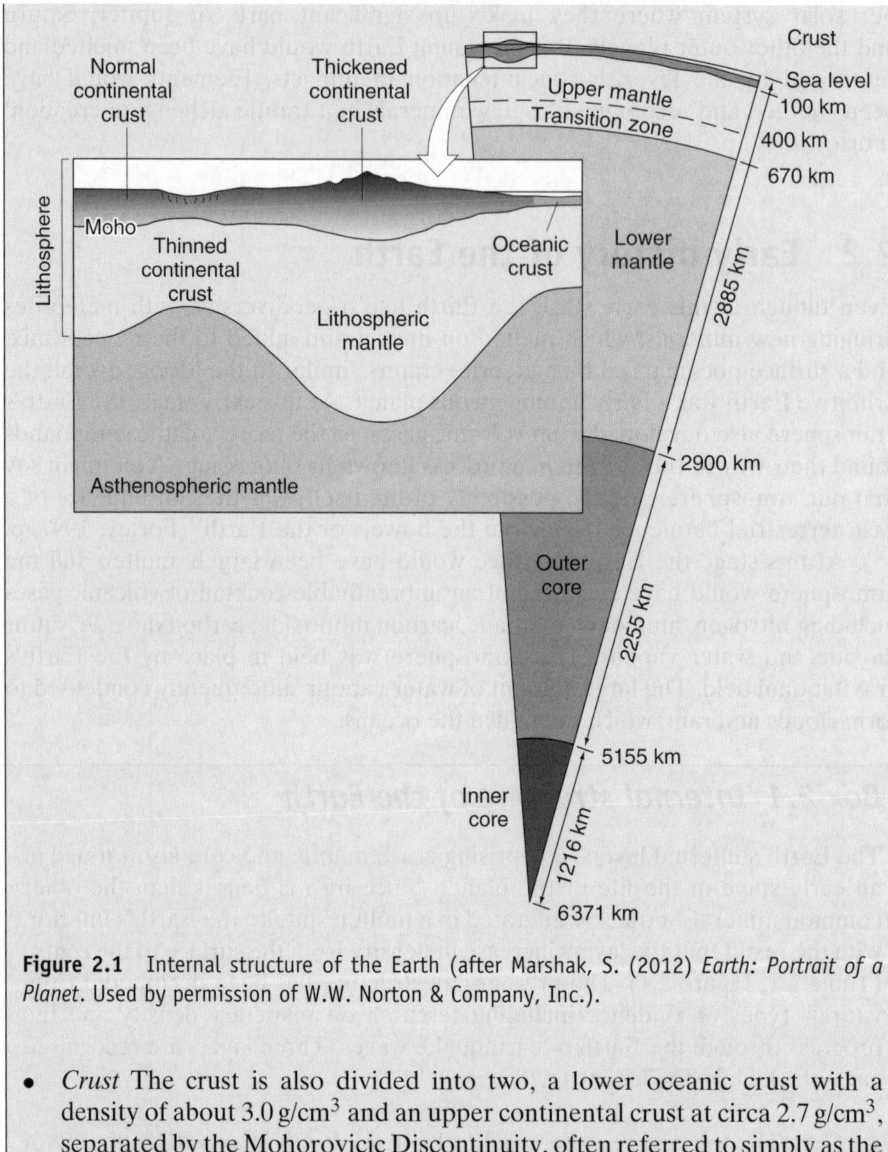

Figure 2.1 Internal structure of the Earth (after Marshak, S. (2012) *Earth: Portrait of a Planet*. Used by permission of W.W. Norton & Company, Inc.).

- *Crust* The crust is also divided into two, a lower oceanic crust with a density of about 3.0 g/cm^3 and an upper continental crust at circa 2.7 g/cm^3, separated by the Mohorovicic Discontinuity, often referred to simply as the Moho. Crustal rocks are dominated by silicon and oxygen, most commonly combined as silica (SiO_2). In turn, silica is combined with other elements (iron, magnesium, aluminium, calcium, potassium and sodium) to form the so-called silicate minerals and silicate rocks that dominate the Earth's crust (see later).

The continental drift hypothesis proposed by the German scientist Alfred Wegener in the early decades of the twentieth century was replaced in the 1960s by the theory of plate tectonics. Much of Wegener's evidence for the matching of continental margins, rocks and fossils was confirmed, but rather than continental blocks drifting across the Earth's surface like supertankers, the plate tectonics theory proposes that the whole crust is broken into pieces like an eggshell, and that these pieces are continuously in motion. In fact, it is now believed that a surface crust began to form on the cooling Earth over four thousand million years ago (4 Ga) and that this crust quickly broke into plates which have been on the move ever since. However, the position of the margins has changed through time. To begin with it is believed that the plates were small, thin and rapidly moving, since at this stage the Earth had cooled only sufficiently for a thin surface crust to form. It also follows that the intense heat below the surface meant that volcanic activity and volcanic gas emission would have been intense, particularly along plate margins. The resultant volcanic arcs were relatively light and therefore were differentiated from the denser crustal materials. Here we begin to see the origin of the continents, a process that was aided by erosion, sedimentation and the uplift of thick sediment sequences and granitic intrusions as in modern mountain building (see later). These blocks of lighter material became the continental crust, segregated from the denser oceanic crust below (see Box 2.1).

Collisions and volcanic suturing created progressively larger plates and larger continents and as the Earth continued to cool these so-called protocontinents became thicker and stronger. By 2.7 Ga the core areas (cratons) of our current continents had formed, comprising rocks dating back to more than 4 Ga discovered in Canada, Greenland, China and Wyoming in the United States. The 4 Ga date is therefore taken as distinguishing the very early phase of Earth history when the earth was too hot for significant solid rock areas to stabilise (Hadean, Figure 2.2), from the Archean period of protocontinent formation. It should be noted however that zircon grains dating back to 4.4–4.2 Ga have been discovered in rocks in Australia and elsewhere, indicating that some minerals were crystallising from the molten rock at this very early date (Wilde *et al.*, 2001).

By 3.8 Ga condensing water vapour and rainfall had caused rivers to run over the lifeless continents and a global ocean had formed. This is deduced from the presence of water-rounded mineral grains from this time indicating that the processes of erosion, transportation and deposition were established early in the age of the continents. It is also likely that the ocean rapidly became salty as surface and groundwater dissolved salts from the rocks and transported the material to the sea. Archean rocks also contain the first record of life, with simple cells, bacteria and algae appearing around 3.8 Ga. It seems likely that the free oxygen produced by these bacteria precipitated the iron out from the seas and rivers, to form vast sedimentary deposits of banded iron deposits, such as those in Western Australia. These 'represent a turning point in the history of life on Earth; they show for the first time the influence of life on the structure of the planet itself.... It is a unique two way interaction' (Manning, 2001, p. 22). Thus by the end of the Archean, 2.5 Ga ago, the Earth had begun to resemble the planet as we know it today with atmosphere, oceans, dry land and the early evolution of life forms.

The Proterozoic (from the Greek meaning 'first life') is clearly misnamed, but this is often the case in geology as new discoveries make older names inappropriate. Nonetheless, there was a significant evolution of multicellular life during the circa 2 Ga of Proterozoic time. In particular, the evolution of photosynthetic organisms changed the atmosphere from an oxygen-poor volcanic

Eonotherm / Eon	Eratherm / Era	System / Period	
Phanerozoic	Cenozoic	Quaternary	2.588
		Neogene	23.03
		Paleogene	66
	Mesozoic	Cretaceous	145
		Jurassic	201.3
		Triassic	252.2
	Paleozoic	Permian	298.9
		Carboniferous	358.9
		Devonian	419.2
		Silurian	443.8
		Ordovician	485.4
		Cambrian	541.0
Proterozoic	Neoproterozoic	Ediacaran	635
		Cryogenian	850
		Tonian	1000
	Mesoproterozoic	Stenian	1200
		Ectasian	1400
		Calymmian	1600
	Paleoproterozoic	Statherian	1800
		Orosirian	2050
		Rhyacian	2300
		Siderian	2500
Archean	Neoarchean	not defined	2800
	Mesoarchian		3200
	Palaeoarchean		3600
	Eoarchean		4000
Hadean (informal unit)			4600

Figure 2.2 The history of the Earth (source: International Commission on Stratigraphy website; http://www.stratigraphy.org/. Reproduced with permission of ICS).

gas mix to something much more like the oxygen-rich air we breathe today, though it was not until the Phanerozoic that today's levels of 21% oxygen were reached. It was the production of this oxygen that allowed the development of energy-producing metabolism and atmospheric ozone to screen ultraviolet radiation. These two developments eventually allowed the great diversification of life to occur as living organisms had the energy and protection to be able to conquer the land.

Also during the Proterozoic eon, the continuing cooling of the Earth brought a slowing of plate tectonic processes, including the accretion of volcanic provinces. By 1.8 Ga most of the large continental cratons that exist today had formed (see Figure 2.3), though parts of the cratons have been buried by later deposition. For example, Figure 2.4 shows that the North American craton is a collage of different plate fragments, volcanic belts and accretionary provinces fused together between 1 and 2 Ga ago. The evidence of collision between Archean blocks is particularly well exposed along the Trans-Hudson belt in central Canada. In the United States most of the Precambrian craton complex is buried below Phanerozoic rocks which form a so-called cratonic or continental platform. The Phanerozoic mountain belts and coastal plain are also the result of subsequent events (see later).

Finally, during the Proterozoic, the Earth also experienced a number of global glaciations. The so-called 'Snowball Earth' theory is still being debated but it is now generally believed that the planet was completely or almost completely encased in ice at these times. It is also possible that the 'Snowball Earth' event that terminated at circa 635 Ma may have driven the subsequent evolution of complex life now traced back to at least 600 Ma with the discovery of the Lantian biota in south China (Yuan *et al.*, 2011).

2.3 Plate tectonics

By the beginning of the Phanerozoic, many of the continents had begun to resemble their present-day outlines, including North America, South America, Africa, Antarctica and Australia. The Palaeozoic Era saw many further changes to the continental masses as they drifted about, but by the end of the Palaeozoic they had come together in the supercontinent of Pangaea (Figure 2.5). It is with the break up of Pangaea over the last 200 Ma that the modern era of plate tectonics begins.

Some basic principles of plate tectonics were mentioned earlier, particularly in relation to the movement of plates. This is driven by convection currents in the Earth's mantle and aided by the penetration of water (Boulton, 2001). The internal areas of plates remain relatively rigid and intact, but there is a high level of geological activity along the plate margins. In fact, three types of plate margin can be recognised:

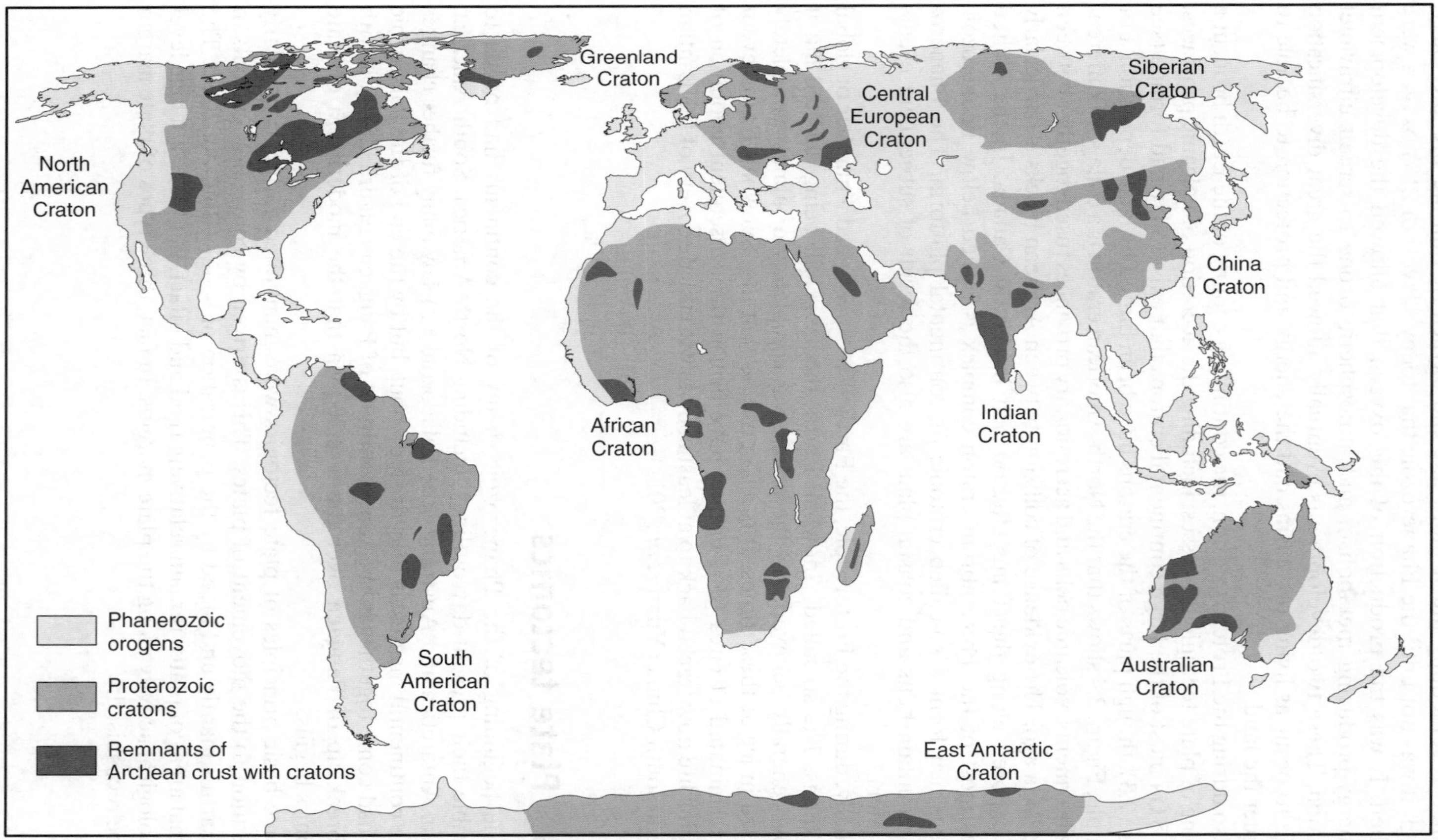

Figure 2.3 Proterozoic cratons incorporating remnants of Archean crust. Note that many parts of these cratons are buried by later sediments (after Marshak, S. (2012) *Earth: Portrait of a Planet*. Used by permission of W.W. Norton & Company, Inc.).

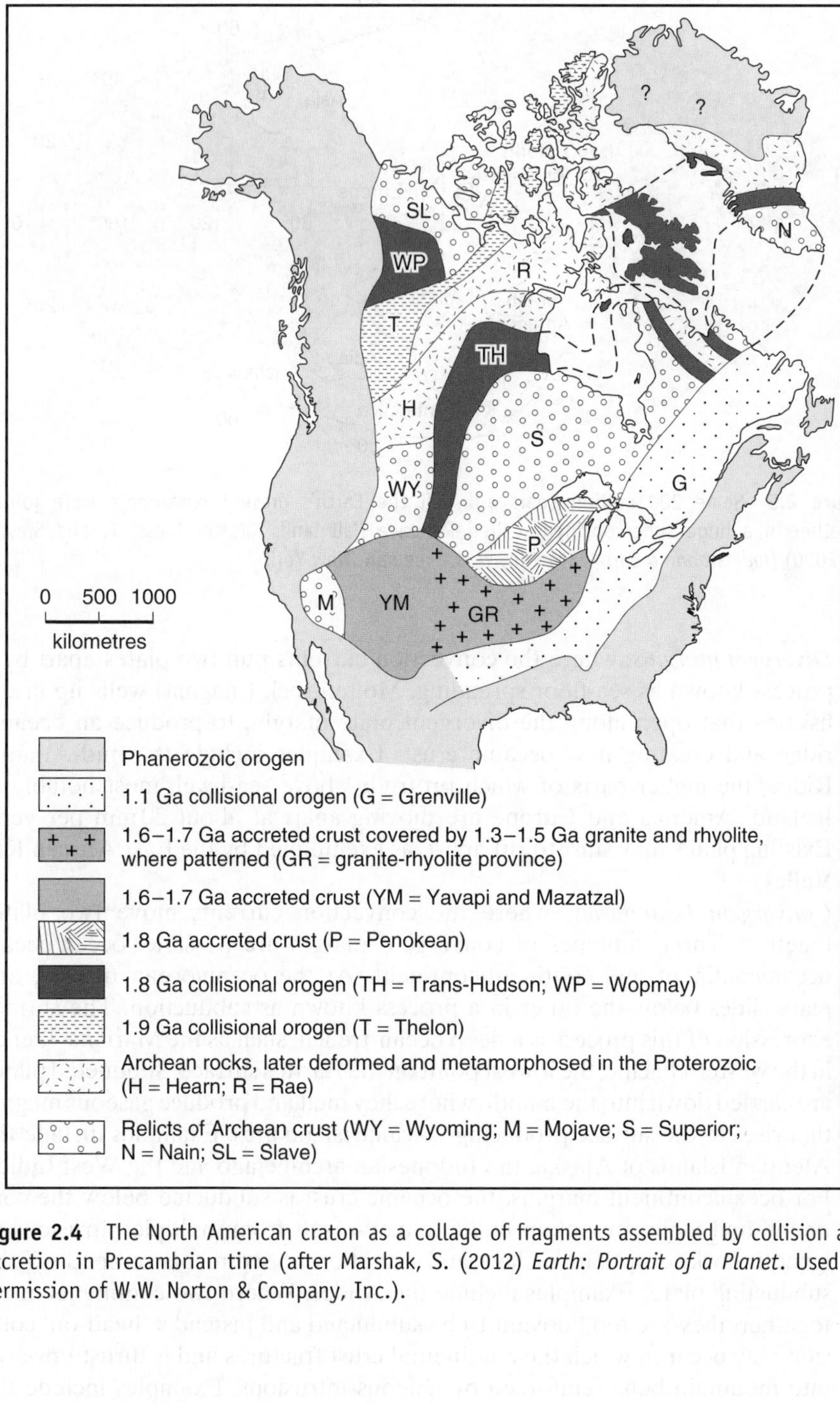

Figure 2.4 The North American craton as a collage of fragments assembled by collision and accretion in Precambrian time (after Marshak, S. (2012) *Earth: Portrait of a Planet*. Used by permission of W.W. Norton & Company, Inc.).

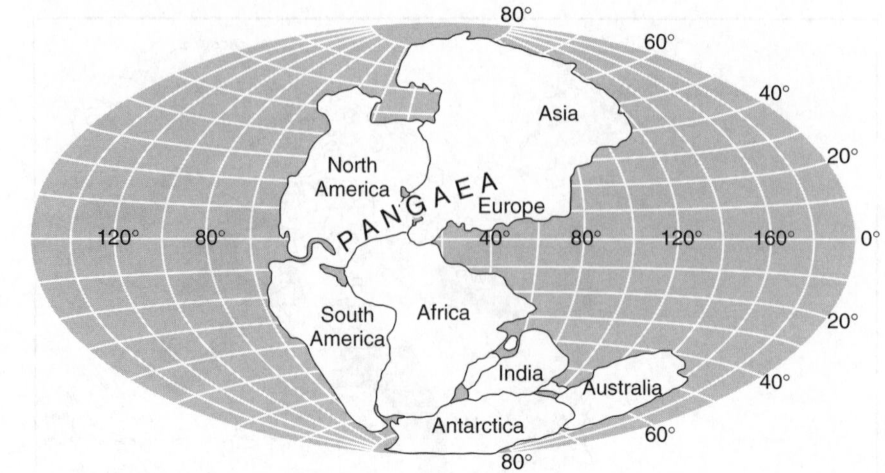

Figure 2.5 Some 200 million years ago, all the Earth's present continents were joined together in a huge supercontinent called Pangaea ('all lands') (after Press, F. and Siever, R. (2000) *Understanding Earth*, 3rd edn. W.H. Freeman, New York).

- *Divergent margins*, where the convection currents pull two plates apart by a process known as sea-floor spreading. Molten rock (magma) wells up in the fissures that open along the divergent plate margin, to produce an oceanic ridge and creating new oceanic crust. Examples include the mid-Atlantic Ridge, the higher parts of which protrude above sea-level, most notably in Iceland. America and Europe are moving apart at about 20 mm per year. Existing plates may start to rift apart, as exemplified by the East African Rift Valley.
- *Convergent boundaries*, where the convection currents move two plates together. Three subtypes of convergent margin are possible: ocean/ocean, ocean/continent and continent/continent. At the ocean/ocean margin, one plate slides below the other in a process known as subduction. The surface expression of this process is a deep ocean-trench, such as the Mariana Trench in the western Pacific, the lowest point on the Earth's surface. Subducted plates are carried down into the mantle where they melt and produce gaseous magma that rises to the surface producing volcanic island arcs. Examples include the Aleutian Islands of Alaska, the Indonesian archipelago and the West Indies. For ocean/continent margins, the oceanic crust is subducted below the continent, and major mountain building may result from multiple intrusions of igneous rock, eruptions of lava and accretion of sediments scraped off the subducting plate. Examples include the Andes. Where two continents move together, they are too buoyant to be subducted and instead a 'head-on' collision may occur in which the continental crust fractures and is thrust upwards into mountain belts reinforced by igneous intrusions. Examples include the Himalayas, where India is moving into the Asian continental plate, and the Alps where the African and European plates collide.

- *Transform boundaries*, where one plate slides laterally past another along a fault line, without the creation or loss of plate material. Transform boundaries or transform faults link sections of oceanic ridges but also occur in other locations. For example, much of the famous San Andreas Fault in North America is a transform fault where the North American and Pacific plates slide past each other.

From these descriptions it should be clear that crustal movement produces a diverse range of processes, materials and morphologies depending on the type of plate margins. As a simplification, divergent margins are typified by shallow-seated earthquakes and relatively quiet, fissure-type eruptions of basalt, resulting in low angle, shield volcanoes. Convergent margins, on the other hand, are typified by deep-seated, destructive earthquakes, explosive volcanic eruptions of more viscous lavas and pyroclastic emissions resulting in steep-sided volcanic cones. They are also typified by violent mountain building, a process known as orogenesis. Summerfield (1991, 2000) describes the development of major mountain ranges including the Andes, the Himalayas, the Tibetan Plateau and the Southern Alps of New Zealand.

2.4 Landscapes of plate interiors

Within this overall pattern there are, of course, exceptions and complications. For example, the Hawaiian Islands are volcanically active, but lie in the middle of the Pacific Plate, far from a plate boundary. The explanation is that they lie above a so-called 'hot spot' where a plume of magma rises from deep in the mantle and is able to penetrate the crust and erupt on the surface. Since these hot spots remain in the same place as the crustal plates drift over them, it is not surprising to find that the Hawaiian Islands form a linear chain of volcanic islands with the oldest being furthest from the current position of the hot spot. Another well-known hot spot occurs at Yellowstone in Wyoming, where no lava has erupted for 600 000 years, but where hot water and steam heated by an underground magma chamber reach the surface as hot springs and geysers (see Figure 6.1 and Figure 6.2). Another hot spot lying away from a plate margin occurs below the western Spanish Canary Islands, but some hot spots lie on plate boundaries as indicated by the presence of more intense volcanic activity there than elsewhere on the same margin, e.g. Iceland.

Another complication involves the accretion of microplate terranes onto larger continental plates. For example, North America has received several of these through its history. Most of Florida is probably a fragment of Africa! The Appalachians comprise several microplate remnants of ancient Europe, Africa and oceanic islands welded onto to an older eastern seaboard. And most of the western seaboard and western part of the North American Cordillera comprise accreted terranes. As a result, the physiography of North America has what Marsh (1997) describes as a 'startling variety', celebrated as part of the national heritage in both Canada and the United States (Figure 2.6). Summerfield (1991) describes the landform and structure of other plate interiors and continents.

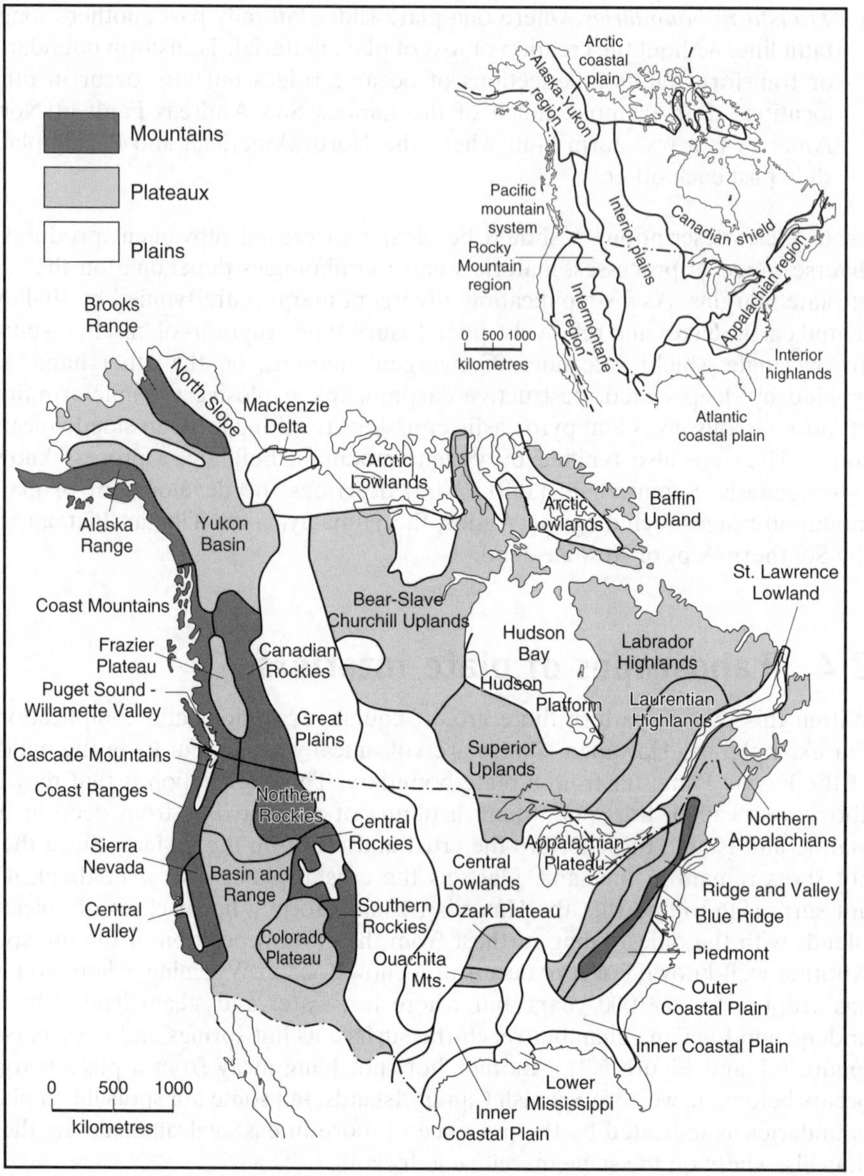

Figure 2.6 The physiographic diversity of North America (after Marsh, W.M. (1997) *Landscape Planning: Environmental Applications*, 3rd edn. This material is reproduced with permission of John Wiley & Sons, Inc.).

It should be clear from this discussion that a knowledge of Earth history and plate tectonics leads to an understanding of the distribution and nature of the major structural features of the Earth's surface—continents and oceans, mountain ranges and ocean trenches, volcanic island arcs and rift valleys, ocean ridges, continental shields and transform faults. All these help to give the planet a variety of form and structure—a global geodiversity.

2.5 Evolution of biodiversity and geodiversity

The world's biodiversity is the result of billions of years of evolution and extinction. Some authors have attempted to analyse the change in biodiversity through geological time, for example by plotting the number of fossil families against time. Figure 2.7 shows a compilation by Benton and Harper (2009). The curve shows a slow development of biodiversity from 3.5 to 0.6 Ga, then a rapid expansion during the Phanerozoic, punctuated by five major mass extinctions and several smaller events during which a significant proportion of living forms disappeared within a brief period in geological terms. For example, it is estimated that 50% of families and 80–96% of species disappeared during the Permo-Triassic event (Benton and Harper, 2009). Alvarez *et al.* (1980) proposed that the Cretaceous–Tertiary extinction was caused by a 10-km-diameter asteroid impacting with the Earth and creating a huge dust cloud that blocked the sun's radiation and prevented photosynthesis. This led initially to the loss of plants, followed inevitably by herbivores and finally carnivores including the dinosaurs. The diversity of the fossil record is therefore not one of simple linear or exponential increase.

So we know quite a lot about the evolution of biodiversity through geological time. But how has geodiversity (minus the fossil record—see Hart, 2012) evolved over the history of the planet. Did it follow the same pattern as biodiversity? Or did geodiversity increase steadily over geological time? Or was there a rapid initial phase followed by an exponential decline? Very little research has been done on this topic, so I will have to speculate a little.

In its very early state, the indications are that the molten Earth can be regarded as largely homogeneous and chondritic in composition (Rollinson, 2007), whereas today it is extremely diverse. Internal diversity resulted from sinking of the heavier elements towards the Earth's interior so that the core, mantle and crust structure developed early in the Earth's life (see Section 2.2). It is suggested that the surface and near-surface geodiversity of the Earth did not develop progressively, but

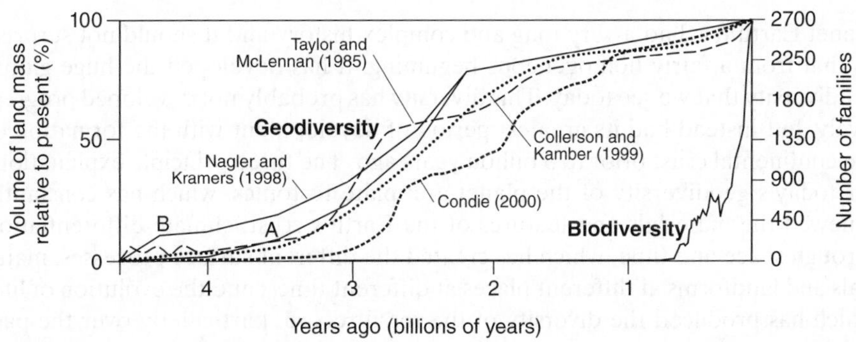

Figure 2.7 The evolution of biodiversity (after Benton & Harper (2009) *Intoduction to Palaeobiology and the Fossil Record*. Reproduced with permission of John Wiley & Sons), and suggested S-curves for the evolution of geodiversity based on several authors' proposals for the development of the continental crust with acceleration circa 3 billion years ago.

instead followed an S-shaped curve, with the maximum expansion of geodiversity occurring during the early development of the continental crust. The related processes, along with the evolution of the Earth's atmosphere, not only created a range of mineral and rock types but also led to the rapid development of a diversity of mountain building episodes, crustal accretion, surface processes and landform evolution. Although there is disagreement on the detailed development of the continental crust, several authors (e.g. Taylor and McLennan, 1985; Nagler and Kramers, 1998; Collerson and Kamber, 1999; Condie, 2000; Rollinson, 2007) have dated the early period of continental growth to the Archean, approximately 3.5–2.5 Ga (see Figure 2.2) and the early part of this is likely to be the time of maximum development of global geodiversity. Furthermore, unlike biodiversity, the Earth's geodiversity has probably had few major setbacks that brought mass extinctions to the biological world. Although these impacts no doubt destroyed some rocks, minerals, landforms or soils they also created new materials and features that added geodiversity to the planet. For further discussion of the geological evolution of the Earth see Windley (1995) and Kusky *et al.* (2010).

In summary, it is suggested that the major factors explaining global geodiversity are:

- plate tectonics—absent on all other planets, with the possible exception of periods of plate growth on Mars;
- climatic differentiation through space and time—with related diversity of physical processes, sediments and landforms;
- evolution and extinction—creating the diversity of the fossil record.

No other known body in the solar system approaches the geodiversity of the Earth. And given the influence that geodiversity has on biodiversity and its evolution, it may be no coincidence that the most geodiverse body in the solar system is also the one where complex life has developed (Ward and Brownlee, 2000).

2.6 Conclusions

Planet Earth has had a very long and complex history and it should not surprise us that from a fairly homogeneous beginning, it has developed the huge global geodiversity that we see today. This diversity has probably not developed progressively, but instead had its greatest period of development with the formation of the continental crust prior to 3 billion years ago. The three principle explanations for today's geodiversity of the planet are: plate tectonics, which has constantly renewed the materials and features of the Earth's crust; climate differentiation through space and time, which has created the different surface processes, materials and landforms at different places at different times; and the evolution of life, which has produced the diversity of the fossil record, particularly over the past 600 000 years.

3

Geodiversity: the Local Scale

'I need your help' said Wallander, 'I'd like you to compare some stones'...

'I'll see what I can do, but it sounds difficult. Do you have any idea how many different species of rock there are?'

from *The Troubled Man* (2012) by Henning Mankell

This chapter will outline geodiversity at the local scale—minerals, rocks, fossils, landforms, processes, soils, and so on. First, the materials constituting the Earth's crust will be described, beginning with minerals, the building blocks of sediments and rocks and therefore of the planet itself. Those with a good knowledge of geology and geomorphology may wish to skip this chapter.

3.1 Earth materials

Earth materials comprise minerals, rocks, sediments fossils and soils. A comprehensive description of these is given in the books by Heffernan and O'Brien (2010) and Klein and Philpotts (2012).

3.1.1 Minerals

Geological minerals are defined as naturally occurring, crystalline, solid, inorganic substances with a specific chemical composition and an internal structure characterised by an orderly arrangement of atoms, ions or molecules in a lattice. This orderly arrangement occurs during the process of crystallisation, when the constituent atoms, ions or molecules come together in the appropriate chemical proportions and alignments related to chemical bonding and electron sharing. During crystallisation, original microscopic crystals grow larger with well-formed crystal faces if they are free to grow (see later).

Geodiversity: Valuing and Conserving Abiotic Nature, Second Edition. Murray Gray.

Table 3.1 Some common minerals and their characteristics (after Marshak, 2012).

Amphibole	Dark-coloured, non-metallic mineral
Calcite ($CaCO_3$)	White, pink or clear with milky lustre; fizzes in acid
Chlorite	Dark green; composed of thin flakes
Clay	Group of minerals occurring as very thin flakes
Corundum (Al_2O_3)	Very hard—used as an abrasive
Diamond (C)	Clear and glassy; hardest mineral known
Dolomite ($CaMg[CO_3]_2$)	Similar to calcite, but less reactive to acid
Feldspar	Group of common, light-coloured, hard minerals
Orthoclase ($KalSi_3O_8$)	Pinkish, common feldspar also known as K-feldspar
Plagioclase	Common feldspar containing Na or Ca; white, grey or blue
Galena (PbS)	Dark grey; metallic cubic crystals
Garnet	Dark brown to maroon, glassy, roundish grains
Graphite	Very soft, grey and metallic
Gypsum	Clear and glassy, very soft
Halite (NaCl)	Clear or grey, cubic grains
Haematite	Either earthy coloured or shiny bluish grey and metallic
Kyanite	Bluish, blade-shaped
Magnetite (Fe_3O_4)	Dark grey to black, metallic and magnetic
Mica	Group of minerals occurring in thin flakes or sheets
Biotite	A black mica
Muscovite	A clear to light brown mica
Olivine ($[Mg, Fe]_2SiO_4$)	Olive green, glassy; occurs in clusters
Pyrite (FeS_2)	Golden bronze; metallic; cubic form
Pyroxene	Dark-coloured; non-metallic mineral
Quartz (SiO_2)	Glassy and hard; typically white, grey or clear
Serpentine	Group of minerals forming thin, thread-like needles
Talc	Very soft mineral

Minerals may crystallise in a variety of ways. When magma cools, minerals will crystallise when the temperature drops below their melting point, which varies from mineral to mineral. Crystallisation also occurs when water evaporates, leaving behind any minerals that were dissolved in the water. For example, salt crystals form when seawater evaporates and this is still used as a method of producing crystalline sea salt in many countries. Finally, new minerals may form when minerals are heated. As the atoms and ions become more mobile, they may rearrange themselves to become new minerals with different crystal structures, a process known as solid-state diffusion.

With 92 elements in the Periodic Table, it should not be surprising to find that there are 4000–5000 known minerals, though many are quite rare and only about 1% of them commonly occur in rocks (see Table 3.1). Minerals have diversity in several respects, and these properties are not only vital in identifying minerals, they also determine the uses to which the minerals can be put (see Chapter 4). Anyone requiring more information on minerals and crystals may wish to consult Putnis (1992), Perkins (2002), Klein (2002), Wenk and Bulakh (2004) and Brocx and Semeniuk (2010). In outline the variations include:

- *Chemical composition*. We have already noted that the most common earth material is silica (SiO_2) which, in its crystalline form, occurs as the mineral quartz, but other silicates have very complex formulae. For example, biotite, a black mica, has the formula $K(Mg,Fe)_3(AlSi_3O_{10})(OH)_2$. The proportions of iron and magnesium in biotite can vary, so that even within a single mineral, we find diversity. Silicates are built from the basic silicate tetrahedral ion $(SiO_4)^{4-}$ in which four oxygen ions surround and share electrons with a silicon ion. These can then be combined into rings, chains, double chains, sheets and frameworks to form several common minerals including feldspars (Deer, Howie and Zussman, 2001), mica, amphibole and pyroxene. Apart from the silicates, other important mineral groups include carbonates, such as calcium carbonate ($CaCO_3$), oxides, such as the common iron oxide haematite (Fe_2O_3), sulphides such as pyrite (FeS_2), sulphates such as gypsum ($CaSO_4.2H_2O$), halides such as fluorite (CaF_2), and native metals much as copper (Cu).

- *Crystal size*. Large, well-formed crystals occur in conditions where they are able to grow slowly and in an unrestricted environment. For example, this may be in open spaces in rocks, such as cavities and fractures, as exemplified by the giant crystals of gypsum discovered in underground cavities in Spain and Mexico (Garcia-Guinea and Calafarra, 2001; Brocx and Semeniuk, 2010). However, if space is restricted or growth is rapid, then crystal growth is limited and they may coalesce to form a solid crystalline mass composed of mineral grains. Natural glass such as obsidian is formed from molten material that solidifies so quickly that there is no internal atomic order and crystals have not had time to form.
- *Crystal form and habit*. Minerals are defined as crystalline solids, which means that the atoms, ions and molecules form a regular pattern known as a crystal lattice. Galena (PbS) for example, consists of lead and sulphur atoms packed together in a regular array. Given the freedom to grow (see above), this configuration results in galena crystals with a cubic form. Common salt crystals (NaCl) also have a cubic form. However, crystals can come in a very wide diversity of shapes For example, they may be plate-like as in the micas, trapezoid as in calcite, pyramidal as in diamond, etc. Some substances with exactly the same chemical composition have more than one crystal structure and therefore form more than one mineral. For example, carbon exists in two mineral forms as diamond and graphite. Diamond, which has a closely packed tetrahedral crystal structure and a density of $3.5\,g/cm^3$ forms at very high temperatures and pressures, whereas graphite, which has hexagonally arranged atoms in weakly bonded sheets and a density of only $2.1\,g/cm^3$, is formed at lower temperatures and pressures. Crystal habit refers to the general shape or character of a crystal. For example, asbestos has a fibrous habit, a characteristic that is now known to cause lung disease.
- *Hardness*. Mineral hardness varies from very hard, like diamond, to very soft, like talc. In 1822 the Austrian mineralogist, Friedrich Mohs, devised a scale of mineral hardness based on the ability of one mineral to scratch others (Table 3.2). Hardness is related to the strength of the chemical bonding. Most silicate minerals lie in the range 5 to 7, except for sheet silicates like mica which lie in the range 2 to 3.
- *Cleavage* refers to the presence of splitting planes within mineral crystals. Muscovite mica, for example splits into thin sheets less than a millimetre thick very easily, whereas quartz and garnet lack cleavage planes. The explanation

Table 3.2 Mohs' scale of hardness.

10	Diamond
9	Corundum
8	Topaz
7	Quartz
6	K-feldspar
5	Apatite
4	Fluorite
3	Calcite
2	Gypsum
1	Talc

again lies in crystal structure. Mica is a sheet silicate, whereas quartz and garnet are bonded strongly in all directions. Some minerals have several cleavage directions. Calcite and dolomite, for example, each have three.

- *Fracture* refers to the nature of break surfaces other than cleavage planes. The different fracture styles include conchoidal, fibrous and splintery.
- *Lustre* refers to the light-reflecting qualities of minerals. Terms such as metallic, vitreous, resinous, greasy, pearly, silky and adamantine are used to describe the lustrous properties of different minerals.
- *Colour and streak* of minerals may be characteristic of minerals or may be the result of impurities. Pure magnesium olivine is white, but with iron impurities it is green. This is unusual since iron normally gives a red or brown colouration to minerals. Quartz may be clear, white, purple, grey or rose. Gemstone colour is discussed in Section 4.5.5. Streak refers to the colour produced by scraping a mineral against a hard, unpolished white surface and again varies.
- *Internal features* of crystals are described by Brocx and Semeniuk (2010) and include zoning, fluid, gas or solid inclusions and reaction rims.
- *Specific gravity* varies greatly in minerals. The difference between the two carbon minerals, diamond and graphite, was referred to earlier. Specific gravity depends on the atomic weight of the constituent elements and how closely the atoms are packed together. Lead, for example, has a specific gravity of 11 g/cm^3 because it has a high atomic weight, whereas for quartz the value is 2.65 g/cm^3.
- *Chemical properties*. Many minerals have distinctive chemical properties, some of which are valuable in their economic use (see Section 4.5.4).

Minerals, therefore, display a huge diversity of physical and chemical characteristics that generally results from their chemical composition and atomic structure. These properties give minerals their practical values and in themselves contribute to the diversity of rocks, as I shall now describe.

3.1.2 Rocks and sediments

Rocks are a natural aggregation of minerals and fall into three groups—igneous, metamorphic and sedimentary—each of which has its own incredible diversity.

Igneous rocks form by the solidification of molten rock, metamorphic rocks have been transformed under intense heat and pressure from pre-existing rocks, and sedimentary rocks are formed by the consolidation and cementation of sediment deposited at the Earth's surface. Together they provide us with a fairly comprehensive history of the planet, they give us an insight into its internal processes and they even suggest explanations for the origin of the solar system. Rocks therefore deserve to be better understood and valued by society for their amazing diversity, for the clues they can give us about past environments and for their practical uses (see Section 4.5).

The theory of plate tectonics allows us to see the three great rock groups as part of the rock cycle originally proposed by James Hutton (1795). Igneous rocks formed during plate collisions together with sediments that have accumulated in the vicinity, are uplifted at convergent margins during mountain-building episodes (orogenies). The heat and pressure exerted during these episodes also create metamorphic rocks. Once uplifted, mountains become susceptible to weathering and erosional processes, and the debris produced is transported to the oceans by rivers, glaciers and mass movement (see later). Here, the sediments accumulate in thick sequences that are hardened and cemented into sedimentary rocks. At the base of these sequences, the temperature and pressure may be sufficient to form further metamorphic rocks. Finally these materials may be transported to a convergent plate margin where the cycle recommences as part of a perpetual recycling process. Of course, this is a generalised representation of a much greater natural complexity and diversity.

Igneous rocks As well as proposing the basis for the rock cycle, James Hutton was also responsible for first identifying a solidified igneous rock at Salisbury Crags in Edinburgh, Scotland, where an underlying block of sedimentary rock had been baked and partially incorporated into the rock above. Hutton argued that the latter must have been molten when emplaced in order to bake the sediment and surround the partially detached block. After more than 200 years of further study geologists now have a very detailed knowledge of igneous rocks and their formation. For detailed information readers are referred to Best and Christiansen (2001), Best (2003) and Gill (2010). Not surprisingly there is a huge diversity of igneous rocks and even those known by a single name can have several subtypes, an example being the eight subtypes of larvikite (Heldal *et al.*, 2008). The classification of igneous rocks is based principally on texture and chemical and mineral composition (Figure 3.1).

- *Texture*, in this context, refers to the size of the mineral grains. 'Coarse-grained' refers to a rock with large crystals, while 'fine-grained' is applied to a rock with small crystals. It has already been noted that rapid cooling of magma produces a non-crystalline natural glass (obsidian). On the other hand, slow cooling in restricted environments results in the growth of large crystals. The relevance of this to igneous rocks is that magma that is intruded deep in the Earth's crust cools only very slowly and becomes a coarse-grained intrusive igneous rock, whereas magma that erupts at the surface as lava (extrusive igneous rock), is subject to rapid cooling and therefore develops a fine-grained texture. Even here, however, there may be differences in crystal size between

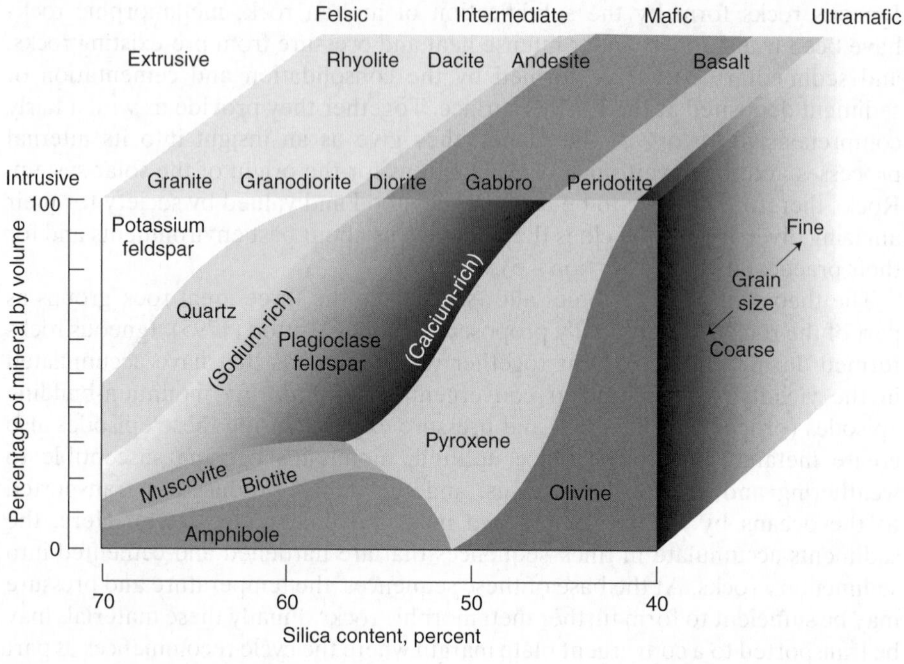

Figure 3.1 Basic classification of igneous rocks (after Press, F. and Siever, R. (2000) *Understanding Earth*, 3rd edn. W.H. Freeman, New York).

the interior of the lava flow, which cooled more slowly than the margins. Special names are given to other types of texture. For example, the term 'porphyritic' refers to an igneous rock in which large crystals (phenocrysts) sit in a finer-grained matrix. A final type of igneous rock comprises materials that have been ejected during violent igneous eruptions when everything from fine-grained ash to large bombs may accumulate around a volcano. These are subsequently hardened into so-called pyroclastic rocks, such as pumice, scoria and tuff.

- *Chemical and mineral composition* is the other important characteristic in igneous rock classification (Figure 3.1). Chemical composition is usually based on silica content, which normally varies between 40 and 70%. Rocks at the upper end of this range like rhyolite and granite are termed 'silicic', or sometimes 'acidic', whereas those at the lower end of this scale, like basalt and gabbro, contain higher percentages of magnesium (Mg) and iron (Fe) and are therefore given the name 'mafic' or sometimes 'basic'. In terms of mineralogy, silicic rocks contain quartz, potassium- and calcium-rich feldspars and lower percentages of mica and amphibole. Mafic rocks, on the other hand, are dominated by sodium-rich plagioclase, olivine and pyroxene. Mineral composition also produces differences in colour and specific gravity. Silicic rocks are lighter in both specific gravity and colour, usually with light- or mid-grey colourations or pink where orthoclase feldspar dominates. Mafic rocks are denser and darker, with black, dark grey and green predominating.

These descriptions are a gross oversimplification of the variations in rock composition and texture that occur in nature and readers should refer to the specialist texts for a full appreciation of the diversity of igneous rocks.

The explanation of the variation in the chemical composition of igneous rocks lies in the chemistry of the magma. At divergent margins, where the magma rises from the mantle and erupts at the surface without passing through continental crust, the lava is usually mafic in composition, contains less gas and is less viscous. This leads to predominantly basaltic lavas, less-violent eruptions and low-angle lava flows and volcanoes. Current examples include Iceland and Hawaii, while past examples include the flood basalts of Washington and Oregon in the United States and at the Deccan Plateau in India. On the other hand, at convergent margins, the magma rises through continental crust and as it does so may melt large quantities of silica-rich sedimentary rocks. The molten mix is therefore much more silicic in composition, contains more gas and is more viscous. This results in violent eruptions, rhyolitic and andesitic lavas and pyroclastic rocks, and steep-sided cones such as Fujiyama in Japan. Igneous processes and morphologies are considered further later in this chapter.

Before leaving igneous rocks, it is also important to consider briefly the possible outcomes of fractional crystallisation, a process proposed by Canadian geologist N.L. Bowen early in the twentieth century (Bowen, 1928). This involves the separation and removal of successive fractions of crystals as magma cools. It occurs because the temperature at which minerals crystallise from a melt varies. For example, olivine has a very high melting point of about 1800 °C. As magma cools, olivine will crystallise at this temperature, and as the olivine crystals form they may begin to settle to the base of the magma chamber. Further cooling may result in new minerals crystallising with the result that a mineralogically differentiated body is produced. While some igneous intrusions do exhibit this character, including the Palisades cliff on the west bank of the Hudson River, facing New York, others do not and it is now recognised by geologists that Bowen's proposals were oversimplified. Geological reality produces a much more diverse range of outcomes.

Sediments and sedimentary rocks Sediment is derived from the weathering, erosion, transportation and deposition of rock fragments (clastic sediments), or from the precipitation from water of chemical and biochemical minerals (chemical and biochemical sediments). The former are much more abundant than the latter and many variables combine to produce a large diversity in clastic sediments. Readers requiring full information on the classification of sedimentary rocks are referred to Prothero and Schwab (1996), Selley (1996), Reading (1996) and Tucker (1996, 2001).

- *Particle size distribution and sorting.* Sediments vary in particle size from massive boulders, through cobbles, pebbles, sand and silt to clay, and several classification systems of particle sizes exist. Not surprisingly, this variability has much to do with the processes of weathering, erosion, transportation and deposition that will be discussed in more detail later. Clays are often the result of chemical weathering of minerals and, because of their fine-grained nature, are easily transported by rivers to lakes or oceans where they settle out in the low-energy, deep-water environments. Boulders, on the other hand, reflect high-energy erosion, transportation and sedimentation, perhaps associated with catastrophic floods. Size is also controlled by bedrock characteristics. Closely jointed rocks such as slates are easily eroded into small rock fragments compared with a massively jointed granite where only large blocks can be detached. But whatever the original size of the entrained particles, they are reduced in size the further the distance of transport due to attrition. Particle sorting refers to the range of particle sizes in a sediment. A sediment with a wide range of sizes from clays to boulders, as in many glacial sediments, is described as poorly sorted, whereas one that is dominated by a single size range, for example a beach sand, is termed well sorted. Some rock units, termed graded units, show a change in particle size vertically within the unit, usually becoming finer upwards, due for example, to larger grains settling out of water faster.

- *Particle composition*. Sediments may be composed of a variety of minerals, depending on the mineralogy of the source rock and subsequent processes. Clays are usually composed of the minerals produced by chemical weathering such as kaolinite, montmorillonite and illite. The beaches people usually find the most attractive are sand coloured and generally composed of quartz grains stained orange by iron impurities. This composition is the same for most desert sands. But sands derived from mafic igneous rocks are black, as in Iceland and Hawaii. Coarse clastic sediment particles are usually composed of more than one mineral or rock type and it is not unusual to see rock particles brought together and cemented as a new rock, as for conglomerates (rounded particles) and breccias (angular particles). Sedimentary rock cements also vary in composition, some being silica based and others being calcitic.
- *Particle shape*. Particle shape is described in various ways. Terms such as blades, rods, discs and spheres are used to describe the three-dimensional shape of coarse clastic particles. Sphericity is a measure of how equal the three axes are, whereas roundness refers to two-dimensional rounding of the edges. Hence a cube has a high sphericity but a low roundness. Mineral grains in an igneous rock are highly irregular where they become interlocked during crystallisation. On the other hand, after erosion and transportation in water they may become highly rounded as they come into contact with other particles and the riverbed. Detailed shape may reveal particle history. For example, frosted grains result from wind transportation whereas glacial transport produces characteristic fractures on mineral grains.

Although there are other sedimentary variables such as colour, fossil content, compaction, etc., these three characteristics—particle size, shape and composition—combine to produce the three main factors used in classifying sedimentary rocks, and Table 3.3 shows some of the main types.

Chemical and biochemical rocks also exhibit significant geodiversity. The main chemical sediments are evaporites, including rock salt and gypsum. Rather like igneous crystallisation, there may also be fractional differentiation of evaporite

Table 3.3 A common classification of sedimentary rocks (modified after Greensmith, 1989; Press and Siever, 2000).

Sediments	Particle size (mm) or minerals	Rocks
Boulder		
	256	Conglomerate
Cobble		Breccia
	64	Agglomerate
Gravel		
---------------	2 ------------------------------	------------------------------
Sand		Sandstone, coarse tuff
---------------	0.062 ------------------------------	------------------------------
Silt		Siltstone, fine tuff
---------------	0.0039 ------------------------------	------------------------------
Clay		Mudstone (blocky fracture)
		Shale (bedding fracture)
Carbonate sand and mud	Calcite	Limestone
	Dolomite	Dolostone
Iron oxide sediment	Haematite	Iron formation
	Limonite	
	Siderite	
Evaporite sediment	Gypsum	Evaporite
	Anhydrite	
	Halite	
Siliceous sediment	Opal	Chert
	Chalcedony	
	Quartz	
Carbonaceous sediment	Coal, etc.	Organic
Phosphatic sediment	Apatite	Phosphorite

minerals with gypsum, halite, anhydrite and finally salts of potassium and magnesium crystallising as evaporation progressively concentrates the salt solution. All of these minerals are commercially valuable as we shall see in Chapter 4. Examples of significant evaporite deposits include the Permian rocks of the Zechstein Sea that stretched from Russia to England, but the evaporite sequences are particularly thick in eastern Germany and the Netherlands. Important lake evaporites occur at the Great Salt Lake in Utah, United States, and in the East African Rift Valley. Phosphorite, ironstone and coal are other examples of chemical and biochemical rocks, while oil and gas are geologically formed organic fluids. Chert is a chemically or biochemically precipitated form of silica and is similar to flint, layers of which are abundant in the English and French chalk rocks.

Chalk is in fact one of a series of biochemical sedimentary rocks, other examples being limestone and dolomite. Chalk and limestone are calcium carbonate rocks made up of lithified carbonate shells and tests of micro-organisms such as foraminifera. Coral reefs are a particular form of calcium carbonate and give rise to reef limestones. Dolomite is a chemically altered form of limestone by the addition of magnesium ions shortly after deposition. Chemically precipitated

Figure 3.2 Dragon Mount: one of Iran's most scenic and rarest structural phenomena is the folding in north Garmsar. (Photo by permission of Alireza Amrikazemi.)

calcium carbonate also occurs, including cave stalactites, stalagmites and other features collectively termed speleothems.

A variety of sedimentary structures may also occur within individual sedimentary rock units. These include horizontal bedding and lamination, cross-bedding, dune bedding, ripple marks, sole structures, channel scours and mud cracks. Deformational structures commonly seen within rock sequences, and which themselves display great diversity, include slump structures, load casts, dewatering structures, organic structures, joint patterns, faults and folds (Figure 3.2).

Metamorphic rocks Metamorphic rocks result from solid-state changes to other rocks induced by high temperatures and pressures. The changes may result in changes in texture, mineralogy or chemical composition of a rock, and sometimes involves all three so that the original nature of the rock is unrecognisable. In these instances, geologists refer to the changes as high-grade metamorphism, whereas lesser changes brought about by lower temperatures and pressures are referred to as low-grade metamorphism. Details of metamorphic rocks can be found in books by Best (2003), Fettes and Desmons (2007), Vernon and Clarke (2008) and Bucher and Grapes (2011).

Temperature in the crust rises by about 30 °C/km depth. Thus at a depth of 10 km the temperature will be circa 300 °C and at 30 km depth it will be circa 900 °C. Most metamorphic rocks form within this depth and temperature range, with high-grade rocks forming in the deeper and hotter parts of the crust. Heat can bring about profound changes to the mineralogy and texture of rocks by breaking chemical bonds, altering the crystal structure of minerals and initiating recrystallisation of new or altered minerals.

The high confining pressures deep in the crust are capable of producing new minerals with denser crystal structures. Furthermore, the directed pressures frequently experienced at convergent plate margins cause minerals to change size, shape and orientation. The combined effect of heat and pressure is critical in producing the foliated appearance of many metamorphic rocks since these influences cause minerals to segregate into planes and grow perpendicular to the applied stress. Since heat also makes rocks more pliable, these mineral bands often display severe folding and deformation structures.

Other metamorphic processes include chemical and physical changes to the rocks surrounding an igneous intrusion. First, contact metamorphism bakes the rocks along the contact zone, with rocks such as hornfels being formed close to the contact and other mineralogical changes occurring farther away within a so-called metamorphic aureole. Secondly, it is normal for magma intrusion to be accompanied by hydrothermal solutions that permeate into the intruded rocks and react with them, changing their chemical and mineral compositions and sometimes completely replacing one mineral with another without changing the rock's texture, a process known as metasomatism. Dissolved minerals may be carried away to reprecipitate as veins in rock fissures. Quartz veins are very common where silica-rich hydrothermal fluids have permeated through a rock, and many valuable metal ores such as copper, lead and zinc are also formed in this way (see Figure 4.15). In the case of divergent, submarine plate margins, heated seawater plays an important role in altering the chemical composition of the ridge basalts.

Like other rock types, metamorphic rocks exhibit a great diversity of types, whose characteristics are summarised here.

- *Cleavage*. This refers to the finely spaced, planar partings in rocks such as slates and phyllites. This fracture cleavage should not be confused with mineral cleavage discussed earlier. It results from low-grade metamorphism of fine-grained rocks such as shales, where the pressure causes a realignment of the mineral grains perpendicular to the applied pressure. This produces parallel cleavage planes unrelated to the original bedding, which may still be visible. Slates are generally dark grey or black, but may be coloured red or purple where iron oxide minerals are present or green where chlorite is present.
- *Schistosity*. This is the name given to a coarser foliation produced by the lineation of platy minerals such as biotite and muscovite micas, but also including quartz, feldspars and other minerals. It gets its name from schist, one of the most common metamorphic rock types, which frequently exhibits wavy splitting planes. Schists are frequently named from the dominant minerals present, and thus we have mica schists, chlorite schists, quartz schist, garnet schists and a diverse range of others.
- *Banding*. This is a still coarser type of foliation common in high-grade metamorphic environments where the micas and chlorite are lost and quartz, feldspars and mafic minerals dominate. These rocks frequently display spectacular, deformed, light (quartz and feldspar) and dark (mafic minerals) segregated bands of minerals, and are termed banded gneisses (Figure 3.3).

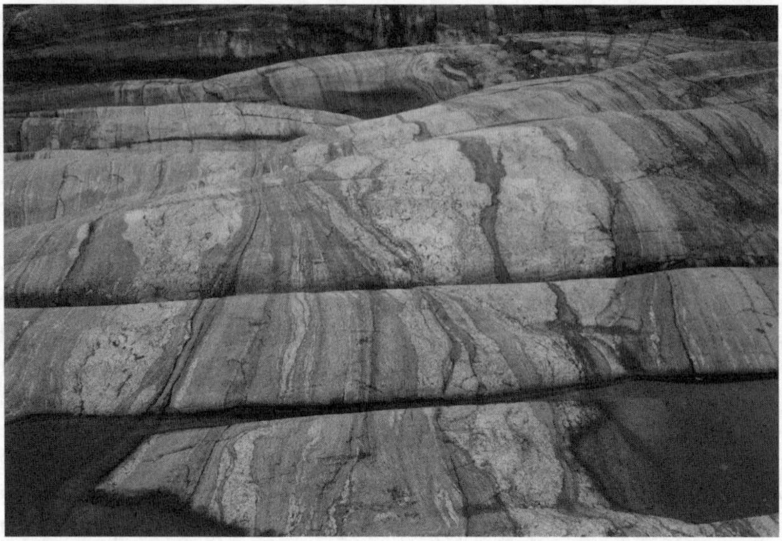

Figure 3.3 Banded gneiss, Rognstranda, Norway.

- *Non-foliated rocks*. Non-foliated metamorphic rocks include quartzites (metamorphosed quartz-rich sandstones) and marbles (derived from limestones and dolomites). Some, like the famous Italian Carrara marble, are pure white, but most contain irregularly coloured bands and streaks, giving the marbles an endless diversity. Other non-foliated metamorphic rocks include argillites (produced by low-grade metamorphism of shaly sedimentary rocks) and greenstones (metamorphosed mafic-rich volcanic rocks, often produced by seawater reactions at divergent oceanic plate margins and receiving their colouration from chlorite).
- *Porphyroblasts*. These are large crystals that have grown faster than the finer grained matrix in which they are set. The crystals may vary between a few millimetres and several centimetres in diameter, and examples include cubic iron pyrites crystals in slate and garnet crystals in schist.
- *Shear textures*. These form when two rock surface slide past each other. Rocks known as mylonites form at the shear surfaces as rocks are crushed and sheared under high pressures deep in the crust.

Where metamorphism has affected a large area, for example in the orogenic belts of the Alps and Appalachians, systematic changes in the metamorphic grade may be seen, indicating a transition from high to low grade. A sequence of index minerals from chlorite, biotite, garnet, staurolite, kyanite to sillimanite is produced by increased metamorphism in a uniform shale, though there are many local variations caused by compositional variations in the parent rocks (see Figure 3.4a). Metamorphism of basalt, for example, produces a zeolite, greenschist, amphibolite, pyroxene granulite series, but at very high pressures blueschist and eclogite form instead (see Figure 3.4b). Migmatites form at very

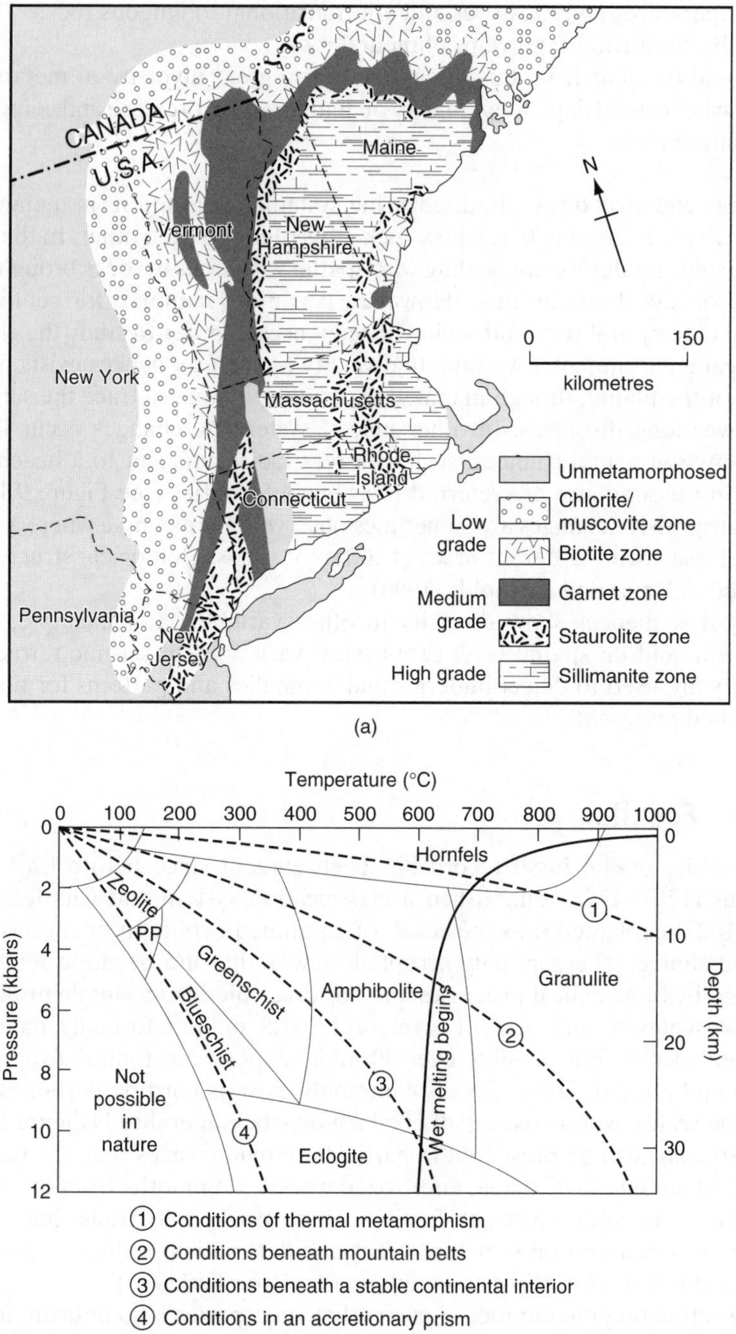

Figure 3.4 Metamorphic diversity: (a) metamorphic zones in New England USA and (b) common metamorphic facies at varying temperature and pressure conditions (after Marshak, S. (2012) *Earth: Portrait of a Planet*. Used by permission of W.W. Norton & Company, Inc.).

high temperatures and pressures and are transitional to igneous rocks. They are typified by contortions and veins of melted rock.

It should be clear from this discussion that a great diversity of metamorphic rocks can be formed dependent mainly on temperature, pressure and composition of the parent rocks.

Sequences and structures In discovering systematic variations in metamorphic grades, we are beginning to see how rock units relate to each other. In the case of metamorphic grades we are dealing with spatial variations in rocks brought about at more or less the same time. However, it is more common for geologists to investigate temporal rock and sediment sequences in order to study the changing geological environments. An infinite diversity of such sequences exists near the surface of the planet, though in many places it is possible to trace the same rock strata over long distances. In other places, systematic changes occur laterally due to environmental changes, such as a reef facies adjacent to a beach facies. Breaks in the sequence are referred to as unconformities (see Figure 9.19), and time-transgressive changes are sometimes observed, for example where a relative sea-level rise covers different areas at different times. Geological structures are described in Davis and Reynolds (1996).

Vertical sequences show diversity in other variables. For example, palaeomagnetism, and on spatial level geophysical variations in seismic refraction or resistivity are used to detect underground anomalies and patterns for both pure and applied research.

3.1.3 Fossils

The diversity of the fossil record has been evident since before Carl Gustav Linnaeus (1707–1778) established a classification system and Charles Darwin (1809–1882) explained it as the result of continued evolution of species from a common source. The generally accepted view is that life began around 3.8 Ga years ago by biochemical processes and has diversified from simple prokaryotes and eukaryotes to the present level of 1.5–1.8 million formally named and described species, but possibly over 30 million species in total (Lovejoy, 1997; Benton and Harper, 2009). Fossils commonly exist as hard parts (bones, shells, teeth, fish scales, woody tissue) all of which may be mineralised (Figure 3.5), but soft parts may also be preserved in particular circumstances (e.g. the Cambrian Burgess Shale Fauna, Canada, and frozen woolly mammoths from the Siberian permafrost). In addition, trace fossils include tracks and trails, burrows and borings, root penetration structures, faecal pellets and coprolites, regurgitation pellets and vomit, gastroliths and teeth marks (Bromley, 1996).

Since estimating the number of currently living species is so difficult, it should not be surprising that it is impossible to say how many species have existed during the history of the Earth. Not all species will have survived in fossil form and many that have survived have yet to be discovered by geological field research, particularly in the remoter parts of the world. Benton and Harper (2009) believe that extinct plants and animals make up 99% of all species that ever lived. They also

(a)

(b)

Figure 3.5 Fossil diversity (a) mineralised wood, Petrified Forest National Park, USA and (b) fossil fish, Miguasha World Heritage Site, Canada.

believe that biotic diversity may have taken 100 million years to re-establish itself from the major Permo-Triassic mass extinction, but was probably of the order of some 10 million years for the Cretaceous–Tertiary event. Figure 3.6 shows the huge expansion of the placental mammals around the Palaeocene-Eocene boundary some 10 million years after the Cretaceous-Tertiary event and illustrates just one fragment of the diversity of life. More details about the evolution

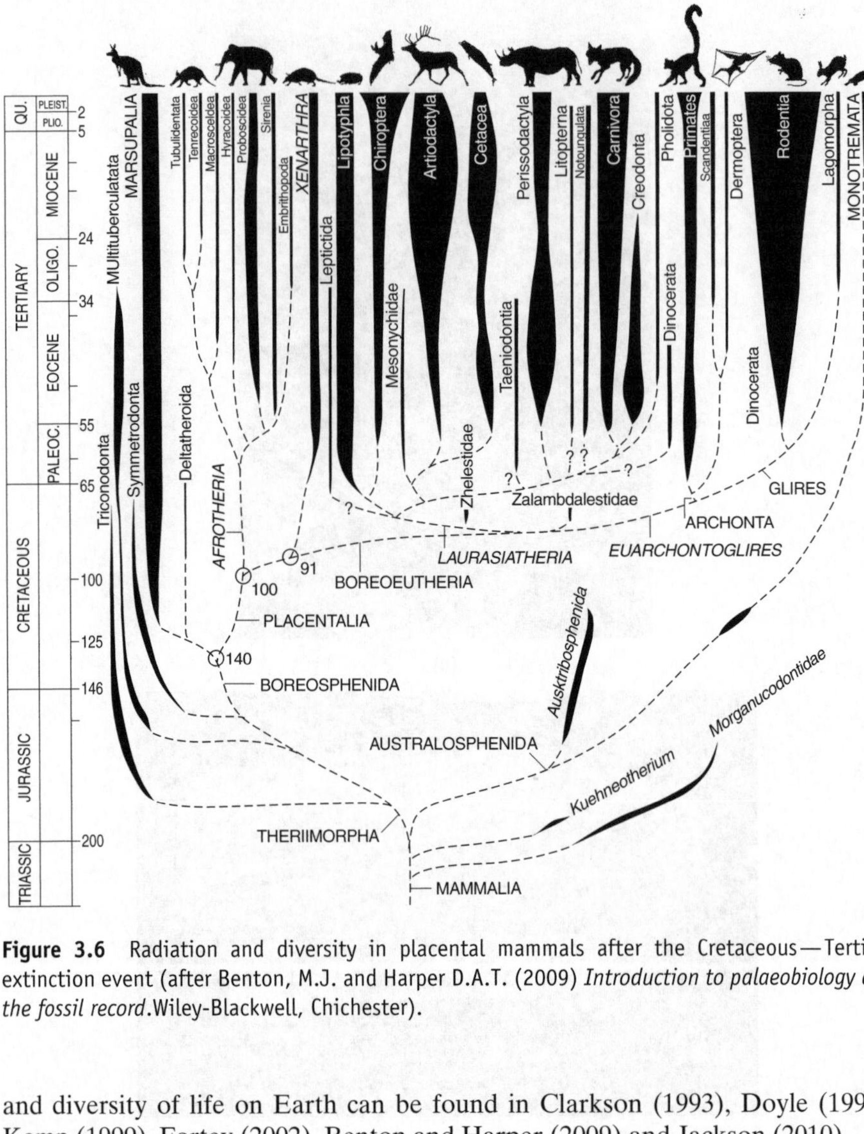

Figure 3.6 Radiation and diversity in placental mammals after the Cretaceous—Tertiary extinction event (after Benton, M.J. and Harper D.A.T. (2009) *Introduction to palaeobiology and the fossil record*.Wiley-Blackwell, Chichester).

and diversity of life on Earth can be found in Clarkson (1993), Doyle (1996), Kemp (1999), Fortey (2002), Benton and Harper (2009) and Jackson (2010).

3.1.4 Soils

Soil is the product of rock and sediment weathering at the Earth's surface, and 'is highly variable from place to place on Earth. In fact, the soil is a collection of individually different soil bodies' (Brady and Weil, 2002, p. 11). Several properties of soil are subject to variability:

- *Colour*. This varies considerably in soils from red to yellows to blacks. Colour is usually described by reference to the internationally recognised Munsell colour chart. Three components of colour are used: hue, chroma (intensity or brightness, grey = 0) and value (lightness or darkness, black = 0).

- *Particle size distribution*. This was discussed with reference to sediments (see earlier) and the same systems are used to describe particle sizes of soils. The percentages of clay, silt and sand present can be used to provide descriptive soil names.
- *Structure*. This refers to the cohesive arrangement of soil particles into groupings called aggregates or peds, and these have a great influence on properties such as water movement, aeration and porosity. The four principle structural types are spheroidal, platy, prismlike and blocklike.
- *Density*. Soil density is usually measured as the bulk density of dry soil. Clays and clay loams usually have lower bulk densities than sandy soils because the former usually form aggregates in which pores exist both between and within the granules. Bulk density is easily affected by a range of human activities (see Chapter 5).
- *Pore spaces*. Pore sizes in soil vary, with macropores, mesopores, micropores, ultramicropores and cryptopores being recognized in descending size order (Brady and Weil, 2002).
- *Other properties*. Several other properties of soils are used, sometimes in particular circumstances. For example, tillage and crusting properties vary and are relevant to agriculture.
- *Horizonation*. Soil profiles are characterized by horizons with five major types being recognised (O, A, E, B and C) plus transition zones. In addition, subordinate descriptions are indicated by lower case letters (see Brady and Weil, 2002, Figure 2.35 and Table 2.6).

The huge diversity of soils is a function of five main factors (Jenny, 1941), all of which are closely interrelated.

- *Parent material*. The nature of the surface rocks and sediments has a profound influence on soil characteristics and properties. For example, weathering of a quartz-rich sandstone will invariably produce a sandy-textured soil with strong vertical drainage properties leading to translocation of fine soil particles and plant nutrients. We have already seen that there is a rich geodiversity of rocks and sediments, and therefore it should not be surprising that this factor alone creates an immense variety of soils.
- *Climate*. This has a major influence on soils because it determines the nature and intensity of weathering processes. In turn this influences several soil characteristics and the rate of soil formation. Water is essential for all chemical weathering processes, and in areas of high rainfall, water is able to percolate deep into the rocks to sustain these weathering processes, particularly where temperatures are also high. Water also leaches soluble and suspended materials and thus leads to soil horizonation. However, in arid areas the lack of water limits soil formation and may lead to the build-up of soluble salts in the surface layers. In turn, these lead to restricted plant growth and limited organic matter in the soil. Thin soils also occur in cold areas of

the world and these can be contrasted with the deeply weathered profiles, sometimes reaching over 50 m depth, of the humid tropics.

- *Biota*. Plants and animals have a strong influence on several soil processes including accumulation of organic matter, biochemical weathering, nutrient cycling, aggregate stability, soil mixing and rates of soil erosion. Since vegetation varies greatly across the Earth's surface, there is a resultant variation in soils. There are also differences in biogeochemical processes between different tree types. The acidic needle litter from coniferous trees decomposes very slowly and recycles only small amounts of Ca, Mg and K. The result is an acidic soil with a thick organic horizon (O-horizon). In contrast, the leaves of deciduous trees are more readily broken down, releasing large amounts of Ca, Mg and K that are then recycled by the trees. The result is a less acidic soil and a thinner forest floor with litter mixed into the A-horizon (Brady and Weil, 2002). Soil organisms (earthworms, termites, etc.) are extremely diverse and they play a diverse set of roles within the soil (Brady and Weil, 2002, Table 11.1).
- *Topography*. Elevation, slope angle, aspect and landscape setting all play a role in soil development, particularly through their impact on other variables. For example, steep slopes encourage water runoff and greater erosion, and therefore tend to have relatively thin soils. On the other hand, thick soils may accumulate at the base of slopes. In low points in the landscape the soils may be waterlogged so that aeration is limited and so-called gleyed soils result. A group of soils developed in a systematic way across a topographically varied landscape, is referred to as a catena.
- *Time*. Rock weathering and soil development take time, so that we would expect that rocks and sediments recently exposed (e.g. those in front of retreating glaciers) would have very thin profiles compared with those where deep weathering has been ongoing for millions of years (e.g. much of the humid tropics). Rates of soil formation clearly depend on other factors, such as climate, but are generally very slow, so that soil can be regarded as a non-renewable resource (see Chapter 4 and Chapter 12).

There is an almost infinite diversity of soils across the surface of the Earth, but in order to make sense of this variation and provide a common language to describe soils, systems of soil classification have been devised. However, different countries have devised their own schemes and there is as yet no fully internationally applied scheme, though Table 3.4 gives the UN classification used on the World Soils Map (Brady and Weil, 2002). For further details, readers are referred to White (1997), Gerrard (2000), Ashman and Puri (2002) and Brady and Weil (2002). The latter give detailed descriptions of the US soil types. As well as the main soil orders, the US system recognizes suborders (many of which are indicative of the moisture regimes), great groups, subgroups, families and, finally, series. There are 19 000 soil series in the United States (Brady and Weil, 2002) and even these do not fully describe the soil variability of the country.

Ibáñez, De-Alba and Boixadera (1995) and Ibáñez *et al.* (1995, 1998) have studied the global diversity and distribution of major soil groups.

Table 3.4 Diversity of World Soils according to the UN Food and Agriculture Organisation (FAO)/UNESCO Soil Map of the World (after Brady and Weil, 2002).

Acrisols	Low base status soils with argillic horizons
Andosols	Soils formed in volcanic ash that have dark surfaces
Arenosols	Soils formed from sand
Cambisols	Soils with slight colour, structure or consistency change due to weathering
Chernozems	Soils with black surface and high humus under prairie vegetation
Cryosols	Soils of cold climates with permafrost
Ferralsols	Highly weathered soils with sesquioxide-rich clays
Fluvisols	Water-deposited soils with little alteration
Gleysols	Soils with mottled or reduced horizons due to wetness
Greyzems	Soils with dark surface, bleached E horizon and Textural B horizon
Histosols	Organic soils
Kastanozems	Soils with chestnut surface colour under steppe vegetation
Lithosols	Shallow soils over hard rock
Luvisols	Medium to high base status soils with argillic horizon
Nitosols	Soils with low cation exchange capacity clay in argillic horizons
Planosols	Soils with abrupt A-B horizon contact
Phaenozems	Soils with dark surface, more leached than Kastanozems or Chernozems
Podozols	Soils with light-coloured eluvial horizon and subsoil accumulation of iron, aluminium and humus
Podzoluvisols	Soils with leached horizons tonguing into argillic B horizons
Rankers	Thin soils over siliceous material
Regosols	Thin soils over unconsolidated material
Rendzinas	Shallow soils over limestone
Solonchaks	Soils with soluble salt accumulation
Solonetz	Soils with high sodium content
Vertisols	Self-mulching, inverting soils, rich in smectite clay
Xerosols	Dry soils of semi-arid regions
Yermosols	Desert soils

3.2 Processes and landforms

This section aims to provide a brief insight into the diversity of processes and landforms at the surface of the Earth, but does not attempt a full description of all processes and landforms. Readers who require more information are referred to some excellent general geomorphological texts such as Summerfield (1991), Bloom (1998), Huggett, (2003) or to the more specialised books referred to later.

3.2.1 Igneous processes and forms

We have already discussed some aspects of igneous processes and landforms in describing plate tectonic processes and igneous rocks earlier. Here we shall concentrate on the diversity of types of eruption, and morphologies of volcanoes and extrusive and intrusive products. For more details, see Francis (1993),

Decker and Decker (1998), Ritchie and Gates (2001), Francis and Oppenheimer (2004) and Lockwood and Hazlitt (2010).

Table 3.5 is a commonly used classification of volcanic eruptions that relates the nature of effusive activity to magma type, explosiveness and volcano morphology. It will be appreciated that this is a simplified system and that the complexity in nature is greater than shown. As already explained, explosiveness is related to magma viscosity and gas content. A basaltic magma with low viscosity and gas content leads to unexplosive eruptions, whereas an acidic, viscous magma typical of a rhyolite leads to gas retention and explosive eruptions as the gas pressure is released at the surface. An example of the latter was the Mount St Helens eruption in Washington, United States, of 18 May, 1980.

Lava types also vary significantly. Low viscosity lava usually forms a ropy surface, given the name 'pahoehoe', where cooling of the surface forms of crust

Table 3.5 Diversity of volcanic eruptions (after MacDonald, 1972; Summerfield,1991).

Type of eruption	Type of magma	Nature of effusive activity	Nature of explosive activity	Structures formed around vent
Icelandic	Basic, low viscosity	Thick, extensive flows from fissures	Very weak	Very broad lava cones; lava plains with construction of cones along fissures in terminal phase
Hawaiian	Basic, low viscosity	Normally thin, extensive flows from central vents	Very weak	Very broad lava domes and shields
Strombolian	Moderate viscosity; mixed basic and acid	Flows absent, or thick and moderately extensive	Weak to violent	Cinder cones and lava flows
Vulcanian	Acid, viscous	Flows frequently absent; thick if present	Moderate	Ash cones, explosion craters
Vesuvian	Acid, viscous	Flows frequently absent; thick if present	Moderate to violent	Ash cones, explosion craters
Plinian	Acid, viscous	Flows may be absent, variable in thickness when present	Very violent	Widespread pumice and lapilli; generally no cone construction
Peléan	Acid, viscous	Domes and/or short, very thick flows; may be absent	Moderate plus nuées ardentes	Domes; cones of ash and pumice
Krakatauan	Acid, viscous	Absent	Cataclysmic	Large, explosion caldera

that is dragged into rope-like forms by further flow (Figure 3.7a). Viscous lava, on the other land forms a blocky, angular lava termed 'aa' (Figure 3.7b). Among the diversity of other features associated with vulcanicity are lava tunnels or tubes which form where the lava drains out from below a solidified crust. Costa *et al.* (2008) have produced a beautifully illustrated book showing the location of over 270 volcanic caves in the Azores, Portugal and describing the phenomena associated with these lava tubes. Pillow lavas are formed by rapid cooling of

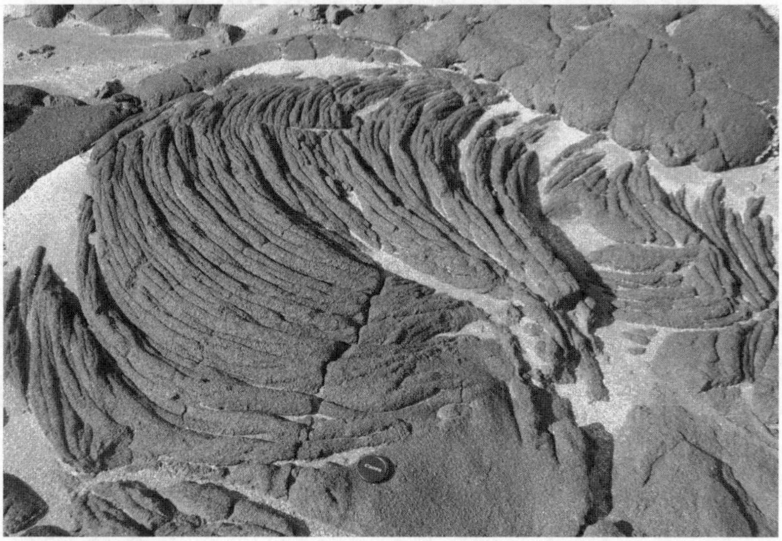

(a)

(b)

Figure 3.7 Lava diversity: (a) pahoehoe lava, Pico, Azores, Portugal and (b) aa lava, Terceira, Azores, Portugal.

Figure 3.8 Pillow lavas, Emilia Romagna, Italy.

lava underwater (see Figure 3.8). Igneous intrusions include batholiths, stocks, laccoliths, lopoliths, sills, dykes, cone sheets and ring dykes and many of these have impressive surface topographic expressions (e.g. Figure 3.9).

3.2.2 Slope processes and forms

Slopes are an integral part of all but the flattest of landscapes and they evolve in a diversity of ways and variety of rates. Much depends on the strength and stability of the slope concerned (Selby, 1993). The downslope movement of slope material under the force of gravity is referred to as mass movement. It occurs where the shear stress exceeds the shear strength of the slope, which is described as actively unstable. Since the operation of many factors potentially leading to instability may vary through time it should be clear that some slopes are described as conditionally stable, depending on the cumulative magnitude of the factors.

The diverse types of mass movement processes have been classified in various ways but generally six types are recognized—creep, flow, slide, heave, fall and subsidence—each of which has a number of sub-groups. Table 3.6 is an attempt

Figure 3.9 Dyke cutting across bedded ash layers, La Palma, Canaries, Spain.

to relate various mass movement types to materials, moisture content, type of movement and rate of movement. However, as Summerfield (1991, p. 168) notes, classifications of this type are valuable in indicating the range of mechanisms and forms of motion, but 'it must be appreciated that most movements in reality involve a combination of processes. Debris avalanches, for example may begin as slides consisting of large masses of rock but then rapidly break up to form flows as the material is pulverised in transit'.

Mass movement processes also result in a range of landforms. For example, Figure 3.10 shows some flow-type morphologies and these do not include major ice avalanches like those in 1962 and 1970 at Huascarán Norte in the Peruvian Andes. Slides occur when movement takes place as blocks without internal deformation. Some are translational slides taking place on planar surfaces, but rotational slides are common in homogeneous clays. Falls involve the downward movement of

Table 3.6 Diversity and characteristics of major mass movement types (after Varnes, 1978; Summerfield, 1991).

Primary mechanism	Mass movement type	Materials in motion	Moisture content	Type of strain and nature of movement	Rate of movement
Creep	Rock creep	Rock (esp. readily deformable types such as shales and clays	Low	Slow plastic deformation of rock or soil producing a diversity of forms including	Very slow to extremely slow
	Continuous creep	Soil	Low	Cambering, valley bulging and outcrop bedding curvature	
	Dry flow	Sand or silt	Very low	Funnelled flow down steep slopes of non-cohesive sediments	Rapid to extremely rapid
	Solifluction	Soil	High	Widespread flow of saturated soil over low to moderate angle slopes	Very slow to extremely slow
Flow	Gelifluction	Soil	High	Widespread flow of seasonally saturated soil over permanently frozen subsoil	Very slow to extremely slow
	Mud flow	>80% clay size	Extremely high	Confined elongated flow	Slow
	Slow earthflow	>80% clay size	Low	Confined elongated flow	Slow
	Rapid earthflow	Soil containing sensitive clays	Very high	Rapid collapse and lateral spreading of soil following disturbance, often by initial slide	Very rapid
	Debris flow	Mixture of fine and coarse debris (20–80% coarser than sand-sized)	High	Flow usually focused into pre-existing drainage lines	Very rapid
	Debris (rock) avalanche	Rock debris, sometimes with ice and snow	Low	Catastrophic low friction movement of up to several kilometres, usually started by a major rock fall and capable of overriding significant topographic features	Extremely rapid

(continued overleaf)

Table 3.6 *(continued)*

Primary mechanism	Mass movement type	Materials in motion	Moisture content	Type of strain and nature of movement	Rate of movement
	Snow avalanche	Snow and ice (plus debris)	Low	Catastrophic low friction movement started by fall or slide	Extremely rapid
	Slush avalanche	Water-saturated snow	Extremely high	Flow along existing drainage lines	Very rapid
Translational slide	Rock slide	Unfractured rock mass	Low	Shallow slide approx. parallel to ground surface of coherent rock mass along single fracture	Very slow to extremely rapid
	Rock block slide	Fractured rock	Low	Slide approx. parallel to ground surface of fractured rock	Moderate
	Debris/earth slide	Rock debris or soil	Low to moderate	Shallow slide of deformed masses of soil	Very slow to rapid
	Debris/earth block slide	Rock debris or soil	Low to moderate	Shallow slide of largely undeformed masses of soil	Slow
Rotational slide	Rock slump	Rock	Low	Rotational movement along concave failure plane	Extremely slow to moderate
	Debris/earth slump	Rock debris or soil	Moderate	Rotational movement along concave failure plane	Slow
Heave	Soil creep	Soil	Low	Widespread incremental downslope	Extremely slow
	Talus creep	Rock debris	Low	movement of soil or rock particles	Slow
Fall	Rock fall	Detached rock joint blocks	Low	Fall of individual blocks from vertical faces	Extremely rapid
	Debris/earth fall (topple)	Detached cohesive units of soil	Low	Toppling of cohesive units of soil from near vertical faces such as river banks	Very rapid
Subsidence	Cavity collapse	Rock or soil	Low	Collapse of rock or soil into underground cavities such as limestone caves	Very rapid
	Settlement	Soil	Low	Lowering of surface owing to ground compaction or shrinkage on withdrawal of water	Slow

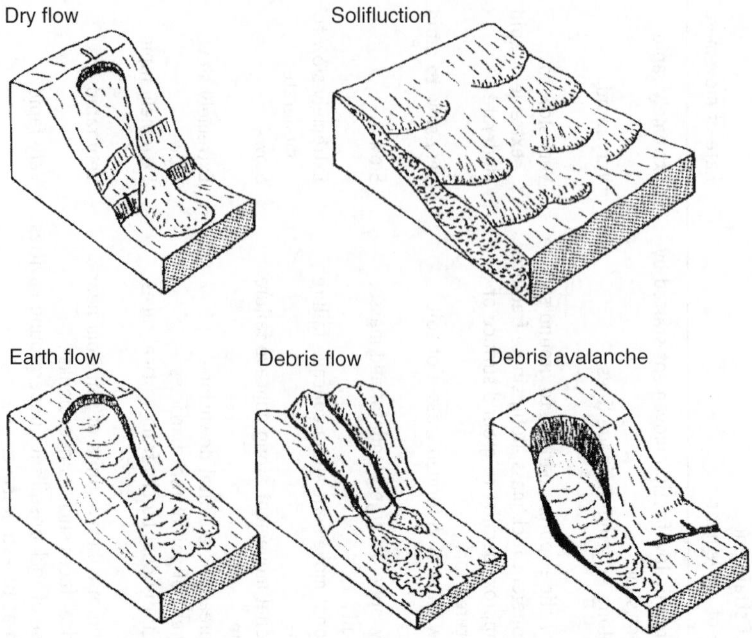

Figure 3.10 Some flow-type morphologies (after Varnes (1978) Special Report 176: Landslides: Analysis and Control, Transportation Research Board, National Research Council, Washington, DC. Reproduced with permission from the Transportation Research Board).

detached rock or soil masses. An example is the Frank Slide in south-west Alberta Canada which occurred in 1903 (Kerr, 1990), while more recently a large section at the summit of Mt Cook in New Zealand collapsed. On a more local scale, gradual fragment detachment from cliffs causes the accumulation of talus slopes and cones at the base of cliffs. As well as mass movement processes, slopes are also affected by water. For example, rain falling on bare soil or loose sediment dislodges particles by rainsplash erosion while flowing water may produce surface rills or gullies.

Much has been written regarding slope form and slope evolution. Figure 3.11 shows the nine possible three-dimensional slope forms and special slope forms include inselbergs, mesas and buttes (Figure 3.12).

3.2.3 River environments

In the previous section we noted that water has an impact on slopes, and here we extend this to outline the diversity of fluvial processes and landforms. For more details see Knighton (1998), Bridge (2002), Robert (2003) and Charlton (2008). Rivers also transport sediment and Bloom (1998), amongst others, believes that 'Water flowing down to the sea and immediately beneath the land surface, is the dominant agent of landscape alteration'. Garrells and Mackenzie (1971) calculated that over the planet as a whole, rivers are responsible for 85–90% of sediment transport to the sea, glaciers for about 7%, with wind, volcanoes and

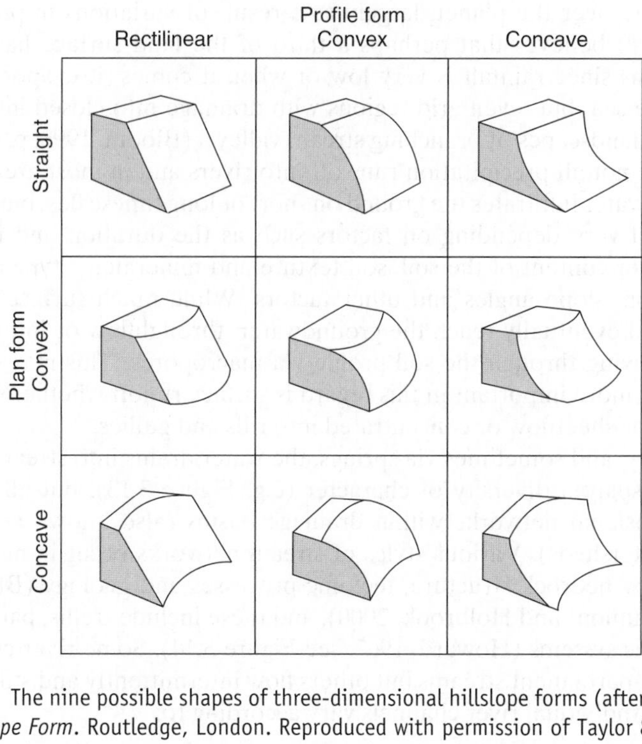

Figure 3.11 The nine possible shapes of three-dimensional hillslope forms (after Parsons, A.J. (1988) *Hillslope Form*. Routledge, London. Reproduced with permission of Taylor & Francis).

Figure 3.12 Buttes at Monument Valley, USA. See plate section for colur version.

other processes responsible for the remainder. Of course, the impact of rivers varies greatly over the planet, largely as a result of variations in precipitation. Bloom (1998) believes that perhaps a third of the land surface has no runoff to the oceans since rainfall is very low or when it comes, it evaporates before reaching the sea, but ‘even arid regions with drainage into closed intermontane basins have landscapes of branching stream valleys’ (Bloom, 1998, p. 198).

Of course not all precipitation runs off into rivers and in most areas the large majority of water infiltrates the ground on short or long timescales, but infiltration capacity will vary depending on factors such as the duration and intensity of rainfall, water content of the soil, soil texture and mineralogy, type and density of vegetation, slope angles and other factors. While much surface water may infiltrate and eventually reach the groundwater, throughflow or interflow refers to water moving through the soil profile via macropores. This may lead to soil erosion, but more important in this regard is surface runoff whether spread over the surface as sheetflow or concentrated into rills and gullies.

Eventually, and sometimes via springs, the water drains into streams or rivers, that also display a diversity of character (e.g. Figure 3.13), but all operate as three-dimensional networks within drainage basins (also known as catchment areas or watersheds). Various styles of stream networks occur in nature, partly controlled by bedrock structure, tectonic processes and geology (Bloom, 1998; Schumm, Dumont and Holbrook, 2000), and these include trellis, parallel, radial and dendritic systems (Howard, 1967; see Figure 3.14). Some river channels are occupied by permanent streams but others flow intermittently and still others are ephemeral. Individual river channels vary according to:

- long profile—normally concave between headwaters and the sea in graded profiles, but can be linear or sometimes convex. At a smaller scale, gradient may vary between and within reaches, and may be interrupted by pools and riffles, waterfalls, rapids, cataracts and lakes.
- planform—may vary between single fairly straight channels, through strongly meandering rivers to the highly complex braided channel networks. In turn, these river forms are related to gradient, sediment load, flow velocity and other factors. Meandering rivers are typified by neck cutoffs, oxbow lakes and point bar deposits on the inside of meander curves.
- channel cross-section in meandering rivers is asymmetrical with deeper sections on the outside of meanders, but in straight streams more or less symmetrical cross-sections occur. In the latter case, river banks may be vegetated and stable, whereas in other cases they are actively eroding.

Other fluvial landforms include incised meanders where downcutting into bedrock has occurred (Figure 3.15), natural arches, floodplains with levees, alluvial fans and river terraces. Miller and Gupta (1999) and Schumm (2005) describe the diversity of river forms and processes. For example, ‘Rivers differ among themselves and through time. An individual river can vary significantly downstream, changing its dimensions and pattern dramatically over a short distance’ (Schumm, 2005). Special features occur in limestone areas where dry valleys, sinkholes and underground caves with a range of speleothems may be

(a)

(b)

Figure 3.13 River diversity: (a) rock channel river, Killin, Scotland and (b) braided outwash river, Exit Glacier, Alaska.

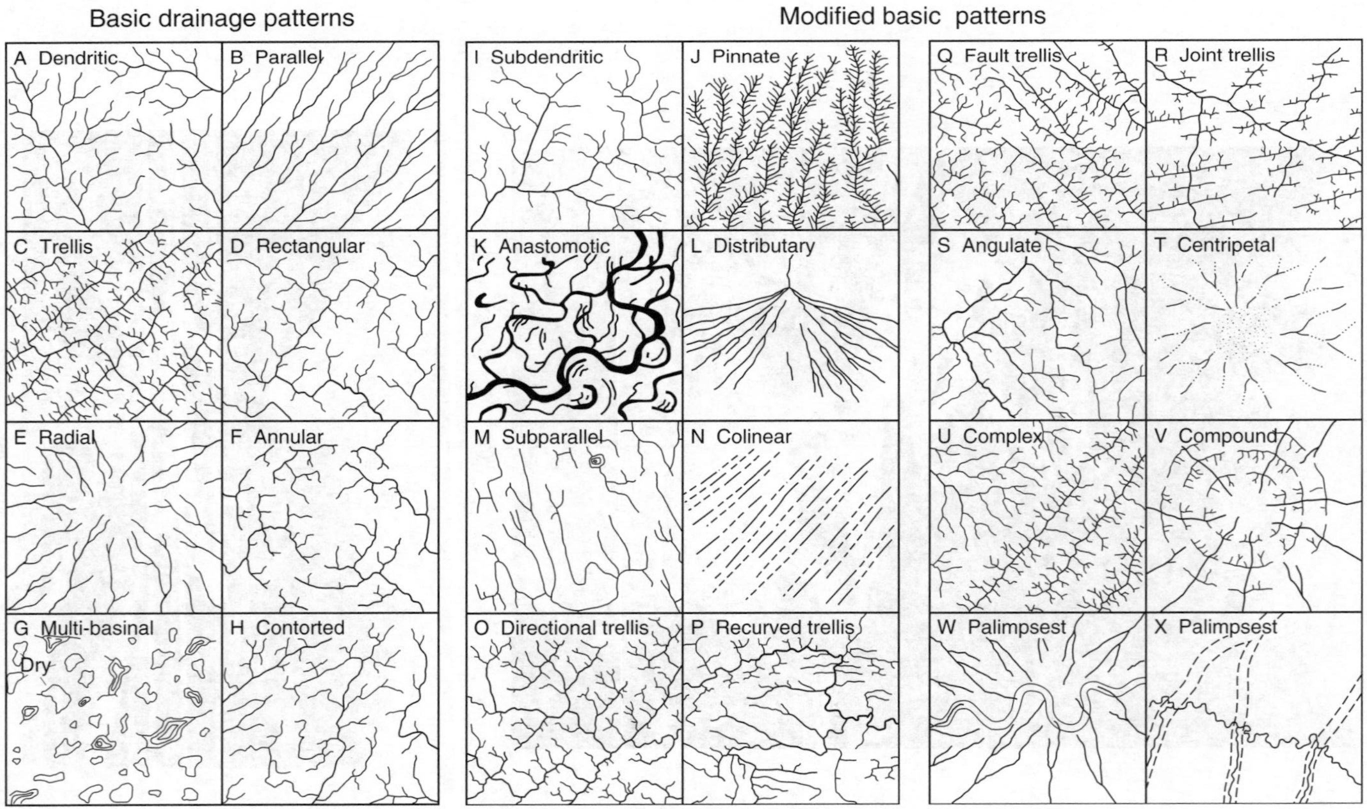

Figure 3.14 Diversity of basic and modified drainage systems (after Howard, A.D. (1967) Drainage analysis in geologic interpretation: a summation. *American Association of Petroleum Geologists Bulletin*, 51, 2246–2259. AAPG © 1967. Reprinted by permission of the AAPG whose permission is required for further use).

Figure 3.15 Incised meanders of the San Juan River, USA. See plate section for colur version.

present (see also Section 3.2.8, Weathering environments for a discussion of karst landforms).

The sediment load of rivers is also diverse, and includes a dissolved load, a suspended load of fine particles and a bed load of coarser debris that rolls, slides or saltates along the channel bed. Stream discharge varies through time as does the sediment load and transport regime. There are also longer timescale changes in rivers induced by climate changes so that some fluvial landforms and deposits are relict features.

Various systems have been used to describe and classify rivers including Geomorphic Characterisation (Rosgen, 1996), River Habitats Surveys (Raven *et al.*, 1997), Stream Reconnaissance Surveys (Thorne, 1998), the River Styles approach (Hardie and Lucas, 2001; Thomson *et al.*, 2001; Brierley and Fryirs, 2005) and Geomorphic Complexity (Bartley and Rutherfurd, 2001). 'Classifications are nonetheless arbitrary divisions of continua and cannot be expected to provide a complete understanding of river functioning' (Soulsby and Boon, 2001, p. 100).

3.2.4 Coastal environments

At a global scale, the world's coastlines can be classified according to their plate tectonic setting and Figure 3.16 shows Davies' (1980) attempt to present such a classification. Passive margin coasts occur on the trailing edge of the continents and are typified by tectonic stability and wide continental shelves. Examples occur along the eastern coast of both North and South America. On the other hand, convergent margin coasts occur near to subduction zones or continental collisions and are typified by coastal instability and narrow shelves. Examples occur on the western coast of South America and Mediterranean. Island arcs are also common

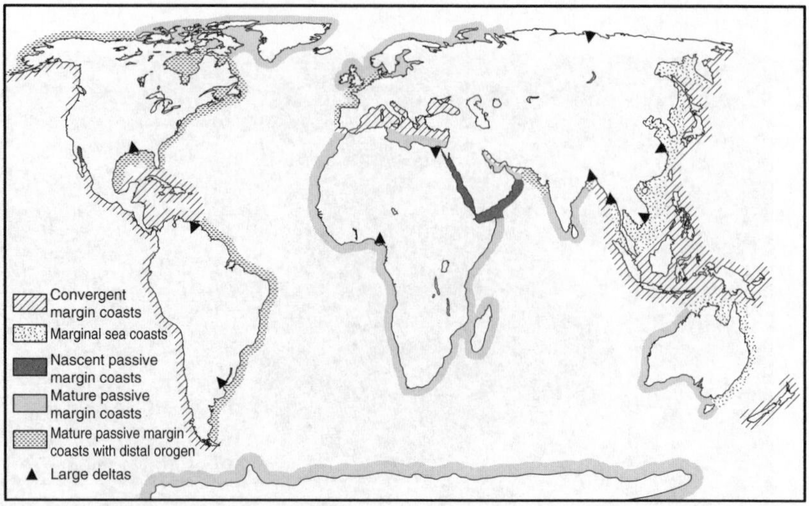

Figure 3.16 Tectonic setting of the world's coastlines and major deltas (after Davies, J.L. (1980) *Geographical Variation in Coastal Development*, 2nd edn. Longman, Harlow; Summerfield, M.A. (1991) *Global Geomorphology*. Longman, Harlow © M.A. Summerfield, 1991, by permission of Pearson Education Ltd).

as exemplified in the western Pacific and continental margins protected by these are termed marginal coasts and occur, for example on coastal China, Vietnam and eastern Australia. Summerfield (1991, p. 315) has noted that large river deltas such as the Mississippi, Niger and Huang Ho are, not surprisingly, more or less confined to passive margin coasts and marginal coasts 'since it is only these that provide the outlets for the world's great drainage systems'.

Davies (1980) notes several other aspects of global-scale coastal diversity. For example, major wave environments vary according mainly to the nature of the atmospheric circulation, with storm wave environments in the westerly zones and monsoonal influences in some tropical locations. High latitude coasts are often low-energy coasts because of the protection of sea ice. Between these locations swell environments predominate. Tidal regime and range are also highly variable.

At a smaller scale, Kiernan (1997a) attempted to assess the factors contributing to coastal geodiversity:

- bedrock variables—lithology, structure, orogenic and tectonic movements, bedrock preparation, sediment characteristics;
- topographic variables—submarine topography, hinterland topography, shoreline topography;
- oceanographic variables—waves, tsunamis, tides, storm surges, currents, sea-level change;
- coastal process variables—shore weathering, sediment transport;
- temporal variables—duration of processes, rate of sea-level change, number of change cycles.

Kiernan (1997a) went on to classify Tasmania's coastal landforms and to specifically identify their geodiversity.

We can recognise both erosional and depositional processes and landforms (Bird, 2000; Davis and Fitzgerald, 2002; Haslett, 2008; Masselink, Hughes and Knight, 2011). Coastal erosion commonly results in cliffs and shore platforms whose detailed morphology will largely be determined by geological structure and lithology. Platforms on dipping sedimentary rocks form series of mini-escarpments compared with the flat platforms that form on more homogeneous rocks. The variety of lithologies, joint patterns, folds and dip angles mean that cliffs and platforms, as well as related features such as caves, undercuts, natural arches and sea stacks, display a very large morphological diversity. Wave erosion is not the only important process either. Sub-aerial processes and mass movements operate on the cliffs, solutional processes operate on carbonate rocks and a range of biochemical processes operate on the shore platforms. Bloom (1998) notes that 'Animals from at least 12 phyla, and numerous kinds of plants and microbes, will graze, browse, burrow, or bore into rocks. ... Many use chemical secretions to dissolve rocks, especially limestone. Many of the animals have abrasive appendages or teeth by which they remove surface layers of rocks or bore into them.' The net effect may be rapid bioerosion and a range of micromorphological features.

Biological processes are also responsible for the development of coral reefs and atolls that grow into morphological features and are represented in the geological record. Barrier reefs, such as the Great Barrier Reef in Australia, are distinguished from fringing reefs, which are attached to a coast and extend seaward, and island reefs of which Davies (1980) identified various types. Detailed morphology depends partly on the reef-building coral genera present, of which there are over fifty.

Depositional coastal processes and morphologies also vary. In some places, sandy beaches occur but these vary in particle size, gradients, morphology and planform. Bars, berms, beach cusps, sand waves, ripples and stream channels are examples of the morphological forms found on sandy beaches. They may or may not be backed by storm ridges and coastal dunes of which there are several types (Davies, 1980). In places offshore barriers separated from the mainland by lagoons form, while spits, baymouth bars, cuspate forelands and tombolos are other well-known landforms. On lower energy coasts, mudflats and saltmarshes with dendritic tidal creeks and ponds occur, sometimes backed by cheniers (beach ridges of sand and shell debris). Passive tropical coasts are often dominated by mangrove swamps. Delta morphology is also diverse (see Figure 3.17).

Since sea-level changes, relict coastal features occur both above and below present sea-level and also at present sea-level which may have reoccupied a much older coastline. Most of the coastal landforms described above occur in relict form (e.g. Figure 3.18 and Figure 4.20).

3.2.5 Glacial environments

Glaciers currently cover about 10% of the Earth's land surface, but during full glacial periods the percentage rose to around 30%, mainly by expansion

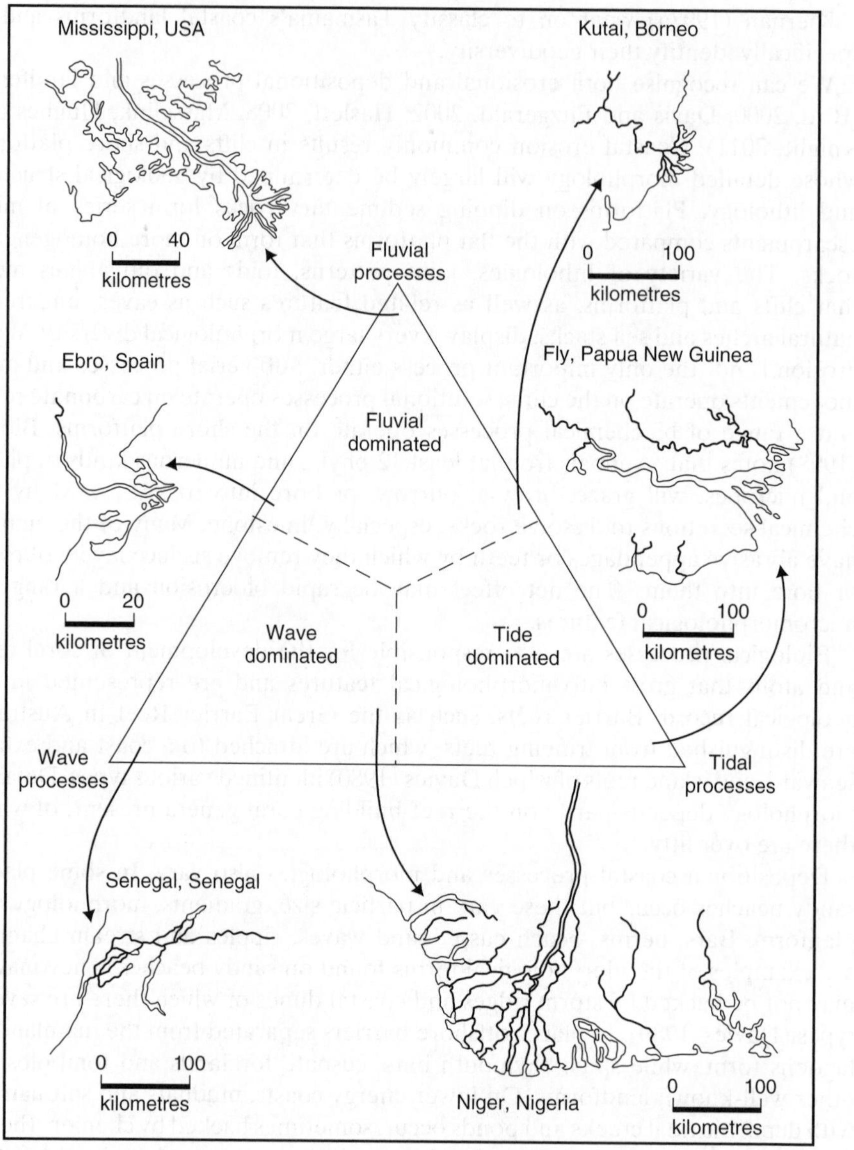

Figure 3.17 Delta diversity related to fluvial, tidal and wave influences (after Summerfield, M.A. (1991) *Global Geomorphology*. Longman, Harlow © M.A. Summerfield, 1991, by permission of Pearson Education Ltd).

of North American and European ice-sheets. Thus, in considering the diversity within glacial environments we must consider both the glacier ice itself (as a solid component of the lithosphere) and the areas of former glaciation where the geomorphological impacts still dominate the landscape. Fuller details of the diversity of glacial environments and land systems can be found in Martini

(a)

(b)

Figure 3.18 Diversity of relict coastal erosional features: (a) underct cliffline and (b) natural arch, Argyll, Scotland. See plate section for colour version.

et al. (2001), Benn and Evans (2010) and Bennett and Glasser (2010). Knight and Harrison (2009) describe paraglacial processes in the unstable landscapes following deglaciation.

Kiernan (1996) has specifically described the geodiversity of glacial landforms, describing both glacial landform 'species' and glacial landform 'communities'.

Table 3.7 Diversity of ice mass types (modified after Martini, Brookfield and Sadura, 2001).

Ice-sheets	Continental ice-sheets	Shield-like domes > 25 000 km^2
	Lowland ice caps	Smaller ice domes in lowland areas
	Plateau glaciers	Flat plateau ice fields with ice cascades
	Highland ice caps	Mountain ice fields with many nunataks
Valley Glaciers	Ice streams	Fast moving ribbons within ice-sheets
	Reticular glaciers	Transitional from ice-sheets and glaciers
	Outlet glaciers	Glaciers descending from ice-sheets/ice caps
	Alpine glaciers	Glaciers originating in cirques
	Cirque glaciers	Glaciers confined to cirques
	Cliff or niche glaciers	Small glaciers on steep slopes or cliffs
Lowland Glaciers	Piedmont glaciers	Low-angle ice-lobes laterally unconstrained
	Expanded foot glaciers	Smaller lobes where valley glaciers fan out
	Ice shelves	Floating sections of glaciers or ice-sheets

He sees the major controls on glacial landscape evolution and therefore on geodiversity as being:

- glacier variables—ice temperature, glacier morphology, glacier constraint, glacier gradient; glacier movement and velocity, ice thickness, glacial processes;
- bedrock variables—lithology, structure, orogenic and tectonic history, bedrock preparation, glacial sediment;
- topographic variables—preglacial topography, contemporaneous topography, postglacial topography;
- temporal variables—duration of glaciation, number of glacial stages.

Ice masses can be classified in various ways but the one most relevant to this book is the so-called morphological classification that considers size, shape and environmental location (Table 3.7). Only two ice-sheets exist at present (Greenland and Antarctica) but there are thousands of cirque and valley glaciers. Surface features may include a variety of crevasse patterns, ogives, supraglacial stream channels and ponds, moulins, lateral moraines, medial moraines and broader debris spreads. Glaciers terminating in the sea or a lake may calve to form impressive iceberg arrays or large rafts.

Glacial erosion produces a very large range of features at several scales (Figure 3.19). At the continental scale, areas such as the Canadian Shield are scoured to produce a complex interplay of rock and freshwater lakes. Mountain areas, on the other hand, are carved into horns, aretes, cirques and glacial troughs (or fjords if flooded by the sea). Smaller features such as crag-and-tails and roches moutonnées occur, while at the smallest scale, striae, polished surfaces and a range of friction cracks are observed.

Glacial depositional processes produce a range of till types which in turn vary in sedimentological properties. Glacial depositional landforms include till sheets and a range of moraines and drumlinoid forms. Drumlins themselves are highly

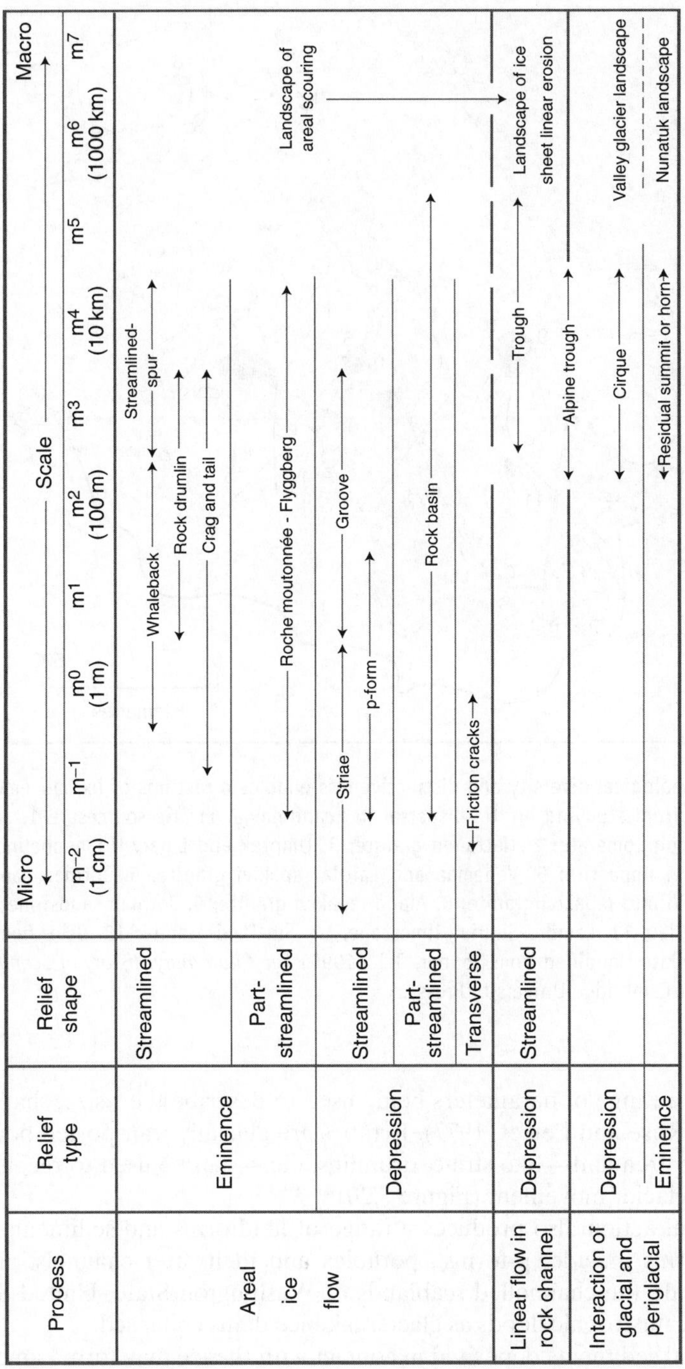

Figure 3.19 Landforms of glacial erosion indicating the diversity of forms and scales (modified after Sugden, D.E. and John, B.S. (1976) *Glaciers and Landscape: A Geomorphological Approach*. Edward Arnold, London. Reproduced by permission of Hodder Education).

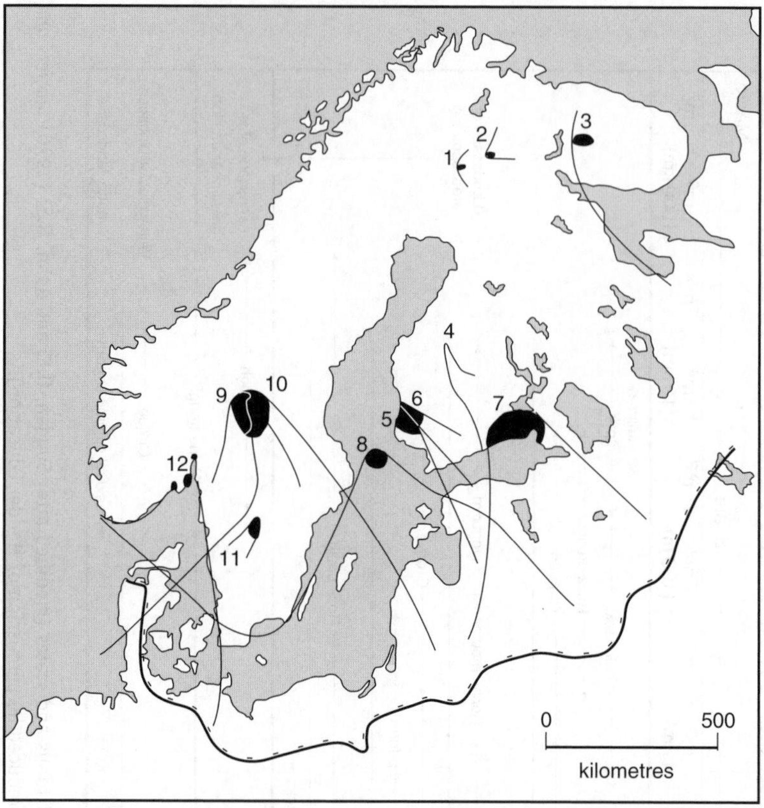

Figure 3.20 Geological diversity and distinctiveness enables directions of ice movements to be estrablished from studying erratic dispersal in Scandinavia. Erratic sourcesare 1. Jatulian sandstone and conglomerate; 2. Nattanen granite; 3. Umptek and Lujarv-Urt nepheline syenite; 4. Lappajarvi impactite; 5. Vehemaa and Laitila rapakivi granite; 6. Jotnian sandstone (Satakunta); 7. Viipuri rapakivi granite; 8. Aland rapakivi granite; 9. Jotnian sandstone (Dala); 10. Dala porphyries; 11. Cambro-Silurian limestone; 12. Smaland granite; 13. Oslo rhomb porphyries and larvikite (modified from Donner, J.J. (1995) *The Quaternary History of Scandinavia*, by permission of Cambridge University Press).

variable with a range of parameters being used to describe their size, shape and distribution (Rose and Letzer, 1977). Erratics are glacially transported boulders foreign to the area and—like striae, drumlins, etc.—can be used to reconstruct directions of glacier movement (Figure 3.20).

Glaciofluvial action also produces a range of landforms and sediments. Erosional landforms include 'p-forms', potholes and meltwater channels, and, at the largest scale the channelled scablands of Washington State, United States, produced by catastrophic floods as glacier lake ice dams collapsed.

Glaciofluvial sediments deposited in contact with the ice may form kames and kettle holes, eskers and kame terraces, while beyond the ice, braided meltwater streams deposit outwash fans, trains or plains (also known as sandar; Figure 3.13) which may become terraced by fluvial downcutting to produce outwash terraces. On reaching the sea or lake, the traction load is usually deposited as glaciofluvial deltas, while the finer material settles as rhythmite sediments called varves. Lakes are frequently formed around the margins of glaciers and ice-sheets including around the southern margin of the North American ice-sheet, for example Lake Agassiz west of Lake Superior. These lakes were frequently large enough to have shoreline features such as beaches and spits, but other lakes have erosional shorelines as seen at the Parallel Roads of Glen Roy in Scotland.

3.2.6 Periglacial environments

Like glacial environments, periglacial activity also exists in both active and fossil forms. The periglacial zone has been defined in various ways. French (2007), for example, has defined it as the area where frost action processes predominate, but Bloom (1998) believed it should be restricted to those non-glacial areas of high latitude or altitude that are underlain by permafrost.

Permafrost is ground that remains frozen for at least two years. Its thickness increases with latitude from sporadic patches, through a discontinuous zone to a continuous zone where depths of over 1000 m are possible. Unfrozen areas in, on, below or between permafrost are termed talik. Above the permafrost is the active layer that freezes and thaws annually, but this also varies in thickness, reaching a maximum of 2–3 m in the discontinuous permafrost zone. Permafrost ice may exist in several forms including pore ice, ice layers and lenses, vein ice, intrusive ice and needle ice.

Periglacial processes include frost weathering responsible for the disintegration of rock by ice expansion upon freezing of water in pores, joints, etc. The effects will vary depending on the nature of the rock. For example, porous rocks such as chalk will be reduced to a paste, whereas a well-cleaved rock like slate will split into thin rock fragments. On horizontal ground this results in blockfields or felsenmeer, but on steep slopes the material collects as talus slopes or protalus ramparts in cases where the material accumulates below a snowpatch. Rock glaciers are tongue-shaped masses of frozen debris (Giardino, Schroder and Vitek, 1987). Involutions result from sediment distortion when coherent beds freeze.

Frost heaving and thrusting are responsible for vertical and horizontal movement of coarse material and the formation of many types of patterned ground, such as sorted polygons and stone stripes. Other types, such as frost wedge polygons are the result of frost cracking in which the ground fractures due to contraction at low temperatures. Washburn (1979) related patterned ground morphologies to formative processes. Solifluction lobes form where thawed material moves slowly downslope, and they may be either stone-banked or turf-banked.

Pingos are ice-cored hills up to 70 m high and 700 m in diameter that are particularly common on the Mackenzie delta area of the Canadian arctic. These are classified as closed system types since they form by sealing of talik below shallow lakes and subsequent freezing and upward expansion of the water/ice. Open system types form where spring water rises to the surface and freezes and they often occur in groups. A variety of other frost mounds have been described including palsas which are smaller mounds of peat or lens ice typical of the discontinuous zone.

Melting of the permafrost creates a variety of features collectively known as thermokarst. The features include thaw lakes, thaw mounds where ice wedge polygons thaw, and alas depressions and lakes where large-scale collapse of the tundra surface occurs (Czudek and Demek, 1970). In riverbanks and coastlines, ice lenses may melt when exposed to the atmospheric temperatures and water action in summer. Lateral slope retreat is often the result of this thermal erosion. Fluvial regimes in periglacial areas show very dramatic annual and diurnal cycles. On the coast, shore platform formation may be rapid due to frost shattering of the cliffs.

The presence of many of these features have been recognised in fossil form south of the former North American and European ice sheets, as well as in other parts of the world. In China, Eastern Europe, parts of the United States and other areas, large thicknesses of wind-blown silt (loess) have accumulated under periglacial conditions when there was abundant fine debris, little vegetation and strong wind action. However, the particle size of loess varies depending on the local sources of dust.

3.2.7 Arid environments

'The processes of landscape development in dry regions differ only in frequency and intensity, rather than in kind, from those in humid regions. ... The largest single identifiable climatic region on earth is the dry region, where either seasonal or annual precipitation is insufficient to maintain vegetational cover and permit perennial streams to flow' (Bloom, 1998, p. 277). Nonetheless, there have been various attempts to identify the dry region, some of which distinguish hyperarid, arid and semi-arid areas. Together they currently cover over 25% of the earth's land surface principally in two belts coinciding with the subtropical anticyclonic belts around 15–30° N and S.

In these areas, fluvial processes may be limited, but when rainfall does occur the normally dry river channels (wadis) are filled with sediment-laden water (Graf, 1988; Bull and Kirkby, 2002). Bullard and Livingstone (2002) describe the interactions between fluvial and aeolian systems in dryland environments. In mountainous desert areas, alluvial fans coalesce to form bajadas, which grade into playas or salt lakes.

Aeolian erosion produces coarse debris lag deposits termed desert pavements often with faceted stones (ventifacts). Yardangs are wind-abraded hills and ridges with steeper stoss slopes, for example at Qaidam in China. Deflation hollows and larger depressions such as the Qattara in Egypt also occur.

Aeolian sediment transport occurs by suspension of fine dust, saltation or surface creep. The sediment often accumulates in ripples, dunes or megadunes. Several types of dune have been recognised, but the basic classification is into free dunes, whose form is related to the wind regime, and impeded dunes, whose

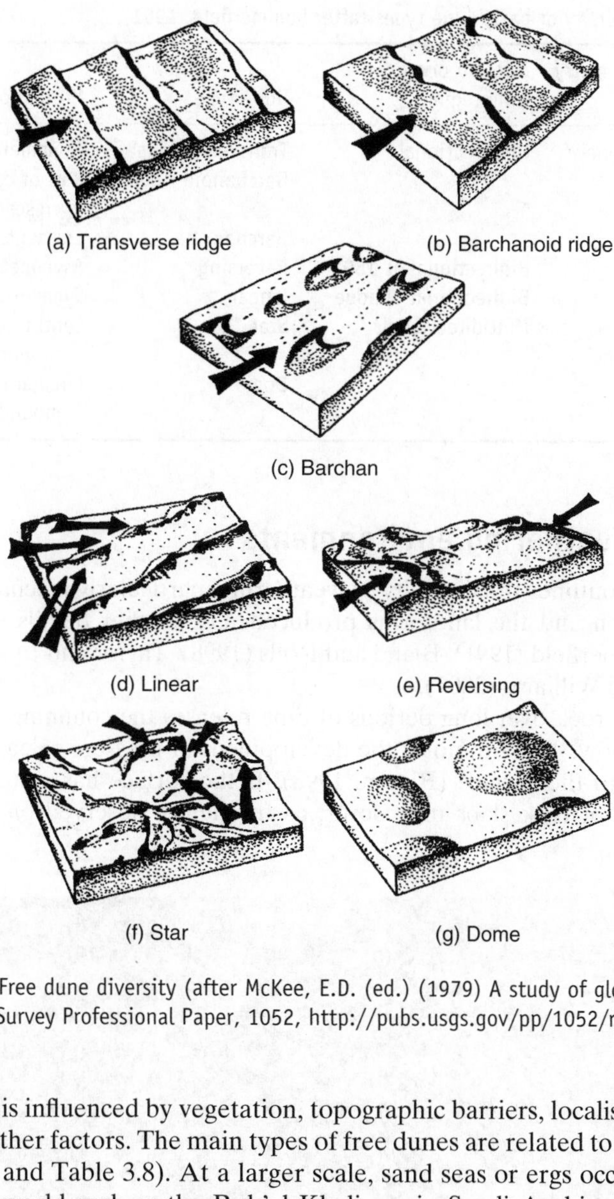

Figure 3.21 Free dune diversity (after McKee, E.D. (ed.) (1979) A study of global sand seas. US Geological Survey Professional Paper, 1052, http://pubs.usgs.gov/pp/1052/report.pdf).

morphology is influenced by vegetation, topographic barriers, localised sediment supplies or other factors. The main types of free dunes are related to wind regime (Figure 3.21 and Table 3.8). At a larger scale, sand seas or ergs occur in certain parts of the world such as the Rub'al Khali erg in Saudi Arabia which covers 560 000 km^2. Summerfield (1991) shows the distribution of ergs both at the present day and relict, illustrating the fact that aeolian processes were more extensive in the past as demonstrated by stabilised dunes, for example, in many parts of Africa. Further details on aeolian environments, sediments and landforms can be found in Cooke, Warren and Goudie, (1993), Abrahams and Parsons (1994), Lancaster (1995), Livingstone and Warren (1996), Goudie, Livingstone and Stokes, (1999), Laity (2008) and Thomas (2011).

Table 3.8 Diversity of basic dune types (after Summerfield, 1991).

Number and geometry of slip faces	Inferred primary wind regime	Dune type	Morphology
One; unidirectional	Unidirectional	Transverse ridge	Asymmetric ridge
		Barchanoid ridge	Row of continuous crescents
		Barchan	Crescentic form
Two; opposing	Bidirectional at 180°	Reversing	Asymmetric ridge
Two; opposing	Bidirectional oblique	Linear	Symmetric ridge
Three or more; multidirectional	Multidirectional	Star	Central peak with three or more arms
None		Dome	Circular or elliptical mound

3.2.8 Weathering environments

This section outlines the diversity of weathering, particularly mechanical weathering, solution and the landforms produced. For further details see Jennings (1985), Summerfield (1991), Bland and Rolls (1998), Taylor and Eggleton (2001) and Ford and Williams (2007).

Erosion of rock over long periods of time releases the confining pressures on the rocks below. This promotes the development of pressure release joints that are parallel to the surface (Bloom, 1998). If the surface has a steep gradient, slope failure of rock slabs may occur, as frequently observed on the sides of

Figure 3.22 Honeycomb weathering, Rab Island, Croatia.

Table 3.9 Diversity of solutional microforms developed in limestone (after Jennings, 1985; Summerfield, 1991).

	Form	Typical dimensions	Comments
Forms developed on bare limestone by areal wetting	Rainpit	<30 mm diameter <20 mm deep	Formed by rain falling on bare, gentle rock slopes. Coalescence gives irregular carious appearance
	Solution ripples	20–30 mm high can be >100 mm long	Wave-like form transverse to dowslope water movement
	Solution flutes (rillenkarren)	20–40 mm across 10–20 mm deep	V-shaped or semi-circular channels formed by concentrated flow down steep slopes
	Solution bevels	0.2–1.0 m long 30–50 mm high	Flat, smooth elements found below flutes. Flow over them occurs as a thin sheet
	Solution runnels (rinnenkarren)	400–500 mm across 300–400 mm deep 10–20 m long	Larger channels formed by increased water flow. May have meandering form
Forms developed on bare limestone concentration of run-off	Grykes (kluftkarren)	500 mm across; up to several metres deep	Formed by the solutional widening of near-vertical joints or bedding
	Clints (flackkarren)	Up to several metres across	Tabular blocks separated by the development of grikes
	Solution spikes (spitzkarren)	Up to several metres	Sharply pointed projections between grikes
Forms developed on partly covered limestone	Solution pans	10–500 mm deep 0.03–3.0 m wide	Dish-shaped depressions usually floored by thin soil, vegetation or algae
	Undercut solution runnels (hohlkarren)	400–500 mm across 300–400 mm deep 10–20 m long	Like runnels but become larger with depth. Recession at depth probably associated with accumulation of humus or soil that keeps sides and base constantly wet
	Solution notches (korrosionkehlen)	1 m high and wide 10 m long	Produced by active solution where soil abuts against projecting rock, giving rise to curved incuts
Forms developed on covered limestone	Rounded solution runnels (rundkarren)	400–500 mm across 300–400 mm deep 10–20 m long	Runnels developed beneath a soil cover that becomes smoothed by more active corrosion associated with acid soil waters
	Solution pipes	1 m across 2–5 m deep	Funnels becoming narrower with depth. Found on soft limestones such as chalk as well as on stronger and less permeable types

glacial troughs, e.g. Yosemite Valley, United States. At a smaller scale, thermal expansion and contraction of rocks aided by the presence of moisture and salts may lead to the spalling of thin sheets of rock in a process known as exfoliation or onion weathering. The end product tends to be a rounding of the weathering mass, sometimes producing exfoliation domes or rounded inselbergs known as bornhardts. Tafoni and honeycomb weathering pits are also believed to be related to salt weathering since they commonly, though not exclusively, occur in the coastal spray zone (Figure 3.22).

Karst landscapes are predominantly the result of limestone solution (Ford and Williams, 2007). The features of karst terrain range from a huge diversity of small-scale solutional forms called karren in Germany or lapies in France (Table 3.9), to major landforms such as solution hollows known collectively as dolines. Sub-types include collapse dolines, solution dolines and subsidence dolines. Uvalas are larger hollows typically resulting from coalescing of doline complexes, while poljes are major depressions sometimes covering over 100 km^2 of landscape. Cockpit or cone karst describes the landscape resulting from major limestone solution and collapse with only residual rounded hills remaining. Similar hills are called mogotes in Cuba and pepino hills in Puerto Rico. Tower karst is a spectacular form of this landscape, seen for example near Guilin in China, where the residual hills are steep-sided and often over 100 m high. In some places in the humid tropics thin pinnacles of limestone produce a remarkable pinnacle karst landscape. Caves and chambers are very common in karst areas and redeposition of calcium carbonate in these caves is common. The resulting speleothems take a diversity of forms including stalactites, stalagmites (Figure 9.9), columns, helictites, curtains, tufa rims, terraces and delicate mineral flowers growing in cave pools.

Finally, duricrusts are hard surface or near-surface layers resulting from the accumulation of one or more materials. Several types of these crusts exist, including ferricretes (iron), alcretes (aluminium), silcretes (silica), calcretes (calcium carbonate) and gypcretes (gypsum). They are typically 1–10 m thick and may form capping layers on residual hills or survive as boulder lags called gibber plains, seen for example in central Australia.

3.3 Conclusions

In this chapter I have tried to outline the huge diversity of materials, landforms and processes of the abiotic world. The aim has not been to describe the full geodiversity of the planet—there are enough books already available that do this—but simply to give a flavour of the factors that have produced this amazing diversity. The chapter has hardly done justice to the real diversity of nature, and readers with limited knowledge of geology and geomorphology are encouraged to delve more deeply into the extensive reading available on any of these aspects, or simply to travel with eyes open within the physical environment that is all around, even in our cities. It is to the value of this geodiversity that I now turn.

Part II

Values and Threats

4 Valuing Geodiversity in an 'Ecosystem Services' Context

there can be no conservation if there is no interest and thus no sense of value.
Barry Thomas and Lynda Warren (2008)

Why is it important to conserve and manage the geodiversity of the planet? The next two chapters aim to answer this question by first discussing the many values of geodiversity and the reasons for treating the physical basis of our environment with care and respect. Then, in the next chapter, I examine the threats to geodiversity, the overall aim of this part of the book being to justify the need for geoconservation by demonstrating that the geodiversity of Planet Earth is hugely valuable but seriously threatened by human impacts.

4.1 Introduction

Several authors have tried to outline the value of nature or the rationale for nature conservation in general (e.g. Huxley, 1947; Nature Conservancy Council, 1984; de Groot, 1992; Daily, 1997a; Costanza *et al.*, 1997; English Nature, 2002) or earth science conservation in particular (Wilson, 1994; Kiernan, 1996; Doyle and Bennett, 1998; Page, 1998; Sharples, 2002a; Webber, Christie and Glasser, 2006; Gordon and Barron, 2011; Gray, 2012). Wilson (1994) recognised two main types of value in the earth's physical resources. First, the economic value in exploiting the physical resources of the planet, and second, the cultural or heritage value in protecting the aesthetic and research resource of the physical environment. This twofold division is a useful one, but other writers have expanded

Geodiversity: Valuing and Conserving Abiotic Nature, Second Edition. Murray Gray.

the classification into four groups (e.g. Bennett and Doyle, 1997; Doyle and Bennett, 1998):

- intrinsic value;
- cultural and aesthetic value;
- economic value;
- research and educational value.

In the first edition of this book (Gray, 2004), I added the functional value of geodiversity for both physical and ecological processes. It is also true to say that there is a difference between the value of a resource and the value of the diversity of a resource, and it is the latter that I shall try to focus on in this chapter.

A crucial concept that has been understood for many years but has recently become central to nature conservation is that of the 'ecosystem services' approach. Ecosystem services are the goods and services provided by ecosystems that benefit society. Daily (1997b, p. 3), in the foundation work on this topic, described ecosystem services as:

> the conditions and processes through which natural ecosystems, and the species that make them up, sustain and fulfil human life. They maintain biodiversity and the production of ecosystem goods, such as seafood, forage, timber, biomass fuels, natural fiber, and many pharmaceuticals, industrial products, and their precursors.... In addition to the production of goods, ecosystem services are the actual life support functions, such as cleansing, recycling, and renewal, and they confer many intangible aesthetic and cultural benefits as well.

More recently, a proposal by the UN Secretary General Kofi Annan in 2000 led to the *Millennium Ecosystem Assessment* (MA) (2005) being carried out between 2001 and 2005 by 1300 international scientists. It demonstrated the importance of ecosystems for human well-being and found that many of the services that ecosystems provide are being lost or degraded. The MA classified ecosystem services into four categories: regulating, supporting, provisioning and cultural services. Some authors have widened the definition into natural resource management and see an ecosystem services approach as vital to human sustainable occupation of the planet. In response to the MA, in 2007 the UK House of Commons Environmental Audit Committee produced a report that called on the UK Government to carry out a full MA-style assessment for the country to enable the development of an effective policy response to ecosystem service degradation. The result was a UK National Ecosystem Assessment (UKNEA) (2011), through a 'wide-ranging, multi-stakeholder, cross-disciplinary process' carried out between 2009 and 2011 using the same four-category classification system. It was 'the first analysis of the UK's natural environment in terms of the benefits to society and the nation's continuing prosperity' (Brown *et al.*, 2011, para 1.1). However, both the MA and the UKNEA grossly undervalue the goods and services associated with geodiversity (Gray, Gordon and Brown, 2013), in particular through:

- the omission of non-renewable resources;
- the lack of long-term perspectives;

- the uneven treatment of geomorphological processes;
- the general lack of integration of the dynamic functional links between geodiversity and biodiversity.

Thus this chapter is concerned with describing 'abiotic ecosystem services' (Gray, Gordon and Brown, 2013) using the four existing categories:

- regulating services;
- supporting services;
- provisioning services;
- cultural services;

plus the additional category:

- knowledge services,

given the value that geoscience knowledge brings to society. A fundamental point is that it is the geodiversity of the planet that delivers all these values. The issue of 'ecosystem services' is further discussed at the end of this chapter where Figure 4.25 and Figure 4.26 are provided as summary diagrams. However, before discussing the ways in which abiotic nature benefits human society, we first need to consider the issue of intrinsic or existence value.

4.2 Intrinsic or existence value

Intrinsic value refers to the ethical belief that some things (in this case the geodiversity of nature) are of value simply for what they are rather than what they can be used for by humans (utilitarian value). This is the most difficult value to describe since it involves ethical and philosophical dimensions of the relationship between society and nature. These have been discussed by a wide range of authors, and Beckerman and Pasek (2001, p. 129) refer to the issue as 'one of the most recalcitrant problems of environmental ethics'. Some have argued that there is no such thing as intrinsic value since the value of nature depends on whichever ethical or belief system that we adopt. Other philosophers argue that nature is not a social construction but has a value in itself and this is not dependent on any uses of nature that humans might adopt (e.g. Norton, 1988).

One view is that the resources of the planet should be freely available for human exploitation and there should be no curbs or restrictions on the use of these physical or biological resources. This 'technocentric' or 'anthropocentric' view of the human place in the environment is one that demonstrates a 'lack of concern for anything non-human: "nature" is seen as an "external" environment with no worth or value, except its ability to be manipulated or exploited by society' (Phillips and Mighall, 2000, p. 14). Under this view, the environment is regarded as a commodity like any other for which there is a market. This view has been adopted by different societies at different times and was promoted by early scientists including René Descartes in France and Francis Bacon in England.

Political systems, too, have been associated with technocentric attitudes. For example, both capitalism and communism have been blamed for promoting economic growth at the expense of the environment (Barry, 1999; Phillips and Mighall, 2000). The technocentric philosophy therefore, in its most extreme form, is not one that would readily recognise the intrinsic or existence value of physical or biological nature or the need to conserve and manage it.

At the other end of the nature/society philosophical spectrum is 'ecocentrism', which sees value in non-exploitative relationships between society and nature. This view has been shared by many movements, cultures and religions over the years though there are many shades of ecocentrism. There are clearly links with the traditional Judaeo-Christian notion of human responsibility for the stewardship of the planet, though there has been intense debate about the interpretation of human–environment relationships in Christian thought (Attfield, 1999; Barry, 1999; Phillips and Mighall, 2000). For example, L. White (1967) used the Book of Genesis to argue that Christian thought has encouraged technocentric views:

> and God said unto them 'Be fruitful, and multiply, and replenish the earth, and subdue it: and have dominion over the fish of the sea, and over the birds of the air, and over every living thing that moves upon the earth (Genesis: 1: 28)

though others such as Passmore (1980) and Hillel (1991) have pointed to the fact that the second chapter of Genesis gives a contradictory view: 'God took man and put him in the Garden of Eden to work it and take care of it' (Genesis: 2: 15).

In this view 'Man is not set above nature . . . his power is constrained by duty and responsibility . . . man is a custodian, entrusted with the stewardship of God's garden, and he can enjoy it only on the condition that he discharges his duty faithfully' (Hillel, 1991, p. 13).

Similarly, many eastern religions espouse the principle of empathy with, and respect for, the natural world. Buddhism displays a marked respect for the natural environment and Islam has its own particular set of rules, taken from the Koran, about the proper way of thinking about, and relating to, the environment (Khalid and O'Brien, 1992; Barry, 1999). For example, there is provision in Islamic law for 'himas', tracts of land set aside to remain undeveloped in perpetuity, of which thousands remain to this day. Aboriginal thought is perhaps the least anthropocentric and more inclined to emphasise the continuity rather than separation between the human and the nonhuman worlds. Ehrlich (1988, p. 22) believed that, given the current state of the planet, a 'quasi-religious transformation leading to the appreciation of diversity for its own sake . . . may be required to save other organisms and ourselves'.

Arguably the most extreme form of ecocentrism is 'Gaianism' first promoted by James Lovelock (1979, 1995) and named after the Greek Earth goddess. This philosophy is not only holistic in seeing the essential integration between the biological and geological elements of the planet, but also sees nature as the primary object of concern. According to Lovelock it is the health of the planet that matters, not that of some individual species of organism, including humans.

It is not difficult to perceive that a central tenet of ecocentrism is the intrinsic or existence value of nature in its own right or for its own sake. As Attfield (1999,

p. 27) puts it, 'it is difficult to credit that nothing but our own species matters' or that 'absolutely everything exists for the sake of humanity, and it alone'. This is a well-understood principle in relation to wildlife, in which there is a strong belief in many societies of the rights of other creatures. Ecocentrism regards ecosystems and the biosphere as having moral significance. Alternative strains of thought include 'sentientism', which accords moral recognition to all creatures with feelings and only to such creatures, and 'biocentrism', which recognises the moral standing of all living things (Attfield, 1999). It is less clear that society holds the same view of geodiversity or would acknowledge a 'geocentrist' ethic. 'Save the Dolphin' is always likely to have greater appeal to the public than 'Save the Drumlin' (Gray 1998a, p. 273). Indeed, since sentientists do not even extend recognition to all creatures, they are hardly likely to attach much value to abiotic nature. But there is no reason, in principle, for separating animate and inanimate nature in this respect and it would certainly be contrary to the Gaia philosophy to do so. Sharples (1993, p. 7) believes that 'it is simply "biocentric chauvinism" to hold that only living things have intrinsic value whilst non-living things do not'.

Another line of argument for intrinsic or existence value relates to natural and human timescales. Bronowski (1973, p. 91), with usual eloquence, described how 'The hidden forces within the earth have buckled the strata, and lifted and shifted the land masses. And on the surface, the erosion of snow and rain and storm, of stream and ocean, of sun and wind, have carved out a natural architecture'. This architecture has taken thousands of millions of years to evolve yet can be destroyed or altered within days. Given the potential of this asymmetric cycle of creation and destruction, it is arguable that if we understand the lengths of time and complexity of the processes involved we may conclude that the end result has some intrinsic value.

A further approach is promoted by Goodwin (1992) who suggests that value of natural processes and landscapes comes about precisely because they are not the work of human hands. According to Goodwin, a 'natural' landscape is more valuable than a 'humanised' landscape, in the same way that a 'fake' or reproduction is never as valuable as the original. In response to Gray's (1997a, p. 313) assertion that unlike the natural vegetation of Britain, 'the natural landforms remain generally intact... substantially unaltered since the major changes of the Pleistocene and modification of the Holocene', Adams (1998, p.168) argues that 'a "natural" landform is valuable because it is unchanged, or minimally changed by human action'.

Belshaw (2001) makes the distinction between the 'intrinsic' value of an object, i.e. that a rock, landform or soil should exist for its own sake, and an 'existence' value, meaning a non-instrumental value to humans, i.e. that humans value the existence of a rock, landform or soil irrespective of any other value to those humans, including an aesthetic value. Beckerman and Pasek (2001, p. 130) agree that 'values cannot exist without a valuer', and they accept (p. 131) that a 'subjective approach to valuation still allows the valuer to attribute intrinsic value to something'.

This leaves the question of whether diversity, as opposed to uniformity, has intrinsic value. Cuomo (1998, p. 132) argued that:

to claim that something is ethically valuable merely because it is unlike something else is incoherent—to be ethically valuable something must itself have a certain quality or status, even if that quality or status is contextually determined. Also, claiming that difference itself renders something morally valuable fails to give attention to the content and origins of the thing itself.

As we saw in Section 1.1, opinions have varied in past centuries about the value of landscape diversity (Midgely, 2001) and the conclusion must be that diversity has no inherent value in itself but can have a subjective intrinsic value, i.e. an existence value.

Other arguments must also be considered in relation to both biodiversity and geodiversity. Attfield (1999, p. 135) rightly reminds us that 'Arguments for preserving whatever is natural assume (implausibly) that whatever is natural is desirable'. Yet there is much in the biological world that is certainly harmful to humans. For example, should we protect and enhance the smallpox virus or the anthrax bacterium, when medical science has spent centuries attempting to eradicate these and other diseases? Should we value spiders or rats as much as pandas and tigers? Do we need to preserve all of the 30 million or so species estimated to exist on earth, even if we ever manage to identify them all? Since carnivorous species hunt their prey, might they not themselves endanger these species? And since extinction itself is a natural process should we attempt to prevent it occurring (Passmore, 1980; Attfield, 1999)? These questions raise important ethical and philosophical issues about whether humans should have the right to decide the fate of species, and if so, how priorities should be identified and resources allocated in conserving biodiversity.

Similar questions can be applied to geodiversity. For example, is geological and geomorphological diversity always beneficial? As was indicated in Section 1.1, the answer must be 'no'. For example, to the civil engineer the endless variety of rocks, sediments, slopes and drainage courses makes life exceedingly difficult and raises the cost of building projects through the need for site investigations, material testing and geomorphological mapping (Cooke and Doornkamp, 1990; Griffiths, 2001). Projects would be much simpler and cheaper to complete if there was greater uniformity and predictability of rocks, sediments, landforms and processes.

Also, hazards such as earthquakes, tsunamis, volcanic eruptions, floods, avalanches and landslides kill thousands of people every year and damage property to the tune of millions of pounds (e.g. Decker and Decker, 1998; Bell, 1999; Smith, 2000a; Abbott, 2009; Hyndman and Hyndman, 2010). Should we conserve such damaging processes or try to eradicate them? By and large human society, perfectly understandably, tries to prevent these hazards and disasters by various means including sensible planning, predicting events, evacuating populations and engineering solutions. But it would be unfortunate if all potentially hazardous processes were eradicated from the planet since they are part of its natural evolution.

Several other questions need to be asked about the aims and principles of geodiversity. For example, since erosion is a natural process, should we be concerned if it removes an element of geodiversity? Do we need to preserve all the world's geodiversity even if we could identify it. If not, how do we decide what is sufficiently significant to conserve. How should priorities be identified and what resources should be allocated to conserving geodiversity relative to biodiversity? If we accept Cuomo's (1998) premise that diversity can only have a subjective intrinsic value, then it also allows us to support Sharples (2002a) proposal (see Section 1.3) for a distinction to be made between 'geodiversity' as a value-free quality and 'geoheritage' as those elements of geodiversity that are seen as significant according to particular subjective values.

For further discussion of these issues see Fox (1990), Nash (1990), Cuomo (1998), Attfield (1999), Belshaw (2001) and Beckerman and Pasek (2001).

4.3 Regulating services

4.3.1 Atmospheric and oceanic processes

The atmosphere and oceans play important regulating roles at various scales, but predominantly at the global scale. For example, the atmosphere plays a crucial role in allowing life to survive by keeping the planet warm (carbon dioxide), shielding us from harmful radiation (ozone), providing water and oxygen for animal life as well as carbon dioxide for plant growth, and by transferring heat away from low latitudes. The latter is also achieved, though to a lesser extent, by oceanic currents. These transfers are important because without them the tropics would get steadily hotter and the poles steadily colder. The dynamic circulations of the atmosphere and oceans therefore act as vital regulating mechanisms.

The hydrological cycle is no less important as a regulating service. For example, the huge quantities of water stored in the world's oceans play an enormous role in regulating global climates, in influencing evaporation and the return of water to the atmosphere and, together with the Antarctic and Greenland ice-sheets and the groundwater stores, have a stabilising influence on the hydrological cycle itself. The need for water for animal and plant growth is well known, but it is also crucial to several human activities including domestic life, industry and agriculture.

4.3.2 Terrestrial processes

The carbon cycle is also an important regulator, linking land, oceans and atmosphere. As we know, much carbon is locked up in fossil fuels but also in other carbonate rocks and sediments including limestones and organic soils. The total carbon stored in the world's soils amounts to 2100–2400 Gt, about 25% of this being held in peatland soils, which cover only 3% of the land area. The burning of fossil fuels is well known as a mechanism for releasing carbon dioxide into the atmosphere but there are also important natural processes that do so. As part of plate tectonics, carbonate sediments and limestones are subducted and melted, releasing carbon dioxide through volcanic activity. In turn, carbon dioxide is returned to terrestrial stores directly by terrestrial plant growth. Another abiotic process responsible for extracting carbon dioxide from the atmosphere is rock weathering and, because weathering increases with temperature, it provides an important regulator of both atmospheric carbon dioxide and therefore temperature. Other biogeochemical cycles include the nitrogen, phosphorus and sulfur cycles.

The rock cycle involves erosion, transportation, deposition and uplift. This not only helps to regulate atmospheric carbon, as explained above, but also renews the planet's relief and provides a constantly renewed supply of fresh rock on which weathering processes can act. Geomorphological processes are part of the rock cycle but have their own regulating functions. Rivers perform the function of transporting water and sediment from land towards the sea and their capacity is adjusted to the stream discharge. They process large fluxes of energy and materials from upstream areas. Beaches act to protect the coastline from erosion under normal conditions and they perform the dynamic function of moving sediment along the coast. Salt marshes are efficient in trapping fine sediment that helped them to accrete vertically and laterally in response to the Holocene sea-level rise (French and Reed, 2001). Many of these geomorphological systems are in dynamic equilibrium and their continued functioning is important in regulating environmental systems and mitigating the impacts of climate change (see Gray *et al.*, 2013 and references).

4.3.3 Flood control

Flood control is often listed as an ecosystem service but many of the processes involved are physical as well as ecological. For example, soil and subsurface sediments absorb large quantities of rainwater and thus reduce surface runoff, i.e. they delay and smooth the delivery of rainwater to river channels and thus reduce flooding. These processes add to similar roles played by tree canopies and vegetation cover so that ecosystems and geosystems operate together in this regard. In fact, soil represents the classic interface between abiotic and biotic systems. Both organic matter and inorganic clay strongly affect soil hydrology and susceptibility to erosion. Clay can block soil pores and thus reduce infiltration rates leading to surface runoff and erosion, while the presence of organic matter in soils increases their water holding capacity.

Flood control is also provided by natural physical barriers such as river levees, shingle beach ridges, salt marshes and barrier islands, while offshore sandbanks play a useful role in reducing wave energy reaching the coast.

4.3.4 Water quality

In terms of water quality, soil, sediment and rock perform a role in attenuating polluting substances and therefore in helping to maintain the quality of surface water and groundwater. This attenuation can occur by adsorption, ion exchange, microbial decomposition or dilution, but the effectiveness of these processes is variable. For example, the thickness, composition and structure of the soils, sediments and rocks will strongly influence the susceptibility of an area to pollution. A thin capping of soil and sediment above a water table will provide little protection from surface pollution by agrichemicals or spills, whereas a thick cover of massive clay will generally be very effective in attenuating pollution.

4.4 Supporting services

4.4.1 Soil processes

As we saw in Section 3.1.4, soil is a mixture of weathered rock debris and organic matter, organised into the structured layers of a soil profile. The weathering essential for soil formation begins when fresh rock or sediment is exposed to the atmosphere and may then be subject to mechanical and/or chemical processes that change the structure and/or composition of the parent material. This weathered material is known as regolith and as it becomes colonised by plants and animals, organic material is added. Thus, as stated earlier, soil is partly organic and partly inorganic, and as Bridges (1994) states, soils are the vital link between the inanimate world (geosphere) and the living world (biosphere). As we saw in Section 3.1.4, there is great diversity of soil types but they all provide important services.

The usual perception of soil is as a medium for plant growth, and there is no doubting the importance of this function. As Rachel Carson (1962) wrote: 'The thin layer of soil that forms a patchy covering over the continents controls our own existence and that of every other animal of the land. Without soil, land plants as we know them could not grow, and without plants no animals could survive'. Plants need sunlight and carbon dioxide from the atmosphere but they rely on soil for the nutrients and water that are essential to life. Soil is therefore an absolutely vital part of terrestrial environmental systems. In fact, a more diverse series of important functions of soil is increasingly being recognised (e.g. Taylor, 1997; Brady and Weil, 2002). For example, soils:

- interact with many other parts of the environment (atmosphere, hydrosphere, biosphere, lithosphere);
- act as a habitat for soil biota and as a gene reserve;
- filter and bind substances from water and receive particulates from the atmosphere;
- act as a store of water and carbon and as a recycler of organic matter;
- support ecological habitat and biodiversity;
- regulate the flow of water from rainfall to watercourses, aquifers, vegetation and the atmosphere;
- act as a growing medium and nutrient supply for food, timber and energy crops and are the basis for livestock production (see earlier);
- act as environmental archives.

Soil provides the growing medium for agriculture, viticulture and forestry.

Agriculture. About 90% of our food comes from the land and only 10% from water and air. Although population has been increasing at a very high rate, food production has been increasing even faster due to the expansion of cropland areas, drainage and irrigation, and the increased use of machinery, fertilisers, pesticides and high-yield varieties (Mather and Chapman, 1995). Soil diversity is important since different crops favour different soils. For example, in the East Anglia region of England, the sandy loam soils are favoured by root crops, and sugar beet, potatoes, parsnips and carrots are grown in rotation with cereals and peas. Rotation is essential for soil health and irrigation is required because of the light soil and low summer rainfall. On the medium soils, winter wheat is grown in rotation with oilseed rape and a legume crop. Wheat is grown on the heavy clay soils derived from the glacial till.

Viticulture. Along with grape variety, climate and manufacturing processes, geodiversity accounts for some of the diversity of wine (Wilson, 1998; Haynes 1999; Gillerman *et al*., 2006; Sommers, 2008). According to Lugeri *et al*. (2012, p. 26) 'Vineyards are linked to the ground more than other kinds of cultivation'. The most important variables are:

- rock and soil mineralogy, which control the nutrient and mineral supply to the vines. Potassium is particularly important and is found in the potash feldspars of many igneous and metamorphic rocks, in arkosic and glauconitic sands and in illitic and chloritic clays (Selley, 2002). Trace minerals give wine diversity of taste but it has not yet proved possible to identify, for example, a Maastrichtian wine, from its taste.
- rock and soil structure/texture, which control water content. Vines do not thrive in either drought or in waterlogged soils, but they do require a steady supply of water. This is best supplied where the soil is well drained but the roots are able to penetrate deep into the rock via fractures. An example of this occurs in the Douro Valley east of Porto, Portugal, where the vine roots penetrate deep into the schists for coolness and moisture. The English and French chalk also provides good conditions given the porosity and fracture characteristics.
- topographic variables for providing appropriate growing conditions. Altitude and aspect affect the degree of solar warmth. In marginal climates for vine growth such as Britain, south-facing slopes at low altitude will maximise solar warmth, particularly where there is a lake or river at the base of the slope to reflect the sunlight. In hot climates like Spain or Italy, vines often grow better at higher and therefore cooler elevations.

Forestry. About 4000 million hectares of the Earth are covered in forest, though this is declining at an alarming rate (10–12 million hectares per year). About 60% of the resource is in America or the former Soviet Union and only 29% being in Africa and Asia. These last two continents have high populations and a long history of use of wood for fuel. Forestry also protects soil from erosion, particularly in areas of high rainfall and sloping terrain.

4.4.2 Habitat provision

The physical environment generally, plays a huge role in providing habitats for biodiversity, yet this appears to be rarely recognised by ecologists. Fortunately, Warren and French (2001) note that the situation is changing and engineers and ecologists have become aware of the need to understand the processes and patterns in the physical environment if they are to manage processes and habitats successfully (see Chapter 14). For example, 'ecologists working in tropical rainforests have suddenly become aware that slope dynamics have a vital role to play in the patterning and dynamics of the ecosystem, yet have only just begun to collaborate with geomorphologists' (Warren and French, 2001, p. 3). Similarly, in the case of the management of coastal dunes, the importance of allowing the operation of shifting sands and dynamic physical processes is being recognised; an understanding of the latter is critical to population dynamics of plant and animal species, and both need attention if dunes are to be managed in a way that conserves their diversity (Warren and French, 2001). Bartley and Rutherfurd (2001, p. 15) believe that 'Biologists increasingly acknowledge that geomorphological surfaces form the template for development of both flora and fauna communities'.

MacArthur and Wilson's (1967) prediction that species numbers were strongly related to land area is clearly undermined by the important variable of topography. McCoy and Bell (1991) and O'Connor (1991) found that heterogeneous habitats have an important effect on measurements of species diversity, and it is often geomorphology, geology and soil diversity that provides habitat heterogeneity. For example, in Rhode Island, United States, geomorphology is a more important control on plant species diversity within woodlands than woodland size (Nichols, Killingbeck and August, 1998). Warren (2001, p. 41) believes that this type of topographical control is even clearer in very dry areas where 'the slightest hollow can support many times more plant production than a nearby slope'. Warren also argues that geomorphology is more of an ecological factor in high-energy than in low-energy environments.

There are similar relationships with animal populations. Warren (2001) quotes the case of the Bay Checkerspot butterfly in California, United States, which needs warm slopes where larvae can develop quickly and contiguous cool slopes where the females can emerge and reach diapause before their host plants senesce as summer proceeds (Murphy, Freas and Weiss, 1988). Moreover, in wet years more of these butterflies survive on warmer slopes, while in dry years there are more on the cooler slopes. Similarly, Belsky's (1995) study of butterflies in East Africa demonstrated that they needed different facets of the landscape at different times of year. He concluded that there was greater animal species diversity and biomass where there was more landscape diversity. Thus the influence of geodiversity on biodiversity is increasingly being recognised by biologists (e.g. Burnett *et al.*, 1998) as well as geoscientists. The degradation of landforms, soils and waters will inevitably adversely impact on the biological species and communities living in or on them.

Among the physical factors that influence biodiversity, altitude and aspect must rate as two of the most important. Most mountains display an altitudinal zonation of natural or semi-natural vegetation, where this exists, linked to soil and climatic gradients. Aspect, by influencing topographic shelter and shade, can have an important effect on plant distributions, by affecting, for example, insolation, temperature, evapotranspiration, soil moisture and duration of snow-lie.

Other environments can also be cited. Salt marshes provide the saline conditions enjoyed by several specially adapted plants. In the intertidal zone, several species of mollusc and seaweed cling to rocks and boulders, sometimes deriving nutrients, and always physical support, from their rocky home. Mudflats and shallow ponds provide excellent environments for wading birds that often have long beaks adapted for probing into the mud for food.

Acid peatlands provide the homes for many flowering plants (e.g. *Calluna vulgaris*), mosses (e.g. *Sphagnum*) and even insectivorous plants (e.g. Sarranaceniaceae). Fenlands, on the other hand, are dominated by grasses or sedges often with rich herbaceous floras and colonised by fen-carr shrubs (Maltby and Proctor, 1996). According to Daly, 1994, p. 18), peatlands are unique ecosystems 'representing extreme habitats where waterlogging and, in many cases, restricted nutrient supply, are important features; and they support unique combinations of plants and animals adapted to these environmental conditions'. They are

Table 4.1 Extract from an analysis of the relationships between National Vegetation Classification communities and the soils and rocks that support them (after Rodwell, 1991; Usher, 2001).

NVC number	NVC name	Notes on habitats and soils
W1	*Salix cinerea—Galium palustre* w	Wet mineral soils on the margins of standing or slow-moving open waters and in moist hollows
W2	*Salix cinerea-Betula pubescens-Phragmites australis* w	Topogenous fen peats, especially flood plain mires
W3	*Salix pentandra-Carex rostrata* w	Peat soils kept moist by moderately base-rich and calcareous groundwater
W4	*Betula pubscens—Molinia caerulea* w	Moist, moderately acid, though not highly oligotrophic, peaty soils
W5	*Alnus glutinosa—Carex paniculata* w	Wet or waterlogged organic soils, base-rich and moderately eutrophic
W6	*Alnus glutinosa—Urtica dioica* w	Eutrophic moist soils
W7	*Alnus glutinosa—Fraxinus excelsior-Lysimachia nemorum* w	Moist to very wet mineral soils, only moderately base-rich and usually only mesotrophic
W8	*Fraxinus excelsior—Acer campestre-Mercurialis perennis* w	Calcareous soils with mull humus
W9	*Fraxinus excelsior—Sorbus aucuparia-Mercurialis perennis* w	Permanently moist brown soils derived from calcareous bedrock and superficial deposits
W10	*Quercus robur—Pteridium aquilinum-Rubus fruticosus* w	Base-poor, brown soils
W11	*Quercus petraea—Betula pubescens-Oxalis acetosella* w	Moist, but free-draining, and quite base-poor soils
W12	*Fagus sylvatica—Mercurialis perennis* w	Free-draining, base-rich and calcareous soils
W13	*Taxus buccata* w	Moderate to steep limestone slopes with shallow, dry rendzina soils
W14	*Fagus sylvatica—Rubus fruticosus* w	Brown earths, with low base status and slightly impeded drainage
W15	*Fagus sylvatica—Deschampsia flexuosa* w	Base-poor, infertile soils

internationally important for many bird species and peat swamp forests are the home of the orang-utan. The UK's National Vegetation Classification illustrates the relationship between woodland and scrub vegetation on the one hand with soils and geology on the other (Rodwell, 1991; Usher, 2001) and an extract is shown in Table 4.1.

Zovodovski Island in the South Sandwich Group, Antractica, has the world's largest colony of chinstrap penguins. Two million of them nest on the island and this is attributed to the high geothermal heat flux from the island's volcano. The penguins can lay their eggs on the snow-free, warm soil, thus giving them an advantage over most other land and penguin colonies of the Antarctic. However, in other parts of the world it is quite common for snakes and reptiles to escape the heat of the day by sheltering under rocks or in rock crevices. Female turtles come ashore to lay their eggs in the beach sands in places like Ascension Island in the Atlantic, Costa Rica in the Caribbean and Crab Island of the Queensland coast in Australia. In summer, Beluga whales rid themselves of their old skin by vigorously rubbing themselves against the gravel beds of Arctic estuaries, and walruses do the same on coastal rocks. The spawning grounds of the Atlantic salmon are well-oxygenated, pool-and-riffle sequences in gravel-bed rivers. Seabirds like the gannet, guillemot and puffin nest on ledges on mainland or offshore island sea-cliffs. Funk Island off the Newfoundland coast in Canada has the world's largest guillemot colony (over a million birds), attracted there by the offshore fish shoals. The gannet (*Sula bassana*) gets its name from the Bass Rock, an island off the coast of East Lothian, Scotland where there is a colony of 20 000 birds (Figure 4.1).

Burrowing animals such as badgers, moles and rabbits rely on the soil and subsoil for their living quarters. Bare soil and sediment, including actively eroding cliffs, provide important habitat for bees, butterflies and other invertebrates.

Figure 4.1 The Bass Rock, Scotland, home to a large colony of gannets (*Sula bassana*).

In fact, soil provides a living environment for thousands of smaller faunal species. As Bridges (1994, p. 12) puts it '... not only does soil support life, it is itself an ecosystem teeming with living creatures belonging to many different phyla. All these different forms of life, from bacteria to mammals, participate to some degree in the recirculation of chemical elements in which the soil plays a central role.... It provides all terrestrial ecosystems with physical support, nutrients and moisture...'. Bats, on the other hand, find amenable living conditions in caves.

Adams (1996, p. 165) argues that 'The link between process and ecology is particularly clear in the case of rivers, where characteristic species and communities are maintained within different parts of the channel and the floodplain by processes of erosion and deposition, and by patterns of over-bank flooding and groundwater recharge. If those processes are changed, ecological changes are likely to follow'. Hughes and Rood (2001, p. 105–107) believe that 'Floodplains are unique linear landscapes. They are among the most important ecosystems on the planet because they are highly productive and have a high species diversity.... Floods promote regeneration of plants... by creating open, moist sites necessary for many riparian species'. Both Hughes and Rood (2001) and Richards, Brasington and Hughes (2002) believe that flood attenuation has led to a loss of biodiversity on floodplains. In parts of the Amazon basin, it has been suggested that the high species diversity is the result of long-term disturbance through flooding and channel migration (Salo *et al.*, 1986). Floodplains also act as biological corridors through landscapes that would otherwise be inhospitable to both plant and animal species (Hughes and Rood, 2001).

Lithology and geochemistry also play a vital role. Calcareous rocks provide the special environments for a range of plants and animals (see Box 4.1). The mortar-filled joints of walls also provide a calcareous habitat for plants

Box 4.1 Limestone pavements, Ingleborough, England

The Yorkshire Dales area of England is an example of an area where geology can be seen to be an inseparable part of the natural world. The development of vegetation on the limestone rock faces, crevices and screes is influenced by the stability of the rock, slope angle, aspect and shelter. The less stable and more exposed the habitat, the more specialised the flora.

Limestone pavements sustain a rich range of habitats for botanical diversity (Ward and Evans, 1976). The crevices (grykes) in the Ingleborough limestone pavements provide the right sheltered and shady conditions for relatively rare plants such as purple saxifrage, yellow saxifrage, alpine-meadow grass, hoary whitlowgrass, lesser meadow-rue, wall lettuce and baneberry. The microclimate in grykes varies depending on their orientation (Burek and Legg, 1999) but in general is more like that of a woodland than an exposed rock surface. Snails and butterflies, including some rare fritillaries, live in these habitats and birds such as wheatear and wren sometimes nest in limestone pavements.

Elsewhere in the area, metallic ores, including mining waste, provide the conditions for a distinctive range of species able to cope with high concentrations of heavy metals.

Figure 4.2 Plants growing in a wall, Porto, Portugal.

(Figure 4.2), while 'on walls built of mixed materials . . . every change in the substrate is reflected by a corresponding adjustment in the covering of lower plants (lichens and mosses)' (Gilbert, 1996, p. 7). In the United Kingdom, heavy metal mineralisation produces toxic conditions for most plants, yet a small number of rare and local plants are more or less restricted to such situations, including leadwort (*Minuartia verna*), alpine penny-cress (*Thlaspi alpestre*) and several rare mosses and lichens (Hopkins, 1994). Similarly distinct plant communities occur on serpentine rocks in many parts of the world including the Klamath-Siskiyou Ecoregion on the Oregon/ California border, United States. Of the 200 endemic plant types in the ecoregion, 141 are either rare or uncommon and have evolved in the toxic and dry conditions. Huggett (1995) refers to these as 'lithobiomes' and gives several examples.

Coastlines and streams, bogs and moors, deserts and mountains, glaciers and volcanoes: the infinite variety of life on Earth is adapted to its physical environment and these diverse physical systems therefore have a functional value for biological systems and biodiversity. As Warren (2001, p. 60) puts it 'Plant and animal species have adapted to and therefore now need the spatial variety and scale that geomorphological processes provide'. It is now fairly clear as a generality that areas of high geodiversity lead to high biodiversity (Hopkins, 1994), though the reverse is not always the case, and certainly more research is needed (see Chapter 14).

4.4.3 Platforms

The land surface provides a platform or foundation on which not just biological habitats but also all human development and activities take place. Different activities require different types of platform. Airport runways require flat sites whereas ski slopes need to be of varying steepness and complexity to cater for the different abilities of the skiers (see Section 4.6.2). South-facing slopes in the northern hemisphere provide solar gain, while north-facing slopes give shade. Solid rock is required for the foundation of many engineering structures. Diversity of platforms is therefore important.

4.4.4 Burial and storage

The physical resources of the land have long been used for human burial, by placing bodies either into the land (as in graves) or into monuments constructed above ground, such as the pyramids or—at a smaller scale—cairns or dolmens. A diverse variety of rock types are also used by modern stonemasons for making gravestones (Figure 4.3), though an important property here is durability, particularly in retaining inscriptions. Older gravestones in sandstone or softer limestones are sometimes unreadable, and the preferred rock types are durable limestone, marble, slate, granite, gabbro and other igneous rocks. In Copenhagen, Denmark, the most suitable headstone materials have traditionally been the rounded erratic metamorphic and igneous boulders found in the glacial tills of the area. This local style has been preserved even today when rectangular blocks imported from around the world are fashioned into rounded boulders (Parkes, 2004). In Iceland, sections of columnar basalt are frequently used as headstones. Portland limestone was almost universally adopted for the graves of the British Commonwealth war dead (Bennett and Doyle, 1997). Some writers mourn the diversity of stone used in modern graveyards. Clifton-Taylor (1987, p. 146), for example believes that 'there

Figure 4.3 Diversity of gravestones in an Irish cemetery.

can be little room for doubt that . . . no stone has contributed so much as polished granite to converting many English churchyards, once so attractive, into ghoulish eyesores'. Some church authorities, such as the Norfolk Diocese, England, agree and have banned the use of stones other than sandstones and limestones.

Waste materials are also buried in the ground (landfill sites) or above ground (landraising). Sites have been selected because of the *in situ* properties of local rocks for waste disposal. This includes municipal waste disposal based on clay soils like the London or Oxford Clays in England, though usually with a reworked floor or lining system of plastic and/or bentonite. The low permeability of glacial till means that it can be used as an aquaclude for use in lining landfill sites or reservoirs, but its fissured nature and sand lens inclusion means that some reworking is generally necessary (Fredericia, 1990; Gray, 1993).

Radioactive waste disposal is an increasing problem, with many countries stockpiling waste in temporary surface stores. One solution is to use underground chambers as retrievable locations for this waste. Finland has already built an underground repository at Olkiluoto. Several other countries are exploring similar options (National Research Council, 2001) and while some are in quite stable parts of the world (e.g. Canadian Shield), others are in tectonically active zones (e.g. Japan and Switzerland). Other disposal suggestions include use of subduction zones where the waste will be carried down into the Earth, the interiors of low-permeability salt domes or the use of underground clay formations, since clay can more readily absorb and trap radioactive products.

The physical environment acts as an important store, for example of water, oil, natural gas and carbon. As regards the latter, much research is underway on the feasibility of using old oil and gas fields as repositories of carbon dioxide emissions from power stations burning fossil fuel, a process known as 'carbon capture and storage' (CCS). The intention is to mitigate the contribution of carbon dioxide to global warming by preventing these emissions from entering the atmosphere. Other suggestions for storage sites include saline aquifers and layered basalts.

4.5 Provisioning services

Provisioning services are those where the physical environment supplies material goods that are valued by human societies. Some geological materials are extremely valuable. The 'Millennium Star', a 203-carat diamond, and 11 rare blue diamonds, were together valued at over £200 million when attempts were made to steal them from the Millennium Dome in London in November 2000 (*The Guardian*, 9 November 2001). On the other hand, a tonne of gravel may be worth only a few pounds. The diversity of these resources has been exploited with ingenuity from the Stone Age through the Bronze Age and Iron Age and now there is a huge range of materials on which our modern sophisticated societies depend. The average car contains steel, aluminium, carbon, copper, silicon, zinc and more than 30 other mineral commodities including titanium, platinum and gold. And a modern mobile phone relies on over 10 different geological materials including plastics, copper, glass, aluminium, iron, silicon, nickel, tin, lithium, cobalt and graphite as well as rarer elements such as tantalum, neodymium and indium (Wilkinson, 2008, p. 86–87).

4.5.1 Food and drink

The physical environment provides sources of drinking water and natural stores of fresh water that can be drawn on as necessary. These include groundwater aquifers, rivers, lakes, glaciers and ice-sheets. In many places the natural topography has been suitable for the creation of artificial reservoirs by damming rivers or increasing the size of lakes.

Bedrock geology affects the chemistry of water circulating at depth. This affects the chemistry of springs many of which are utilised for bottled mineral water. Whisky is also made from water, and the diversity of whiskies (there are about 100 different single malt whiskies distilled in Scotland) is partly due to the different sources of water (see Box 4.2).

Other drinks are made from surface waters or groundwaters and have variations in taste or other attributes as a result of geological diversity (Maltman, 2003). For example, Lloyd (1986) examined the relationship between hydrogeology,

Box 4.2 The whiskies of Islay, Scotland

Islay is known for its heavy, pungent, phenolic whiskies, but in fact the whiskies of Islay vary depending partly on the geology and local source of water. The geology of Islay is shown in Figure 4.4 but in general terms is an anticline dipping to the north-east. Quartzites and greenstones form the limbs of the anticline in the north and south-east with older sedimentary rocks from the core of the anticline exposed by erosion in the centre of the island.

Three distilleries Laphroaig, Lagavulin and Ardbeg are situated in bays on the south coast of the island. The water used by these distilleries originates as rainfall on the quartzite hills to the north. The water descends to the coast crossing phyllites and greenstones and their peaty cover on the way, but may be intercepted by small natural or dammed lochs. The quartzite rocks of the hills and the peaty land make the waters highly acidic, but because they are surface waters they have a low mineral content. These are the heavy whiskies of Islay though their phenolic taste derives more from the peat smoke allowed to permeate the barley prior to distilling.

On the north-east coast of Islay the distilleries of Caol Ila and Bunnahabhain are situated on older rocks in the core of the Islay anticline. The waters here are derived from springs from the Port Askaig tillite (Caol Ila) and dolomite (Bunnahabhain) and the whiskies are light and clean.

At Bowmore, in the centre of the island, the water is extracted from the river Laggan having flowed westwards towards Loch Indaal over quartzite hills, limestone outcrops, glacial deposits, alluvium, peaty lowlands and grey Moinian sandstones. The water is therefore a blend of Islay's geology and the whisky likewise is intermediate between those of the south and north-east coasts (Cribb and Cribb, 1998).

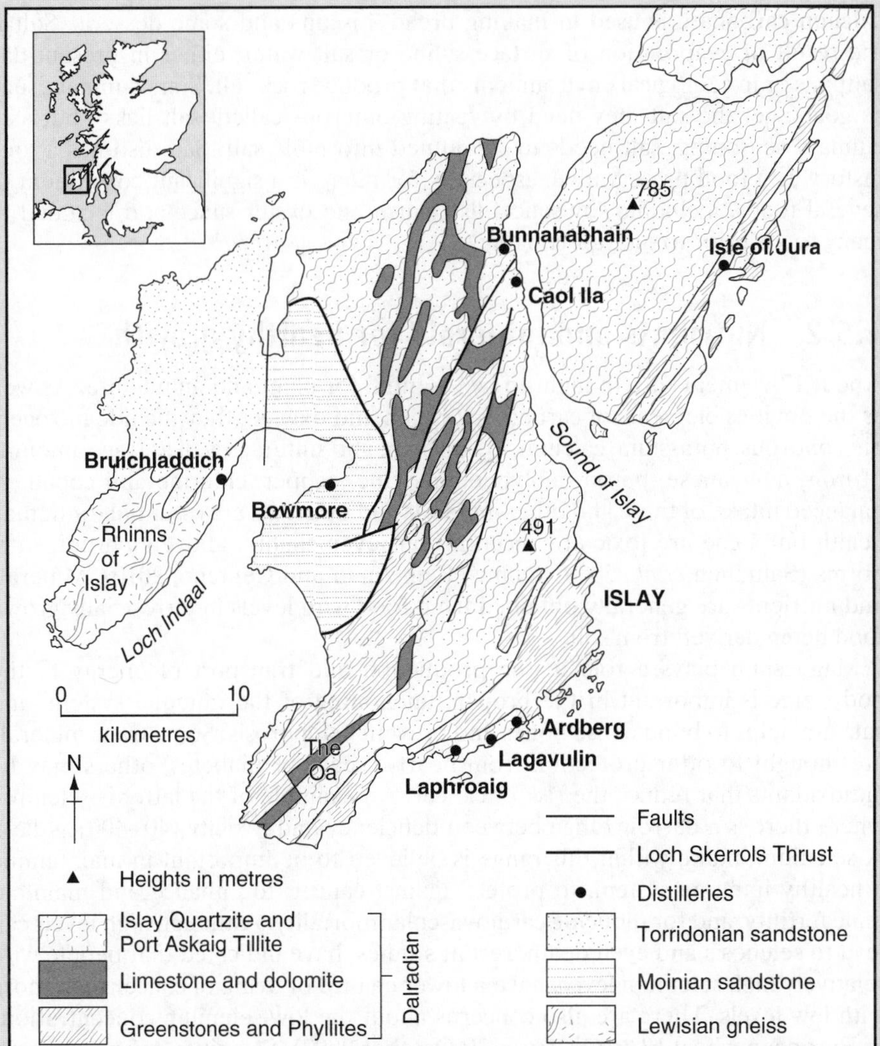

Figure 4.4 Diversity of rock types and water chemistry on Islay, Scotland, helps to explain the diversity of malt whiskies from the distilleries on the island (modified from Whittow, 1992).

brewery location and beer taste in England. He found that English breweries were mainly concentrated on the Triassic Sandstone and Chalk aquifers though there were important breweries on other major aquifers. He also found that calcium, bicarbonate and sulfate are important in the brewing process with the latter ion having the most direct impact on taste. He quotes sulfate values of 820 mg/l for the Triassic Sandstone but only 65 mg/l for the Chalk, with Jurassic Limestone, Magnesian Limestone and Coal Measures at intermediate values.

Almost all foods are biotic but there are some that are abiotic. For example, calcium carbonate is used in making bread, biscuits and some desserts. Salt is the result of evaporation of surface saline or salt water, either in present-day saltpans or in geological environments that produce rock salt. Some animals, such as goats, get the salt they need by visiting outcrops called 'salt licks', but as a culinary seasoning, salt needs to be refined into table salt. Salt also has a long history as a food preservative, especially for meat, is a significant constituent of several foods, including soy sauce, fish sauce and oyster sauce and is added to many foods prior to or after cooking.

4.5.2 Nutrients and minerals for healthy growth

About 17 elements are thought to be essential for plant and animal life. As well as the obvious elements of carbon, hydrogen and oxygen, they include nitrogen, phosphorous, potassium, calcium, magnesium and sulfur, as well as trace amounts of iron, manganese, boron, molybdenum, zinc, copper, chlorine and cobalt. A balanced intake of these and other minerals and nutrients are essential for human health but some are toxic at concentrations even slightly above environmental norms (Salminen *et al.*, 2008; Selinus, Finkelman and Centeno, 2010). Minerals and nutrients are generally obtained from food with levels in terrestrially grown food being derived from soil.

Magnesium plays a role in the production and transport of energy in the body, zinc is important in the proper functioning of the immune system, and calcium helps to build dense bones and prevent osteporosis. Some trace minerals are thought to offer protection from heart disease or diabetes, others may be antioxidants that reduce the risk of cancer. An example of the latter is selenium where there is a narrow range between deficiency and toxicity (40–400 μg/day). A selenium intake within this range is believed to be important in maintaining a healthy immune system, to protect against cancer, to enhance and maintain male fertility, and for reducing cardiovascular mortality. Although high levels can lead to selenosis and even death, recent studies, have indicated that people with relatively high selenium levels have a lower chance of dying of cancer than those with low levels. There are also concerns about the low selenium concentrations in European wheat (*The Observer*, 20 October 2002). Selenium deficiency leads to Keshan disease, named after a Chinese district where it was discovered in 1935 after causing damage to the heart muscles and eventually death of many local people. Low soil selenium levels have since been discovered in Croatia, New Zealand and Slovakia. Selenium supplements, fertilisers or crop sprays can be used to counter deficiency (Fordyce and Johnson, 2002).

The recommended intake of iodine is circa 100 μg/day and iodine deficiency causes a swelling of the thyroid known as goitre, which affects 190 million people worldwide. Chromium is a trace mineral that is essential for good health and is also believed by many to be usefully supplemented for protection against or treatment for Type 2 diabetes. Adequate chromium is necessary for insulin to be effective. The sodium in salt (sodium chloride) is important for maintaining the correct concentration of body fluids, helping cells to take up nutrients and stimulating electrical impulses in the nerves.

Geophagy is the human or animal consumption of inorganic minerals such as clay, soil or chalk. Clays have been used for a variety of purposes including being taken internally, for example in the kaolin and morphine medicine for diarrhoea and indigestion. Soil eating is quite common in tribal and traditional rural societies including some African countries (Uganda and Kenya), perhaps because it contains magnesium, iron and zinc that are essential during pregnancy, early childhood and adolescence (Smith, 2002). It is common practice in Haiti where clay mud is worked into cake called 'bon bon de terres'. Some South American Indians neutralise poisons in wild potatoes by cooking them with clay. In the animal world, geophagy is very common. For example, parrots have been observed to gorge themselves on river clay in the Amazon rainforest and this is thought to be a way of neutralising their diet of poisonous berries and fruits (*The Guardian*, 4 April, 2002).

4.5.3 Mineral fuels

The three main types of mineral fuels are coal and peat, petroleum and uranium.

Coal and peat These are the products of accumulation of mainly terrestrial plants in a humid environment. Peat that has accumulated over the past few thousand years has undergone only limited compaction and alteration so that it is still moist and relatively spongy. It is the traditional fuel source in parts of Finland, Sweden, Ireland, Scotland, Russia, Belarus, Ukraine, China, Indonesia and the Falkland Islands (Asplund, 1996) and in places is still dug, dried and burnt today. The International Peat Society (IPS) estimates that the world's peatlands cover an area of almost 4 million km^2, representing about 3% of the Earth's land surface, and contain 5000–6000 Gt of peat. In addition, there are over 2.4 million km^2 of wetlands where the amount of accumulated peat or other organic matter is unknown (Lappalainen, 1996). Some countries have large areas of peatland. Finland, for example, has 10 million ha of peatlands covering 30% of the land area and both it and Ireland use peat for electricity generation (Kelk, 1992; Asplund, 1996). Other economic uses of peat include agriculture, forestry, horticulture and health therapy and a diverse range of minor, but locally important, uses.

There is great diversity in peat deposits depending on the types of source water and organic materials. For example, Daly (1994) and Maltby and Proctor (1996) describe raised bogs, blanket bogs, tropical forest peats and fen peats. The first two of these form ombrogenous bogs in cool temperate climates where the only water source is rainfall. Consequently, they form deep acid peats with low mineral content. Fen peats, on the other hand, are fed by mineral rich springs or other sources and have near-neutral pH, high cation concentrations, and higher bulk densities and mineral content. The Geological Survey of Finland maintains a peat data bank with information on all bogs covering more than 20 ha and includes over 200 variables of bog and peat characteristics (Kelk, 1992).

Coal is the compacted and chemically matured form of peat in which increasing pressure and temperature produce physiochemical changes that in turn define a continuous spectrum of coal maturity or rank (see Figure 4.5). This spectrum

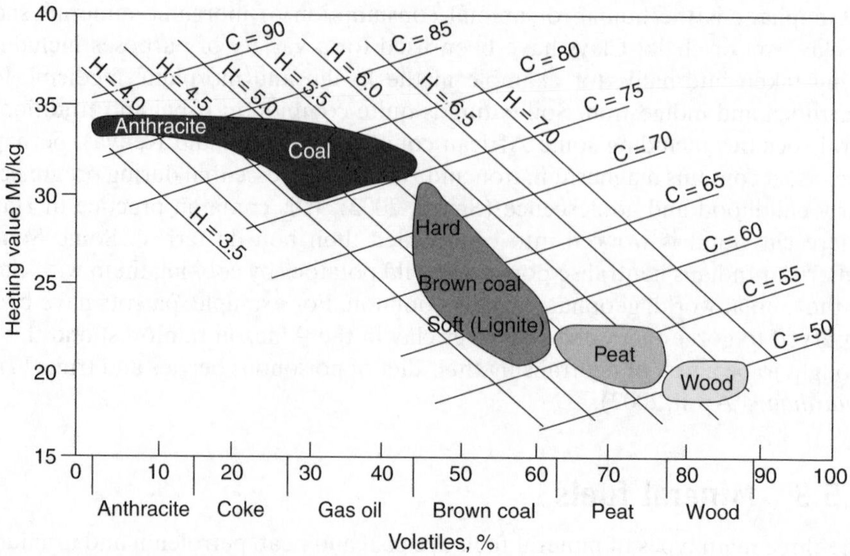

Figure 4.5 Diversity of fuel characteristics from wood to anthracite (after Asplund, D. (1996) Energy use of peat. In Lappalainen, E. (ed) *Global Peat Resources*. International Peat Society, Finland, 319–325, by permission of International Peat Scociety).

extends from lignite (soft brown coal) through sub-bituminous coal (hard brown coal), bituminous coal, anthracite to graphite. This succession is produced in anaerobic conditions by progressive loss of water, carbon dioxide, methane and volatiles. During the process of conversion, oxygen and hydrogen content decrease and the percentage of carbon increases until it reaches 100% in the case of graphite. The calorific value also increases, bringing greater economic value to the higher ranking coals such as anthracite. Coal may contain a variety of impurities, including clay minerals (which increase ash residues), pyrite (which gives emissions of sulfur dioxide) and sodium chloride (which accelerates corrosion of boilers) (Woodcock, 1994), so diversity is not always beneficial. Rarer types of coal include sporinite (coalified spores and pollen) and alginite (coalified algae) also present in oil shales (Evans, 1997; Thomas, 2002).

Petroleum This name covers a diverse range of natural, solid (e.g. bitumen), liquid (e.g. crude oil) and gaseous (e.g. natural gas) hydrocarbons. For an economically exploitable resource to occur there must be a coincidence of appropriate source rock lithologies, processes and rock geometries. The source rock must have a significant organic content (0.5–10%), as would occur through the steady accumulation of dead organisms (algae or plankton) in lake or seabed muds which, by diagenesis, are transformed into shales. Other source rocks include limestones such as the Jurassic Hanifa Formation of Saudi Arabia and the Cretaceous La Luna Formation of Venezuela. In the early stages of burial in anaerobic conditions, the organic matter is converted to kerogen, but the type of kerogen produced will have different proportions of hydrogen, oxygen and carbon depending on the source of the organic matter. Impurities in crude oils include sulfur, nitrogen, oxygen and organic compounds of some heavy metals such as nickel. Up to 3% sulfur may be present but there is a high demand for low sulfur fuels (sweet crudes) (Evans, 1997).

The processes are also important in producing diverse end products. At low pressures and temperatures the decay of organic matter results in methane, much of which escapes to the atmosphere, though some is trapped by frozen ground in permafrost areas, by water pressure on the deep seabed or within coal beds. It is sometimes exploited for local heating as in Anchorage, Alaska, United States. The more economic natural gases as well as oils are produced at temperatures of 50–200 °C, with crude oil, thermogenic wet gas and thermogenic dry gas being released successively as the temperature rises. Oil will itself be transformed to gas if heated further, and the transformations can occur at slightly lower temperatures if timescales of millions of years are involved.

Once released, the fluids will migrate along pressure gradients, usually towards the surface and will escape if nothing prevents them from doing so. Thus in Venezuela and some other oil-rich areas it is not unusual to find 'tar ponds' where liquid hydrocarbons are seeping out at the surface. However, in some cases the rock geometry has been such that low permeability barriers (caprocks or seals) trap the upward migration of oil and gas. Shales and evaporites form most of the oil and gas seals and Figure 4.6 shows some common examples of stratigraphic and

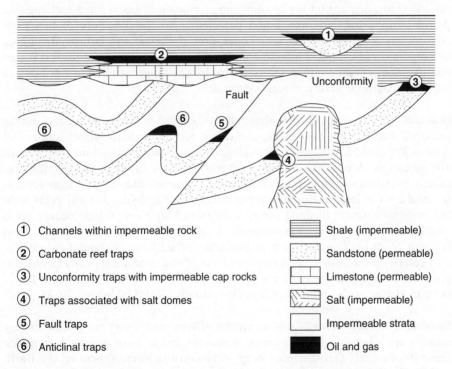

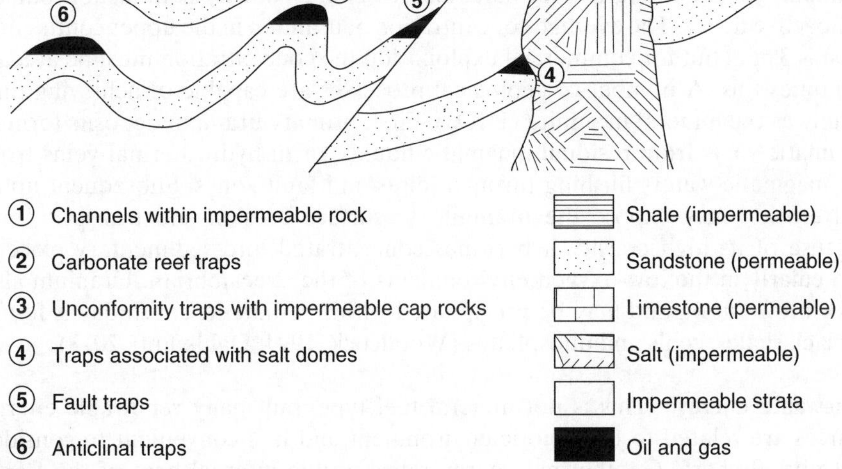

Figure 4.6 Diversity of oil and gas traps (after Bennett, M.R. and Doyle, P. (1997) *Environmental Geology*. John Wiley & Sons Ltd, Chichester, by permission of John Wiley & Sons Ltd).

structural traps. Below these traps, the oil will accumulate in suitable reservoir rocks (predominantly sufficiently thick sandstones and limestones) that have high porosity or permeability or preferably both. High porosity is important since it provides the pore spaces in which the oil and gas can accumulate, while high permeability is important in allowing fluids to migrate freely into the reservoir and out again via extraction boreholes (Bjorlykke, 2011).

As well as crude oil and natural gas mentioned above, a range of other hydrocarbons occur, reflecting the diversity of source materials and maturation processes. These include heavy oils, such as occur in the Orinocco Oil Belt of Venezuela, which are the result of low maturity or alteration processes within the reservoir. Bitumen has an even higher viscosity than heavy oils and acts as a tar cement within sandy reservoir rocks. The bitumen is extracted either by *in situ* steam or gas injection or, in the case of the famous Athabasca Tar Sands in Canada, by mining and heating in retorts. Shale oil occurs where shallow burial has resulted in low-maturity kerogen that has not migrated from the source shale. Other forms of natural gas occur either as 'geopressure natural gas', in deeply buried source rocks, or as 'tight-reservoir gas' in low-permeability sedimentary rocks where it can be released by hydraulic rock fracturing also known as 'fracking'.

This diverse range of hydrocarbons is valuable in providing a huge range of end uses, though about 90% is used as fuel. Crude oil is distilled to produce fractions such as bitumen for road surfacing, wax and naptha for chemical manufacture (detergents, textiles, rubber, agrichemicals, plastics, cosmetics), fuel oil and kerosene/paraffin for heating and aircraft fuel, gasoil for diesel engines, petrol/gasoline for cars, and liquid petroleum gas for a variety of fuel uses. Natural gas is used for electricity generation, domestic and industrial cooking and heating, and chemical manufacture (methanol, ammonia, fertilisers). Waxes and heavy oils can be converted to petrol by cracking, thus giving flexibility in meeting geographical or temporal demands.

Uranium This is mineral fuel that releases energy not by combustion but by radioactive decay. The average concentration of uranium in the upper continental crust is 3 ppm but for commercial exploitation the concentration must be at least 100 times this. A number of geological processes are capable of achieving this, mainly as the mineral uraninite (U_3O_8). First, primary uranium deposits form in magmatic veins from residual magmatic liquids, or in hydrothermal veins from hot magmatic waters flushing through joints and fault zones. Subsequent uplift and erosion may expose the uraninite to surface erosion and transport, and because of its high density it becomes concentrated into sedimentary placers, particularly in the low-oxygen environments of the Precambrian. Uranium also goes into solution and may be precipitated within sediments and today is found in black shales, coals and phosphates (Woodcock, 1994; Dahlkamp, 2013).

Renewable energy This is not mineral fuel type, but many renewable energy sources are related to the abiotic environment and it is convenient to consider these briefly here. Geothermal energy relies on the internal heat of the Earth and has been used for some time in active volcanic provinces such as Iceland and New Zealand (Figure 4.7). More recently, schemes that circulate water

Figure 4.7 A geothermal enery plant in Iceland.

into the crust to be heated at depth and pumped to the surface have been explored. Hydroelectric power relies upon a vertical fall of water and therefore requires certain topographic situations. Similarly, wave and tidal power can exploit particularly coastal locations where conditions are favourable, and wind power is greatest in upland and coastal situations.

4.5.4 Construction minerals

The greatest volume of geological materials is used in construction work and there is a very diverse range of geomaterials used in a huge number of applications (e.g. Figure 4.8). Although biomaterials are used in building construction (e.g. timber frames, thatched roofs, wooden cladding), urban environments are dominated by bulk minerals (Figure 4.9) in ways that should make the value of geodiversity very evident to their inhabitants. However, most urban dwellers probably never give their mineral surroundings a second glance or second's thought.

Construction minerals are usually classified into:

- building stone—quarried blocks, rough or cut, used in wall construction or as facing slabs, floor tiles or roofing slates;
- aggregates—either coarse clastic sediments or crushed stone used in concrete manufacture and many other uses;
- limestone—used in the manufacture of cement;
- structural clay—used in the manufacture of bricks and tiles;
- gypsum—used in plaster;
- sand—for glass manufacture;
- volcanic products;
- bitumen—used in asphalt (see petroleum earlier).

Figure 4.8 Diversity of wall materials: mixed, rounded glaciofluvial cobbles, brick and metamorphic stone cladding in juxtaposition, Banff, Canada.

In some cases rock is used *in situ* to provide shelter (as in the case of the cave dwellings at Cappadocia in Turkey) or form the walls of a building (as in the Temppeliaukio Church in Helsinki, Finland), but in most cases the geological material is excavated and processed in some way. Pohl (2011) provides a comprehensive description of economic geology.

Building stone This has a long history of use dating back to at least 8000 BP when stone was built into the walls of Tell-es-Sultan near Jericho (Prentice, 1990). Not all stone is suitable for use as load-bearing or facing material. The five key factors that determine whether a rock type is viable as a building stone are (Bennett and Doyle, 1997):

Figure 4.9 A part wood, part stone clad building, Pennsylvania, USA.

- structural strength (load-bearing capacity—determined by mineralogy and presence/absence of discontinuities);
- durability (ability to withstand exposure, weathering, salt crystallisation, etc.);
- appearance (e.g. colour and texture);
- ease of working (e.g. whether the rock can be worked into precisely shaped blocks);
- availability (e.g. cost of transport).

Sedimentary rocks are often thinly bedded and in these circumstances may be usable when thin sheets or slabs of rock are required, e.g. roofing slabs, floor tiles or cladding. Where sedimentation has been continuous, there may be thicker beds suitable for load-bearing wall construction. Intrusive igneous rocks, such as granite batholiths, often give massively jointed rocks suitable for cutting into variously shaped and sized blocks. Metamorphic calcareous rocks are relatively easily recrystallised to produce massive marble deposits, whereas granitic rocks require high-grade metamorphism to be converted to usable gneisses (Prentice, 1990).

Several types of building stone can be recognised. Drystone walls are constructed from roughly broken rock and rely on the skill of the builder to produce an interwoven structure. The diversity of the United Kingdom's drystone walls is brilliantly illustrated by the 19 different sections of the Millennium Wall, at the National Stone Centre in Derbyshire (Figure 4.10).

Rough cut blocks of various sizes and shapes are referred to as 'rubble' though the final effect in buildings is usually very attractive. 'Dimension stone' (ashlar to the architect) is shaped and dressed into regular sized and shaped blocks, slabs or tiles which are then used to build load-bearing walls, cladding for walls, roofing materials (Figure 4.11), floor tiles, pavement flagstones or road sets and

Figure 4.10 Diversity of drystone walling materials illustrated by a section of the Millennium Wall at the National Stone Centre, England.

Figure 4.11 Diversity of roof tiles used to create patterns in the roof of St Vitus Cathedral, Prague Castle, Czech Republic.

cobblestones. This type of use has a long history, and reached its peak in the late nineteenth century when a host of public buildings and private houses were constructed in stone. In developed countries, apart from the use of building stone in the construction of monuments, e.g. in Washington, DC (Withington, 1998), its use declined in the twentieth century as cheaper, composite materials became available. However, in most parts of the world the vernacular architecture can be seen to reflect the local geology, and there has been a welcome recent revival in the use of building stone. The local variations in building stone often reflect the geological diversity of particular countries and nowhere is this relationship between geology and traditional building stone clearer than in Britain (see Box 4.3). In general, local stone was used in traditional buildings because of the high cost of transport. The major exception is Caen stone, which was easily shipped to south-east England from northern France.

> But usually no materials look so well as the local ones, which belong organically to their landscape, harmonize with neighbouring buildings and nearly always give the best colours. There are still many old houses scattered over our countryside which convey the impression of having grown out of the soil untouched by the hand or mind of Man. (Clifton-Taylor, 1987, p. 23–24)

Box 4.3 Building stone in the United Kingdom

In Scotland, grey granites dominate in Aberdeen where local sources have been actively quarried, for example at Rubislaw Quarry. In Edinburgh and Glasgow, on the other hand, Carboniferous and Devonian sandstones dominate the traditional buildings. 'To see the best of all Britain's Carboniferous sandstones, it is generally considered necessary to go to Edinburgh' (Clifton-Taylor, 1987, p. 132). McMillan, Gillanders and Fairhurst (1999) list 97 sandstone quarries in Scotland and northern England which have supplied stone for the city's buildings and the stone from each has its own distinctive rock texture, colour and structure that provides a rich and diverse heritage of stone buildings.

Throughout Britain there are examples of villages where distinctive Old Red Sandstone dominates the buildings and field walls (e.g. East Lothian and Devon). Clifton-Taylor (1987, p. 138) describes the diversity of this stone as follows:

They may be a dense and even rather depressing red; but they can also be pink, purple, brown, greenish-grey, pure grey, or grey with a blush of pink, a delicious colour. The tint may change slightly almost from block to block, and sometimes within the single block of stone, yielding effects of much charm.

However, the most intensively worked sandstone in Britain has been the Carboniferous buff-coloured 'Yorkstone' of the Pennine region of west Yorkshire. The term 'Yorkstone' covers a variety of types. Some is massively bedded with blocks weighing several tonnes offered by some quarries. Others are finely bedded at spacings of 2–3 cm making these types ideal for splitting into flagstones, copings or sills. The mica grains in many Yorkstone types give them an attractive sparkle.

From Humberside to Dorset, Jurassic limestones are used as a major building stone and have been quarried in the Cotswolds, Bath area and Isle of Portland. Many public buildings in London were constructed of Portland stone, including the British Museum. In the Lake District, Wales and south-west England, Lower Palaeozoic slates and slatey mudstones are extensively

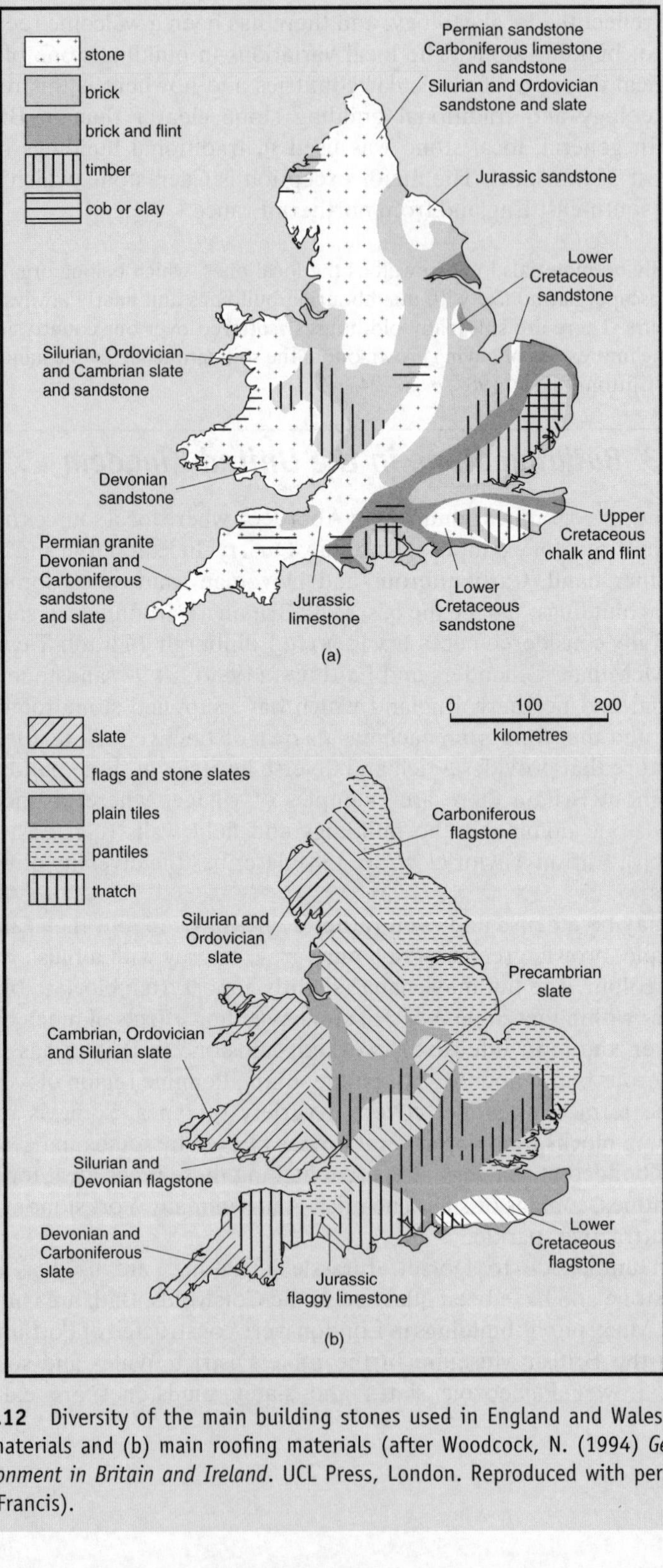

Figure 4.12 Diversity of the main building stones used in England and Wales: (a) main walling materials and (b) main roofing materials (after Woodcock, N. (1994) *Geology and the Environment in Britain and Ireland*. UCL Press, London. Reproduced with permission of Taylor & Francis).

used as a building material for walls, floors and particularly roofs. About 20 standard sizes of roofing slates are recognised, many named after the female aristocracy. Therefore we have princesses, duchesses, marchionesses, countesses, viscountesses and ladies, not to mention the wide countesses and narrow ladies (Clifton-Taylor, 1987)! In many places the fashion is to place the largest tiles at the base and decrease the size with each course towards the ridge.

Eastern England is generally poor in building stone, though locally, Ragstone, Greensand, Chalk and Carstone have been used (Clifton-Taylor, 1987). There is also extensive use of flint derived from the Chalk as a walling material, but 'not used for building in any other country on so extensive a scale' (Clifton-Taylor, 1987, p. 193). In the Neolithic, there were 200 flint mines up to 12 m deep at Grimes Graves in Norfolk, England, exploiting flint seams in the Chalk. It can be worked in various ways, for example as rounded facing pebbles derived from local beaches or glaciofluvial deposits, or as knapped or squared flints producing 'flushwork' blocks in churches or important buildings. It is sometimes mixed with chalk or brick rubble in walls.

The local distinctiveness of vernacular architecture in Britain has much to do with the diversity of geological materials in use in different parts of the country (Figure 4.12). A superbly illustrated map of building stones and quarries of the United Kingdom, compiled in collaboration with various other groups, has been published by the British Geological Survey (2001).

In developed countries, dimension stone is now rarely used in a load-bearing role, but instead there has been a major expansion in its use as facing material on steel-framed or concrete buildings. It is not just major companies that have become globalised; their geological edifices provide dazzling geological displays of colourful facades and interiors assembled from around the world. This includes polished cladding slabs for use as external or internal facing materials, service counters or bathroom tiles. As Robinson (1996, p. 39) puts it: 'all hell broke loose in the post-war years as buildings became clad in Namibian gneiss or Brazilian yellow granite. The important fact remains, that geology is all around us if we care to look'. Robinson (1984, 1985) has described geological walks in London to view the geological materials but there have been many further additions in the past 30 years. Stone has become 'the symbol of prestige, and it is not surprising that it finds favour with the banks and financial houses of today, as it did with the church builders of medieval times, in conveying a sense of permanence and solidity' (Prentice, 1990, p. 43). Burek (2012a) has coined the term 'rock miles' (in comparison to 'food miles') to convey the issue of the increasing global transport of stone.

The use of the attractively coloured and textured geological materials led to their early use in artistic contexts. The Romans used the diversity of stone colour to create beautiful mosaics. And traditionally in Portugal, white durable limestones are used in conjunction with black basalts to produce an impressive set of geometric pavement patterns or pictures (Figure 4.13).

In some cases a great variety of ornamental stones have been used in building construction. For example, Robinson (1987) gives a detailed account of the 10

Figure 4.13 Diversity of pavement patterns produced by alternating black basalts and white limestones, Portugal.

main building stones used in the construction of the Albert Memorial in London in the 1860s. This includes one Irish and two Scottish granites as well as Campanella Marble from Italy. In addition, in the upper parts, there are bronze statues, several other stones and what the architect George Gilbert Scott described as 'an elaborate fretwork of exquisite workmanship, and inlaid with polished, gem-like stones. These gems and inlays are formed of vitreous enamel, spar, agates and onyxes, upwards of 12 000 in number; of these, 200 are real onyxes, many of which are 0.75 inches in diameter' (quoted in Robinson, 1987, p.30).

The use of stone in gardens also has a long history in many parts of Europe and the Far East, though different approaches are used. In Europe, a common aim has been to use natural stone to mimic the natural bare rocky habitats of alpine plants. However, the removal of weathered limestone from natural settings is of concern (Section 9.4.5), as is the increased use of water worn cobbles and pebbles if derived from active beach or river environments. In the Far East, rock colour, structure and texture are used to give artistic and structural qualities to gardens.

Finally, armourstone is used to protect sea walls, breakwaters and vulnerable structures from wave impact and erosion. In high-energy environments very large blocks (up to 20 tonnes) are necessary to withstand wave attack and therefore require durable rock types with widely spaced joints. In the East of England an accessible source of such blocks is by boat from Scandinavian coastal quarries.

Aggregates These are collections of rock particles, produced either by natural processes of erosion, transportation and deposition, or by the mechanical crushing of larger rock masses. Aggregates vary in particle size from sand to cobbles and are by far the greatest volume of exploited material in most countries of the world.

Natural sand and gravel aggregates come from either active or inactive sources and of course sediment can be naturally recycled through various geological deposits. Glacial, fluvial, coastal and aeolian processes all produce aggregates, but the major sources are from present-day, active river or beach environments or from unconsolidated Quaternary glaciofluvial sources, either quarried onshore or dredged offshore. Subsidiary quantities may come from older, weakly cemented rocks such as the Cretaceous Greensand of south-east England and Triassic Bunter sandstones and pebble beds of the English Midlands (Woodcock, 1994). Glacial till is rarely used as a gravel aggregate because of its high fine sediment (clay and silt) content, but it is widely used as a fill material, for example, in dam or road construction.

Crushed aggregates are usually coarse, angular limestones, granites, dolerites or other igneous rocks. In the United States, 70% of crushed rock is limestone, 20% is granite or basalt, and 10% is sandstone and quartzite. There is an increasing use of coastal superquarries to provide large quantities of easily transported (by boat) crushed aggregate and armourstone blocks. Glensanda, a granite quarry in the western Highlands of Scotland is currently the only UK superquarry, but several exist in Scandinavia. In Ireland, a coastal quarry at Arklow Head, south of Dublin, produces dolerite for export to Germany as well as armourstone for coastal defences in Wales.

Once quarried, these materials are sorted into different size fractions that are then used in a diverse series of applications. For example, sand is used in the production of cement, concrete and mortar, as a base in road construction (Box 4.4 and Figure 4.14) or pipe laying, and in the manufacture of bricks, tiles and drainage pipes. Because the packing behaviour of grains is important, grain size and shape analysis are important in assessing the suitability of sand for particular uses (Prentice, 1990; Evans, 1997). So called 'soft sands', often from marine or fluvial sources, have rounded grains and are more easily worked than 'sharp sands', for example from fluvioglacial sources or crushed rock which have rougher textured particles, which is more acceptable in concrete or road aggregate.

Gravels of various sizes are also tested for their suitability for various functions. If they are to be combined with cement or bitumen they must react favourably with these materials to ensure the structural stability of the end product in its construction position. Resistance to heavy loads, high impacts and severe

Box 4.4 Road construction

Roads have been constructed since Roman times, but it was John McAdam who developed the modern bitumen or 'black-top' road in the nineteenth century. Modern roads are still constructed in layers though the details vary from country to country and depend on the type of traffic loading expected. The sub-base comprises a drainage layer of unbound coarse aggregate. This is overlain by a road base which is generally sand and gravel, compressed to give a firm bed, followed by a base course which is usually a bitumen bound coarse, crushed aggregate. Finally the wearing course is an asphalt with 30% aggregate sometimes with a top-dressing of fine gravel chippings. This wearing course is the most important since it must be able to withstand frictional wear, provide grip for vehicles with tyres and result in low surface noise. Much testing of particle size, shape, strength and durability is carried out to ensure that only suitable materials are used. In Finland, studded tyres are permitted in winter but this imposes special demands on the quality of the aggregates in the asphalt (Kelk, 1992).

Concrete roads are still constructed but are criticised for their surface noise. Gravel roads are also common in some countries, for example the United States, Canada, Norway and Sweden, and a variety of aggregate materials is used in their construction, including glacial till (Knutz, 1984). The Denali Highway in central Alaska, United States, is about 200 km of mainly glacial till-surfaced road open only in summer. The till matrix makes a good compacted road surface but the clasts are often loose on the surface with the result that most drivers allow at least 5 hours to drive it! Scoria, a natural, fused volcanic cinder is extensively used as a road-base for minor roads in recently active volcanic areas, e.g. Cascade Mountains in the United States, and Iceland.

abrasion may be important as, for example, in its use as roadstone. Strong, durable, crushed aggregates are commonly used as the bed for rail track sleepers. But crushed aggregates result in less easily worked and denser concrete mixes and natural aggregates are usually preferred. The most suitable aggregates are hard, quartz-rich lithologies such as quartzite and durable sandstones. Aggregate is also used in roughcast or pebble-dash wall renders, which can be produced in a diverse series of colours and textures.

Limestone Obtaining limestone for use in cement and concrete manufacture is a major quarrying industry in most countries. Evans (1997, p. 230) refers to it as 'probably the most important industrial rock or mineral used by man' and lists its uses (Table 4.2). Prentice (1990, p. 171) believes that 'It is difficult to visualise what our present-day urban landscape would be like without concrete; or to imagine how today's major engineering structures—bridges, tunnels, roads, high-rise buildings—could have been constructed without it'. Over 1 billion tonnes of cement and concrete are produced annually and they are clearly of major importance to both the global construction industry and the global economy.

Cement is a mixture of 75–80% crushed limestone and 18–25% clay or ash (silica and alumina) which is fired in a kiln, cooled to produce clinker and finally mixed with about 5% gypsum. Cement, with sand and water added, is

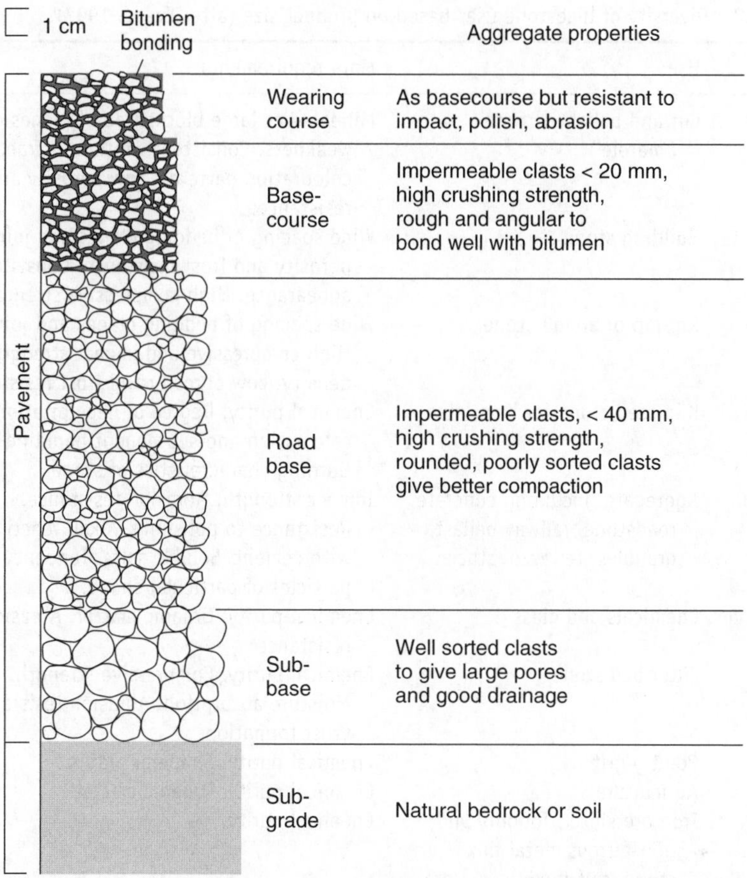

Figure 4.14 Diversity of aggregates and other geomaterials used in road construction (after Woodcock, N. (1994) Geology and the Environment in Britain and Ireland. UCL Press, London. Reproduced with permission of Taylor & Francis).

worked as a floor screed or wall render which sets hard when dry. In the United Kingdom, Lower Carboniferous or Jurassic limestones and Cretaceous chalks are frequently used to provide the carbonate content, while Jurassic, Cretaceous and Tertiary clays and shales or pulverised fly ash supply the silica and alumina. In some countries, for example Italy and the Rhine rift valley, volcanic ash has a long history of use. Some rock lithologies, for example cementstones, come 'ready mixed' in the correct proportions or can be readily made up to them. The Rosendale Formation on the Hudson River upstream from New York contains this type of mixture and has been mined and quarried since 1820 to supply New York, particularly its concrete sidewalks. At Dunbar in Scotland, Carboniferous limestone and shale are interbedded so that quarry design has allowed the correct mix to be produced on site. Most countries have limestone and clay that can be used for cement production even if the carbonate has to come from Precambrian marble, Holocene coral sands, ultrabasic igneous rocks (Greece) or carbonatite intrusions (Malawi) (Prentice, 1990). Mixed with sand, and sometimes lime, cement becomes mortar for use in bricklaying.

Table 4.2 Diversity of limestone uses based on product size (after Evans, 1997).

Size	Use	Main requirements
>1 m	Cut and polished stone ('marble')	Mineable as large blocks free of planes of weakness. Consistent white or attractive colouration patterns. Low porosity and frost resistance.
>30 cm	Building stone	Wide spacing of bedding planes and joints. Low porosity and frost resistance. Consistent appearance. High compressive strength.
>30 cm	Rip-rap or armourstone	Wide spacing of bedding planes and joints. High compressive and impact strength. High density. Low porosity and frost resistance.
1–30 cm	Kiln feed stone for lime kilns	Chemical purity. Degree of decrepitation during calcination and subsequent handling. Burning characteristics of stone.
1–20 cm	Aggregate, including concrete, roadstone, railway ballast, granules, terrazzo, stucco	Impact strength. Abrasion resistance. Resistance to polishing. Alkali reactivity with cement. Soluble salts. Tendency to form particles of particular shape.
0.2–0.5 cm	Chemicals and glass	Chemical purity. Organic matter. Abrasion resistance.
3–8 cm	Filter bed stone	Chemical purity. Compressive strength. Moisture absorption. Abrasion resistance. Crust formation.
3–8 mm	Poultry grit	Chemical purity. Shape of grains.
<4 mm	Agriculture	Chemical purity. Organic matter.
<3 mm	Iron ore sinter, foundry and non-ferrous metal flux stone, self-fluxing iron ore pellets.	Chemical purity.
<0.2 mm	Filler and extender in plastics, rubber, paint, paper, putty	Chemical purity. Whiteness and reflectance. Oil, ink and pigment absorption. pH. Nature of crushing and grinding circuits.
<0.2 mm	Asphalt filler	
<0.2 mm	Mild abrasive	Near white colour. Low quartz content.
<0.2 mm	Glazes and enamels. Funcicide and insecticide carrier	Chemical purity. Near white colour. Organic matter.
<0.1 mm	Flue gas desulphurisation	Chemical purity. Surface area. Microporosity.
Various	Bulk fill	Customer requirements. Size gradation.

When mixed with a 60–80% gravel aggregate and water, cement becomes concrete. The cement minerals react with water to form interlocking crystals of hydrated silicates and aluminates, binding the aggregate into a solid material (Woodcock, 1994). As stated earlier, the characteristics of the aggregate are very important to the strength of the concrete. There have been some notable failures due to aggregate shrinkage during drying or adverse reactions between alkali solutions from the cement and water, and silica or silicates in the aggregate, particularly when crushed types have been used (Prentice, 1990; Bennett and

Doyle, 1997). Failure may also occur where reinforcing rods rust or salt crystallisation occurs within the concrete. Many examples of the latter have been experienced in the Middle East due to rising saline capillary waters in coastal locations (e.g. Cooke and Doornkamp, 1990).

Structural clay This is clayey sediment used in brick or tile making. The processes have a long history, stretching from the time of the Egyptians, through the impressive innovations by the Romans, to the sophisticated variety of modern bricks and tiles (Prentice, 1990). Brick clays are usually sedimentary deposits though some are of residual (weathering) origin. But whatever their source, they are a diverse mixture of four clay minerals—kaolinite, illite, smectite and chlorite—each of which reacts differently to the various stages of brickmaking and many interact in complex ways. Quartz sand or silt grains may also be present or are added as 'grog' (inert additive) in order to provide a more open texture, assist in the processes and provide strength and durability. In fact quartz can make up to 90% of the total. Micas, vermiculite and sepiolite may also be present in brick clays to affect their properties.

Clifton-Taylor (1987) describes brick colours in England and believes that 'There is scarcely a limit to the colours that have been obtained' (p. 236). The diversity of colours is partly related to the mineralogy of the clay and the nature of the firing, but nowadays, mineral colorants such as ochres, are often used. The iron minerals, such as haematite, limonite, pyrite, siderite and magnetite will provide the natural red colour of most bricks with depth of colour increasing at firing temperatures of over 1000 °C. Blue colours, such as Staffordshire Blue bricks made from the iron-rich Etruria marl of the West Midlands of England, are produced by producing reducing conditions, either by setting the bricks close together in the kiln so reducing the passage of air around the bricks or by controlling the fuel supply so that all the oxygen is burnt. 'In a reducing atmosphere, the iron combines with the silicates in the clay to form ferrous silicates, which, unlike haematite, become liquid at kiln temperatures, thus forming a dark blue skin on the surface on cooling' (Prentice, 1990, p. 149). Black colourations can be produced by creating reducing conditions in the interior of individual bricks. This is done by sealing the brick with a high temperature pulse, thus producing a vitrified external skin. This prevents the escape of gases that burn internally producing a black core which may extend in patches to the brick surface. Pale colours (buff or yellow) can be produced by a complex series of processes that suppress free haematite, or by adding calcite to produce the lighter colours of iron carbonates such as ankerite, or dicalcium ferrite. However, control of clay mineralogies and firing conditions are not always easy so the precise control of colour is in fact quite difficult (Prentice, 1990). Brick texture also varies. In hand-made bricks, creases or wrinkles may sometimes be seen on the surface where the clay has been pushed together. Texture is often added by dusting with sand and modern bricks can be patterned or glazed in a variety of colours. Engineering bricks are produced for their durability and resistance to frost and damp.

Most commercial brick-making clays are post-Devonian because of the degree of alteration of older clays. This explains why there is relatively little brick making in the western United States but major industries in the centre and eastern states

(Prentice, 1990). In the United Kingdom nearly half of all bricks are produced from the 70-m thick, Upper Jurassic Oxford Clay which has a relatively high hydrocarbon content. In Belgium there are thick and extensive deposits of clays that form the basis of a large-scale industry. Included are the late Oligocene 'Boom clay', a 50-m thick deposit of smectite-illite rich rhythmites. Quaternary rythmites (varves) and wind-blown loess are also used in brick making. Indeed the wind-blown deposits of south-east England are often referred to as 'brickearths' and produced the yellowish-grey bricks (due to high calcite content) common in London. However, the most common source of brick clays are recent alluvial sediments where for centuries the deposits of the Yangste Kiang, Hwang Ho, Indus, Ganges, Mekong, Nile and others have been extracted from shallow excavations to produce bricks, albeit of relatively low strength (Prentice, 1990).

For underground pipes that must be able to withstand attack by internal fluids, external groundwater and overlying loads, particular clay qualities are required. In western Europe the kaolinite-smectite Westerwald clays from Germany are often used and exported overseas, while in North America carefully blended coal-measure shales are commonly used. The high strength is achieved by firing to 1200 °C.

Floor, roof and wall tiles are also made from clay, though often of a high quality, more carefully prepared and baked harder (Clifton-Taylor, 1987). Like bricks, the colour of traditional tiles is linked to clay composition and a great diversity of clays have been used in ceramic tile production. The German Westerwald clays are certainly used as are the ball-clays of south-west England. A greater variety of clays is used in Spain and Portugal, the latter having a famous hand painted, azuleju tile tradition. In the United States, extensive kaolinite deposits occur in South Carolina and Georgia where they are derived from Cretaceous erosion of earlier weathered rocks. The clays were deposited in a series of lakes and deltas and subsequently had iron removed by leaching processes during the Tertiary so that the clays are very pure. Tile morphology also varies with flat floor and roof tiles, half-cylinder tiles, pantiles and wall-hung tiles of various shapes being popular in different parts of the world.

Porcelain and other ceramics are made from kaolinite or china clay, while high-grade ceramics use ball clay (plastic kaolin) or halloysitic clay (see Table 4.3). To produce porcelain, the clay is partially melted at temperatures above 1250 °C. Just as it begins to melt, the potter cools it quickly to create a translucent, vitreous appearance. South-west England is the major source of china clay which is used in paper, plastics and paint industries as well as in ceramics throughout Europe (Bristow, 1994).

Gypsum This is the main raw material used in the production of plaster for use as a finishing render on internal walls and ceilings. It originates as marine or terrestrial evaporite deposits but nowadays is also produced as a by-product of the flu-gas desulfurisation process in coal-burning power stations, where a single large plant can produce a million tonnes of gypsum in a year.

Plaster of Paris gets its name from the Tertiary gypsum deposits around the city and there are also extensive deposits in Italy, Spain and many other Mediterranean countries derived from the time during the Miocene when the Mediterranean

Table 4.3 Diversity of clay types, mineralogy and uses (after Evans, 1997).

Type	Mineralogy	Uses
Kaolin (also called *china clay*)	Kaolinite of well-ordered crystal structure plus very minor amounts of quartz, mica and sometimes anatase. Highest quality material will be almost pure kaolinite and will contain >90% of particles <0.002 mm. Low to moderate plasticity.	Paper filler and coating; porcelain and other ceramics; refractories; pigment/extender in paint; filler in rubber and plastics; cosmetics; inks; insecticides; filter aids.
Ball clay (plastic kaolin)	Kaolinite of poorly ordered crystal structure, with varying amounts of quartz, feldspar, mica (or illite) and sometimes organic matter. Kaolinite particles usually finer than in kaolin. Moderate to high plasticity.	Pottery and other ceramics; filler in plastics and rubber; refractories; insecticide and fungicide carrier.
Halloysitic clay	Halloysite with varying amounts of quartz, feldspar, mica, carbonate minerals and others. Similar particle size distribution to ball clay. Moderately plastic.	Pottery and other ceramics; filler in plastics and rubber.
Refractory clay or fire clay	Structurally disordered kaolinite with some mica or illite and some quartz. Very plastic.	Refractories (firebricks); pottery; vitrified clay pipes.
Flint clay	Kaolinite and diaspore. Plastic only after extended grinding	Refractories (firebricks); pottery.
Common clay and shale	Kaolinite, and/or illite, and/or chlorite, sometimes with minor montmorillonite, quartz, feldspar, calcite, dolomite and anatase. Moderately to very plastic.	Bricks; tiles; sewer pipes.
Bentonite and Fullers' earth	Montomorillonite (smectite) with either Ca, Na or Mg as dominant exchangeable cation. Minor amounts of quartz, feldpar and other clay minerals. Forms thixotropic suspension in water. Forms a very plastic, sticky mass.	Iron ore pelletising; foundry sand binder; clarification of oils; oil well drilling fluids; suspending agent for paints; adhesives; absorbent; landfill site lining.
Sepiolite and attapulgite	Sepiolite or attapulgite with very minor amounts of montmorillonite, quartz, mica, feldspar. Thixotropic and plastic as montmorillonite clays.	Absorbent; clarification of oils; some oil well drilling fluids; special papers.

was landlocked. Other evaporites include: the Cambrian of Pakistan; Silurian of New York and Ohio, the Carboniferous of Utah and the Permian deposits of New Mexico and Texas in the United States; and the famous Permian deposits of the Zechstein Sea in Holland, Germany and Poland. Plaster is produced by dehydrating gypsum at about 107 °C and mixing the end product with other materials to provide a range of plaster types with different uses. One is render for walls and ceilings, with both base-coat bonding and finishing plaster being produced. Fine plaster can be worked into intricate raised ceiling or wall patterns, the latter being referred to as pargeting in England. It is also found in plasterboard and used for setting broken bones.

Glass sand This is a particularly pure type of sand deposited in iron-free environments or subsequently iron-leached. The sands should also ideally be clean (clay free) and well size-sorted to ensure even melting during production. Such sands are found extensively in the United States, Holland and Argentina and in smaller quantities in many other countries. Very pure sand is needed because impurities would produce discolourations (iron) or imperfections (e.g. zircon, chromite) in the glass. Optical glass must be low in alumina (Evans, 1997). But even the purest quartz sands usually have some impurities and these are removed by washing, sieving or flotation methods. Additives may include boron (from evaporites) or feldspar (from igneous rocks) as flux materials, lithium (from the mineral spodumene) for use in ceramics, and fluorite or arsenic for heat-resistant glass. Another product is glass fibre, used as insulation or incorporated with resin to produce fibreglass.

The ideal conditions for deposition of pure, iron-free sand occur where granitic rocks have been weathered (to reduce the silicates to clays), eroded (to produce the quartz) and very well sorted (to remove the iron clays). Such conditions occurred during the Upper Cretaceous at Loch Aline in Scotland and Provodin in the Czech Republic and during the Tertiary in Guyana, where very large reserves exist. Iron leaching has produced suitable sands and sandstones, for example in the Cretaceous of the northern Paris Basin in France.

Volcanic products We have already mentioned the use of volcanic ash to provide the clay fraction in cement production in, for example, Italy. Other volcanic rocks used in construction include pumice, a highly porous, solidified volcanic froth. Moulded with cement and cured slowly, it produces a lightweight block with good insulating properties. Pumice is worked on Lipari in Italy, the Greek islands and Turkish mainland, and in Oregon, California, Arizona and New Mexico in the United States. Large reserves occur on the Caribbean islands. Other low-density volcanic materials with similar properties include perlite and vermiculite.

4.5.5 Industrial and metallic minerals

These minerals form by a wide variety of mineralisation processes (Woodcock, 1994; Evans, 1997) some of which mirror those discussed above for uranium (Figure 4.15). For example, endogenic mineralisation may occur by early

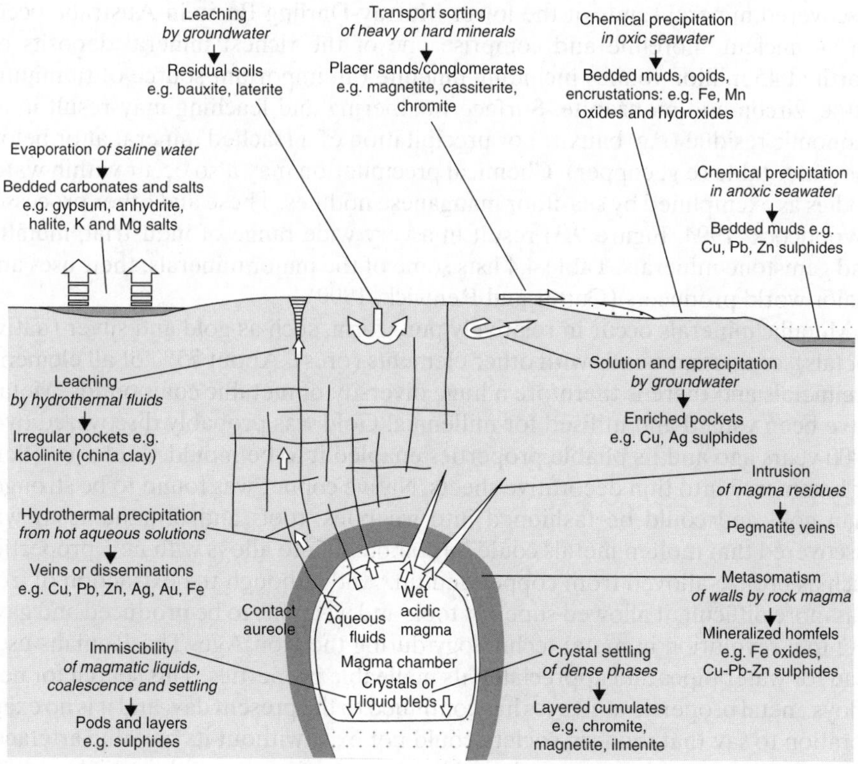

Figure 4.15 Diversity of mineralisation process (after Woodcock, N. (1994) *Geology and the Environment in Britain and Ireland*. UCL Press, London. Reproduced with permission of Taylor & Francis).

crystallisation and differential settling of heavy minerals within a magma chamber, resulting in economically exploitable concentrations of, for example, magnetite, ilmenite and chromite. In other cases, immiscible liquids may develop in these magmas with the separation of, for example, nickel sulfide liquid. On crystallisation, pods of pyrite (iron sulfide), chalcopyrite (copper sulfide), and sometimes gold and platinum may form economic concentrations. Metamorphism associated with heat and water transfer from acidic magma intrusions may cause mineral deposits to be concentrated in the surrounding country rock. Magnetite, haematite, cassiterite, pyrite, chalcopyrite, galena, sphalerite and molybdenite all form in this way. Hydrothermal precipitation occurs when superheated water travels through joints and fissures carrying dissolved compounds that are then precipitated and concentrated as the water cools to form mineral veins. Mineralisation also occurs by igneous and hydrothermal processes (e.g. black smokers) on the ocean floor or by the flow of saline water through basin floor sediments (Bennett and Doyle, 1997).

Exogenic processes also result in economic concentrations of minerals, including fluvial or coastal transport, density sorting and deposition in placers of heavy metals and some gemstones such as diamonds (Sutherland, 1984). The recently discovered mineral sands of the lower Murray-Darling Basin in Australia occur on an ancient shoreline and comprise one of the richest mineral deposits on Earth at 45 million tonnes including ilmenite (an important source of titanium), rutile, zircon and monzanite. Surface weathering and leaching may result in an economic residue (e.g. bauxite) or precipitation of a leached mineral at or below the water table (e.g. copper). Chemical precipitation may also occur within water bodies as exemplified by sea-floor manganese nodules. These and other processes (Woodcock, 1994, Figure 9.3) result in a very wide range of industrial, metallic and gemstone minerals. Table 4.4 lists some of the major minerals, their uses and major world producers (Cutter and Renwick, 1999).

Metallic minerals occur in relatively pure form, such as gold and silver (native metals), or as compounds with other elements (ores). About 75% of all elements are metals and there is therefore a huge diversity of metallic compounds. Metals have been valued and utilised for millennia. Gold was probably discovered over 8000 years ago and its pliable properties enabled it to be moulded into jewellery or hammered into thin decorative sheets. Native copper was found to be stronger than gold and could be fashioned into weapons, tools and ornaments. It was discovered that molten metals could be combined into alloys with new properties such as bronze alloyed from copper and tin. And although the extraction of iron was more difficult, it allowed superior tools and weapons to be produced and gave rise to a revolution in metal technology during the Iron Age. The Romans used lead for water pipes and appreciated its malleable properties. This search for new alloys, metal properties and uses has continued to the present day, and it is no exaggeration to say that modern society could not exist without its metallic artefacts, from aluminium with its strength and lightness to titanium with its high melting point, toughness and corrosion resistance for use in rockets and aeroplanes. Iron has by far the greatest production volumes and its use in the production of steel also utilises manganese, chromium, nickel and molybdenum (Woodcock, 1994).

The rare earth elements (REE) are a group of 17 metals, including scandium (atomic number 21), yttrium (39), lanthanum (57) and lutetium (71). REEs are in increasing use in the production of magnets, catalysts, alloys, glass, electronics, etc. REEs occur in a number of minerals but deposits are rarely of sufficient size, type and concentration to exploit commercially. Of the current global REE production, 95% comes from China, particularly Bayan Obo in Inner Mongolia, but other significant reserves occur, for example, at Mountain Pass in California and at Mount Weld in Western Australia. Nonetheless, some REEs are expected to be in short supply, at least in the short term, and price instability is possible (Henderson *et al.*, 2011). In 2012, the United States, European Union and Japan made a formal complaint to the World Trade Organisation that China was illegally restricting exports of REEs, thus deliberately creating a scarcity (*The Independent*, 20 March 2012).

4.5.6 Gemstones

These are valued for their rarity, durability and aesthetic beauty, though some have more mundane uses (e.g. as abrasives or drill-bits). There is, in fact, a very wide range of gemstones, displaying a huge diversity of chemical composition,

Table 4.4 Major minerals, their uses, major producing nations and reserves (after Cutter and Renwick, 1999).

Mineral	Major uses	Major producing nations	Reserves (yrs)
Antimony	Flame retardants, transportation, batteries	China, Russia, S. Africa, Bolivia	39
Arsenic	Wood preservatives, glass, agricultural chemicals	China, Chile, France	20
Asbestos	Insulation, friction products, gaskets	Russia, Canada, Kazakhstan, China	Large
Barite	Well drilling fluids, chemicals	China, USA, India, Morocco	40
Bauxite	Packaging, building, transportation, electrical	Australia, Guinea, Jamaica, Brazil	211
Beryllium	Alloys for aerospace and electrical equipment, computers	USA, China, Brazil, Russia	n/a
Bismuth	Pharmaceuticals, chemicals, machinery	Peru, Mexico, China	34
Boron	Glass, agriculture, fire retardants	Turkey, USA	57
Bromine	Fire retardants, agriculture, petroleum additives	USA, Israel, UK	n/a
Cadmium	Coating and plating, batteries, pigments	Japan, Canada, Belgium, USA	29
Caesium	Electronic and medical applications	Canada, Namibia, Zimbabwe	n/a
Chromium	Metallurgical and chemical industries, refractory industry	S. Africa, Kazakhstan, India	349
Cobalt	Superalloys, catalysts, paint driers, magnetic alloys	Canada, Zambia, Russia, Australia	205
Columbium	High-strength low-alloy steels, carbon steels, superalloys	Brazil, Canada	259
Copper	Building construction, electrical and electronic products	Chile, USA, Canada, Russia	32
Diamond	Machinery, abrasives, stone and ceramic products, minerals	Australia, Russia	20
Diatomite	Filter aid, fillers	USA, France, former USSR	533
Feldspar	Glass, pottery	Italy, USA, Thailand, S. Korea	Large
Fluorspar	Metal processing	China, Mexico, S. Africa	52
Gallium	Optoelectronic equipment, integrated circuits	Germany, Russia, Japan	n/a
Garnet	Petroleum industry, filters, transportation	USA, Australia, China, India	Moderate
Germanium	Fibre-optics, infrared optics, detectors	USA	n/a
Gold	Jewellery, electronics, medicine	S. Africa, USA, Australia	20

(*continued overleaf*)

Table 4.4 (*continued*)

Mineral	Major uses	Major producing nations	Reserves (yrs)
Graphite	Refractories, brake linings, packings	China, S. Korea, India	29
Gypsum	Cement retarder, agriculture, plaster	USA, China, Canada, Iran	Large
Helium	Cryogenics, welding, pressurising, controlled atmospheres	USA, former USSR, Algeria	n/a
Ilmenite	Titanium pigments	Australia, S. Africa, Canada	82
Indium	Coatings, solders, alloys, electrical, semiconductors	Canada, Japan, France	17
Iodine	Animal feeds, catalysts, chemicals	Japan, Chile, USA	n/a
Iron ore	Steel	China, Brazil, Russia, Australia	150
Kyanite	Refractories	S. Africa, France, India	Large
Lead	Batteries, fuel additives	Australia, USA, China, Peru	24
Lime	Steel furnaces, water treatment, construction, agriculture	China, USA	Moderate
Lithium	Ceramics, glass, aluminium, lubricants	Chile, Australia, Russia	349
Magnesium	Metal refractories, aerospace, auto components	USA, Canada, Norway, Russia	Moderate
Manganese	Construction, machinery, transportation	S. Africa, China, Ukraine	93
Mercury	Electronic and electrical applications, paints, chemicals	Spain, China, Algeria	42
Molybdenum	Machinery, electrical, transportation	USA, China, Chile, Canada	47
Nickel	Metal alloys, stainless steel	Russia, Canada, New Caledonia	51
Perlite	Building construction, filters, horticulture	USA, Turkey, Greece	412
Phosphates	Fertiliser	USA, China, Morocco, W. Sahara	80
Platinum	Automotive, electronic, chemical, jewellery	S. Africa, Russia, Canada	431
Potash	Fertiliser	Canada, Germany, Belarus, Russia	321
Pumice	Building blocks	Italy, Turkey, Greece	n/a
Quartz	Electronics	USA, Brazil	Large
Rare earths	Petroleum catalysts, metallurgical, ceramics	China, USA, former USSR	Large
Rhenium	Petroleum catalysts, super alloys	USA, Chile, Peru, Canada	86
Rutile	Titanium pigments, titanium metal, welding	Australia, S. Africa, Sierra Leone	83
Scandium	Metallurgical research, halide lamps, lasers	China, Kazakhstan, Madagascar	n/a

Table 4.4 *(continued)*

Mineral	Major uses	Major producing nations	Reserves (yrs)
Selenium	Electronics, glass, chemicals, pigments	Japan, USA, Canada, Belgium	37
Silicon	Metal alloys, stainless steel	China, USA, Russia, Ukraine	Moderate
Silver	Photography, electronics	Mexico, Peru, USA, Australia	20
Strontium	Colour TV tubes, magnets	Mexico, China, Turkey, Iran	36
Sulphur	Fertiliser, chemicals	USA, Canada, China, Mexico	27
Talc	Ceramics, paint, paper, plastics	China, USA, Japan	Large
Tantalum	Electronics, machinery	Australia, Brazil, Canada	65
Tellurium	Iron and steel, catalysts, chemicals	Canada, Japan, Peru	n/a
Thallium	Electronics, pharmaceuticals, alloys	USA, Canada	25
Thorium	Nuclear fuel, electrical	Australia, Brazil, Canada, India	n/a
Tin	Cans and containers, electrical, construction	China, Indonesia, Brazil, Bolivia	39
Titanium	Aerospace, chemicals	Japan, Russia, Kazakhstan	4
Tungsten	Lamps, electrical, metalworking	China, Kazakhstan, Russia	105
Vanadium	Transportation, machinery, tools, building	S. Africa, Russia, China	286
Yttrium	Television monitors, lasers, alloys, catalysts	China, former USSR, Australia	699
Zinc	Metal plating, alloys	Canada, Australia, China	20
Zirconium	Foundry sands, refractories, ceramics	Australia, S. Africa, Ukraine	36

n/a = not available

crystal structure, colour and light properties. O'Donoghue (1988), Read (1999) and Thomas (2008), for example, list about 200 gemstone types, but there is a huge variety even within each gemstone type (see Box 4.5). An important characteristic of gemstones is their inclusions, whose composition allows geologists to predict further occurrences and gemmologists to say where a particular stone has come from. 'Not only can characteristic inclusions identify a natural stone, distinguish it from its synthetic counterpart and provide valuable information on the formation conditions of an unknown specimen, they can also in many cases establish exactly where the stone came from—sometimes down to the actual mine' (O'Donoghue, 1988, p. 3). Burmese rubies and Kashmir blue sapphires are regarded as high quality and therefore more valuable than other sources, but certification can only be given by a gemmologist from the characteristic inclusions.

Gemstones are used in an infinite and dazzling variety of metallic and other settings. For example, Fabergé made several ornamental Easter eggs for Tsar Nicholas II, the most expensive of which is embellished with more than 3000 gemstones and was sold in 2002 for circa £4 million. The engraved rock crystal shell sits on a rock crystal base in the form of a melting ice block with platinum-mounted rose diamond rivulets (*The Independent*, 21 March 2002). René Lalique's Dragonfly Pectoral made in 1897–1898 is a masterpiece of design and colour in gold, enamel, chrysoprase, chalcedony, moonstones and diamonds, and can viewed in the important Lalique collection in Lisbon's Gulbenkian Museum. Gemstones are one of the most remarkable demonstrations of the principle and economic value of geodiversity.

Box 4.5 Gemstone diversity

Beryl has the formula $Be_3Al_2Si_6O_{18}$ with some iron, manganese, chromium, vanadium and caesium. It displays a wide range of colours from green in emerald, blue to greenish-blue in aquamarine, golden yellow, pink or peach in morganite and an orange- to near ruby-red in bixbite from Utah, United States. Beryl can also be colourless (goshenite). The colour in emerald arises from replacement of aluminium by chromium and/or vanadium. The colour of green aquamarine is due to iron, while morganite is believed to acquire its pink colour from manganese. A dark blue beryl (Maxixe or Maxixe-type) was discovered in Brazil in 1917, and derived its colour from nitrate or carbonate impurities. Beryl is found in a variety of geological environments and rock types around the world. It is found in mica schists (particularly emerald), metamorphic limestones (emerald) and hydrothermal veins. It is common in granitic rocks, particularly pegmatites. The finest emeralds come from Colombia and were known to Colombian Indians as far back as 1000 AD.

Corundum (Al_2O_3 with some iron, titanium and chromium) comes in a number of forms but the best known are ruby and sapphire. Ruby colours range from a strong purple-red to orange-red. The world's finest rubies come from Burma and show a strong red fluoresecence, whereas those from Thailand, which make up 70% of the world's fine ruby production, are less intensely red due to the presence of iron. Sapphire comes in a range of blues, greens, oranges and purples. Blue sapphires of industrial and gem quality were mined from dykes in Montana, United States, and those recovered from the Yogo Gulch mines between 1865 and 1929 were worth $25 million. Placer sapphires have also been quarried from Missouri River terrace gravels where they occur along with mined gold deposits. Some corundum shows 12-pointed or 6-pointed stars due to rutile inclusions.

Quartz (SiO_2) comes in diverse forms and colours including amethyst, rock crystal, milky, smoky and rose quartz, chalcedony, cornelian, onyx, jasper and agate, but each type itself exhibits a range of forms and colours (O'Donoghue, 1988) due to impurities and formation processes.

4.5.7 Fossils

Fossils can have significant value if they are rare, well preserved and/or well known. Dinosaur fossils, for example, can command large sums given their popularity with the general public. A complete skeleton of *Tyrrannosaurus rex* ('Sue') collected by a commercial company on federal lands in South Dakota and subsequently confiscated by the FBI was sold at auction by Bonhams in New York for $8.36 million (Fiffer, 2000). The Staatliches Museum in Stuttgart, Germany offered £180 000 for the world's earliest fossil reptile ('Lizzie'), discovered at East Kirkton Quarry in Scotland. The *Caloceras* beds at Doniford Bay in Somerset, England are valued at £825 per m^3, but 65% has been removed by irresponsible collecting (Webber, 2001). Even material already collected can be sold off for commercial gain. The famous collections of the Moscow Palaeontological Museum have been broken up and have started turning up in the commercial markets of Europe and the Americas (Karis, 2002).

The market for fossils is not just scientific. So-called 'décor fossils' are sold to a general market on the basis of their aesthetic appeal, examples being insects encased in amber and the brightly coloured (green, red, blue) and iridescent 'Ammolite' jewellery range derived from particular species of Upper Cretaceous ammonites found around the Alberta–Montana border. In fact, the value of most fossils is assessed according to aesthetic appeal or rarity rather than on scientific grounds (Norman, 1994). Forster (1999) lists fish, crustaceans, crinoids, ammonites, trilobites and large leaves as the most popular specimens, with prices ranging from £5 to £5000 dependent on size, colour and quality of preservation. Fossil and mineral shops have opened to cater for public demand (Figure 4.16).

Fossiliferous ornamental stone is also common, coralline and crinoidal rocks being particularly popular. In Scandinavia, Ordovician red limestones with straight nautiloids provide an attractive stone, while in southern Europe, Cretaceous limestones containing rudists are commonly polished as a facing stone (Fortey, 2002).

Fossils *in situ* have great economic value in stratigraphic correlation and thus in oil and mineral exploration worldwide. Particular use is made of a diverse range of microfossils (e.g. conodonts, ostracods, foraminifera, coccoliths, pollen and spores).

Figure 4.16 Mineral & fossil shop, Quebec City, Canada.

4.6 Cultural services

By cultural services we mean the value placed by society on some aspect of the physical environment by reason of its social or community significance. It is not difficult to find examples of these attachments in both past and present societies, and it follows that because the physical environment is valued in this way, it is appropriate to consider conserving the landscapes and features involved.

4.6.1 Environmental quality

Environmental quality refers quite simply to the visual appeal (and those of other senses) provided by the physical environment. This may be through landforms at all scales from mountain ranges to local ponds, from coastlines to river banks, but all have value because of the diversity of topography they provide for residents or travellers. There are also psychological and physiological health benefits of having access to natural areas, sometimes referred to as therapeutic landscapes and well known for raising morale and benefiting mental health.

Many landscapes have aesthetic appeal. Daly (1994) argued that peatlands are major landscape features in countries like Finland, Ireland, Scotland and Poland. 'As such, they are not only of geomorphological importance but are part of the beauty and scenery of these countries...'. With reference to limestone pavements, Goldie (1994, p. 220) believed that 'the urge to conserve these landforms for the common good derives largely from local and personal appreciation of their scientific and aesthetic worth'. Jarman (1994, p. 41) argued that 'Human perception values variety, intricacy, pattern and regional character.... The contribution of landform type and wealth of surface detail to the popularity of tourist areas such as the Yorkshire Dales or residential areas (Kent v Essex), is under-rated. Erosion of this character and detail will reduce the attractiveness of an area'.

Norton (1988) refers to this as 'amenity value' where the existence of a natural feature improves our lives in some non-material way. This is certainly true of many physical features such as mountains, beaches, cliffs, rivers, lakes, glaciers and waterfalls. Landform is undoubtedly undervalued as an element of landscape (see Chapter 10) and it is the ever-changing nature (diversity) of the natural world that creates much of its visual interest and beauty. And aesthetic value also bestows economic value and social status. As Jarman (1994, p. 42) argues 'Advanced societies will pay a premium for landscape attributes of a property, starting with a sea view'. Hilltop locations often acquire more than topographic height since they are associated with greater social and economic status than adjacent lowland.

4.6.2 Geotourism and leisure activities

Just as 'geodiversity' has been developed as the abiotic equivalent of 'biodiversity', so 'geotourism' has become a popular topic in recent years as the abiotic parallel of 'ecotourism' (e.g. Larwood and Prosser, 1998; McKirdy, Threadgold and Finlay, 2001; Dowling and Newsome, 2006, 2010; Newsome and Dowling, 2010;

Dowling 2011; Hose, 2008; Hose 2012a, 2012b), though the two may, of course, be linked. Many definitions of ecotourism exist including:

- 'A tourism market based on an area's natural resources that attempts to minimise the ecological impact of the tourism;
- 'Tourism supported by natural ecological attributes of an area, for example bird watching.

There are also several existing definitions of geotourism, such as:

- 'The provision of interpretative facilities and services to promote the value and societal benefit of geological and geomorphological sites and their materials and to ensure their conservation, for the use of students, tourists and other recreationalists' Hose (2000, p. 136).
- 'A form of natural area tourism that specifically focuses on geology and landscape. It promotes tourism to geosites and the conservation of geo-diversity and an understanding of earth sciences through appreciation and learning. This is achieved through independent visits to geological features, use of geo-trails and view points, guided tours, geo-activities and patronage of geosite visitor centres' (Newsome and Dowling, 2010, p. 4).

Despite the inclusion of geodiversity in this definition, the one preferred here is simpler:

- 'Tourism based on an area's geological or geomorphological resources that attempts to minimise the impacts of this tourism through geoconservation management' (Gray, 2008a, p. 295).

Hose (2008, p. 55) quotes James Boswell's remark to his wife in 1773 as he prepared to travel to Scotland with Dr Samuel Johnson: 'Madame, we do not go there as to paradise. We go to see something different from what we are accustomed to'. This encapsulates the key principle of tourism in that places are different and provide different experiences and a diversity of environments. Clearly then, geotourism is based on geodiversity, in the sense that it provides the opportunity to experience different geologies, geomorphological environments and landscapes and/or take part in geological activities. With reference to landscapes, it must be understood that they are always at least partly abiotic whether through rock outcrops, sediment deposits, topographies or the operation of geomorphological or geological processes and this is a topic that I shall explore in some detail in Chapter 10. Larwood and Prosser (1998, p. 99) conclude that 'Tourists, whether they are aware or not, will in some way all be geotourists'.

As well as differences between places, differences also occur within places. One of the most popular geotourism locations in the world is the Grand Canyon (Figure 4.17), resistances to erosion bringing diversities of slope morphologies, processes and talus materials. If the geology and geomorphology of the canyon had been uniform so that a simple V-shaped canyon had resulted it would still have been impressively large, but it is the internal diversity of rocks, topographies

Figure 4.17 Geotourism at the Grand Canyon. See plate section for colour version.

and processes that gives the canyon much of its character and aesthetic appeal to the tourist.

One of the aims of geotourism is to improve the economy of parts of the world, many of which may be economically poor but geologically rich. I shall deal with geoparks, one of whose aims is to promote economic development through geotourism in Chapter 7, and in Chapter 13 I shall outline some examples of geoeducation, geotourism facilities and other geological initiatives for the general public. The aim here is to give some examples of geotourism and its role in economic development.

In 2007, in the western part of the Grand Canyon, the Indian Hualapai Tribe opened the 'Skywalk', a U-shaped, glass-bottomed bridge that takes visitors out over the canyon 1200 m above its base. Built at a cost of over $30 million it attracted over a million visitors in its first three years of operation, each paying $25 for the 15-minute thrill of walking out over the canyon. It provides much-needed income and employment for the Hualapai people who previously had an unemployment rate of 70%. Future development of a museum, gift shop, restaurants and bars is planned.

Another geotourism walk occurs at Hopewell Rocks in New Brunswick, Canada, where visitors are invited to 'walk on the ocean floor'. Of course we all do this on the beach when the tide goes out, but the difference at Hopewell Rocks is that it lies on the coast of the Bay of Fundy where the maximum tidal range is circa 15 m. One of the main attractions of the site at low tide is the opportunity to explore the spectacular natural arches, caves and sea stacks in the sandstone and conglomerate rocks. Many of the stacks are topped with vegetation and are known as 'flowerpot rocks' (Figure 4.18).

The general value of scenery, wilderness and environment are often promoted as part of national tourism campaigns. There is increasing interest in touring

Figure 4.18 Geotourism at Hopewell Rocks, New Brunswick, Canada.

and walking holidays and the general attraction of rural landscapes for day trips and short breaks by increasingly urban populations. This is in addition to the worldwide attraction of beach visits and holidays. There are also specific geological/geomorphological wonders that are highly attractive to tourists. As well as the Grand Canyon, examples include Niagara Falls, Old Faithful geyser in Yellowstone National Park, the Norwegian fjords, Uluru/Ayers Rock in Australia, swimming in the geothermal 'Blue Lagoon' in Iceland, etc. Sometimes the use of scenery in films can lead to major boosts for tourism as occurred with *Lord of the Rings* and *The Hobbit* filmed in New Zealand. The countless 'Westerns' filmed in Monument Valley on the Utah–Arizona border, United States, continue to draw tourists to these landscapes (Figure 3.13). A survey by the BBC of 20 000 viewers, concluded that the top place 'to see before you die' is the Grand Canyon, while other places in the top 50 include Uluru (Australia), the Matterhorn (Switzerland), Lake Louise (Canada) and no less than four waterfalls (Angel Falls, Niagara Falls, Iguacu Falls, Victoria Falls). The list suggests that abiotic features are able to attract tourists at least as much as biotic ones and that the value of geotourism ought to be better understood.

Global Geotourism conferences have been held in recent years in Australia (2008), Malaysia (2010) and Oman (2011) with the fourth in Iceland (2013) (Dowling, 2012), and an International Association for Geotourism (IAGt) was established in Krakow, Poland in 2007 with the related journal *Geoturystyka* published since 2004. Several edited books (e.g. Dowling and Newsome, 2006, 2010; Newsome and Dowling, 2010) and journal special issues (e.g. Hose, 2012a) on geotourism have been published and have helped to develop geotourism principles as well as providing many case studies of geotourism from around the world. Dowling and Newsome (2006) also consider some of the general issues and challenges for geotourism. These include the issue of *National Geographic's* definition of geotourism as 'geographical tourism' which Hose (2008, p. 39) dismisses as simply 'superficial re-branding of existing tourism activities'.

Another issue is the potential adverse impact of geotourism, including fossil collecting, which is permitted in some locations but strictly prohibited in others. The key to preventing adverse impacts lies in the management of both the geological resources and the visitors and this is an issue I shall return to in Section 5.10. And geotourism is also an activity with huge economic potential. According to *The Guardian* (13 December 2012), the number of world travellers will reach 1.8 billion by 2020, growth from China and Russia being particularly strong. By then it is predicted that one in ten people will be employed in tourism, so there are significant opportunities for geotourism, if marketed correctly.

Many sports and leisure pursuits also depend on geodiversity. Climbers, in particular, value the diversity of rock types and structures for the variety of challenges they bring (Mellor, 2001), from the granite slabs of the Yosemite Valley in California, United States, to the sandstone pinnacle of the Old Man of Hoy in Scotland, to the columnar phonolite porphyry of the Devils Tower in Wyoming (Figure 4.19).

Most ski resorts provide ski runs with a wide diversity of slope steepness and complexity for different skiing abilities from children and beginners to experts. The North American classification of ski runs identifies:

Green Circle Runs, easiest, 6–25% slopes;
Blue Square Runs, intermediate, 25–40% slopes;
Black Diamond Runs, difficult, >40% slopes;
Double Black Diamond Runs, expert, very steep + jumps, etc.

Thus, diversity of topography is of great importance to the development of ski runs and ski resorts. In some cases the topography is mechanically altered to provide required features.

However this is more common in golf course design. Nonetheless, Price (1989, 2002) was able to classify Scottish golf courses by their geomorphological characteristics. These include the Open Championship links courses of St Andrews, Carnoustie, Troon and Turnberry, all of which are built on stabilised sand dunes developed on raised beaches.

Table 4.5 is a list of some sports and leisure activities that value topographic diversity. But other sports require a 'level playing field'. Also included in leisure activities related to geology are fossil and mineral hunting, lapidary and geocaching.

4.6.3 Cultural, historical and spiritual meanings

Folklore (geomythology) The term 'geomythology' was first introduced by Vitaliano (1973) who identified that 'primarily, there are two kinds of geologic folklore: that in which some geologic feature or the occurrence of some geologic phenomenon has inspired a folklore explanation, and that which is the garbled

Figure 4.19 Rock climbing at the Devils Tower, Wyoming, USA.

explanation of some actual geologic event, usually a natural catastrophe' (Piccardi and Masse, 2007). It is the former that I am mainly concerned with in this section since it is the myths and legends associated with the origin of landforms, geomaterials and physical processes that gives these places or materials cultural value and hence creates a reason for their geoconservation.

This field of study has developed over the decades since Vitaliano's book and culminated in 2004 with a special conference session held at the 32nd International Geology Conference in Florence, Italy, the papers from which subsequently formed the basis of an edited volume on *Myth and Geology* (Piccardi and Masse, 2007). Among the many examples cited in the book is the story of Pelé, the

Table 4.5 Some of the sports and leisure activities that value topographic diversity.

Skiing (downhill & cross-country)
Snowmobiling
Mountaineering
Rock climbing (also values rock diversity)
Hill walking
Hill running
Cross-country running
Motor cycle scrambling
Mountain biking
Road race cycling (e.g. Tour de France)
Off-road driving
Whitewater rafting
Whitewater canoeing/kayaking
Canyoning
Caving/potholing
Golf

Hawaiian volcano goddess. All the major landforms in Hawaii are attributed to the actions of Pelé including the craters and the smooth, glassy lapilli that are reputed to be her tears. Even the alignment of the Hawaiian island chain has a mythological explanation. Pelé came to Hawaii in flight from her older sister. She first came to the most northwestern island where she dug a crater in search of fire, but her sister chased her onto the next island where another pit was dug. By this repeated process, the chain of volcanic islands was created (Vitaliano, 2007). This legend illustrates how some stories contain an impressive level of understanding, in this case that the volcanoes become younger from north-west to south-east.

Many other igneous rock formations or landforms are attributed to supernatural forces and many still bear the names of these associations. For example, according to Californian Indian legend, Glass Mountain was created when Ground Squirrel dropped a load of obsidian that had been stolen from Adder. We also have famous explanations of columnar jointing such as the Giant's Causeway in Northern Ireland, supposed to be the stepping stones of giants, and the Devils Tower in Wyoming, United States, where native American legend has it that a giant grizzly bear trying to reach a group of people on its summit, left its claw marks on the sides of the tower (see Figure 4.19).

In fact the Devil is very common in the names of physical features around the world. *The Times World Atlas* identifies mountains or hills (Devil Mountain, Devil's Peak, Devils Paw, Devilsbit, Devil's Mother, Devil's Riding School, Devil Post Pile, Devil River Peak, Devil's Elbow), hollows (Devil's Beef Tub, Devil's Gorge, Devil's Hole, Devil's Kitchen), passes (Devil's Gate), lakes (Devil's Lake) and deserts (Devil's Playground) with associations with the Devil. In Serbia, the Devil's Town is a series of over 200 earth pillars, reputed to be petrified humans or beings from another planet (Vasileva, 1999). The Devils Marbles are huge rounded residual boulders in the Northern Territories, Australia, though Aboriginal belief is that they are eggs of the Rainbow Serpent (see also Box 4.8

later). The Devil may even have been held responsible for geomorphological sounds. *The Guardian* (27 August 2001) reported that a cavern in Derbyshire, England has changed its name back to its Anglo-Saxon one of the Devil's Arse, because of the noise made when water which has built up in the cavern drains away! As a result, visitor numbers have increased by 30%! Duffin and Davidson (2011) have recently described in detail the supernatural folklore traditions of geology, including those associated with the Devil, fairies, elves, pixies and witches. They describe how some dinosaur footprints and other tracks have been interpreted instead as made by the Devil.

Other examples of mythical explanations for landforms can be related. Several examples occur in Scotland (see Box 4.6) but many countries have similar folklore. In the United States there is a Chippewa Indian legend to explain the Manitou Islands in Lake Michigan, Wisconsin. A mother bear and her two cubs were driven into the lake by a raging forest fire. They swam and swam but the cubs became tired and lagged behind. The mother bear eventually reached the opposite shore and climbed to the top of a sand dune to await her offspring, but they had drowned. Today the cubs are the Manitou Islands, while the sand dune is called Sleeping Bear Dune.

Box 4.6 Scottish landform legends

The Dog Stone, a raised, undercut sea stack near Oban, Scotland, is said to be where the Irish Giant Fingal tied up his dog whilst he went hunting in the Hebrides (Figure 4.20). Not far away, in Glen Roy, there is a series of near-horizontal ice-dammed lake shorelines called the 'Parallel Roads of

Figure 4.20 The Dog Stone, Oban, Scotland named from the legend that this is where the giant Fingal tied up his dog Bran.

Glen Roy', because they are said to be hillside hunting roads used by Fingal. Also near Oban, a complex shaped kettle hole on the Loch Etive kame terraces (Gray, 1975) is said to be where a giant witch's cow lay down, while nearby a perfectly circular kettle hole represents her cheese mould. At Morebattle in the Scottish Borders, the well-sorted sands of the local kames are reputed to have been sifted by nuns as a penance. There are also tales of what we now know to be glacial erratics reputedly being heaved around the country by giants, witches or apocryphally strong men. By the shores of Loch Etive lies 'Rob Roy's Putting Stone'. Two other boulders on either side of the Kyle of Durness are reputedly the result of a local witches' feud in which the matter was settled by hurling boulders across the Kyle with the furthest throw being acclaimed the victor. Loch Bran in Wester Ross is reputed to be the result of digging by Bran, the legendary hound of Ossianic legend (Robertson, 1995).

Islands are often associated with legends. The small Norwegian island of Torghatten is in the shape of a hat with a crown and a brim, the latter having been eroded by a higher relative sea level. This sea level eroded a cave that passes all the way through the island, but legend has it that the cave was produced when an arrow was shot through the hat which then fell to the ground and turned to stone. In Vietnam, the steep karst islands of Halong Bay are reputed to be a defensive wall spat out by dragons sent by the gods to defend the country from invaders.

In New Zealand, geologists believe the famous Moeraki Stones to be giant septarian nodules eroded from Palaeocene claystones, but Maori legend has it that they are bread rolls dropped from a giant explorer's food basket. Further south, a Maori god, Tu Te Raki Whanoa, is said to have been responsible for carving Doubtful Sound, the largest of New Zealand's fjords, as he tried to create a route from the sea to the interior.

Fossils are also the subject of folklore and legends. For example, Mayor (2005) has documented in great detail the myths surrounding the fossils observed and collected by native Americans long before contact with Europeans. In Alberta, Canada, large dinosaur bones were thought to come from 'giant' or 'grandfather' buffalo, this being the largest animal known to local people (Mayor, 2005). Invertebrates have been interpreted in even more imaginative ways. Ammonite fossils are reputed to have once been living serpents or horns of an ancient Egyptian god. Amber is reputedly the hardened tears of the daughters of the sun god. Species of the genus *Gryphaea*, particularly *G. incurve*, has the nickname 'Devil's toenail' and belemintes have been called 'fairies' fingers' and 'elf arrows' (Duffin and Davidson, 2011).

Archaeological and historical value Our early ancestors had a very close relationship with their physical surroundings, and geology and landscape must have played an absolutely crucial role in their lives. As Jacob Bronowski wrote in *The Ascent of Man* (1973. p. 40), when man first made rudimentary stone tools by using a simple blow to put an edge on a pebble 'He had made the fundamental invention, the purposeful act which prepares and stores a pebble for later use . . . he had released the brake which the environment imposes on all other creatures'.

During the Palaeolithic, the increasing sophistication of stone implements is evident. From initial, crude chopping tools, through well-crafted hand axes to delicate arrow heads, the history of the human use of stone for hunting, butchery and warfare is clear, as is the ability of humans to seek out the best rock types for these purposes such as flint, obsidian, quartzite or other hard rock displaying a conchoidal fracture. Here is an example of geodiversity being sought and exploited in innovative ways. Lynch (1990) describes a 0.7-m thick, naturally baked, Ordovician shale at Mynydd Rhiw in Wales which was used for axe and chisel manufacture. The Alibates flint quarries in Texas, United States, date back at least 12 000 years and tools and spear points made from the flint are found in many places in the Great Plains and South West United States, indicating either trading or extensive nomadic hunting. According to Evans (1997, p. 3), 'by 3000 B.C. large underground flint mines were in operation at Grime's Graves in Norfolk, UK, and it is clear that the miners had noted that particular horizons in the chalk host rock carried the best and most numerous flints'.

Later, when agriculture became important and milling equipment was developed, millstones or 'querns' were manufactured with grooves to allow the flour to escape. Clay and other materials began to be used in pottery manufacture, while stone and subsequently metals were used to produce art and coinage. Pigments such as natural ochres (iron oxides), umbers (iron-rich clays), cinnabar (red mercury sulfide), wad (black manganese oxides) and galena (grey lead oxides) were extracted.

Early humans used the diversity of the natural environment to their advantage. They sought out natural caves or excavated them in suitable rock types. In western North America they selected cliff sites as 'buffalo jumps' where herds would be driven to their death over suitable precipices. But they also discovered the power of construction using rock materials. They soon began to build crude stone shelters and houses, defensive structures, stone monuments or burial sites, and they began decorating the rock surfaces with carvings (Figure 4.21) and paintings. As they did so they utilised the natural diversity of the geomaterials that they found all around them. Important examples include the beautiful temples and massive pyramids of the Nile Valley in Egypt, the huge stone monuments at Borrobodur and Prambanan near Yogjakarta on Java, Indonesia and Angkor Wat in Cambodia, the giant Buddhas at Bamiyan in Afghanistan (destroyed by the Taliban), the Great Wall of China, the granite city and agricultural terraces of Machu Picchu in Peru, the carved rock city of Petra in Jordan, and the cliff-houses of the Mesa Verde, Colorado and Canyon de Chelly, Arizona, United States (Box 4.7).

The use of stone as part of ritual acts is exemplified in many parts of the world. For example, in Scotland, ancient carved footprints have been discovered in Orkney and Shetland, and examples also occur in Ireland and France. Breeze and Munro (1997, p. 12) believe that 'The act of a king or chief standing on a special stone to be invested can be seen as symbolic of a relationship with an object of great antiquity—rock. . . . and also with the land from which his people earned their food'. But Breeze and Munro (1997) go further in noting associations between rock and investiture: 'It is but a short step to remove a piece of such rock and make it into an object on which a king sat to be invested'. And they quote

Figure 4.21 Stone relief carving used to portray the union of the crowns of Upper and Lower Egypt.

Box 4.7 Canyon de Chelly, United States

The Canyon de Chelly is located in a Navajo Reservation in north-east Arizona and was established as a National Monument in 1931. The vertical red sandstone walls of this and adjacent Canyon del Muerto rise to heights of 300 m and contain several large caves both at the cliff base and within the walls. These caves were occupied as cliff dwellings between the fourth and fourteenth centuries, the later ones involving construction of houses within the caves sometimes rising to five or six storeys, and made up of hundreds of rooms. Notable examples are the White House (occupied from about 1060–1275 (Figure 4.22) and the Mummy Cave dating from 1253. The walls of the canyon and caves are decorated with fine examples of rock carvings and paintings including antelope, fish and abstract patterns.

> There is a great intellectual step forward when man splits... a piece of stone, and lays bare the print that nature had put there before he split it. The Pueblo people found that step in the red sandstone cliffs... The tabular strata were there for cutting; and the blocks were laid in courses along the same bedding planes in which they had lain in the cliffs of the Canyon de Chelly. (Bronowski, 1973, p. 95)

Figure 4.22 The 'White House' in Canyon de Chelly, Arizona. See plate section for colour version.

examples from Kingston upon Thames, England, several sites in Ireland, Zollfeld in Austria, Uppsala in Sweden and Aachen in Germany. But the most notorious example must be the 'Stone of Destiny', a block of Perthshire sandstone on which the Scottish Kings had been inaugurated for centuries until it was removed from the abbey at Scone by King Edward I of England in 1296. He took it to Westminster Abbey in London where, for 700 years it sat within the Coronation Chair where the kings and queens of England, and subsequently of the United Kingdom, were crowned. The stone was returned to Scotland in 1996 and today it is on display in Edinburgh Castle.

Edinburgh Castle is, of course, an impressive defensive structure, built as it is atop a volcanic neck, and there is also no doubting the role that the physical environment often played in the siting of settlements for defensive, resource or other reasons. Other examples include the hilltop towns of Tuscany and Umbria in Italy, the 'Gold Rush' towns of the American west or Canadian Yukon, and the clustering of Egyptian settlements in the Nile Valley.

Bennett and Doyle (1997, p. 168) point out that the outcome of many famous battles was influenced by the local landscape: 'Waterloo (1815) was fought on a clay plain in Belgium, while the Somme (1916) took place in the dissected chalk upland of northern France. Both of these are commemorated on the ground and are intimately associated with the local landscape'. The outcome of the Battle of the Boyne (1690), so important in Irish history, was strongly influenced by the topography of the area (Stout, 2002). The Falklands War (1982) was fought between the United Kingdom and Argentina over the sovereignty of a set of islands (Falklands/Malvinas) in the South Atlantic, though like many conflicts, there may have been a background issue of rights to mineral resources in the South Atlantic. Similarly, the Gulf War and the threats to Iraq are alleged to have been at least partly about oil resources. For further examples of geologically induced conflicts and terrain affecting the outcome of warfare see Woodcock (1994, Figure 3.8) and Doyle and Bennett (2002). It has also been predicted that future wars may well be fought over water.

Spiritual value Many human societies place spiritual or religious value on the physical environment. 'Adam' is derived from the Hebrew 'adama' meaning earth or soil. Thus 'Adam's name encapsulates man's origin and destiny: his existence and livelihood derive from the soil, to which he is tethered throughout his life and to which he is fated to return at the end of his days' Hillel (1991, p. 14). And since 'Eve' is derived from the Hebrew 'hava' meaning life, together Adam and Eve signify soil and life.

This is reflected in the views of Jomo Kenyatta, former president of Kenya who is reported as describing the value of land to the Gikuyu people in 1938 as follows:

> It supplies them with the material needs of life through which spiritual and mental contentment is achieved. Communion with ancestral spirits is perpetuated through contact with the soil in which ancestors of the tribe lie buried... it is the soil that feeds the child through lifetime; and again after death it is the soil that nurses the spirits of the dead for eternity. Thus the earth is the most sacred thing above all that dwell in or on it. (Mackenzie, 1998, p. 24)

North American Indian tribes each have their own stories about the origin of the world. The Blackfoot, for example, believe that the landscape was made by Old Man or Naapi. Consequently they regard the Earth as sacred and some even see ploughing as violating the land. Many natural sites are regarded as sacred where people can communicate with spirits, both good and bad. 'What is important for traditional Indian religious believers is... a location made holy by the Great Creator, by ancient and enduring myth, by repeated rituals such as sun dances, or by the presence of spirits who dwell in deep canyons, on mountaintops, or in hidden caves' (Gulliford, 2000, p. 69). Kelley and Francis (1994) undertook a survey of 13 Navajo communities to find out what places they regarded as sacred and concluded that these sacred places, that remain integral to tribal histories, religions and identities, should not be disturbed. Gulliford (2000) classifies sacred places into several types, many of which are related to physical features, for example vision quest sites, group ceremonial sites or burial sites and Figure 4.23

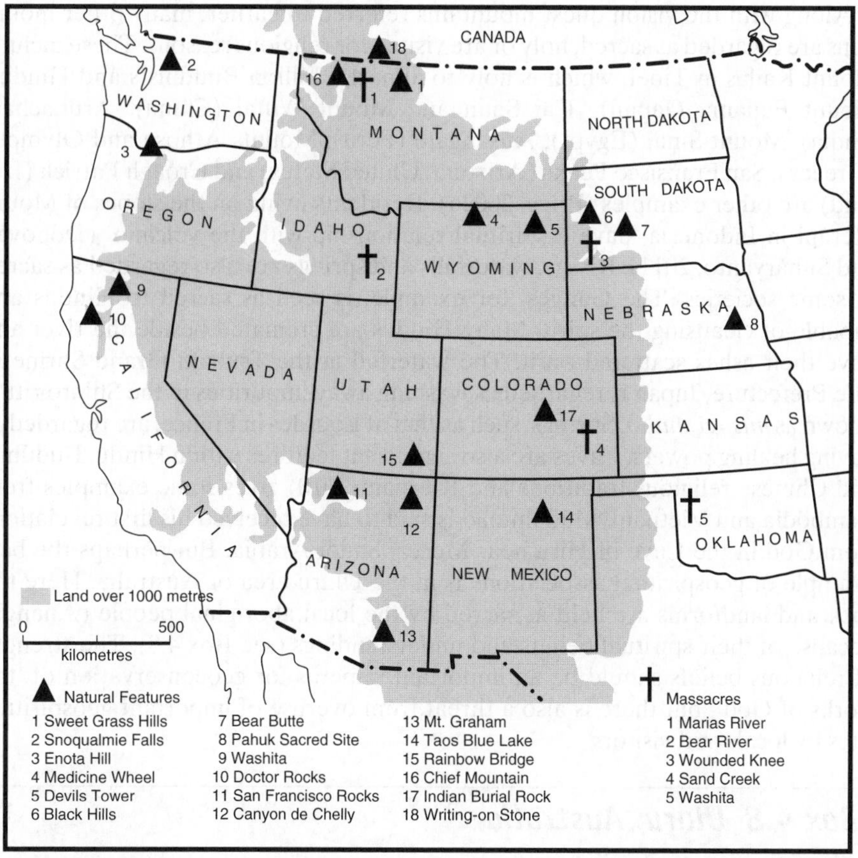

Figure 4.23 Selected Native North American sacred sites in Western USA and Canada (after Gulliford, A. (2002) *Sacred Objects and Scared Places: Preserving Tribal Traditions*. University Press of Colorado, Boulder, by permission of Andrew Gulliford).

shows the locations of selected sacred sites in the western United States. Vision quest sites, such as the Sweet Grass Hills and Chief Mountain in Montana, are usually isolated and remote peaks where individuals remained without food or water until the arrival of a spiritual bird or animal to give them guidance.

There is a North American Indian story related to pipestone, a red, durable but easily carved rock type found particularly in Minnesota where the quarry is a National Monument. The origin of pipestone is said, in Sioux accounts, to date from an ancient time when the Great Spirit, in the form of a large bird, broke off a piece of red stone, formed it into a pipe and smoked it. He told the gathered tribes that this red stone was their flesh, that they were made from it, that they must all smoke to him through it, and that they must use it for nothing but pipes. The ground was sacred to all tribes and no weapons must be used on it. Consequently, the quarrying of the pipestone has always been carried out with respect for the earth and for what it yields.

Along with the vision quest mountains referred to earlier, many other mountains are regarded as sacred, holy or are visited for religious reasons. These include Mount Kailas in Tibet, which is holy to almost a billion Buddhists and Hindus. Mount Fujiama (Japan), T'ai Shan and Mount Wutai (China), Arunachala (India), Mount Sinai (Egypt), Ausangate (Peru), Mounts Athose and Olympus (Greece), San Fransisco Peaks (Arizona, United States) and Croagh Patrick (Ireland) are other examples (Price, 2002a). Residents living on the slopes of Mount Merapi in Indonesia 'have a spiritual relationship with the volcano' (Donovan and Suharyanto, 2011). Rivers, waterfalls and springs are also regarded as sacred in some societies. The Ganges, for example, is seen as sacred to Hindus and capable of cleansing the spirit. Many Hindus are cremated beside the river and have their ashes scattered on it. The waterfall at the Tsubaki Grand Shrine in Mie Prefecture, Japan is regarded as washing away impurities in the Shinto ritual known as *misogi shuho*. Springs, such as that at Lourdes in France, are regarded as having healing powers. Caves are also significant features within Hindu, Buddhist and Chinese religious traditions and Kiernan (2010) gives some examples from Cambodia and Vietnam. Muhammad is said to have received his first revelations from God in the Cave of Hira near Mecca, Saudi Arabia. But perhaps the best example of geospiritual associations is in the Uluru area of Australia. Here the rock and landforms are held as sacred by the local aboriginal people (Ananga) because of their spiritual beliefs and understandings (see Box 4.8). The strength of religious beliefs should be an important impetus for geoconservation of 'the works of God', but there is also a threat from overuse of important geospiritual sites by locals and visitors.

Box 4.8 Uluru, Australia

Uluru (formerly Ayers Rock) in Australia still has spiritual significance for the local aboriginal people (Anangu). This is hardly surprising given the dramatic appearance of the hill rising over 300 m from the flat desert floor and the near vertical bedding in the arkosic sandstone bedrock (Figure 4.24).

The Aborigines believe that there is a hollow below ground and an energy source known as 'Tjukurpa', a word that has several meanings, including the creation period. The Anangu believe that the world was once a featureless place. However, during the creation period the area around the rock is inhabited by dozens of ancestral beings whose activities produced the physical features of the area, such as rocks, sand hills or caves. This gives each of these features, including Uluru itself, eternal spiritual significance as the living presence of Tjukurpa.

For example, in the creation period, Tatji, the small red lizard, came to Uluru. He threw his kali, a curved throwing stick similar to a boomerang, at the rock but it became stuck in a joint in the rock. Tatji used his hands to try to scoop it out, but in his efforts to retrieve his kali, he left a series of bowl-shaped hollows in the rock. Unable to recover his kali, he finally died in a nearby cave and his implements and bodily remains survive as large boulders on the cave floor. Elsewhere a fractured slab of sandstone is believed to represent large joints of meat from an emu killed and butchered by two lizard men, Mita and Lungkata. In several caves at Uluru, rock paintings believed to date back many thousands of years but regularly refreshed, relate the many stories of the Tjukurpa.

Figure 4.24 Uluru, Australia.

The spiritual significance of the rock, caves and other features give the physical features of the area a particular value to aboriginal society that is not generally appreciated by tourists to the area. They certainly value the spectacular geomorphology of the rock, its ever-changing colour and the opportunity to climb to the summit, but they are largely ignorant of the sacred lands that lie below or the offence they cause to aboriginal beliefs by doing so (not to mention the footpath erosion). Tourists have also been removing pieces of rock from the site though some are now being returned because of the bad luck they are reputed to bring.

Sense of place Many other present-day societies also feel a strong bond with their physical surroundings and value these ties for cultural, as well as economic, reasons. Agricultural communities are dependent on soil quality and have long valued the material on which their living depends, hence phrases such as 'son of the soil' or 'mother earth'. Trudgill (2001) sees the cultural links with 'soil', 'land' and 'earth' as representing yield and fertility, provision and abundance, or ownership, patriotism and nationality. The latter is exemplified in phrases such as 'on American soil' or demonstrated by the significance to the British of 'the white cliffs of Dover' or to Gibraltarians of 'the Rock'.

In the United States each state has a State Fossil (Table 4.6). Maine's State Fossil, for example, is the primitive and rare Devonian plant *Pertica quadrifania*

Table 4.6 State fossils of some states in the United States.

State	State fossil	Year adopted
Alabama	*Basilosaurus cetoides* (whale)	1984
Alaska	*Mammuthus primigenius* (woolly mammoth)	1986
Arizona	*Araucarioxylon arizonicum* (petrified wood)	1986
Arkansas	None	
California	*Smilodon fatalis* or *californicus* (sabretooth cat)	1973
Colorado	*Stegosaurus* (dinosaur)	1991
Connecticut	*Eubrontes giganteus* (dinosaur track)	1991
Delaware	*Belemnitella americana* (belemnite)	
Florida	*Eupatagus antillarum* (sea urchin relative)	1979
Georgia	Shark tooth	1976
Hawaii	None	
Idaho	*Equus simplicidens* (Hagerman horse)	1988
Illinois	*Tullimonstrum gregarium* (Tully monster)	1990
Indiana	Limestone	1971
Iowa	Crinoid	
Kansas	None	
Kentucky	Brachiopod	1986
Louisiana	Petrified palmwood	1976
Maine	*Pertica quadrifaria* (plant)	1985
Maryland	*Ecphora gardnerae* (marine snail)	1984
Massachusetts	Dinosaur tracks	1980
Michigan	Petoskey stone (fossilised coral)	1966
Minnesota	None	
Mississippi	*Zygorhiza kochii* (whale)	1981
Missouri	*Delocrinus missouriensis* (crinoid)	1989
Montana	*Maiasaura peeblesorum* (dinosaur)	1985
Nebraska	Mammoth	1967
Nevada	*Shonisaurus popularis* (dinosaur)	1977
New Hampshire	None	
New Jersey	*Hadrosaurus foulkii* (dinosaur)	1991
New Mexico	*Coelophysis* (dinosaur)	1981

which was first discovered in the state. The town of Dudley in England has had the trilobite *Calymene blumenbachii*, as part of its emblem. Norway's national rock is larvikite. Raudsep (1994) notes that Estonia is famous for its large erratics that have a special place in the hearts of Estonians, particularly in the Lahemaa National Park.

Coastal communities often feel a strong relationship with, and respect for, the sea and coastal landscapes, frequently due to their dependence on fishing, and a long history of loss of life of both fishing and lifeboat crews. Coastal topography provides navigational landmarks such as headlands and islands, and thus increases the sense of place. Some societies have complex relationships with the rivers on which they are sited, depending on them for their agricultural and domestic water use but fearing the impacts of flood events on life and property. Similarly, those living on the slopes of Mount Etna in Italy value the fertile volcanic soils but fear the often very real effects of the periodic eruptions.

Developing some ideas outlined by Gray (1997a), Adams (1998, p. 168) argues that societies may see landforms as valuable 'because they form a basic constituent of landscape, and hence share the values associated with its cultural construction, related, for example, to 'distinctiveness' or "familiarity". Landscape gives identity and a sense of place to our surroundings.

> It provides the setting for our day to day lives, gives identity to the places where we live, work, and visit, and supplies us with enjoyment and inspiration. We all have special associations with landscape because it triggers responses, memories and emotions that are very personal to us. Every landscape is important to someone, whether it be a small patch of wasteland or a great wilderness. (Swanwick and Land Use Consultants, 1999, p. 1)

4.6.4 Artistic inspiration

The physical world including its materials, landscapes and processes are an extremely important source of inspiration for artists, musicians, poets, writers and others. After seeing Fingal's Cave on the Isle of Staffa in Scotland, Turner painted it, Jules Verne used it as a location for a short story, Strinberg set a scene in *Dreamplay* on the island and Mendelssohn composed the *Hebrides Overture*. Gordon (2012) has reviewed the influence of geology on artistic creativity in Scotland.

Novelists such as Thomas Hardy founded their writing within distinctive local landscapes and were clearly inspired by them. In Hardy's case the landscape was the rolling downlands and vales of Wessex (Hampshire to Devon) in England, and in Box 4.9 he gives an evocative description of the aesthetic qualities of the landscape structure, colour, atmosphere and local distinctiveness of the 'Vale of Blackmoor'.

Box 4.9 Example of Thomas Hardy's landscape description

The village of Marlott lay amid the north-eastern undulations of the beautiful Vale of Blakemore or Blackmoor... an engirdled and secluded region, for the most part untrodden as yet by tourist or landscape-painter.... It is a vale whose acquaintance is best made by viewing it from the summits of the hills that surround it.... This fertile and sheltered tract of country, in which the fields are never brown and the springs never dry, is bounded on the south by the bold chalk ridge that embraces the prominences of Hambledon Hill, Bulbarrow, Nettlecombe-Tout, Dogbury, High Stoy, and Budd Down. The traveller from the coast, who, after plodding northwards for a score of miles over calcareous downs and corn-lands, suddenly reaches the verge of one of these escarpments, is surprised and delighted to behold, extended like a map beneath him, a country differing absolutely from that which he has passed through. Behind him the hills are open, the sun blazes down upon fields so large as to give an unenclosed character to the landscape, the lanes are white, the hedges low and plashed, the atmosphere colourless. Here, in the valley, the world seems to be constructed upon a smaller and more delicate scale; the fields are mere paddocks, so reduced that from this height their hedgerows appear a network of dark green threads over-spreading the paler green of the grass. The atmosphere beneath is languorous, and is so tinged with azure that what artists call the middle distance partakes also of that hue, while the horizon beyond is of the deepest ultramarine. Arable lands are few and limited; with but slight exceptions the prospect is a broad, rich mass of grass and trees, mantling minor hills and dales within the major. Such is the Vale of Blackmoor.

Thomas Hardy (1891) *Tess of the D'Urbervilles*

Poets have also been inspired by geology and landscapes. The United Kingdom poet laureate Carol Ann Duffy described the White Cliffs of Dover as England's '. . . glittering breastplate . . . the sea's gift to the land . . . ' (*The Guardian*, 7 November 2012). Two centuries earlier, the English poet Lord Byron, on the other hand, preferred the landscape of the Scottish Highlands:

England! Thy beauties are tame and domestic
To one who has roved o'er the mountains afar;
Oh for the crags that are wild and majestic
The steep frowning glories of the dark Loch na Garr.

Lord Byron (1807) 'Lachin Y Gair'

The Geological Society of London held a celebration of *Poetry and Geology* in October 2011 at which a large number of geological poems were read, including the following extract about the Irish karst by the poet/geomorphologist Richard Hayes Phillips:

I love the beauty of a karst
That others find so barren:
I love the pitted, rounded rocks,
Etched with rillenkarren,
The fractures widened into grikes
That open to the ocean,
The sinkholes through which I can see
The churning waves in motion.

Richard Hayes Phillips (2001) 'Where the limestone meets the sea'

Many artists have found inspiration in the physical environment or have used physical materials to create art works. Baucon (2009) has examined the link between geology and art in various fields including painting, music, literature, sculpture and photography.

Finally, fossils can also be presented as art, most notably in the remarkable and beautiful exhibition developed by Dolf Seilacher at the Geological Institute at Tubingen University in Germany, subsequently taken on tour and illustrated and described by Seilacher (2008). Most of the examples are trace fossils (burrows, traces, tracks and trails) inscribed on bedding planes and taken as casts all over the world. Subsequently they were reversed back into stiff positives with epoxy resin and suitably painted by Hans Luginsland in Tubingen.

4.6.5 Social development

Local geological activities are important in promoting community and personal development. These include local geological societies who hold meetings with speakers and run field trips to important geological localities. And voluntary environmental work sometimes brings the volunteers into contact with the geological world, for example in the construction of footpaths and steps, the repair of drystone walls, or creation of ponds and ditches. In turn this leads to greater awareness and appreciation of the character and properties of geomaterials and processes.

4.7 Knowledge services

Although this is the final value to be reviewed, it is in many ways the most important. The physical environment is a laboratory for future research, and 'it is often only field sites which provide a reliable test of many geological theories' (Bennett and Doyle, 1997, p. 161). In this section, the need to conserve geology and geomorphology for research and education purposes will be assessed by citing some examples of the contribution of past research. Damage to physical systems and sites, inevitably damages our ability to undertake research and teaching on the physical environment. Consequently 'we should maintain the means to seek knowledge in the future' (Nature Conservancy Council, 1990, p. 17).

4.7.1 Earth history

The study of the geological record has enabled geologists to reconstruct in considerable detail the history of the earth over the past 4 600 000 000. It is a record of amazing complexity and a tribute to the meticulous work of thousands of geologists over a long period of time. It has been deciphered from rock and sediment outcrops and boreholes in all countries of the world and it continues to be refined by further research. Major discoveries are still being made, particularly in the less well-studied parts of the planet, but even where intensive studies have been made, the geological record needs to be conserved for future study using new techniques and approaches and to allow findings to be checked and reinterpreted (Page, 1998). This geological rock record therefore has enormous research value. In the case of Britain, Ellis *et al.* (1996, p. 8) argued that 'Natural rock outcrops and landforms, and artificial exposures of rock created in the course of mining, quarrying and engineering works, are crucial to our understanding of Britain's Earth heritage. Future research may help to resolve current geological problems, support new theories and develop innovative techniques or ideas only if sites are available for future study'.

The fossil record contained in the rocks has demonstrated the evolution of species from the simplest unicellular organisms to the early history of humans. 'Fossils are not only records of evolution . . . they also allow us to have a look at the construction of living matter of past biospheres' (Wiedenbein, 1994, p. 118).

For example, the Rhynie Chert in Scotland contains some of the oldest known fossils of higher plants, insects, arachnids and crustaceans in such a well-preserved state that microscopic detail and cell structures can be studied. Such sites are a rare and irreplaceable part of the world's geological heritage. The history of the dinosaurs and their extinction has caught the public imagination. Research on the triggers for mass extinctions, including the idea of extraterrestrial impacts, has also been stimulated by the nature of the fossil record, which has great research and educational potential and therefore value.

Geological research has also enabled the reconstruction of the changing geography of the planet as the supercontinent of Pangaea initially fractured and then drifted apart on huge tectonic plates driven by mantle convection currents. Research has demonstrated the close link between tectonic plate margins and volcanic and earthquake activity, and study of the pattern of past natural hazards helps us to predict the location and timing of future disasters.

Other researchers have deciphered the history of the Quaternary Ice Age through studies of geology, geomorphology and biostratigraphy. The record is one of glacials and interglacials and shorter climate changes during which the ice-sheets and glaciers advanced and retreated on many occasions. Figure 3.20 shows how geological diversity has enabled reconstruction of directions of ice movement due to the distinctiveness and limited outcrop of the source rocks.

The conclusion of these ideas is that the rock and sediment record present on the planet has a heritage value in giving us a detailed knowledge of the Earth, its history and workings, and the evolution of life including the origins of our own species. Why would we not want to conserve that record for future generations and future research?

4.7.2 History of research

Ellis *et al.* (1996, p. 7) have drawn attention to the fact that many British sites have value because they 'have played a part in the development of now universally applied principles of geology'. Examples include Hutton's unconformity at Siccar Point in Scotland (Figure 9.19) and his section on Salisbury Crags in Edinburgh, Scotland where he deduced that the dolerite forming the crag must have been molten. Many of the names of periods of geological time are derived from Britain while many others serve as international reference sections. These sites have become international 'standards' that 'must be conserved so that they can continue to be used as the standard references' (Ellis *et al.*, 1996, p.7) (see also Section 6.9 on Global Stratotype Section and Points).

Of course, it is not just Britain that has such sites or where such arguments for conservation apply. For example, the history of fossil hunting is important in several parts of the world, including the dinosaur-rich areas of the western fringes of the American and Canadian Rockies (e.g. Wyoming, Montana and Alberta) (Horner and Dobb, 1997; Gross, 1998; Mayor, 2005). Thus sites where important discoveries have been made in the past are worthy of conservation.

4.7.3 Environmental monitoring

The record of sediments in lakes, bogs and ice cores also provide records of the effects of human activities on the environment through pollution, vegetation clearance, soil erosion, etc. These records are valuable not only in reconstructing the past human impacts on the environment and the history of human use of the land, but also in assessing the effects of current and potential future impacts. For example, research in the 1980s based on the analysis of diatoms, trace metals and fly-ash particles in lake sediment records provided definitive evidence for the causes of lake acidification in the United Kingdom (Battarbee *et al.*, 1985; Battarbee, 1992). This research contributed significantly to UK government decisions to introduce sulfur emission reduction policies in the late 1980s leading to the establishment of the UK Acid Waters Monitoring Network in 1988.

The climate history of the Earth of the past few million years gives us many clues as to the causes of climate change, the role of humans in altering climate and how physical systems will respond to future changes. The evidence for climate change comes from a very wide variety of sources, but they include the record preserved in peat bogs, cave sediments and ice-sheets. Records like these are important in reconstructing recent climate changes and correlating these with other terrestrial and marine locations. Study of such sites can also provide empirical data on which to develop and test global models of climate change and may help to predict future environmental changes (Gray, Gordon and Brown, 2013).

4.7.4 Geoforensics

An interesting application of geodiversity is geoforensics. Ideally, this involves linking suspects to crime scenes through the distinctiveness of soil, sediment and even topography. An overall review of the field is supplied by Pye and Croft (2004) and Ruffell and McKinley (2008). The most common application involves comparing soil or sediment on items such as a suspect's clothing, footwear, implement or car tyres to that of a crime scene (or an alibi location) that then allows the suspect to be either eliminated or linked to the locality. It generally involves searching for exotic minerals, mineral assemblages, distinctive surface textures on mineral grains or other indicators in the soil or sediment samples and linking these with the distinctiveness of the soil mineralogy at the crime scene (Donnelly, 2002), that is, the techniques rely on the geodiversity of soil or sediment. However, in practice, diagnostic evidence only arises in very specific situations and the technique is usually applied as an exclusionary one—to exclude rather than confirm hypotheses—and is usually combined with other independent lines of evidence.

Although the principles have been understood for over a century and included investigations by fictional detectives such as Sherlock Holmes, the first attempt to summarise principles and techniques and bring together several case studies was by Murray and Tedrow (1975, 1992). More recently, Morgan and Bull (2007) and Ruffell and McKinley (2008) have reviewed the philosophy, nature and practice of forensic sediment analysis that now includes sophisticated electron microscopy and biological techniques including DNA and pollen analysis (see Morgan and Bull, 2007, Table 1). Of fundamental importance is the taking of representative samples and interpretation must be carried out with extreme caution given the real possibility of false-positive or false-negative results. Pirrie *et al.* (2009) and Pirrie and Rollinson (2011) describe the application of automated mineral analysis to geoforensic work.

Bull and Morgan (2006) used the results of 738 soil and sediment samples analysed during or following the investigation of 20 forensic cases in England to classify quartz grain surface textures and develop this into an exclusionary mechanism. The importance of this method is that quartz is easily the most common mineral on the Earth, is particularly resistant to weathering and that grains often retain surface textures that reflect their geological history. Hence it is the geodiversity of quartz grain surface textures that gives them their geoforensic value.

Experimental work has also been carried out. For example, Morgan *et al.* (2008) report that during a terrorism trial, the prosecution used quartz grain surface texture analysis of various soils, including some from a vehicle that had been completely destroyed by fire, as part of their evidence. The fire had destroyed all biological evidence (pollen, fibres, DNA) and the defence argued that the fire would have fractured and altered the surface characteristics of the quartz grains so that this geological evidence could be disregarded. Consequently experiments were carried out that concluded that the maximum temperature generated by a fire under natural conditions (810 °C) was insufficient to affect quartz grain surface textures, and thus the use of this forensic evidence recovered from fired cars was confirmed.

4.7.5 Education and employment

The geological record has huge research value, but it also has a role in education and training. Students and teachers need sites and areas that they can use to demonstrate geological principles and processes in the field. Trained geologists, geomorphologists and pedologists are needed to locate and utilise mineral resources, predict natural hazards and ensure the sustainable use of land. Rock exposures, fossil sites, landforms, soil sections and active processes play a valuable role in the education of children, the training of the next generation of geologists and for amateurs with an interest in their environment and the geological history of the planet. We should not lightly allow them to be destroyed. Geology is also an important source of employment, for example in geoparks (see Chapter 8).

4.8 Geodiversity and the 'ecosystem services' approach

In the introduction to this chapter the issue of 'ecosystem services' was introduced. This chapter has outlined the many goods and services related to geodiversity and in this section the aim is to bring these together and present them in an 'ecosystem services' context. The problem is what to call these 'services' as we are faced with difficult issues of terminology and principle.

- First is the fact that many authors equate ecosystems with nature. For example, Daily (1997a), in the seminal work on this topic referred to 'nature's services', yet in her descriptions, she only included those services associated with the biological elements of nature. Similarly, the UKNEA (2011) is subtitled 'understanding nature's value to society' yet focuses on biological nature.
- Second, the Convention on Biodiversity defines an ecosystem as 'a dynamic complex of plant, animal and micro-organism communities and their non-living environment *interacting as a functional unit*' (my emphasis; see http://www.un.org/en/events/biodiversityday/convention.shtml). By definition, therefore, the ecosystem approach includes abiotic elements but only where they are interacting with biotic elements. In this chapter we have discovered that there are many values of geodiversity that do not involve biological systems. Hence, the term 'ecosystem services' is, in strict terms, inappropriate to apply to all the values that nature provides. For this reason I used the term 'geosystem services' in a previous paper (Gray, 2011), to refer to the services associated with geodiversity. Others have used the term 'abiotic ecosystem services' (e.g. Gordon *et al.*, 2012).
- Third, and related to this, is the fact that the Millennium Ecosystem Assessment (2005) excluded non-renewable resources from the assessment. This view was supported by Brown *et al.* (2011) in their introduction to the UKNEA which 'focuses on "ecosystem services" that are derived from ecosystem processes including biotic interactions; as such, it does not provide an assessment of "environmental services" that may be purely abiotic in origin such as minerals extracted from the ecosystem' (Brown *et al.*, 2011, para. 1.3.1). Yet, it is vital that non-renewable resources are included in any assessments of the value that society gains from nature. This is because non-renewable resources are natural goods of fundamental value to society. Indeed, as this chapter has shown, modern societies could not function without them. Furthermore, because they are non-renewable it is even more important that they are conserved and managed sustainably for future generations.
- Fourth, despite the earlier statement by Brown *et al.* (2011), an analysis of the 27 chapters of the UKNEA by Gray *et al.* (2013) has revealed that many of the individual chapter authors have included descriptions of abiotic services that do not involve, or only marginally involve, biological elements. Thus, the current position of geodiversity within the ecosystem approach is confused and inconsistent.
- Fifth, if we could start again, the terms 'natural services', 'environmental services' or even 'Earth system services' would be preferable to the term 'ecosystem services' because they would lead to a more holistic and integrated assessment of the value of all of nature to society. By their limited acknowledgement of the role of abiotic nature, current ecosystem assessments grossly underestimate the total value of nature to society. This cannot be in the best interests of nature conservation. However, the fact is that the term 'ecosystem

services' is now deeply entrenched in both policy and practice throughout the world and will be difficult, if not impossible, to replace.

The choice is therefore to ignore the formal definition of an 'ecosystem' and promote the crucial role of geodiversity in contributing to 'ecosystem services' in the informal sense that includes all of nature including its non-renewable resources. If this is unacceptable to the scientific community, then I propose that we adopt the term 'geosystem services' as the abiotic equivalent of 'ecosystem services' but with the obvious overlaps between them. For example, some functions normally referred to as 'ecosystem services' (Daily, 1997b) are actually shared between biotic and abiotic systems. One example, is 'mitigation of floods and droughts' which is partly accomplished by ecosystems but also by natural physical materials (e.g. infiltration, storage in groundwater aquifers) or structures (e.g. levees, beach barriers). Similarly 'generation and renewal of soil and soil fertility' must be partly due to geosystem processes (e.g. weathering and soil development) and partly due to ecosystem processes (e.g. nitrogen fixing). 'Providing of aesthetic beauty and intellectual stimulation that lift the human spirit' (Daily, 1997b, p. 4) is also a function shared by both ecosystems (e.g. forests, wildlife, arctic and alpine meadows) and geosystems (e.g. mountains, rivers, coasts). It follows from this that sustainable management of natural resources must involve both biotic and abiotic systems. It is interesting to note that the Welsh Assembly has recently consulted on *Sustaining a living Wales: a Green Paper on a new approach to natural resource management in Wales* (Welsh Government, 2012) that adopts an ecosystem approach that includes geodiversity as an essential component. Figure 4.25 is an attempt to summarise the full range of values of geodiversity discussed in this chapter, amounting to over 25 distinct values and is based on a layout developed by de Groot (1992), English Nature (2002) and Webber *et al.* (2006). Figure 4.26 presents Gray, Gordon and Brown's (2013) four-dimensional landscape illustration of some of the goods and services associated with geodiversity. It should be clear from these two diagrams that there are many abiotic values of nature that are not represented in the ecosystem services approach as currently promoted. This is particularly true of most of the provisioning and knowledge services, but there are also many examples from regulating, supporting and cultural categories.

Economists have attempted to put a financial value on all environmental assets (see, e.g. Pearce and Turner, 1990; Foster, 1997; Balmford *et al.*, 2002), including biodiversity and ecosystem services. Values of geological provisioning services (goods) are well known and the value of some fossils and minerals has been referred to earlier. Only a few research examples of economic valuation of other geosystem services have been carried out, putting economic values on geodiversity or geosystem services. For example, Hanley, Mourato and Wright (2001) calculated a consumer surplus (the difference between what consumers would be willing to pay and what they actually do pay) per rock climbing trip of £31.15. More significant is the study by Webber *et al.* (2006).

For geodiversity-based recreation and tourism, Webber *at al.* (2006) carried out a questionnaire survey presenting hypothetical management options for the future management of the Wren's Nest National Nature Reserve at Dudley in the English Midlands. This is a site with good exposures of fossiliferous Silurian limestones and shales in disused quarries and underground caverns. Their results showed that access to the reserve for recreation without any educational material was valued at an average of £7.83 per household per year, but when geological explanation was added, rose to £21.26. Access to the Seven Sisters Cavern rose from £12.22 to £13.95 per household per year if geological educational materials were included. The opportunity to collect fossils from spoil heaps was valued at

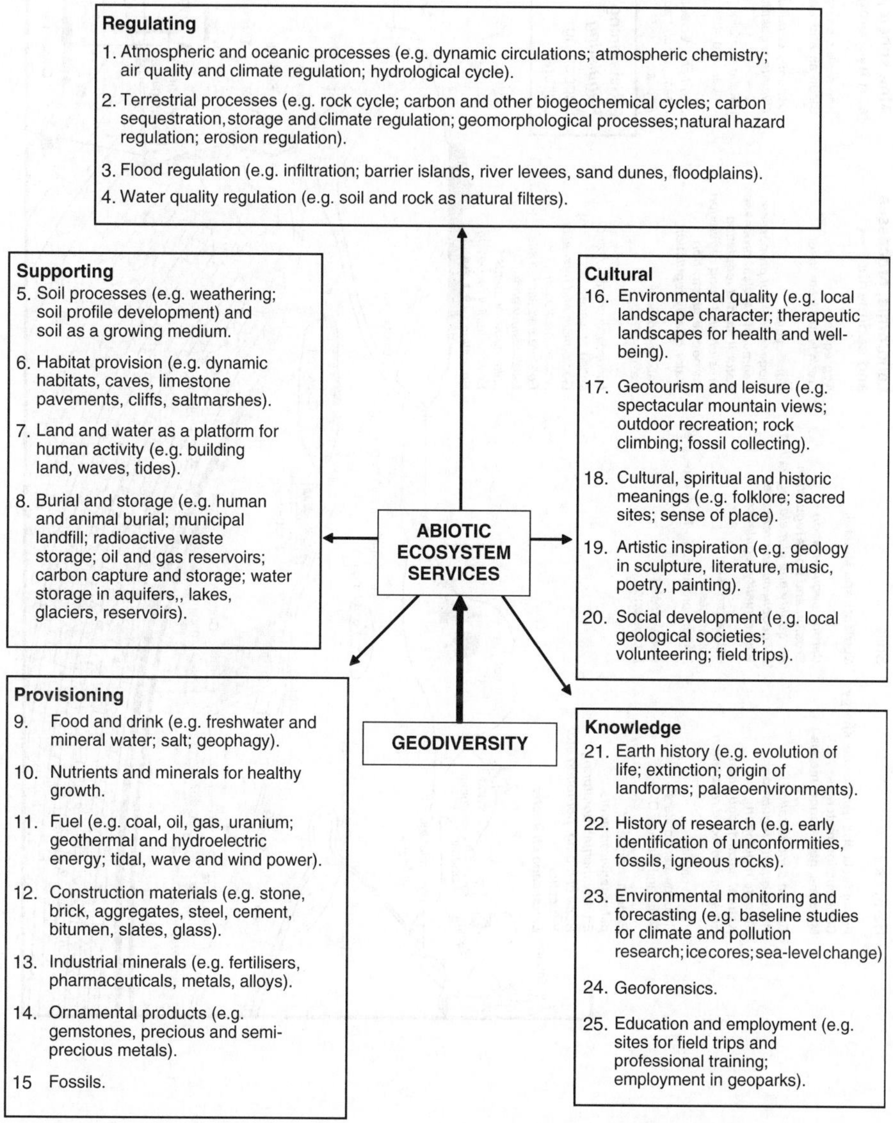

Regulating
1. Atmospheric and oceanic processes (e.g. dynamic circulations; atmospheric chemistry; air quality and climate regulation; hydrological cycle).
2. Terrestrial processes (e.g. rock cycle; carbon and other biogeochemical cycles; carbon sequestration, storage and climate regulation; geomorphological processes; natural hazard regulation; erosion regulation).
3. Flood regulation (e.g. infiltration; barrier islands, river levees, sand dunes, floodplains).
4. Water quality regulation (e.g. soil and rock as natural filters).
Supporting
5. Soil processes (e.g. weathering; soil profile development) and soil as a growing medium.
6. Habitat provision (e.g. dynamic habitats, caves, limestone pavements, cliffs, saltmarshes).
7. Land and water as a platform for human activity (e.g. building land, waves, tides).
8. Burial and storage (e.g. human and animal burial; municipal landfill; radioactive waste storage; oil and gas reservoirs; carbon capture and storage; water storage in aquifers,, lakes, glaciers, reservoirs).
Cultural
16. Environmental quality (e.g. local landscape character; therapeutic landscapes for health and well-being).
17. Geotourism and leisure (e.g. spectacular mountain views; outdoor recreation; rock climbing; fossil collecting).
18. Cultural, spiritual and historic meanings (e.g. folklore; sacred sites; sense of place).
19. Artistic inspiration (e.g. geology in sculpture, literature, music, poetry, painting).
20. Social development (e.g. local geological societies; volunteering; field trips).
ABIOTIC ECOSYSTEM SERVICES
GEODIVERSITY
Provisioning
9. Food and drink (e.g. freshwater and mineral water; salt; geophagy).
10. Nutrients and minerals for healthy growth.
11. Fuel (e.g. coal, oil, gas, uranium; geothermal and hydroelectric energy; tidal, wave and wind power).
12. Construction materials (e.g. stone, brick, aggregates, steel, cement, bitumen, slates, glass).
13. Industrial minerals (e.g. fertilisers, pharmaceuticals, metals, alloys).
14. Ornamental products (e.g. gemstones, precious and semi-precious metals).
15 Fossils.
Knowledge
21. Earth history (e.g. evolution of life; extinction; origin of landforms; palaeoenvironments).
22. History of research (e.g. early identification of unconformities, fossils, igneous rocks).
23. Environmental monitoring and forecasting (e.g. baseline studies for climate and pollution research; ice cores; sea-level change)
24. Geoforensics.
25. Education and employment (e.g. sites for field trips and professional training; employment in geoparks).

Figure 4.25 The role of geodiversity in generating goods and services.

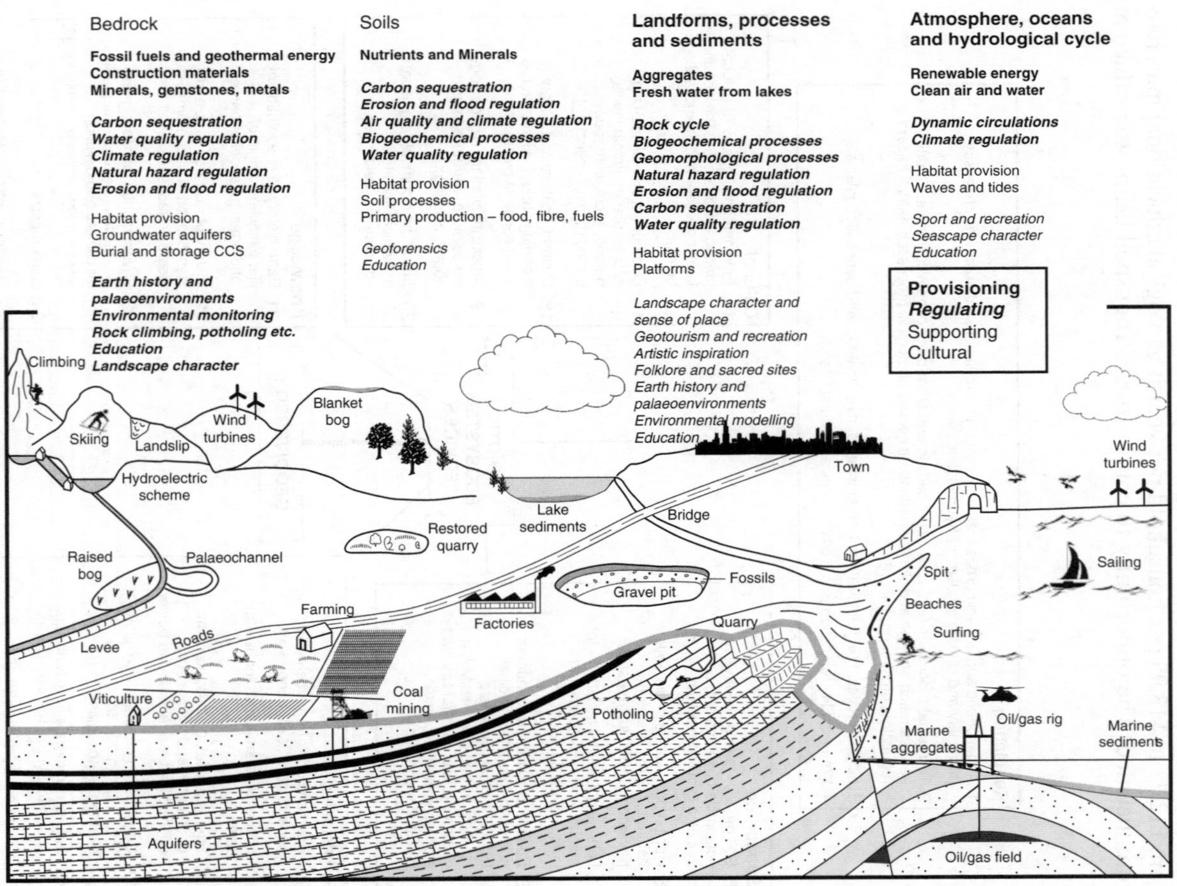

Figure 4.26 Cartoon illustrating some of the goods and services associated with geodiversity (after Gray *et al.*, 2013).

£5.18 per household per year on average. Webber *et al.* (2006) concluded that the provision of geological interpretation and opportunities for fossil collecting clearly add value to people's enjoyment of the site.

A similar study was carried out at the Dorset and East Devon Coast, a World Heritage Site (see Section 7.7.4). On average, people were prepared to pay £23.69 per household per year for access to the coast with some explanation of the geology, but this rose to £62.35 for access and extensive interpretive material. Maintaining the opportunity to collect fossils (via the local code of conduct) was valued at £57.73 per household on average. Webber *et al.* (2006, p. 17) concluded that 'people gain value from recreation in areas of high geological importance. Importantly, people appear to gain greater benefits if the geological features are explained. Also, it is demonstrated that people highly value opportunities to collect fossils.' Webber *et al.* (2006) also carried out an economic multiplier study to estimate the benefits of geodiversity to the economy of the Isle of Wight off the south coast of England and famous for its Mesozoic geology. The conclusion was that geotourism contributes £137 million per year to the economy of the island and supports between 4038 and 5491 full-time equivalent local jobs.

4.9 Conclusions

The combined values of geodiversity are considerable and together provide society with an enormous range of goods and services. They are absolutely essential to sustaining modern societies in both physical and other ways, yet the current 'ecosystem services' approach under-recognises and undervalues the role of geodiversity in contributing to society's well-being. A more holistic and integrated approach is essential if we are to manage our natural systems sustainably in the long-term interests of the planet and its peoples. Yet, the ecosystem approach, as currently promoted, accepts only a limited role for abiotic nature and specifically excludes non-renewable resources. This cannot be in the best interests of nature conservation or sustainable management of the planet, since it results in the value of nature to society being grossly undervalued.

5 Threats to Geodiversity

Eroding cliffs with their extensive exposures of rock sections, the source of sand and shingle beaches elsewhere, are rendered invisible and geomorphologically impotent behind concrete.

Sir David Attenborough (in Nature Conservancy Council, 1990)

There is perhaps a general tendency to think of the biological world as fragile and vulnerable and therefore in need of conservation, whereas the abiotic world of mountains and rock is seen as stable, static and much too prolific ever to be endangered. We use terms such as 'set in stone' or 'written in stone' to express solidity and permanence. This is a gross oversimplification, and many threats to the geodiversity of the planet or local areas are comparable to those facing biodiversity. Furthermore, geoconservation is not just about protecting the static elements of the landscape. It is also about allowing dynamic processes to continue operating within the historical range of natural rates. It should also be noted that disturbances to geological, geomorphological and soil processes can be produced in ways that are not always local or obvious. For example, individually or cumulatively, vegetation clearance, agriculture, water diversion, forestry and urbanisation can all have profound impacts on fluvial landforms, sediments and processes by changing runoff rates and magnitudes, sediment loads, and so on. There are many significant, real and potentially damaging activities that ought to be better understood if we are to properly conserve and manage geodiversity.

5.1 The Nature of the threats

The number of these threats is great (Glasser, 2001; Gordon and MacFadyen, 2001), though only a few are likely to apply in most locations. The types of artificial activities that will degrade geodiversity depend on the types of values concerned (see Chapter 4). Furthermore the impact of an operation will vary depending on the sensitivity, stability or robustness of the site in question

Geodiversity: Valuing and Conserving Abiotic Nature, Second Edition. Murray Gray.

(Brunsden and Thornes, 1979; Schumm, 1979). An operation that would have a devastating effect in one area may be more acceptable in another, more robust, location. This is because some systems are capable of repairing themselves in a relatively short time due to the continued operation of natural processes (e.g. reformation of ripples destroyed by human footsteps), whereas other changes are irreversible because the processes no longer operate or the change to the landscape is fundamental (e.g. removal of an esker by quarrying or loss of soil cover). This concept of landscape sensitivity is a fundamental one in understanding the threats to geodiversity (Werritty and Brazier, 1994; Gordon *et al.*, 2001; Haynes *et al.*, 1998; Werritty and Leys, 2001; Downs and Gregory, 2008), and is considered further in the last section of this chapter.

In general terms, threats to geodiversity are the result of development pressures and land-use change (Gordon and MacFadyen, 2001), but others may result from natural processes (e.g. coastal or river-bank erosion) or from human-induced change (e.g. climate change and sea-level rise), though it is often difficult to separate these effects (Harrison and Kirkpatrick, 2001). The impacts of development can be summarised as:

- complete loss of an element of geodiversity;
- partial loss or physical damage;
- fragmentation of interest;
- loss of visibility or intervisibility;
- loss of access;
- interruption of natural processes and off-site impacts;
- pollution;
- visual impact.

Some of these impacts affect specific sites of geoconservation value while others impact widely across large land areas, but all may lead to loss of or damage to elements of geodiversity. Even protected sites can be subject to damage (Figure 5.1).

The UK's *Wildlife and Countryside Act* (1981) allows for the specification of a standard list of 'Potentially Damaging Operations' (PDOs). Although some are exclusive to biodiversity interests, many can and do affect geodiversity. Table 5.1 is a more explicitly earth science related list of threats (modified after Gordon and MacFadyen, 2001) and will be expanded upon in this chapter. Major reviews of the human impact on the environment have been undertaken by several authors, and readers are referred to Turner *et al.* (1990) and Goudie (2013) for fuller information.

5.2 Mineral extraction

'Mining can be a nasty business; it can be one of the most environmentally damaging activities undertaken by humans. ... Today, more land is devastated due to the direct effects of mining activities than by any other human activity' (McKinney and Schoch, 1998, p. 266). This may not be entirely accurate since

Figure 5.1 Graffiti on a protected site near Krakow, Poland.

Table 5.1 Threats to geoheritage (modified after Gordon and MacFadyen, 2001).

Threat	Examples of on-site impacts	Examples of off-site impacts
Mineral extraction (includes pits, quarries, dunes and beaches)	Destruction of landforms and sediment records Destruction of soils, soil structure, soil biota May have positive benefits in creating new sections	Contamination of watercourses Changes in sediment supply to active systems Extraction from rivers and beaches, leading to erosion and scour
Landfill and quarry restoration	Loss of exposures Loss of natural landform and soil disturbance Detrimental effects of leachate and landfill gases Habitat creation	Contamination of surface watercourses Contamination of groundwater
Land development and urban expansion	Large-scale damage and disruption of landforms and soils Changes to drainage systems Creation of slope instability	Changes to processes downstream Contamination of watercourses
Coast erosion and protection	Loss of coastal exposures Loss of active and relict landforms Disruption of natural processes	Changes to sediment supply and processes downdrift

(*continued overleaf*)

Table 5.1 (*continued*)

Threat	Examples of on-site impacts	Examples of off-site impacts
River management, hydrology and engineering	Loss of exposures Loss of active and relict landforms Disruption of natural processes	Changes to sediment movement and processes downstream Change in process regime; drying of wetlands
Forestry, vegetation growth and removal	Loss of landform and outcrop visibility Physical damage to small-scale landforms Stabilisation of dynamic landforms Soil erosion Changes to soil chemistry and soil water regime	Increase in sediment yield and runoff during planting and deforestation Changes to groundwater and surface water chemistry
Agriculture	Damage or loss of small-scale landforms through ploughing, ground levelling and drainage Soil compaction, loss of organic matter and soil biota Changes to soil chemistry from fertilisers Effects of pesticides on soil biota Soil erosion	Changes in runoff arising from drainage Episodic soil erosion by wind and water Pollution of surface and groundwater from excess agrochemical application
Other land management changes	Loss or degradation of exposures and landforms Loss or contamination of soils Changes to soil-water regime	Changes in runoff and sediment supply
Recreational/tourism pressures	Physical damage to small-scale landforms and soils Localised soil erosion Damage to cave systems	
Removal of geological specimens	Loss of fossil record Loss of mineral specimens	
Climate and sea-level changes	Changes in active process systems Coastal erosion and inundation	Changes in flood frequency Changes in rates of geomorphological processes
Fire	Loss of organic soils Loss of vegetation leading to soil erosion	
Military activity	Loss and damage to soils and small-scale landforms by vehicles Production of craters by bombing	
Lack of information/education	Loss or damage to active processes or static features through ignorance of values	

estimates are that this disturbance amounts to only about 1% of the Earth's land surface, though this is likely to be an underestimate given off-site impacts like acid drainage and the fact that large areas have been disturbed by past mining activities. However, other impacts such as agriculture, deforestation and development may be much more spatially significant.

Extraction of minerals from quarries and other open excavations, including building stone, rock aggregate, metallic ores, sand and gravel, beach shingle, peat and soil is clearly necessary to modern society but always results in loss of some geodiversity since the land surface is disturbed and sediments or rocks below are removed. This may not be significant where the resource being quarried is large. In fact mining has often been invaluable in creating new rock and sediment exposures. But it will become problematic where rare soils, important landforms, limited rock outcrops or important fossil-bearing strata are removed during mineral extraction or where whole landscapes are devastated by strip mining and not restored. According to Vasiljevic *et al.* (2010), the loess site at Cerveny Kopec near Brno in the Czech Republic, one of the key sites for investigating and correlating climatic variations in the Middle and Late Pleistocene (Kukla, 1975, 1977), has been progressively destroyed to provide construction materials.

Underground mining has a less serious impact on the surface, but may still involve subsidence of several metres, for example as a result of coal mining (English coalfields), iron mining (north-east France), brine pumping (e.g. Lüneburg, northern Germany), water abstraction (Mexico City) or oil extraction (Wilmington, California, United States). It is also more expensive and not feasible for many materials such as aggregate. Visual intrusion of the open-cast quarries is therefore a major issue. Jones and Hollier (1997) point out that the extent of visual intrusion from mining depends to a large degree on the local topography but some, like the Bingham Canyon copper mine in Utah in the United States, are so large (4 km wide and 1000 m deep, plus the surrounding spoil heaps) that (allegedly) they can be seen from outer space.

The impacts on the landscape are greatest where area strip mining is involved (US Department of the Interior, 1971). This method is commonly used where there are gently dipping beds within a few tens of metres of the surface, e.g. Carboniferous coals, Jurassic ironstones or sedimentary kaolins. Examples of the latter occur in Georgia and South Carolina in the United States, while Ohio, Pennsylvania, West Virginia, Kentucky, Illinois, Wyoming and Missouri are States with significant coal strip mines. The method is to work a continuous swathe across an area, extracting the mineral and backfilling with the stripped overburden and other wastes. Two other important examples from North America and one from the Czech Republic illustrate the impacts of surface mining (Box 5.1, Box 5.2 and Box 5.3).

Box 5.1 The Athabasca Tar Sands, Canada

The Athabasca tar sands occur in northern Alberta, Canada, around the town of Fort McMurray. Smaller deposits to the south and west are called the Cold Lake and Peace River Tar Sands, respectively, the total economically recoverable resource being estimated at 175 billion barrels of heavy oil. The tar sands are bitumen soaked sands that are 200–300 million years old. According to Nikiforuk (2008), about 20% of the tar sands are shallow enough to be mined by huge (four-storey high, 400-tonne, 4 m diameter tyres) Caterpillar trucks and electric shovels. Nikiforuk describes the impacts as follows: 'To coax just one barrel of bitumen from the Athabasca sand pudding, companies must now mow down hundreds of trees, roll up acres of soil, drain wetlands, dig up four tons of earth to secure two tons of bituminous sand, and then give those two tons a hot wash ... Since 1997, one major mining company has moved enough earth to build seven Panama canals'. In fact, about 5000 tonnes of overburden and sand are excavated every minute (Gillespie, 2008), with the result that large areas of badly scarred landscape are spreading across northern Alberta. But most tar sands are too deep for surface mining and must be steamed or melted out of the ground. This is less disruptive to the landscape but still involves the construction of surface infrastructure including extensive production plants and other buildings, chimneys, pipes, pumps and road systems (Gillespie, 2008).

But there are other impacts, particularly from the more than a dozen toxic ponds of waste water covering over 120 km^2 on both sides of the Athabasca River. Separating the bitumen from the sand requires huge amounts of water (one barrel of bitumen requires about three barrels of water) that is then flushed into these settling ponds. The water contains pollutants including bitumen, phenols, polycyclic aromatic hydrocarbons, cyanide and naphthenic acids, all highly toxic to the environment. Given that production is planned to grow from 1.3 million barrels/day in 2008 to 5 million barrels/day by 2030, the environmental impact is huge and increasing.

Box 5.2 Mountain Top Removal mining, United States

Mountain Top Removal (MTR) mining is practiced in the coal-bearing hills of the Appalachians from Ohio to Virginia, but is particularly prevalent in West Virginia and Eastern Kentucky. It involves removal by blasting of up to 300 m of overburden on mountain tops and ridges in order to access the coal seams running through these mountain tops and ridges. Previously these would have been exploited by mine shafts, but mountain top removal allows cheaper, opencast mining of the coal. There is therefore a significant topographic alteration to the landscape even though some of the overbuden may be replaced on the tops after the coal is extracted. In many cases, however, some of the overburden is deposited in the adjacent valleys or hollows, sometimes impeding headwater streams and thus increasing the topographic and environmental impacts. Over 5000 km^2 had been mined in this way in the Appalachians by 2010, an area greater than Delaware and involving the removal of some 500 mountain tops (*The Guardian*, 5 August 2009). Since coal often occurs in multiple stratified seams, the process of removing overburden may be repeated over 10 times thus increasing the impacts (Figure 5.2). In a debate at the University of Charleston in January 2010, Robert F. Kennedy Jr called this process 'the worst environmental crime that ever happened in our history' (*The Guardian*, 23 January 2010).

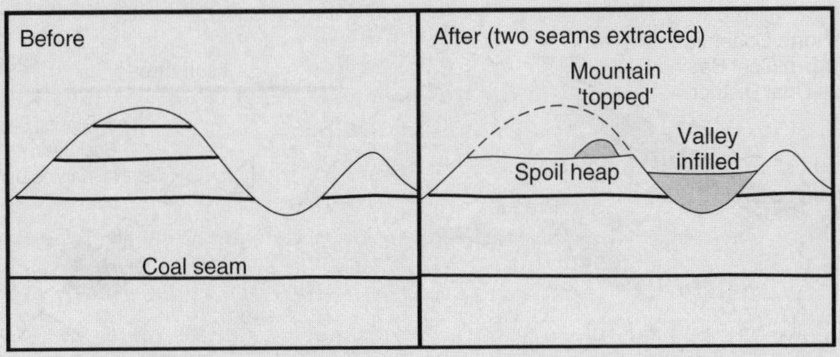

Figure 5.2 Schematic diagram of mountain top removal.

Box 5.3 Impact of Brown Coal Mining in the Czech Republic

Domas (1994) describes the impact of brown coal extraction in the Czech Republic in the 1960s and 1970s. According to Domas (p. 93) 'The government decided to obtain energy from brown coal, regardless of the impact'. Open cast mines were used so that the impact came from both the pit and the spoil. Pits were 1–5 km long, 1–3 km wide and up to 150 m deep. 'Each coal quarry was required to deposit the overburden on its outer reaches at the beginning of mining operations, in the process burying many square kilometres of original landscape ... the new profiles which result dominate the landscape'.

In the North Bohemian Brown Coal Basin, some 260 km^2 of land surface has been influenced by opencast mining, 80 km^2 of which are spoil dumps (see Figure 5.3). More than 70 villages and settlements were destroyed by the advancing quarries. The mining policy of the government at the time was to exploit this resource to exhaustion of the basin's reserves, but a subsequent change of policy imposed obligatory spatial limits on the operating mines, which must not be overstepped by either quarrying or spoil deposition.

In the Upper Silesian Black Coal Basin mining is mainly by underground workings, sometimes on up to 10 floors. This has caused significant land subsidence of up to 40 m in places. The original glacial and periglacial geomorphology of the landscape has been totally changed, with large areas now being flooded. In addition to the constant need to transfer and uplift roads and railways, more than 2500 houses have been destroyed. Settling pits infilled by waste from coal preparation plants contain several metres of thick black mud with admixtures of organic and inorganic pollutants. With no lining systems, polluted water migrates from the base of these ponds into surface and groundwaters. As Domas (1994, p. 97) states 'Elimination of the mining impact, so devastatingly and carelessly caused in the past, will certainly not be easy'.

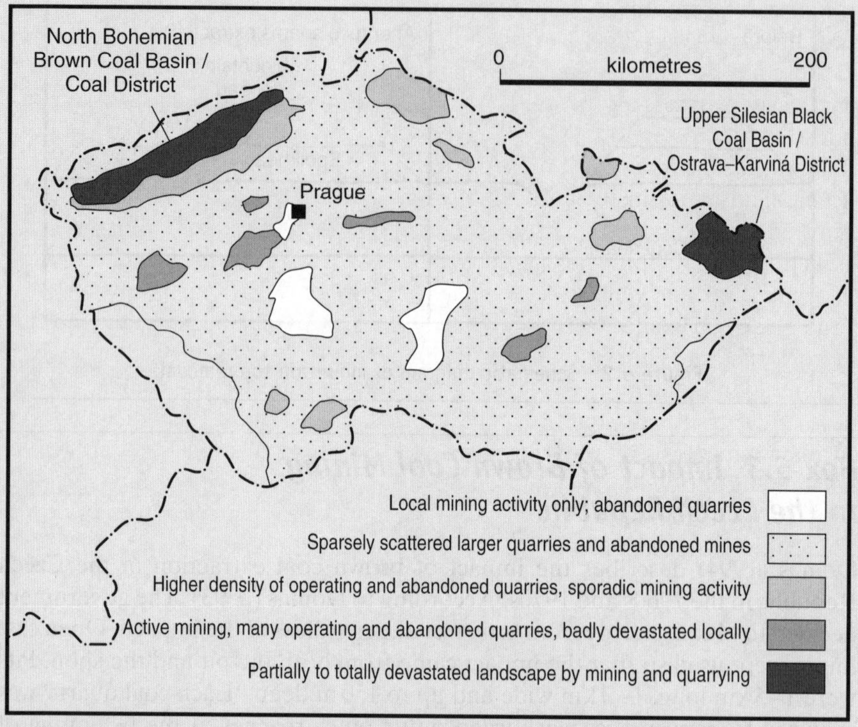

Figure 5.3 Impact of opencast coal mining in the Czech Republic (after Domas, J. (1994) Damage caused by opencast and underground coal mining. In O'Halloran, D., *et al.* (eds) *Geological and Landscape Conservation*: Proceedings of the Malvern International Conference 1993. Geological Society, London, 93–97, by permission of the Geological Society of London).

Sand and gravel extraction provides a more local threat to valued landscapes. Stürm (1994) noted that the impressive moraine landscape around Zurich in Switzerland was under pressure because of the demand for sand and gravel to satisfy the building boom in the area. In Cheshire, England, English Nature (1998) saw an important threat from 'the extensive winning of construction sand from the Delamere Forest so significantly modifying the undulating landscape' which is glaciofluvial in origin.

The excavation of marine, beach and river aggregate needs careful assessment in order to avoid damage to functioning geomorphological systems. Masalu (2002) describes illegal sand mining along the beaches and rivers of the Tanzania coast. This is causing local coastal erosion, threats to coastal properties and instability of bridges due to alteration of streambed morphology. Offshore dredging for sand and gravel alters the shape of the seabed and hence may impact on current patterns, possibly aggravating coastal erosion. At a smaller scale, problems have arisen at Crackington Haven in Cornwall, England where the beach comprises dark grey rounded shale pebbles with attractive quartz veins. These have become the target of collectors as a result of television gardening programmes promoting the use of rounded cobbles and pebbles in gardens. According to Anon (2000, p. 6) 'pebbles are being removed by the bucket load' and there have even been incidents reported of car and trailer loads being taken from the site which is a protected area. The threat is not just the loss of the beach but also to the protection it gives to the cliffs and local settlement.

Thorvardardottir and Thoroddsson (1994, p. 228) described the threat to volcanic landscapes in Iceland. 'If an area is not protected ... Landowners are permitted to mine gravel, rock, scoria, or pumice on their estate. This has led to the production of many mining pits, especially where lava and pumice are found, as these have thin or no vegetation cover, are lightweight, porous, without frost activity and give good insulation'. They are therefore used for road, footpath and driveway construction. Thorvardardottir and Thoroddsson (1994, p. 228) argued that the many abandoned mining pits in Iceland 'look like battlefields in the landscape'. They believe that many interesting volcanic formations have been lost, although nowadays mining is planned with more care, to avoid craters in particular. In New Zealand, Buckeridge (1994) noted that all 48 volcanic cones around Auckland have suffered some damage, and over half have been quarried away or covered over. Similarly, Rosengren (1994) expressed concern about the impact of quarrying on Australia's important Late-Cenozoic volcanic province. Although the quarries often reveal important internal structures, contacts and stratigraphies,

> scoria, and tuff are non-renewable resources and the configuration of the eruption points cannot regenerate or recover until there are more eruptions. Quarry operators may not recognise the significance of uncovered material or structures and may unknowingly destroy specimens or exposures, fill in craters or bury material with over-burden or stockpile. Outcrops or topography significant in displaying volcanic history may be modified, buried or removed ... quarrying produces synthetic landforms. Holes and over-burden mounds alter the form and slope angles of cones and mounds and may confuse future interpretation of original eruption topography and products. (Rosengren, 1994, p. 109)

This extract has been quoted at length because it neatly describes a series of threats to an important geological and geomorphological heritage, from a quarrying operation, and thus implicitly makes the case for control that takes account of the heritage interest.

The extraction of peat for commercial or domestic use as a fuel or garden compost can result in loss of the palaeoecological record of vegetational and faunal change, the archaeological record of human use of the wetland, as well as disrupting wetland landscape, drainage and wildlife (see Box 5.4). Peatlands once covered 8% of western Europe and were a dominant landscape feature. However, they are 'extremely delicate and prone to damage ... [and] their utilization has a considerable economic importance. As a consequence, peat cutting, drainage, afforestation and agricultural development have destroyed most of the peatlands of northwestern Europe' (Daly, 1994, p. 17). The Netherlands have lost almost 100% of their peatlands, Britain has lost 98% of its raised bogs and 90% of blanket bogs, and in Ireland the corresponding figures are 94% and 86%, respectively (Daly, 1994). In February 2002, the UK government paid the company Scotts £17.3 million to buy out its extraction rights at three of England's most important remaining peat bogs (*The Guardian*, 28 February, 2002).

Box 5.4 Peat extraction in Ireland

About 16% of the land surface of Ireland is covered in peat over 1 m thick. This occurs either as blanket bog 1–6 m deep along the western coastal strip and mountain areas throughout Ireland where there are high rainfall amounts and relatively impermeable rock types, or as raised bog up to 15 m thick over the low-lying, poorly drained midland counties. After the Second World War the Irish government established a large-scale peat excavation project both for domestic use and for electricity generation. A state company (Bord Na Mona, BNM) was established to extract the peat from 90 000 ha of bog over the 80 years 1950–2030. Average production is about 3–6 million tonnes/year. However, in recent decades there has been concern from environmentalists about the impact of this extraction, particularly on the raised bogs, 95% of which have been lost, fragmented or affected by human activity. Thus there is currently a dilemma about the exploitation of Irish peat bogs. On the one hand the material is valuable as a fuel and horticultural material as well as providing local employment. On the other hand the bogs are part of the Irish landscape and heritage, attracting leisure and tourist activities (Kelk, 1992).

Similarly to peat extraction, soil excavation can also lead to landscape damage. An example in Slovenia is described by Cernatic-Gregoric and Zega (2010) who document the damage done to karst dolines. Tractor roads are built down the doline sides to get to the soil in the doline floors and the soil is then excavated 'literally emptying the dolines' and making it hard for vegetation to grow on the bare slopes. The soil is used to create new vineyards or as landscaping material but some is sold into Italy.

Silvestru (1994, p. 224) described the effects of limestone and marble quarries on the karst landscapes of Romania, 'the damage does not consist of mere subaerial artificial cuts. ... Sometimes ... underground drainage is diverted or disrupted, which always results in important regional changes and has long-term consequences. Even more destructive, are the auxiliary works like roads, buildings and everyday activities ... '. Silvestru also describes the impact of bauxite mining that takes place in closed hollows up to 60 m deep. 'A closed hollow of this type, takes a very long time to recover, and represents a strikingly unnatural feature in the heart of karstic landforms. In addition, the waste dump left on the slopes is either terra rosa or reddish sediments associated with bauxite, both practically sterile for vegetation. Therefore the recovery time is even longer.' He describes large areas of the northern Apuseni Mountains of north-west Romania as being 'a land of desolation' (Silvestru, 1994, p. 224).

In some developing countries, there may be little by way of restoration and landscapes can become devastated by quarrying operations (Hilson, 2002) sometimes carried out illegally. There are several examples of illegal diamond quarries in Africa, many of which are used to fund civil wars in Sierra Leone, Angola and the Democratic Republic of Congo (*The Guardian*, 12 December 2001). Levy and Scott-Clark (2001) refer to the 'brutalised landscape' at the jadeite quarries at Hpakant in northern Burma, with 'the mountains reduced to rubble'. At Serra Pelada in Brazil, hundreds of miners are destroying hillsides by digging for gold in 6-m^2 plots (*The Guardian*, 23 November 2009).

As indicated by these examples, mineral extraction often produces waste materials that themselves impact on the landscape. This may either be due to the need to remove overburden to reach the mineral, or is due to the low concentration of a mineral within an ore body. Mineral ores are rarely greater than 30% pure and can be under 1% pure. For example at Bingham Canyon, Utah, United States, the copper concentration is about 0.3%. Thus large quantities of waste materials need to be disposed of (Figure 5.4). For deep quarries the backfilling option is excluded by the need to work vertically and not sterilise the remaining resource at depth. Waste material must therefore be dumped beyond the quarry edge, thus increasing the landscape impact. The most serious examples occur where there are high quantities of waste relative to mineral recovered, e.g.

Figure 5.4 Spoil heaps at the Bingham Canyon Mine, Utah, United States.

kimberlite diamond pipes or copper mines. In the United Kingdom the most serious examples are at the north Wales slate quarries and at the china clay pits in south-west England where about 6–8 times as much waste quartz and mica are produced than usable kaolinite (Bristow, 1994).

The dumping of metal mine tailings can have serious pollution impacts. In Papua New Guinea in the 1980s, 130 000 tonnes of metal-contaminated tailings per day from the Panguna copper mine were dumped into the adjacent river. This polluted 30 km of river, destroyed all aquatic life and led to civil war. Some elements, for example cadmium, mercury, antimony and arsenic, are very common in small amounts in many polymetallic sulfide ores, but are toxic to living organisms. These elements may find their way into soils and surface water via mine tailings. An example of this occurred in the Solberg area of western Germany where wastes from the lead-zinc ore mining in the Carboniferous and Devonian limestones, and the smelting of the ores, led to contamination of soils and grassland. Cattle deaths from lead poisoning in the 1970s brought fears that drinking water and foodstuffs might be contaminated. An extensive series of tests was carried out and various remedial measures put in place, including guidance on the cultivation of vegetables in the cadmium-contaminated soils (Aust and Sustrac, 1992).

Toxic chemicals such as mercury and cyanide are also used to extract metals such as gold, and can lead to major land or water pollution, e.g. at Witswatersrand in South Africa (Evans, 1997), Caroni basin in Venezuela (Cutter and Renwick, 1999) and at several sites in Ghana (Hilson, 2002). In October 2010, 700 000 m^3 of toxic sludge, a highly alkaline aluminium by-product also containing heavy metals, burst from a waste lagoon at Ajka in western Hungary. Several people were killed or injured, villages were devastated and wildlife suffered as the sludge flowed into local rivers and onwards towards the Danube. A related problem is acid mine drainage resulting from oxidation of sulfide minerals. Sulfuric acid may be generated which in turn attacks other minerals to produce acidic water that is polluted, for example, by cadmium or arsenic. The problems may only occur after mine closure when the water table rebounds following cessation of pumping (e.g. Bell, Hälbich and Bullock, 2002). Hydraulic mining, the washing out of rock with high-pressure water jets, can also be very damaging in silting rivers and lakes downstream or downslope. It is used to extract gold in countries such as Indonesia, the Philippines, Venezuela and Brazil.

Having examined the impacts of mineral extraction, it is important to re-state the fact that mining frequently also brings geological benefits by creating exposures of rock and sediment that would otherwise be inaccessible. This often results in valuable evidence of the local stratigraphy, sedimentology or geomorphological origin. In these cases, continued quarrying using traditional techniques and carried out in collaboration with geological and conservation organisations makes available otherwise inaccessible rock or sediment sections. Many disused quarries have become important stratigraphical, mineralogical or palaeontological sites. In assessing mineral extraction proposals, it is important to make a balanced assessment on both the benefits and drawbacks of the proposals and to weigh the arguments, which should always include the economic need and geological advantages of extraction as well as the visual, landscape and other impacts.

5.3 Landfill and quarry restoration

Waste management is a major problem in many parts of the world. The waste material of modern societies is of various types (e.g. domestic, industrial, construction, agricultural, quarry) and some of it may be highly toxic (e.g. hospital waste, radioactive waste, contaminated soils). In many countries, quarries are infilled with domestic, industrial or agricultural waste materials, but this may be a major threat to geodiversity since it obscures the geological interest that has been exposed by the quarrying process. An example is Webster's Clay Pit in Coventry, England where Westphalian sandstones and mudstones representing the only available exposure of alluvial plain deposits within the Enville Formation were completely destroyed by landfilling in the 1990s. It was the best site in Britain for studying Walchia-like conifers of the Upper Palaeozoic (Prosser, 2003). Similarly, landfilling of Kirkill Quarry in Aberdeenshire, Scotland buried one of the most important Quaternary sites in Scotland and has prevented further research being carried out on the original exposures (Gordon and Leys, 2001b). Casual tipping of waste can also obscure exposures. A more successful outcome was achieved at Southerham Grey Pit in East Sussex, England where the single most important section through the Chalk Marl and Grey Chalk (Lower-Middle Cenomanian) was threatened by proposals in 1991 to landfill the quarry and trench the main section. Fortunately these proposals were dropped (R. Mortimore, pers. comm).

'But even when sections of the rock face are left uncovered, accumulation of leachate and landfill gases, slumping of the waste towards the face and difficulties of access often seriously interfere with research or education ...' (Nature Conservancy Council, 1990, p. 35). Furthermore, in the absence or failure of a waste containment system, groundwater or surface water contamination by leachate can occur and landfill gases, including methane and volatile organic compounds, are released to the atmosphere and may contaminate the soil capping materials (Gray, 1996).

Overfilling of sites or creation of landraised sites will result in the creation of new landforms, many of which are out of character with the local topography (Gray, 1998b, 2002; see Section 10.5.2). In other cases, natural valleys or depressions may be filled with waste with the resultant loss of landform, rock exposure and soils, and therefore loss of geodiversity.

5.4 Land development and urban expansion

New building works can have large impacts on geodiversity by removing topsoil and damaging soil structure and soil biota, removing and re-profiling land surfaces, thus leading to a loss of landforms, sub-surface sediments and fossils, and obscuring the underlying rocks and sediments. Urban expansion and infilling of undeveloped land in cities is leading to the accelerating encroachment into undeveloped land, and this is resulting in the loss of many important geological sites or semi-natural landscapes. At Gingko Petrified Forest State Park, Washington, United States construction of a pipeline would have cut through the heart of the important fossil site known for the diversity of its 200 genera and species encased in lava flows. Fortunately this threat was recognised and the plan was dropped (Gibbons and McDonald, 2001). But thousands of important geological sites have been lost through development.

Development also often leads to major landscape and pedological changes, e.g. the construction of road or railway cuttings, embankments, dams and reservoirs. Engineering solutions such as the use of gabions are often visually intrusive as are the formation of forest roads across mountain slopes. In the 1960s and 1970s, hundreds of kilometres of vehicular tracks were bulldozed in the Scottish Highlands to improve access to hunting areas and forestry plantations, resulting in unsightly hillside scars (Watson, 1984). Similarly in Iceland, Thorvardardottir and Thoroddsson (1994, p. 228) describe how the development of hydroelectric and geothermal power plants has affected the landscape. 'Even in the preparation stage of a power plant, off-road driving in connection with scientific work to find the best spot for electricity pylons, leaves tracks in sensitive areas. ... roads and powerlines create scars in the open wilderness landscape ...'. In Taiwan, Wang, Sheu and Tang (1994, p. 14) describe the damage to the coastal landscape by expansion of the Suhua Highway. 'Massive drilling and collapses of slopes have severely scarred the landscape'. Kiernan (1996) describes the loss of part of an important section in Tasmania as a result of widening of the Lyell Highway. Jungerius, Matundura and van den Ancker (2002) describe the erosional problems associated with roads in Kenya: 'however carefully the measures against erosion are designed, they become rapidly outdated because a new road attracts settlement'.

However, as in the case of mineral extraction, road, railway and pipeline cuttings often reveal important permanent or temporary rock exposures (Davies and Pearce, 1993). Where 'permanent' sections are left after completion of the project, subsequent maintenance of these rock cuttings can obscure the sections if inappropriate grading or stabilisation techniques are used such as covering with soil of 'shotcrete'. Building development on quarry floors can obscure the faces left by quarrying, while seawalls and marinas can threaten coastal rock exposures.

Related effects may be the disruption of local hydrological systems and the pollution of watercourses. For example, the construction of low-permeability surfacing in urban areas (tarmac and concrete) leads to lower infiltration rates and faster runoff of storm discharge. In turn this may lead to increases in flooding and channel erosion downstream of an urban area, as well as increasing pollution from urban drainage (e.g. oil, salt and pesticides). Pollution from sewage can also occur. Silvestru (1994, p. 222) describes the impact of rural settlement on karst terrain in rural Romania. The karst absorbs all liquid discharges including sewage 'so that practically all karstic aquifers in the area are likely to be polluted'.

Similarly water extraction can threaten river and lake levels. The Everglades National Park in Florida, United States is threatened with drying up as water is abstracted for urban populations and crop irrigation. Hurlstone and Long (2000) report that of five major geyser fields active in New Zealand in the late 1800s, only Whakarewarewa survives. This is due mainly to hydro-electric developments, though the Rotomahana field was destroyed by a volcanic eruption. Of more than 200 geysers active in the 1950s only 40 remain active today. At Whakarewarewa

Figure 5.5 Example of a coast defended by concrete wall and tetrapods, Hong Kong.

there were previously seven large geysers but as a result of more than 800 commercial and domestic wells drilled to tap a cheap source of heat, three have stopped erupting (Hayward, 1989; Buckeridge, 1994; Hurlstone and Long, 2000). Similarly, some geologists fear that exploitation of the Island Park Known Geothermal Resource Area (IPKGRA) in Idaho, United States could affect the famous geothermal activity at nearby Yellowstone National Park in Wyoming (see Box 6.1).

In 1998, the United Nations forecast an increase of more than 50% in world population between 1998 and 2050. This will almost inevitably significantly increase pressure on the world land resource and its ability to sustain social and economic development.

5.5 Coastal management and engineering

The erection of sea defences such as sea walls, cliff stabilisation, rock armouring, flood embankments or coastal slope regrading (Figure 5.5) can adversely affect or destroy the geological and geomorphological interest of coastlines since 'all are designed to counter the natural evolution of the coastline' (Lees, 1997, p. 3). Unfortunately, poorly informed, coastal decision-making can increase demands for coastal defence and stabilisation, e.g. construction of homes very close to eroding coastlines or agricultural land-uses downwind of mobile coastal dunes.

The impacts of coastal defences include:

- loss of or damage to geological exposures, landforms and habitats;
- stabilisation of active coastal landforms and processes such as dune systems;

- interference with natural interchange of sand between beaches and dunes, reducing sand deposition inland;
- increased erosion at the flanks of defences;
- reduction in input of sediment to the beach from an eroding coastline, which may exacerbate erosion downdrift.

There is also a loss of habitat since many specialised flora and fauna flourish on eroding coastal cliffs and intertidal zones (Lees, 1997).

As Hooke (1998) points out, many important geoscience sites for research and teaching occur on the shoreline because of the exposure of geological strata by marine processes and the dynamic landforms created by these processes. 'Continued availability of sites is essential to advance knowledge and understanding of both the past and the present' (Hooke, 1998, p. 2).

Even if not permanently destroyed, rock and fossil sites may be damaged by the works and are lost to research and education for the design life of the structures. There are many examples of loss of geological exposures by coastal defence works. At Barton-on-Sea in Hampshire, England, a proposal to extend the coastal defences and drain a slope threatened to severely compromise the integrity of the international Bartonian stratotype (Doyle, 1989). At Burnie in Tasmania, Australia, coastal reclamation has covered natural shore platform exposures of Precambrian dolerite dykes that were amongst the earliest 'Geological Monuments' in Australia (C. Sharples, pers. comm.).

'Coastal defences can isolate landforms such as shingle bars, beaches, saltmarshes and mud flats from the sediment supply that feeds and maintains them, often causing erosion of these features' (Glasser, 2001, p. 893). There is therefore often a case for allowing erosion to continue, at least at a slow rate, in order to maintain the exposure. Sea defences such as flood embankments also fix the position of the coastline and may result in 'coastal squeeze' on the retreating salt-marsh areas seaward of the embankments. Clearly any interference with sand dunes will alter the morphology of the dunes as well as affecting the stability and processes affecting the dunes.

English Nature has published a report highlighting the degraded condition of the coastline of England, as affected for example by coastal defences, port development and recreational pressures (Covey and Laffoley, 2002). Brocx (2008) describes the anthropogenic impacts on the Pilbara coast of Western Australia where exploitation of local iron ore, oil, gas, sands, gravels, hard rock, limestone and salt has taken place over many years. The most significant impacts have been 'a result of unplanned exploitation of mineral resources in an environmentally unique region. This outcome is the product of some forty years of *ad hoc* planning decisions ...' (Brocx, 2008, p. 144). Particularly damaging have been the development of ports, harbours and causeways. Most recently, Wimbledon (2012) has highlighted the destruction of internationally important coastal exposures on the south of the Crimean peninsula in the Ukraine, with the stratigraphy being buried and much material being pushed into the sea.

5.6 River management, hydrology and engineering

The human impact on rivers has been extensive, and these have come about from both engineering within the channel and floodplain, and from wider land-use changes. For example, deforestation within a drainage basin may increase overland flow and hence the potential erosion and transfer of sediment to the channel. Prosser *et al.* (2001) calculated that at least 127 million tonnes of sediment is delivered to Australia's streams and rivers, and most of this is the result of the extensive land clearance, sheep and cattle grazing and mining activities of the past 200 years. The result is higher water turbidity, lower water quality and sand slugs on the river bed that raise bed levels, increase the risk of flooding and smother aquatic habitats (Bartley and Rutherfurd, 2001).

Clifford (2001) classifies the traditional direct engineering impacts on rivers into:

- flood prevention and mitigation;
- channel stabilisation;
- flow regulation for navigation;
- water supply and quality;
- effluent disposal.

Many natural river channels have been altered in the past, often as part of flood protection or river management schemes. Rivers have been channelised, straightened, embanked, dammed, diverted, culverted, dredged and isolated from their floodplains (Brookes, 1985, 1986, 1988; Graf, 1992, 1996). In turn these works have changed the river characteristics including channel roughness and river dynamics, as well as the natural bank, river bed and floodplain habitats (Hooke, 1994a; Soulsby and Boon, 2001; Richards, Brasington and Hughes, 2002). As Clifford (2001, p. 72) puts it 'Channelisation degrades habitat ... by removing bed and bank sediments and vegetation and by altering flow dynamics' (Figure 5.6).

But perhaps the most radical impacts are from dam construction where there is wholesale impounding and regulation of flow (Graf, 1996; Clifford, 2001). At the large scale, China's Three Gorges Dam on the Yangtze River has certainly changed the downstream flow regime of the river as well as raising water levels in the gorges themselves and 'will forever transform one of the most spectacular river valleys in the world from a natural-flowing stream into a reservoir ' (Cutter and Renwick, 1999, p. 339). The dam is 185 m high, 1.9 km long and impounds a lake 600 km long. One of the most controversial dams in the world is the O'Shaughnessy Dam built in 1913 across the Hetch Hetchy Valley in Yosemite National Park (see Section 9.2). Three major dams on the River Indus in Pakistan have led to water deficiencies in the lower reaches of the river and salinisation of soils (Husain, 2003). It was the flooding of Lake Pedder in Tasmania, Australia

Figure 5.6 Example of a channelised river between two roads.

for a hydroelectric power scheme on the Gordon and Serpentine Rivers that is widely regarded as having initiated environmental politics in Australia as well as the awakening of interest in geodiversity issues in Tasmania (C. Sharples, pers. comm.).

Beavis and Lewis (2001) noted that agricultural expansion and the need for 'drought-proofing' led to the building of 300 large dams in Australia between 1940 and 1983, but even more significant effects come from the construction of farm dams of which there are estimated to be nearly half a million in Australia (Beavis and Lewis, 2001). These have a major impact on catchment hydrology and Australian water resources. The 'tank' landscape of south-east India is a landscape dominated by small earth dams collecting water from myriads of little streams and areas of overland flow (Spate and Learmonth, 1967).

Interbasin water transfers are also associated with major changes to geomorphological and hydrological systems. For example, the Snowy Mountains Hydro-Electric Scheme in Australia consists of 16 large dams, many smaller diversion structures and hundreds of kilometres of tunnels, aqueducts and pipelines (Pigram, 2000). 'Understandably, an undertaking of this magnitude has had far-reaching impacts on the landscape, both in the upland areas immediately affected by construction works and in areas downstream. Perhaps the greatest impact has been on the Snowy River itself'. In reaches of the river downstream from the Scheme there are marked deteriorations of stream flow, riparian vegetation and the river channel itself (Pigram, 2000, p. 343–344).

Despite this, major water transfer schemes are still being planned. For example The Spanish government approved a €18 million scheme in 2008 to divert water from the Ebro northwards to Barcelona. Diversion southwards to the coastal areas of Valencia, Murcia and Andalucia where the tourist complexes, swimming pools, golf courses and vegetable growing consume huge quantities of water has also been proposed. Even bigger is the engineering project recently authorised by the Chinese government to transfer 48 trillion litres of water per year from the Yangtze River to the drier northern provinces including Beijing via three new channels up to 500 km long (*The Guardian*, 27 November 2002). Previous river diversions in the former Soviet Union have caused the drying of the Aral Sea with profound effects on geomorphological, biological and social systems.

Hughes and Rood (2001, p. 106) see floodplains as among the most abused of ecosystems because of competing interests for resources. Dam construction, channelisation, water removal, gravel and sand extraction and forest clearance all affect the timing and quantity of water and sediment delivery to floodplains'.

5.7 Forestry, vegetation growth and removal

5.7.1 Afforestation

The impacts of afforestation and planting relate partly to potential obscuring of landforms or rocks by blanket woodland so that individual features and landform associations are masked by trees or other planting. The Nature Conservancy Council (1990, p. 40) believed that 'Large-scale conifer plantations render location, mapping and inter-relation of outcrops virtually impossible' and also encourage lichen growth on outcrops or bury them in leaf litter. Natural vegetation growth may also lead to loss of visibility. For example, the columnar jointing in a young lava flow in the Organ Pipes National Park near Melbourne, Australia, is now largely hidden behind rapidly growing plantings and native vegetation (Joyce, 1999). Larwood (2003) illustrates by photographs how vegetation growth has obscured the visual impact of the geology of Wren's Nest in the West Midlands, England. In the Yorkshire Wolds, England, English Nature (1998) saw the afforestaion of karst valleys as undesirable and in Breckland, England, afforestation or scrub and tree encroachment has led to the obscuring of fossil pingos. Planting operations are potentially damaging if large-scale mechanical equipment is used. It is easy to disrupt the soil structure, alter soil chemistry or change soil biota, and there may be serious effects in areas where important subtle landforms occur.

Accessibility of landforms and rock exposures or sediment sections are also likely to be affected by blanket planting, while more limited planting or natural vegetation growth can reduce the visual continuity of landforms or rock exposures. Outcrops on open land or in quarries can quickly become degraded and overgrown by scrub vegetation unless site clearance is regularly undertaken.

Root damage to sensitive geological sections is not unknown. It is therefore important to try to locate new forests and woodlands in less sensitive locations in this respect. For example, in the 1970s there were plans by the UK Forestry Commission to plant a new forest in Glen Roy, site of the famous 'Parallel Roads', interpreted as glacial lake shorelines. The shorelines, which are today continuously visible for several kilometres, would have been totally obscured by this plan, which fortunately was abandoned when the impacts were pointed out by the geoscience community. Nowadays much greater account is taken of earth science interests when planning planting schemes.

Soil chemistry can also be affected by afforestation, though this will depend on the species planted. Conifers and eucalyptus can induce soil acidification with a decrease in exchangeable cations and an increase in aluminium (Aust and Sustrac, 1992).

5.7.2 Deforestation

Deforestation or logging operations may also have major impacts on soils and other earth science interests. Trees and their leaf litter generally have the role of intercepting rainfall and reducing the impact of raindrops on the soil so that erosion rates in forests tend to be low. The removal of trees therefore often results in soil erosion by overland flow, particularly from sloping terrain in high rainfall areas. Spectacular examples of gulley erosion following deforestation are found in areas of tropical rain forest areas such as Madagascar. Silvestru (1994, p. 223) describes the effects of local populations felling trees on the karst terrain of rural Romania. 'If the present rate of cutting is maintained ... the soil will rapidly vanish, leaving a stony desert which would eventually exacerbate the karstification process'. Sivestru describes an area in northern Romania where a whole mountain slope has been clear-cut for commercial purposes, 'the soil immediately starting to move down the 35° slope, exposing bare limestone to weathering. Moreover, since the limestone dip roughly corresponds to the slope angle, rock slides occurred within 4 years (after the cutting), because of intense karstification'. Valiūnas (1994, p. 275) urges caution about deforestation of fossil sand dunes in the Dzūkija National Park in Lithuania 'otherwise there is a danger of renewed deflation, soil degradation, etc.'

Commercial felling operations involving mechanical equipment and the uprooting of trees has a major effect in compacting and disrupting the soil. Brady and Weil (2002, p. 141) believe that such practices 'can impair soil ecosystem functions for many years. Timber harvest practices that can reduce such damage to forest soils include selective cutting, use of flexible-track vehicles and overhead cable transport of logs, and abstaining from harvest during wet conditions'. Kiernan (1991, 1996) describes the devegetated and eroded condition of the soils and landforms in the central part of Tasmania's West Coast Range caused by commercial timber cutting.

5.8 Agriculture

As Kibblewhite, Ritz and Swift (2008) have pointed out 'Soil management is fundamental to all agricultural systems yet there is evidence for widespread degradation of agricultural soils in the form of erosion, loss of organic matter, contamination, compaction, increased salinity and other harms'. In general terms, there is likely to be little damage to geodiversity in areas that have had a long and successful history of cultivation. Farmers who have rotated crops, retained fertility and managed the land to prevent soil erosion have been able to conserve and sustain the soil resource. In some cases, however, particular problems may arise. For example, regular ploughing of steep slopes can affect the physical integrity of the slopes and landforms by increasing erosion and downslope movement of soil. The effects are particularly serious under wet soil conditions. In areas where the landforms are small scale and intricate (e.g. low sand dunes or abandoned channels on river terraces) ploughing can reduce the relief and landform detail. In Breckland, England, much fossil periglacial patterned ground has been ploughed out this century (English Nature, 1998). In Northumberland, England, the wintering of cattle on dune areas has led to damage to the dune vegetation, thus increasing the vulnerability to wind erosion. In fact about a third of the arable area of England and Wales is thought to be at risk from wind erosion, with drained peat soils and Quaternary sands being most at risk. Similar threats have been detected in other areas, for example on the coastline of north Germany (Aust and Sustrac, 1992). Stocking levels in semi-arid areas can have serious impacts on vegetation cover and thus soil and gulley erosion potential and delivery of sediment to streams (Caitcheon *et al.*, 2001).

But the biggest threats to geodiversity have come from unsustainable land management, and the history of land clearance and cultivation is littered with examples of the degradation of soil and other resources (see Box 5.5). Bringing areas into agriculture which have not previously been cultivated, will alter many of the soil properties (e.g. soil compaction, organic matter content and distribution, soil biota, soil chemistry from fertiliser or pesticide application) and reduce natural soil diversity. This is particularly true if intensive agriculture is practised year after year, resulting in loss of soil structure and leaching of nutrients. Soil compaction has become an increasing problem as bigger and heavier farm machinery has been introduced. Subsoil compaction may occur below the plough-layer producing a 'plough-pan'. 'Nearer the surface, a combination of compaction and tractor wheel-spin may limit the extent to which plant roots can exploit the soil for moisture and nutrients and the free movement of gases between soil and the atmosphere' (Bridges, 1994, p. 13). Compaction decreases infiltration capacity and porosity resulting in faster saturation and overland flow. In turn soil erosion by wind and water, including transport of pesticide residues, siltation of river courses and increased flooding may result. Figure 5.7 shows soil erosion in the United States based on US Department of Agriculture data (Cutter and Renwick, 1999). In the United Kingdom, Boardman (1988) related increases in soil erosion in the 1980s to increased growing of winter cereals and consequent expansion of bare ground in the autumn and winter when rainfall is highest.

There has been much debate as to whether soil should be regarded as a renewable or non-renewable resource. Although the organic content of soil, derived from dead and living plants and animals, is at least partially replaceable relatively quickly by pioneer vegetation, leaf falls, colonisation by soil fauna and manuring, the mineral components may take thousands or millions of years to weather from the parent rock. Thus it is important that soil is regarded as a non-renewable resource in which soil erosion losses are unsustainable even with the addition of fertilisers (Trudgill, 2001).

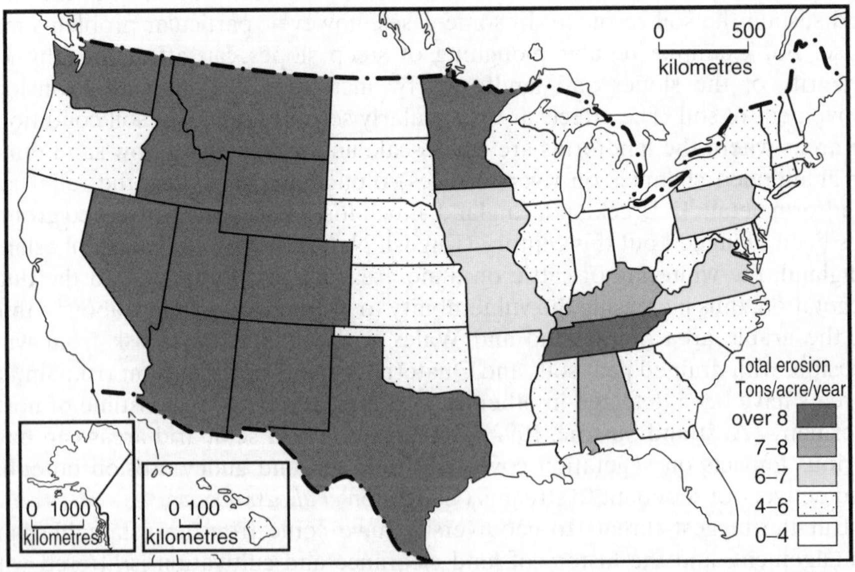

Figure 5.7 Soil erosion rates (1992) in the United States (after Cutter, S.L. and Renwick, W.H. (1999) *Exploitation, Conservation, Preservation: A Geographic Perspective on Natural Resource Use*. 3rd edn. This material is reproduced with permission of John Wiley & Sons, Inc.).

Box 5.5 *Agricultural impacts: Mexico City, the US Great Plains and northern China*

Leopold (1959) recorded the sequence of events that led to the degraded land and denuded soils in the foothills north of Mexico City (see also Dasmann, 1984). Early descriptions described a tall, open forest of pine and oak in the area, but as the population grew the demand for farmland increased and extended farther up the slopes of the foothills. The forests were cleared and turned into charcoal, and the land was then planted with corn or wheat. However, rainfall on the sloping ground of the foothills quickly led to soil degradation resulting from topsoil erosion and mineral leaching. In the end, the soil could no longer support a viable corn crop, but maguey, a cactus-like plant grown for its fibres, could be cultivated on the impoverished ground. The open nature of this plant led to further erosion and gulleying until only an impervious hardpan remained, incapable of supporting the maguey. The remaining scant cover of a few desert-tolerant plants could only support the grazing of goats and donkeys until they too were gone and only a wasteland was left. 'The story illustrates a process that has been repeated again and again throughout the world, an attempt to find food ... resulting in the degradation of the basic and irreplaceable natural resource on which terrestrial life depends' (Dasmann, 1984, p. 179).

Another famous example was 'The Dust Bowl' of the central United States. In the 1880s, the first settlers began to move into the Great Plains of the United States, an area prone to great fluctuations in rainfall. The natural vegetation was short grassland and droughts did little lasting damage to the underlying soils. However, when the brown soils were cultivated for wheat they became susceptible to wind erosion under drought conditions. The most serious erosion occurred in 1933 when severe drought brought dust storms that eroded a topsoil that had lost much of its structure due to ploughing and overgrazing. A large area on the borders of Kansas, Oklahoma, Texas and Colorado was affected by a loss of topsoil of between 50 and 300 mm. The area became known as 'The Dust Bowl' and was affected by drifting dunes which buried roads, fences, dwellings and crops in thick deposits of re-deposited dust (Dasmann, 1984). Whitney (1994, Chapter 10) described this as 'the most rapid rate of wasteful land use in the history of the world' and uses phrases like 'earth butchery' and 'predatory agriculture'.

The most recent example is occurring in northern China where dust clouds testify to wind erosion of soil brought about by over-cultivation, overgrazing, over-cutting and over-pumping of marginal land. This 'Chinese Dust Bowl' is seriously threatening Chinese food supplies and raising world grain prices (Lean, 2003).

The application of fertilizers and pesticides to land has already been referred to and can be very beneficial to the quantity and quality of crop yields. However, residues may remain in the soil and excess amounts may be flushed and leached into surface or groundwaters thus leading to nitrate and other pollution and water eutrophication. Irrigation can also impact on soil by leaching nutrients, changing groundwater levels and increasing salinity if the water used has a high salt content. Salinity levels can also rise as a result of over-abstraction of groundwater in coastal areas. An example is the Llobregat delta near Barcelona in Spain. Irrigation near the Hagerman Fossil Beds in Idaho, United States, has added water to the unconsolidated sediments of the Glenns Ferry Formation, resulting in an increase in frequency and magnitude of landslides and threatening to obscure the exposure of the fossil beds.

O'Halloran (1990), Hardwick and Gunn (1994), Gunn (1995) and Eberhard and Houshold (2001) summarise research on the impact of agricultural operations on limestone cave and karst systems. The potential impacts are illustrated in Figure 5.8 and include:

- farmyard runoff including oil, organic slurry and animal dips can drain into cave systems causing cave and groundwater pollution;
- tipping of farm waste, including animal carcasses, may block cave entrances, infill dolines and lead to underground pollution;
- digging drainage ditches increases peak discharges and may flood caves and erode cave sediments and speleothems. New drainage routes may be opened with consequent impacts on both new and old drainage ways;
- spoil and sediment from surface excavations, ploughing and erosion may be washed into cave systems;
- the use of agrichemicals (fertilisers, pesticides, etc.) on thin soils overlying porous and permeable rocks often leads to pollution of speleothems, water courses and cave sediments.

Buffer zones around cave sites have been suggested and are useful, but changes in the whole catchment area can affect caves and cave water in both obvious and more subtle ways. Hardwick and Gunn (1994) described the impacts of ploughing on cave systems in the Castleton area, England. Clastic sediment input may infill cave entrances and internal passages, bury speleothems, generate back flooding and loss or damage to cave features, alter cave microclimates and transport agrichemicals into underground environments. Changes in agricultural practices can also have an effect on cave hydrology. For example, drainage works and higher stocking densities may increase runoff, leading to accelerated erosion and downstream deposition.

One of the major impacts of agriculture on the landscape has been the construction of terraces. The function of these is to produce flatter areas of land even on steep slopes, to reduce soil erosion and, in the case of rice cultivation, to retain water. Although some spectacular landscapes have been produced (Figure 5.9), the result is a loss of the natural landform and change in the soils. Bayfield (2001, p. 20–21) notes that 'Terraces represent a major disruption of the original soils and have a massive landscape impact'.

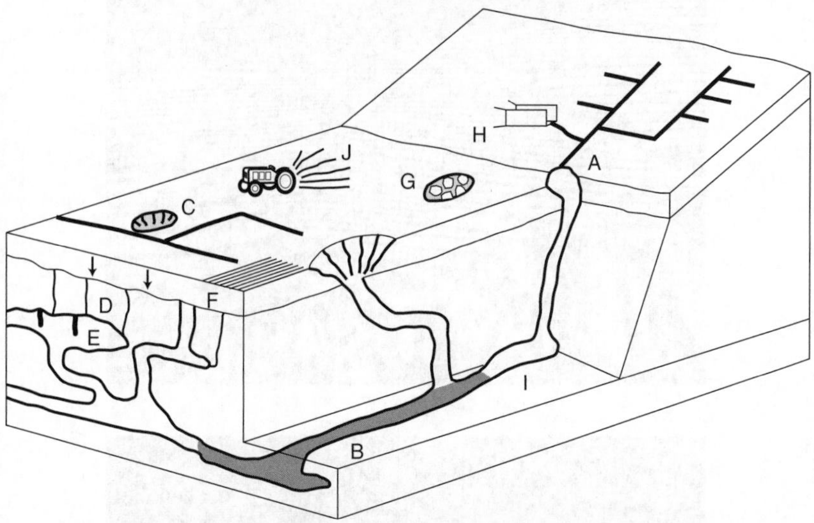

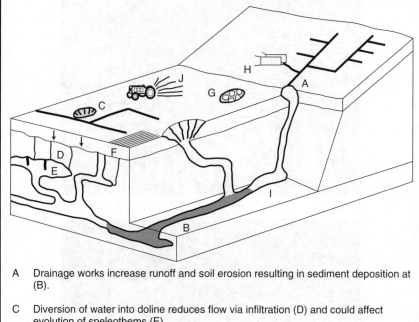

A Drainage works increase runoff and soil erosion resulting in sediment deposition at (B).

C Diversion of water into doline reduces flow via infiltration (D) and could affect evolution of speleothems (E).

F Surface ploughing increases runoff and reduces infiltration and sub-surface recharge.

G Doline infilled by tipped waste to increase area of cultivated land.

H Pollution from animal dips may enter sink at (A) causing contamination of underground channel and groundwater at (I).

J Spreading of manure and agrochemicals results in diffuse pollution to the karst hydrological system.

Figure 5.8 Potential impacts of agricultural operations on karst and cave systems (after O'Halloran, D. (1990) Caves and agriculture: an impact study. *Earth Science Conservation*, 28, 21–23. Reproduced with permission of Natural England).

5.9 Other land management changes

Activities such as cutting, filling and spreading can have very serious effects on many parts of the physical (and biological) environment. For example, the excavation of materials to create agricultural reservoirs or fishing lakes often results in the dumping or spreading of the subsoil on adjacent land, particularly given transport and other costs or regulations involved in removing the material. However, these processes, in badly managed situations, can lead to soil burial and creation of new, artificial landforms.

(a)

(b)

Figure 5.9 Impacts of agricultural terracing: (a) vine terraces, Douro Valley, Portugal and (b) rice terraces, Bali, Indonesia.

The dumping of solid materials can lead to infilling of natural hollows such as kettle holes and dolines (Cernatic-Gregoric and Zega, 2010) and may obscure important rock, mineral and fossil exposures. Infilling of ponds and marshes leads to loss of geomorphological and hydrological character and therefore habitat. Some geomorphological and biological landscape richness is therefore lost. Significant land remodelling, as often occurs in golf course construction, will also change the local geomorphological character and soil conditions.

The dumping of quarry and mine wastes has had an important effect on some landscapes, such as the coastline of Durham, England, where waste was tipped over the cliffs, for example at Easington and Horden. This altered the coastal topography and beach sediments to the detriment of earth science and aesthetic interests (see Section 10.4.3). Deliberate land reclamation will also result in changes to the coastal morphology and can have detrimental effects on shoreline processes by altering the coastal dynamics. The largest scale projects have been in the Dutch polders where land reclamation from the sea has provided new agricultural land and flood control measures but has resulted in changes in groundwater characteristics, ground subsidence and foundation damage in adjacent towns (Aust and Sustrac, 1992).

5.10 Recreation/tourism pressures

Recreational tourism is an increasingly important economic activity but one that can lead to damage to biodiversity and geodiversity. Miller (1998, p. 404) quotes a 12-fold increase in visitors to US National Parks between 1950 and 1990 and there has been a further rapid increase since then.

> Under the onslaught of people during the peak summer season, the most popular national and state parks are often overcrowded with cars and trailers and are plagued by noise, traffic jams, litter, vandalism, deteriorating trails, polluted water, drugs and crime ... Park Service rangers now spend an increasing amount of their time on law enforcement instead of on resource conservation and management.

It is not surprising that Hunt (1988) expressed the view that in the United Statesthere is a danger of people 'loving their national parks and historic sites to death'. While not all of these impacts will directly threaten geodiversity, some of them will. For example, footpath erosion initially tramples the vegetation but ultimately exposes the soil and sub-soil to erosion and can lead to significant gulleying (Bayfield, 2001). Dovedale in the English Peak District National Park receives up to a million visitors every year, including many school and university groups. However, the impact of these is leading to the destabilisation of scree slopes and significant footpath erosion. In the Lake District National Park there have been concerns about the hundreds of hikers who each summer attempt to climb Scafell Pike as part of a three peaks in 24 hours challenge that also includes Ben Nevis in Scotland and Snowdonia in Wales. Slopes climbed by large numbers of people in heavy rain erode very fast and as a result, deep gullies have been cut in the soil next to established footpaths (*The Guardian*, 21 January 2008).

Figure 5.10 Vehicle access to sand dune areas can initiate significant erosion, Donegal Bay, Ireland.

This potential threat and the actual damage to the landscape in rural areas has increased with the use of mountain bikes, motorbikes and all-terrain vehicles (ATVs).

Some areas are particularly vulnerable to recreational pressures including sand dune areas (Figure 5.10) where worn footpaths and ATVs can instigate serious wind erosion and sand movement (ASH Consulting Group, 1994; Scottish Natural Heritage, 2000a; Catto, 2002). Recreational pressures on beaches can affect the build-up of sand on the dune front and change the natural processes of dune formation and renewal. Beach cleansing can also destroy embryo sand dunes before they can fully develop.

Thorvardardottir and Thoroddsson (1994, p. 229) believe that in Iceland one of the major tourist impacts on volcanic landscapes is the creation of multiple paths on lava cones. 'This is especially striking where mosses have covered the crater's slope, but trampling breaks the moss cover, leaving scars open to erosion. Tephra formations and some lava formations are so fragile that they disintegrate underfoot'. Off-road driving creates even greater problems, 'it seems that tourists, Icelanders and foreigners alike, cannot resist the temptation to try out their four-wheel drive vehicles and their ability to drive, or rather to spin their wheels, on slopes of volcanic craters', although steps have recently been taken to encourage more responsible driving attitudes (Thorvardardottir and Thoroddsson, 1994, p. 228).

Another volcanic landscape affected by pedestrian access is brittle lava that may be broken underfoot. An example occurs at Craters of the Moon National Monument in Idaho, United States, where the glazed black lava crust cracks under visitor weight to reveal an orange lava beneath (see Figure 5.11). Not

Figure 5.11 Perceived threats from pedestrian impacts on brittle lava, Craters of the Moon National Monument, Idaho, United States.

far away at Yellowstone in Wyoming, visitors have been found throwing coins, branches, tyres, clothes, and so on into geysers, which can block the geothermal pipes and lead to explosive and damaging events when the pipes are cleared.

Cave systems are another geological environment sensitive to visitor pressure. A count in the Carlsbad Caverns in New Mexico, United States, in 1993 revealed 32 000 damaged speleothems by visitors up to that date. Caving can also lead to damage to the speleothems or floor deposits (Wright and Price, 1994; Hardwick and Gunn, 1994, Kiernan, 2010). 'As soon as a new cave is discovered, there is a gradual, often insidious, degradation of the cave. The rate and degree of this degradation depends on the fragility of the features contained and the care of the cavers involved' (Wright and Price, 1994, p. 196). Touch, breath and light introduced by cavern visitors can encourage algal growth. There are also external threats to caves resulting from human-induced hydrological changes, which may cause erosion or degradation of underground systems (see earlier).

Rock climbing, canyoning and similar pursuits can also damage important geological sections for 'as geotourism grows so does the pressure placed on the resource' (Larwood and Prosser, 1998, p. 98). For example, climbers have been ascending Devils Tower, Wyoming, United States, for over 100 years. Nowadays over 5000 climbers per year use the more than 200 routes identified for scaling the massive columns with the result that the geological integrity of the tower has been affected (Figure 4.19 and see Section 9.2). Particularly damaging has been the permanent placement of bolts and pitons. On the other hand, tourism and outdoor leisure pursuits bring people into contact with natural or semi-natural environment and may lead to a more appreciative public willing to support efforts to reduce the threats and conserve the resource.

Finally, 'High-mountain environments are particularly sensitive to the disturbances caused by recreation use. Steep topography, thin soils, sparse vegetation, short growing seasons, and climatic extremes (e.g. heavy precipitation, cold temperatures, high winds) all contribute to the sensitivity of high-mountain environments. Under such conditions, it takes relatively little use to create long-lasting impacts' (Parsons, 2002, p. 363). The impacts of hiking and camping include construction of campgrounds, soil compaction and erosion, inadequate disposal of human waste, blackening of rocks from campfires and movement of rocks to build fireplaces and wind breaks. Parsons (2002) reports that management proposals to restrict recreational use in urban-proximate wilderness in Oregon and Washington, United States, have led to conflicts between hiking groups and conservation groups.

A particular issue in mountain environments is the development of skiing areas. Removal of boulders and smoothing of slopes may impact on geological, geomorphological and pedological features (Bayfield, 2001). Pfeffer (2003, p. 24) states that 'in areas above the timber line, geomorphologic features, for example morainic arcs, rock glaciers or outcropping bedrock, are the main targets to be destroyed in the landscape'. She gives the examples of the removal of 50% of an important frontal moraine of the 'Gepatsch' Glacier in the Austrian Tyrol by road construction to a skiing centre, and the blasting of a well-developed rock glacier at Sölden to create a steep ski run. 'In both cases evidence of geomorphologic evolution has been destroyed in favour of skiing facilities' (Pfeffer, 2003, p. 25).

5.11 Removal of geological specimens

Collecting appears to meet an innate human need. Casual removal of coloured rocks or pebbles from rivers and foreshores may seem inconsequential, but over long timescales it depletes the resource. At Yellowstone, United States, at least one petrified tree has been completely lost through gradual removal of pieces by visitors, and at Fossil Cycad National Monument in South Dakota, United States, 35 years of neglect, unauthorised fossil collecting and unchallenged research collecting led to the near loss of the resource and its removal as a National Monument in 1957 (Santucci and Hughes, 1998).

Fossils, minerals and rocks are often aesthetic specimens and have long attracted collectors in the same way, though not perhaps to the same extent, as some of the Victorian wildlife collectors of, for example, butterflies or birds' eggs. Gemstones and minerals have long had a commercial value, but fossils have acquired significant economic worth only relatively recently. As Norman (1994, p. 63) points out 'Attaching a monetary value to a natural heritage item such as a fossil raises a spectrum of conflicting assessments ...'. Nowadays, there is even the risk of fossils, including type specimens, being stolen from museums or being forged (V. Santucci, pers. comm.).

The removal of rock, fossil and mineral specimens is not regarded as significant where the geological resource is extensive (Alcalá and Morales, 1994; Forster, 1999; King and Larwood, 2001). It is also recognised that many fossils are more valuable out of the rock than *in situ*, particularly where they may be vulnerable to erosion, quarrying, burial or other loss. In such cases, proper scientific field recording, curation and accessibility in a suitable museum are essential.

However, there can be serious effects on geodiversity where the resource is extremely limited (e.g. within a cave or river channel deposit) or where there are very rare or scientifically valuable specimens (e.g. dinosaur bones or eggs). Some sites that are spatially extensive and contain an abundance of common fossils can nevertheless still be vulnerable if they also contain a few exceedingly rare fossils. Sites like these are vulnerable to irresponsible collecting and it is important that specimens are not removed from sites without the recording of crucial information including stratigraphic position (MacFadyen, 1999). And fossils can be lost from the public domain by sale into private collections or abroad (Norman, 1994; MacFadyen 2001a).

Mechanical excavators, explosives, power tools and sledgehammers have all been used to remove fossil material but have resulted in major damage to important exposures. In general this sort of collecting leads to no scientific or educational gain and is therefore unacceptable. Instead, large amounts of rock are removed in pursuit of the rare or 'perfect' specimen, resulting in loss of other, often important, fossils (Horner and Dobb, 1997; MacFadyen, 2001a). This has a major effect on our fossil heritage and is therefore to be condemned by all responsible geologists. For example, MacFadyen (1999) quotes the example of the amphibian fossil-bearing Cheese Bay 'Shrimp Bed' near North Berwick in Scotland, half of which was removed illegally in a matter of hours by a collector using a mechanical digger. Another Scottish site, Birk Knowes, known for its arthropods and early fish, attracted ruthless fossil collectors during the 1990s. Due to the remoteness of the site, they were able to overcollect from the limited exposures causing huge damage and loss of scientific information (MacFadyen, 2001a). Of the *Caloceras* ammonite beds at Doniford Bay in Somerset, England, 65% have been destroyed by irresponsible collecting (Webber, 2001). Horner and Dobb (1997) make the point that commercial collectors do not have the motivation or time to map, extract and catalogue their finds in a rigorous manner. They 'cannot conduct the sophisticated analyses that enable us to identify exactly what kind of depositional environment the sediments represent ... or the geological events responsible for the formation of the environments. ... when the desire to make money is paramount there's no incentive for conducting scientifically sound excavations' (Horner and Dobb, 1997, p. 241–242).

Other examples of overcollecting of minerals or fossils that have led to the total destruction of sites can be quoted, and these can also be regarded as the equivalent of biological extinctions (see Section 14.2.3). Swart (1994, p.321) believes that overcollecting at the world famous Ediacara Fossil Reserve in South Australia has lead to the virtual destruction of the site. Ironically the establishment of the Reserve, intended to give the site protection, may actually have had the opposite effect by drawing attention to its importance. Collecting is only allowed at the site with permission, but its remoteness means that there is no way of monitoring visitors or activities. In Russia, some important ammonite sites have been mechanically excavated and the material shipped abroad under the label 'of no scientific value' (Karis, 2002). Amateur collecting has threatened several important dinosaur egg sites in Aix en Provence, France (Gomez, 1991).

Murphy (2001, p. 14) quotes the case of the Hope's Nose unique gold-bearing limestone-hosted veins that were very limited in extent. 'Intensive and unconsented collecting, using heavy-duty rock saws, has removed the veins and the site is now effectively destroyed'. Some mine spoil can be of geological importance, for example for supporting rare minerals or fossils, for example Writhlington in England (Jarzembowski, 1989). The indiscriminate depletion of these spoil heaps for minerals or use in track construction or fill can be a threat to this resource. A survey of fossil theft and vandalism in the US National Parks between 1995 and 1998, discovered that there had been over 1400 reported incidents, over 500 citations and 15 arrests resulting in fines of S150 000 (Santucci, 1999).

Even research and educational activities can damage sites. Bayfield (2001) refers to an outcrop of serpentine which was entirely removed by successive visits by geology students (Speight, 1973) and this is not an isolated case. The Upper Ordovician Sholeshook Limestone Quarry in South Wales is one of many sites to have been 'ruined by endless student parties' (Clarkson, 2001, p. 17). Even where sites are not destroyed, the impact of student hammering on rock outcrops can produce unsightly fresh scars. Toghill (1972, p. 514) believed that Shropshire, England, 'is really no place for elementary geology students' who could be taken to less sensitive coastal exposures.

Impacts of palaeomagnetic coring can also be significant. Campbell and Wood (2002) describe how a valuable glacially striated rock surface in Wales first described by Sir Andrew Ramsay in 1860 has been peppered with core holes and MacFadyen (2011) has described the impact of coring on important geosites in Scotland (Figure 5.12). Similar impacts have raised concerns in Tasmania, so that special measures are taken to manage coring and restore core holes (M. Pemberton, pers. comm.).

The concept of landscape sensitivity can therefore be extended to fossil sites. MacFadyen (1999) has classified site vulnerability into:

- *Robust*—sites where the fossil resource is extensive and fossils are common, or the fossils are rare, difficult to collect, unspectacular and perhaps poorly preserved, or the sites are inaccessible to collectors.
- *Vulnerable*—sites where the fossil resource is substantial or of unknown extent and the specimens are generally well preserved, some of them having considerable scientific, aesthetic or commercial value.
- *Very vulnerable*—sites where there is a very small fossil-bearing resource, that could in theory be totally removed in hours using appropriate equipment, and/or where the fossils are of the highest scientific value and accessible to collectors.

As MacFadyen points out, site vulnerability may change over time as a result of resource re-evaluation, research excavation, commercial quarrying or discovery of new fossil material.

Figure 5.12 Irresponsible coring of the Sandwick Fish Bed near Stromness, Orkney, Scotland. Despite there being numerous less well exposed faces to choose from, the holes have been drilled in highly visible surfaces. (Reproduced by permission of Colin CJ MacFadyen, Scottish Natural Heritage.)

5.12 Climate and sea-level change

The most authoritative work on the progress, impacts and mitigation of climate change is the Intergovernmental Panel on Climate Change (IPCC) whose most recent report (IPCC, 2007) predicts that global temperature will warm by 1.1–6.4 °C by 2100 and that globally averaged sea-level will rise by 0.18–0.59 m by 2100. The warming will however vary by region and there would be both increases and decreases in precipitation, changes in the variability of climate, and changes in the frequency and intensity of extreme events. The impacts of these changes on physical systems are not completely known, but are likely to be considerable. There are likely to be changes in the nature or rates of processes (e.g. a rise in precipitation may lead to increases in soil erosion, severe floods and limestone solution). Some processes, and their related landforms and sediments, may be reactivated while others will become fossilised.

It is likely that there will be increases in annual mean stream flow in high latitudes and south-east Asia, and decreases in central Asia, the Mediterranean, southern Africa and Australia. Rivers draining from ice masses in the Alps and the Rockies are likely to have reduced flows in summer due to the disappearance of many of the glaciers in the next few decades. It is estimated that the 37 named glaciers in Glacier National Park, United States, in the 1990s could disappear within the next few decades (D. Fagre, pers. comm.) and glaciers are also shrinking rapidly in Alaska (Arendt *et al.*, 2002). Since many mountain societies rely on these glaciers for water resources, there is concern about their future disappearance (UNEP-WCMC, 2002).

One environment in which climate change is likely to have a serious impact is the periglacial environment. Although quantitative estimates of the temperature rise vary, most authors agree that global warming is likely to have the greatest impact in the world's cold climate areas. A warmer climate may not mean less snow if precipitation increases, but it may mean a shorter snow-lying season. This would have serious implications for the skiing industry. Also potentially important will be melting of significant parts of the permafrost. This would result in disruption of the delicate thermal balance of the permafrost areas and result in surface subsidence and erosion of the melted areas (thermokarst subsidence and thermal erosion). In inhabited areas the result would be disruption to buildings and roads, which have generally been carefully engineered to preserve the permafrost but not to take into account climate change. In particular, there is a significant oil industry on the edge of the Arctic Ocean in Alaska and climate change in this area could reduce sea-ice, accelerate coastal erosion and threaten onshore oil installations (Demek, 1994; Bennett and Doyle, 1997).

Rises and falls in relative sea-level can have important effects by changing coastal processes, landforms or exposures. Changes in wave conditions, oceanic circulation and sea-ice cover are also likely. Coastal erosion of beaches and cliffs, flooding of low lying coasts and islands, loss of salt marshes and mudflats and saline water intrusion into freshwater environments are particular impacts of rising sea-level in some parts of the world (French and Spencer, 2001) and these are likely to increase over the next century due to global warming. This is caused by two sets of processes:

- thermal expansion of the upper ocean;
- melting of land ice.

However, the effect of local land movements must also be taken into account since these may either accentuate or reduce the impacts of sea-level rise (Nichols and Leatherman, 1995). Although sea-level change is a natural process that has been continuous since the first oceans were formed over 4 thousand million years ago, sea-level change is currently being largely driven by human-induced global warming (IPCC, 2007).

The Bruun model has been widely used as a tool for predicting shoreline migration on soft coasts under a given sea-level rise, but has been criticised for being overly simplistic in its initial suggestion that coastal retreat will be 100 times greater than the rise in sea-level. Geological and hydrodynamic factors are likely to be important in controlling responses to level rise. On hard coasts, coastal change will be limited and controlled by inland gradient.

Other parts of the environment may also be affected by climate change. For example, Carvalho and Anderson (2001) state that climate affects lake physics (flushing, stratification), chemistry (dissolved oxygen content, pH, nutrients) and biology (species diversity) and any changes to climate are therefore likely to alter many lake characteristics. Finally, soils are vulnerable to global warming, which may accelerate the decomposition of soil organic matter, potentially releasing large quantities of terrestrial carbon dioxide into the atmosphere (Puri, Willson and Woolgar, 2001). In Tasmania, Australia, the soil fungus *Phtopthera* that kills some vegetation types is now spreading rapidly, probably because of higher soil temperatures (C. Sharples, pers. comm.).

5.13 Fire

Fire is both a natural and human-induced process. It can have advantageous effects by releasing nutrients from the ash, but the burning of peats and organic soils has a significant impact. In Tasmania, Australia, for example, research is being carried out on fire frequencies on the blanket bogs in the south-west of the island over the past 6000 years and the implications for organic soil accumulation and fire management. Other impacts result from the destruction of the vegetation layer and subsequent effects. For example, fire spreading into the Mount Field National Park in Tasmania from adjacent forestry land has killed vegetation that formerly stabilised the slopes and landforms. The result has been soil erosion, landslides, burial of stream sinks, increased stream turbidity, sedimentation in limestone caves and track construction as part of the salvage logging operations.

5.14 Military activity

This can have a serious effect on many aspects of the earth science environment, though many military training areas comprise huge tracts of infrequently disturbed country and thus aid conservation. In sensitive areas, even routine operations can have damaging effects. For example, at Dungeness in England, English Nature (1998) has recognised that 'the use of vehicles and explosives have caused great damage to these sensitive features' which include shingle ridges and sand dunes. Similarly, helicopter landings are damaging the periglacially patterned ground in Snowdonia, North Wales. There is also compression and erosion of soils generated by such military training activities. Kiernan (1997a, p. 14) described how 'One of the more bizarre means by which Tasmania's coastal geoheritage has been diminished involved the use by naval ships of the spectacular freestanding dolerite columns on Cape Raoul, Tasman Peninsula—for target practice'.

At times of war, the landscape impacts may be particularly high. The excavation of trenches and tunnels during the First World War has been the subject of much recent research (e.g. Doyle and Bennett, 2002), though this certainly increased our geological knowledge and many of the trenches themselves are now protected for their historical value. Westing and Pfeiffer (1972) calculated that the bombing of Indo-China (Vietnam, Cambodia, Laos) in the 1960s produced 26 million craters displacing 2.6 billion m^3 of soil. This, together with landscape bulldozing to remove vegetation and soil down to the parent material, destroyed at least 10% of the area's agricultural land. As the authors acknowledge (p. 28), 'It has been a war against the land as much as against armies'. Kiernan (2010) describes the impact of the war in the Mekong Delta area of Cambodia and Vietnam, including damage to rock surfaces, soils and caves by explosives and military infrastructure. In northern Laos, Kiernan (2012) describes in detail the wartime damage to karst caves from impacts including trampling and compaction of cave floors, damage to speleothems, introduction of alien materials, changes to cave architecture and other damage.

Modern bombing methods have become even more effective, including the use of cluster bombs, and will create heavy pitting of the landscape that may do serious damage to important geological interests. Pollution of land from depleted uranium used in some modern bombs is a further issue as are deep-penetrating bombs that can destroy cave systems.

5.15 Lack of information/education

A final threat to geodiversity, and some would say the most important of all, is ignorance. A lack of survey information, documentation and designation of geoheritage has resulted in loss or degradation of sites and landscapes through inappropriate development in the past and this remains a threat in many parts of both the developed and developing world.

As a result of this lack of information, geological conservation measures are often lacking, including their integration into important land-use planning legislation, policy and practice. Joyce (1999) notes that in Australia 'Problems arise because of the general lack of geologically-trained staff in National Parks and similar organisations, the common lack of geological input to management plans, and the lack of a high profile for geological heritage'. Similarly, Gray (2001) noted that a major nature conservation initiative in England (English Nature, 1998) lists 28 organisations that were consulted on the content, but not a single geological, geomorphological or geographical organisation appears in the list of consultees. Gray also noted the need for more geomorphological/Quaternary training amongst English Nature's Local Teams. In Northern Ireland, a premier Jurassic fossil site at Garron Point was lost during a road reconstruction scheme in the 1990s when the Roads Department failed to consult the Environment and Heritage Department about its plans (Doughty, 2002). A greater understanding of the planet's geodiversity, its value, threats and the importance of geoconservation and management are important aims of this book.

5.16 Cumulative impacts and sensitivity to change

This brief review has established that there are many real threats to the geodiversity of the planet, mainly related to increasing development and environmental change. Hooke (1994b) calculated that the average annual transport rate of rock and soil by humans is about 42 billion tons per year, equal to about half that transported by rivers, glaciers, oceans and wind combined, and three times that created by mountain building. The loss of cultivatable soil is now taking place at an faster rate than new soil is being created or brought into cultivation (Pimental, 1993).

An important message of this chapter is that earth surface processes are complex and often sensitive to human interference. An important essential is therefore that we understand the implications of change and development. As Boulton (2001, p. 50) puts it, 'Whether we are engineering a dam or a marine wind farm, we are engineering into the environment, and we should seek to understand what the consequences will be'. It also becomes important 'to identify the significance or otherwise of features prior to potential damage, and to assess the likely significance of that damage, in terms of the significance of the feature itself, the extent to which it is likely to be damaged or the nature of the damage inflicted' (Kiernan, 1996, p. 15).

In some places development pressures are high and significant cumulative losses of geodiversity have occurred. A survey of the Peterborough area of the United Kingdom in 1989 recorded 32 important geological sites, but by 1998, 26 had been lost mainly due to landfilling, vegetation growth and restoration works (Peterborough Environment City Trust, 1999). Box 5.6 lists some of the development impacts and losses of geodiversity in Tasmania, Australia, in recent years (Kiernan, 1996; Pemberton, 2001a; Sharples, 2002a).

Box 5.6 Recent geodiversity impact and losses in Tasmania, Australia

- Loss of 28 Tertiary and Quaternary fossil sites destroyed or buried by quarrying, impoundments, landfill or other developments. This represents 50% of such sites identified in the past 100 years;
- Destruction of significant geological sites in road cuttings. These include an eclogite site in the designated Tasmanian Wilderness Area;
- Removal by uncontrolled collecting of valuable or rare silicified tree stumps and Thylacine subfossils from caves;
- Collection of rare or significant speleothems and minerals;
- Erosion of dunes and Aboriginal middens by off-road vehicles and cattle grazing;
- Only three out of over 50 lunette dune features are left undisturbed;
- Inundation of the globally unique landform assemblage (beach and dunes of pinkish-white quartzite sand, sand bar, spectacular sequence of sub-aqueous megaripples, smaller bedform features, lagoon system, ferromanganese concretions) by hydro-electric development at Lake Pedder (Figure 5.13);
- Damage to dunes and moraines at Lake St Clair by hydro-electric works. A road has been carved along the crest of the most prominent dune, sand

has been quarried from its downvalley flank and rock rip-rap has been plastered across its upvalley;

- Erosion by wake from tourist boats of significant fluvial landforms and deposits on the Gordon River;
- Removal by quarrying of a terminal moraine marking the maximum glacial ice advance in the Mersey Valley;
- Damage to the Exit Cave system from quarrying, destruction of magnesite tower karst in the mid-1980s, and degradation of spring mounds by agricultural and residential development;
- Infestation of coastal dunes by marram grass, altering natural processes and removing mobile sand from the system;
- Soil erosion during and following fire, including impacting on peatlands of international significance;
- Covering of important coastal sections by land reclamation and sea-walls.

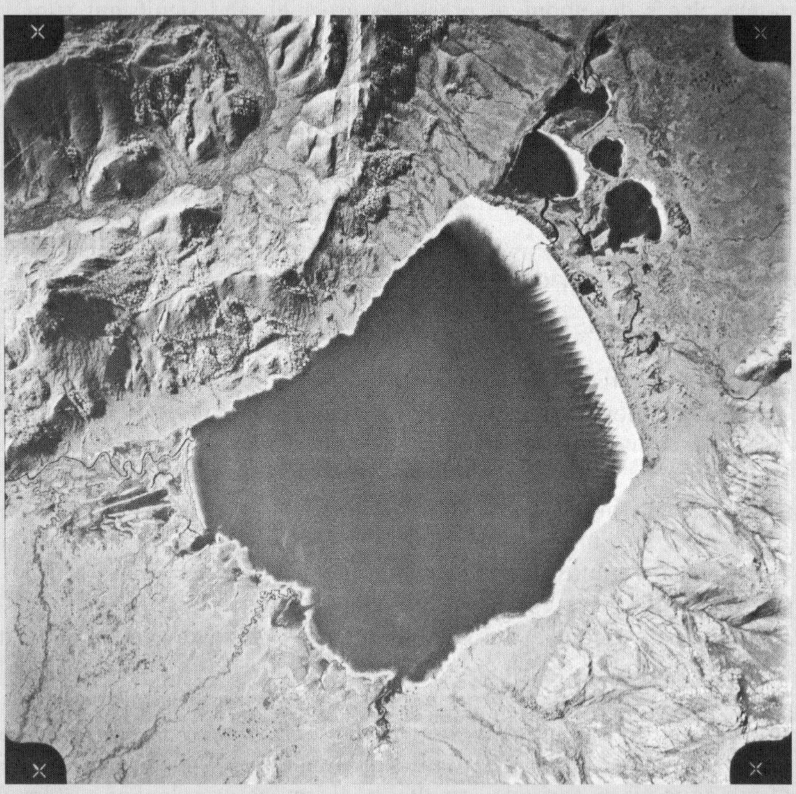

Figure 5.13 Lake Pedder, South West Tasmania, before it was flooded in the 1970s showing the beach, sand bar and megaripples on its north-eastern shore. These and the adjacent glaciofluvial landforms, dunes and blanket bogs were flooded (M. Pemberton, pers. comm.). (Reproduced by permission of Department of Primary Industries, Parks, Water and Environment (DPIPWE), Tasmania.)

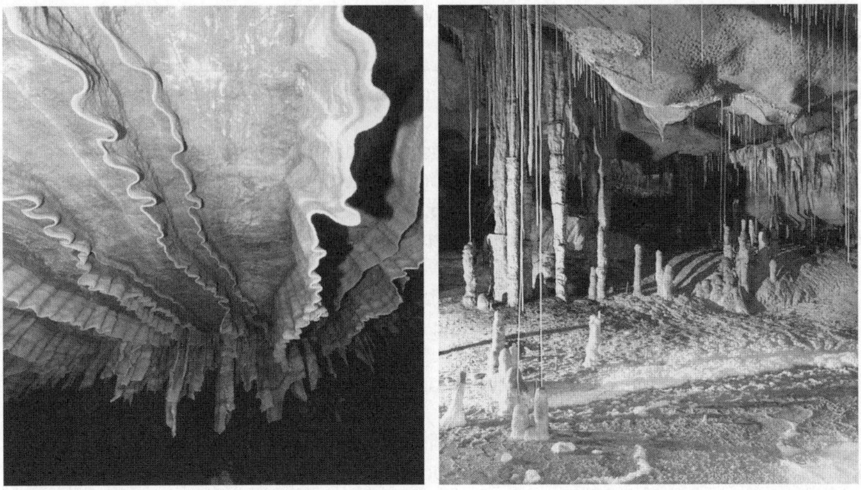

(a) (b)

Figure 5.14 Fragile speleothems in Tasmanian caves, Australia (a) shawl speleothems in the 'Silk Shop' section of Kubla Khan Cave in the Mole Creek karst. Approximately 0.5 m between foreground shawls, (b) straws and stalagmites in the 'Ballroom' section of one of Australia's longest caves, Exit Cave in Tasmania. The longest straws in this view are about 4 m high. (Photos by permission of Ian Houshold.)

Table 5.2 The 5-point Scottish geosensitivity scale (after Werrity and Brazier, 1991).

1.	Activity not generally applicable
2.	Activity obscures or masks the surface and stratigraphic exposures of the landform
3.	Activity causes localised disruption or destruction of part of the landform surface or stratigraphy
4.	Activity causes general disruption or destruction of the surface and/or stratigraphy of the landform
5.	Activity causes destruction of individual or groups of landforms, permanently interrupting morphological and stratigraphic relationships within the landform assemblage

In response to this history of geodiversity degradation, Tasmanian geoconservationists have constructed a 10-point scale of sensitivities indicating the types of activities that might degrade features of a given sensitivity level. In general, it is suggested that a more protective management response is required for highly sensitive features (low on the scale, e.g. Figure 5.14) while those high on the scale will require little or no active conservation management (Kiernan, 1997b; Sharples, 2002a). This scheme is shown in Table 5.3 and follows an earlier, more limited attempt in Scotland (Werritty and Brazier, 1991: see Table 5.2).

Implicit in Table 5.2 and Table 5.3 is the idea that all systems are ultimately sensitive to some level of impact, but some values are robust in the face of impacts that would be highly degrading to other values. For example, the palaeontologically important, Burgess Shale site in the Canadian Rockies is very sensitive to overcollecting, but this damage has little long-term impact on the

Table 5.3 The 10-point Tasmanian geosensitivity scale (after Kiernan, 1997b; Sharples, 2002a).

1. Values sensitive to inadvertent damage simply by diffuse, free ranging human pedestrian passage, even with care.
   Examples: fragile surfaces that may be crushed underfoot, such as calcified plant remains; calcite or gypsum hairs and straws in some karst caves that may be broken by human breath, speech or touch (e.g. see Figure 5.14).
2. Values sensitive to effects of more focussed human pedestrian access even without deliberate disturbance.
   Examples: risk of entrenchment by pedestrian tracks; coastal dune disturbance; drainage changes associated with tracks, leading to gulleying; defacement of speleothems by touching their surfaces.
3. Values sensitive to damage by scientific or hobby collecting or sampling or by deliberate vandalism or theft.
   Examples: some fossil and mineral collecting, rock coring and speleothem sampling.
4. Values sensitive to damage by remote processes.
   Examples: degradation of geomorphic or soil processes by hydrological or water quality changes associated with clearing or disturbance within the catchment; fracture/vibration due to blasting in adjacent areas (e.g. stalactites in caves).
5. Values sensitive to damage by higher intensity linear impacts, depending upon their precise position.
   Examples: vehicle tracks, minor road construction, or excavation of ditches and trenches.
6. Values sensitive to higher intensity but shallow generalised disturbance on site, either by addition or removal of material.
   Examples: clearfelling of forests and replanting, but without stump removal or major earthworks; land degradation such as soil erosion due to bad management practices; vegetation/weathering of exposures.
7. Values sensitive to deliberate linear or generalised shallow excavation.
   Examples: minor building projects, road construction, shallow borrow pits, removal of stumps, construction of small bunds.
8. Values sensitive to major removal or addition of geomaterial.
   Examples: quarrying; dam construction, landraising.
9. Values sensitive only to very large-scale contour change.
   Examples: large opencast quarries; major reservoirs causing inundation; major river channelisation schemes.
10. Values sensitive only to catastrophic events.
    Examples: meteorite impacts; human-induced sea-level change; major landslides and tsunamis.

landscape of Yoho National Park in which it is located. Werritty and Brazier (1994) illustrated the distinction between robust and sensitive behaviour (see Figure 5.15) and applied this to the fluvial geomorphology of the River Feshie.

Sensitivity thus refers to susceptibility to change/damage. Vulnerability, on the other hand refers to the likelihood of damage because of actual or potential human threats. Geoconservation should focus on protecting important interests that are both sensitive and vulnerable.

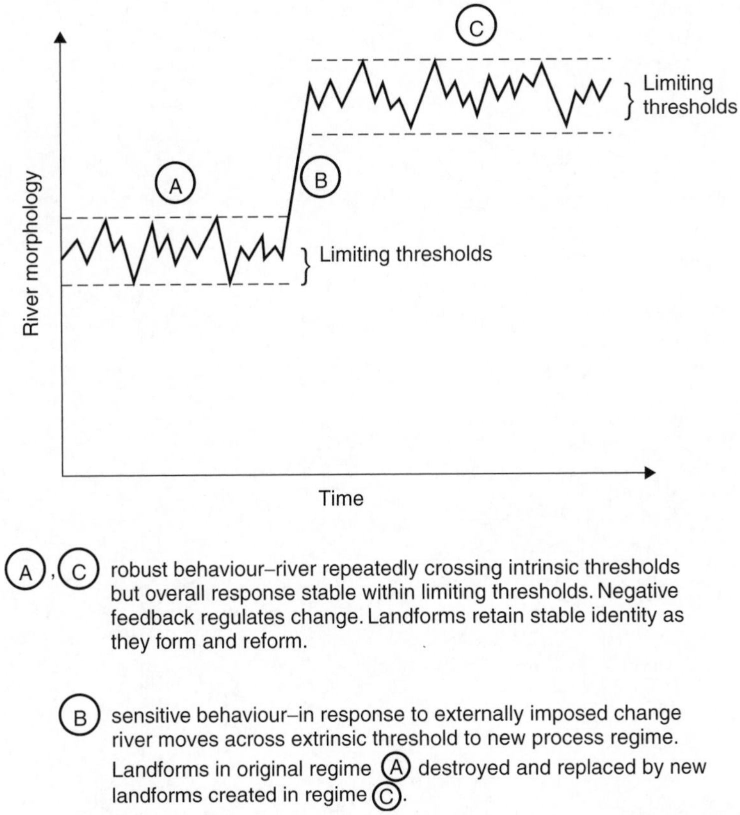

C
Limiting thresholds
A
B
Limiting thresholds
River morphology
Time
A, C robust behaviour–river repeatedly crossing intrinsic thresholds but overall response stable within limiting thresholds. Negative feedback regulates change. Landforms retain stable identity as they form and reform.
B sensitive behaviour–in response to externally imposed change river moves across extrinsic threshold to new process regime. Landforms in original regime A destroyed and replaced by new landforms created in regime C.

Figure 5.15 The distinction between robust and sensitive behaviour in geomorphological river systems (after Werrity, A. and Brazier, V. (1994) Geomorphic sensitivity and the conservation of fluvial geomorphology SSSIs. In Stevens, C., Gordon, J.E., Green, C.P. and Macklin, M.G. (eds) *Conserving our Landscape. English Nature*, Peterborough, 100–109. Reproduced with permission of Natural England).

5.17 Conclusions

In this chapter I have outlined the main threats to geodiversity and noted that these threats may apply to specific important sites or to the wider landscape. We have seen that the cumulative impacts can be considerable and that the threat will depend on the sensitivity of systems and the values placed upon them. Development is of course essential to the improved prosperity of human populations and the essentials of life, but there is a need to balance these against the irreplaceable nature of much of our geoheritage. It is often possible to develop less sensitive sites or to manage the land in less damaging ways. It is only by understanding the value of geodiversity and threats to it that this balance can be achieved.

Part III

Geoconservation: the 'Protected Area' Approach

6 International Geoconservation: an Introduction

Landscape and memory combine to tell us certain places are special, sanctified by their extraordinary natural merits and by social consensus.

Chris Johns (2006)

The start of this chapter represents an important point in the book. Previous chapters have defined, described and valued geodiversity and established that there are significant threats to it. The intention has been to establish that there is a case for geoconservation. Parts III and IV of the book follow directly from Part II, since conservation is an appropriate human response to perceived threats to features regarded as having value. These next seven chapters can therefore be regarded as the most important in the book since it is the embodiment of an environmentally sophisticated society that it acts to avoid losses and protect what is valuable but vulnerable to damage. These chapters therefore move us on to consider how geoconservation has been or can be implemented. In Part III, the current chapter describes the ways in which international organisations attempt to protect areas of geological or geomorphological interest. Chapter 7 and Chapter 8 then describe two specific programmes supported by UNESCO that are used to give significant status, though not in themselves protection, to important geological sites and areas, viz. World Heritage Sites and Global Geoparks. Chapter 9 then moves on to national and local geoconservation schemes.

Geodiversity: Valuing and Conserving Abiotic Nature, Second Edition. Murray Gray.

6.1 Beginnings of the conservation movement in North America

Conservation probably has a long history, including a Forest Ordinance introduced in France in 1669 and the Forest Acts passed in Denmark in 1763 and 1805 (Peterson del Mar, 2012). But the most important organised activity supported by government was probably initiated in the United States as a response to overhunting, overgrazing and soil erosion (Dasmann, 1984). In the 1830s, George Catlin, an artist and naturalist, expressed his concern for the future of the buffalo and the well-being of the Plains Indians who depended on it and who had been careful stewards of its numbers and its rangelands for centuries (www.georgecatlin.org). He was perhaps the first to propose a huge 'national park' across the Great Plains where Indians and wildlife could be left in peace to pursue their traditional way of life. However, the proposal was not taken up and it was another 20 years before the issue raised its head again.

In the 1850s, Henry David Thoreau wrote of his concern for the future of all wild nature and a hope that it could be preserved (cited in Richardson, 1986). He was also wise enough to realise that people have an important relationship with, and obligation to, 'the more sacred laws' of nature, but again few paid much attention. In 1864, George Perkins Marsh, in his book *Man and Nature*, was the first to attempt a systematic description of the human impacts on the natural environment, and formulated the view that man must live with nature. At the same time, urban dwellers came to appreciate the value of city parks and several early environmentalists like John Muir, a Scottish immigrant, argued strongly for the preservation of nature (e.g. Muir, 1901). Their arguments were successful and as a result the Yosemite Valley, California became a protected area in 1864. As the U.S. had no mechanism for managing parks at a national level at that time, it was deeded to the State of California (though it was later incorporated in the Yosemite National Park). This was followed in 1872 by the world's first designated national park at Yellowstone, United States. The growth of the national park network was slow at first—Mackinac Island (1875), Sequioa, Yosemite and General Grant (1890), Mount Ranier (1899) and Crater Lake (1902). Canada's first national park at Banff was established in 1885. Many of the early National Parks were established because of their scenic or geological values (e.g. Yellowstone (see Box 6.1), Yosemite, Mount Ranier, Crater Lake, Banff) though this fact seems to have been lost in the subsequent overwhelming emphasis on wildlife conservation, and needs to be rediscovered.

Box 6.1 Yellowstone National Park, United States

After European colonisation of the east part of North America, the first explorers and travellers to the westernmost reaches began to return with curious tales of the unearthly natural wonders of Yellowstone, particularly of its steaming ground and hot springs. But it was the Washburn–Langford–Doane expedition of 1870 that really awakened public interest in Yellowstone, though there has been much debate as to whether the motives were truly altruistic or, to a greater or lesser extent, commercially driven (Sellars, 1997). It was this party that, on emerging from the forest into what is now known as the Upper Geyser Basin were met with 'an immense body of sparkling water, projected suddenly and with terrific force into the air to the height of over one hundred feet. We had found a real geyser'. And since it repeated its spectacular show at regular intervals they named it 'Old Faithful' (Langford, 1870).

The results of the Washburn expedition were promoted in a series of public lectures and articles by one of the participants, Nathaniel Langford. In the audience at one of these in Washington, DC, was the Director of the US Geological Survey, Ferdinand V. Hayden, whose interest was immediately aroused and who managed to obtain congressional funding to survey the area in the summer of 1871. On his return to Washington, Hayden received the suggestion from Judge William Keeley of Philadelphia that 'Congress pass a bill reserving the Great Geyser Basin as a public park forever' (Schullery, 1999). Hayden was persuaded to take up the cause and the Yellowstone National Park Act (1872) was signed by President Ulysses S. Grant on 1 March 1872. The park covered nearly one million hectares of public land provided for 'the preservation from injury or spoilation, of all timber, mineral deposits, natural curiosities, or wonders within said park, and their retention in their natural condition'. The land was 'reserved and withdrawn from settlement, occupancy, or sale under the laws of the United States, and dedicated and set apart as a public park or pleasuring ground for the benefit and enjoyment of the people'.

The protected area was drawn by Hayden to include all the geothermal features known at the time and the main reason for establishing the park was to protect these geological wonders (Schullery, 1999). It is estimated that there are in the order of 10 000 individual geothermal features in the park, including 200–250 active geysers, though with concentrations into groups (Figure 6.1).

Figure 6.1 Steam emerging from several geysers and vents in the Upper Geyser Basin, Yellowstone National Park, United States.

This makes Yellowstone one of the greatest concentrations of geothermal activity on the planet. It also contains the impressive falls and canyon of the Yellowstone River and one of the world's largest calderas measuring 75 by 45 km (Figure 6.2). However, having protected the area for its geology this also effectively protected the wildlife within it, and over the years this became the main focus for nature conservation efforts and controversies (Pritchard, 1999; Schullery, 1999).

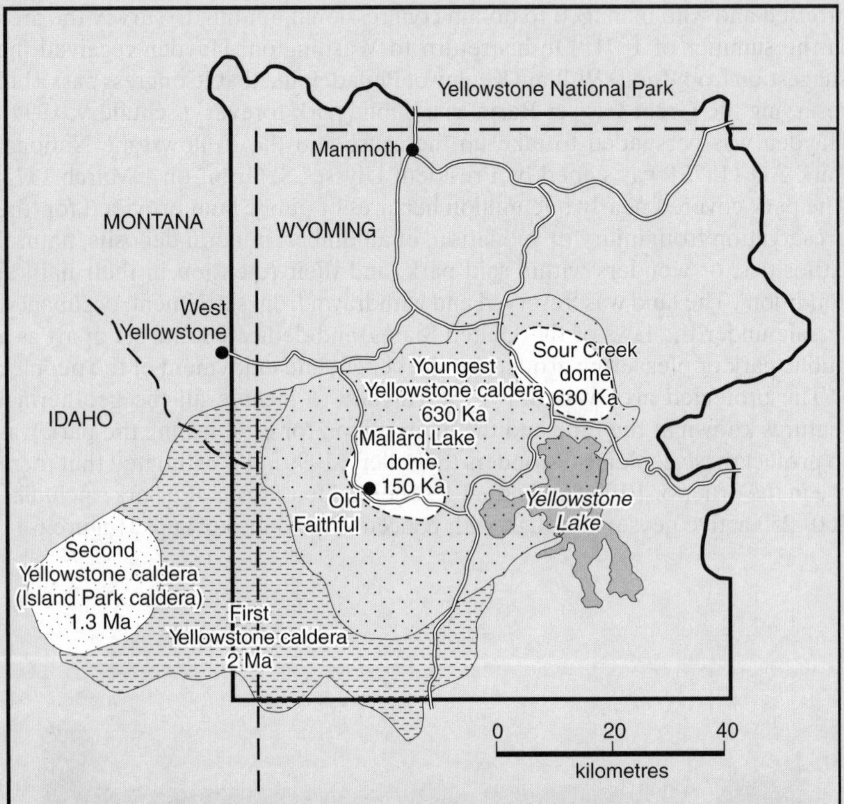

Figure 6.2 Calderas of Yellowstone National Park, United States (after Smith, R.B. and Siegel, L.J. (2000) *Windows into the Earth: the Geological Story of Yellowstone and Grand Teton National Parks*. By permission of Oxford University Press, Inc.).

Since established, the park boundaries in the east have been extended and realigned to conform with the watersheds, it being recognised that rivers flowing into the park ought to be protected from exploitation and pollution. The greatest threat to Yellowstone's geothermal activity lies only 40 km to the west of Old Faithful where, beyond the park boundaries in southern Idaho, lies the Island Park Known Geothermal Resource Area (IPKGRA) (Figure 6.2). It is known that the centre of igneous activity over the past 16.5 million years has shifted eastwards from southern Oregon through southern Idaho to north-west Wyoming as the American tectonic plate has drifted westwards above a volcanic hot spot at a rate of circa 2.5 cm/year (Smith and Siegel, 2000). It is therefore possible that the geothermal activity related to these igneous centres are connected at depth and that any exploitation of the IPKGRA could depressurise the Yellowstone geothermal system with catastrophic results for Yellowstone's geothermal activity. Clearly, more research is needed to understand the underground geothermal interconnectivities of the region (Smith, 2000).

In 1892, Muir founded the famous Sierra Club in California and it is still an influential non-governmental organisation (NGO) over a century later. But it was not until 1908 that US President Theodore Roosevelt used the name 'conservation' to describe the activities and the movement (Dasmann, 1984). Roosevelt's home was in New York but he went to North Dakota in 1883 and decided to develop his interest in cattle ranching. He was brought to the west by the prospect of big game hunting, but when he arrived, the last large buffalo herds were gone as a result of overhunting and disease. Throughout his time in North Dakota Roosevelt became more and more concerned about the disappearance of some species. Overgrazing destroyed the grasslands and with them the habitats for small mammals and songbirds. Roosevelt therefore developed a keen interest in conservation and what would nowadays be called sustainable land use. During his presidency (1901–1909) he established the US Forest Service and set aside land as national forests, signed the Antiquities Act (1906) under which he proclaimed 18 national monuments, and obtained Congressional approval for five national parks and 55 wildlife refuges. There is a memorial to him on Theodore Roosevelt Island, Washington, DC, and a national park in North Dakota named after this major American conservationist whose interest in politics allowed him to implement many of his ideas.

The work he started continued after his presidency had ended. On 25 August 1916 the Federal Government assumed responsibility for the ownership and management of the 40 national parks of the time through the National Park Service Organic Act (1916), the objective of which was to 'conserve the scenery, natural and historic objects and the wild life therein, and to provide for the enjoyment for the same in such a manner and by such means as will leave them unimpaired for the enjoyment of future generations'.

Following the disaster of the mid-west Dust Bowl (see Box 5.5), President Franklin D. Roosevelt established the US Soil Erosion Service in 1933 later becoming the US Soil Conservation Service. Dasmann (1984, p. 9) believes that it was not until the later decades of the twenty first century that environmental concerns amongst the general public led to more active pursuit of conservation goals and establishment of a political environmental agenda. The Alaska Lands Bill (1980) was a major milestone (Cutter and Renwick, 1999), but as the recent debates over further oil exploration in Alaska's National Wildlife Refuge indicates, tensions still exist in American society between economic growth and natural resource exploitation on the one hand and nature conservation on the other.

6.2 Early British experience

In Britain, the Romantic poets of the late eighteenth and early nineteenth centuries, such as William Wordsworth, had extolled the virtues of the Lake District's landscape and talked about it becoming 'a sort of national property, in which every man has a right and interest who has an eye to perceive and a heart to enjoy' (Wordsworth, 1952, p. 127). Although there is a democratising theme here, in seeing areas of beauty as a national resource, there is also an exclusionary tone, since 'if you do not have the necessary perceptual eye and emotional heart, then by implication you have no rights or interests with regard to the Lake District' (Phillips and Mighall, 2000, p. 325). Wordsworth concerns were undoubtedly driven by a fear that the arrival of 'the masses' in areas like the Lake District would ruin their 'natural character'.

> 'A vivid perception of romantic scenery is neither inherent in mankind nor a necessary consequence of even a comprehensive education ... Rocks and mountains, torrents and wild spread waters ... cannot in their finer relations to the human mind, be comprehended without opportunities of culture in some degree habitual ...' (William Wordsworth, quoted in Glypteris, 1991, p. 27)

Those without the necessary insight and emotions would be better off 'taking little excursions with their wives and children among the neighbouring fields within reach of their own dwellings'! Ignoring the snobbish and sexist implications of these remarks, it is interesting to note that Wordsworth's perception of the essence of the Lake District landscape was of its abiotic elements—rocks, mountains and water.

Other writers also expressed distaste for the tourist invasion. James Payne (quoted in Glypteris, 1991, p. 27) was disdainful that a steam boat 'disgorges multitudes upon the pier; the excursions trains bring thousands of curious vulgar people ... our hills are darkened by swarms of tourists'. These fears eventually led to the establishment, in 1883, of the Lake District Defence Society expressly set up to oppose, successfully as it turned out, the extension of the railway line into Borrowdale. This organisation subsequently was incorporated into the National Trust, or to give its full title, The National Trust for Places of Historic Interest and Natural Beauty. The National Trust is one of a very large number of conservation organisations formed in the United Kingdom in the late nineteenth and the early twentieth centuries.

In the 1940s, Alfred Steers carried out a survey of the entire coastline of England and Wales and concluded that he could 'not emphasise too strongly that if we as a nation wish to preserve one of our finest heritages for the good of the people as a whole, we must act now and act vigorously on a national scale' (Steers, 1946). It was not until 1949 that the National Parks and Access to the Countryside Act allowed the establishment of National Parks in England and Wales, though Scotland had to wait until 2002 for its first National Park and Northern Ireland is still waiting. However, even these National Park are very different to the American and international model (see Section 9.4). An interesting review of the history of Nature conservation in the United Kingdom is given by Adams (1996) but his conclusion (p. 99) is that 'Conservation in the UK has grown up without a coherent philosophy, a cultural and scientific rag-bag of passion, insight and good intentions'.

6.3 The 'Protected Area' and legislative approaches

McKenzie (1994, p. 127) commented that 'Rarely, are Earth science features and processes effectively protected, conserved or managed successfully on a site-by-site basis without some supporting institutional framework'. As the national park system exemplifies, the traditional way in which areas for conservation have been identified and defined is by drawing boundaries on a map and then giving particular status to the land, wildlife or features within these boundaries that does not apply outside the boundaries. This 'protected area' approach is identifiable throughout the world though it takes many different forms in different countries as applied to different types of area. Since animal life is not confined by such boundaries, a second traditional approach has been to proscribe the taking, killing or trading of named species or animal products (see later). Thirdly, the removal or trading in historical or archaeological artefacts has been controlled by legislation and this has sometimes been extended to palaeontological objects.

6.4 The UN

The United Nations (UN) is heavily involved in international conservation mainly through its Educational, Scientific and Cultural Organisation (UNESCO) and its Environment Programme (UNEP). The latter has a World Conservation Monitoring Centre (WCMC) based in Cambridge, England, which has a Protected Areas Programme aimed at establishing a Nationally Designated Protected Areas Database. However, it is almost exclusively concerned with biodiversity and ecosystems with a Vision of 'A world where biodiversity counts' and a Mission 'To evaluate and highlight the many values of biodiversity and put authoritative biodiversity knowledge at the centre of decision-making'. On the other hand, UNESCO has a World Heritage Sites network that includes geological sites (Chapter 7) and it supports the Global Geoparks programme (Chapter 8). The remainder of this chapter will outline some of the other current or potential international programmes relevant to geoconservation.

6.5 The IUCN

The International Union for the Conservation of Nature and Natural Resources (IUCN) was founded in 1948. It's Vision is 'a just world that values and conserves nature' and its Mission is 'to influence, encourage and assist societies throughout the world to conserve the integrity and diversity of nature and to ensure that any use of natural resources is equitable and ecologically sustainable' (IUCN website, www.iucn.org). It was the first global environmental organisation and today is the largest professional conservation network comprising more than 1,000 member organisations in 140 countries including over 200 governmental and over 800 NGOs. It employs more than 1,000 professional staff in 60 offices worldwide and has almost 11 000 voluntary scientists and experts grouped into six Commissions. Its headquarters are in Gland, near Geneva, Switzerland. It is the official technical advisory body to the World Heritage Committee on natural heritage (see Chapter 7). In 1980 it published the World Conservation Strategy where the term 'sustainable development' was first widely publicised (Barrow, 1999).

The IUCN has recognised that protected areas come in many different forms in different countries, over 140 names for protected areas being used around the world. It has therefore devised a list of categories (last fully revised in1993 and reviewed in 2008; Dudley, 2008) to allow international comparisons (Table 6.1). At the head of this list are the Strict Nature Reserves, which are areas of land and/or sea 'possessing some outstanding or representative ecosystem, geological or physiographical features and/or species, available primarily for scientific research and/or environmental monitoring', and Wilderness Areas, which are 'large areas of unmodified or slightly modified land and/or sea, retaining its natural character and influence without permanent or significant habitation, which is protected and managed so as to preserve its natural condition'. In these areas, natural processes are allowed to operate without direct human interference, and tourism, recreation and public access are restricted. Ownership and control should be by the national or other level of government. An exception is Antarctica that is governed by International agreement (see Box 6.3). Earlier ideas of keeping these areas inviolate proved to be a somewhat outdated concept since in many cases it excluded indigenous populations with long histories of living off the land and doing so sustainably. However, attitudes have now shifted with the recognition that 'conservation cannot be achieved simply by preservation, but will have to

Table 6.1 The IUCN protected area categories.

I	Strict Nature Reserve/Wilderness Area
	Ia Strict Nature Reserve
	Ib Wilderness Area
II	National Park
III	Natural Monument
IV	Habitat/Species Management Area
V	Protected Landscape/Seascape
VI	Managed Resource Protected Area

acknowledge the values, needs and aspirations of local people' (Mather and Chapman, 1995, p. 129). Thus one of the management objectives for Wilderness Areas is 'to enable indigenous human communities living at low density and in balance with the available resources to maintain their life style'.

The World Commission on Protected Areas (WCPA) is an IUCN Commission that works to help governments and others to plan, manage, strengthen and enhance the world network of protected areas (see Section 9.1). It strongly recommends management planning of protected areas and has produced guidelines for the production of management plans (Thomas and Middleton, 2003).

The IUCN defines a protected area as 'a clearly defined geographical space, recognised, dedicated and managed through legal or other effective means, to achieve the long term conservation of nature with associated ecosystem services and cultural values' (Dudley, 2008). In this context, nature 'always refers to biodiversity, at genetic, species and ecosystem level, and often also refers to geodiversity, landform and broader natural values'. The implication of these definitions is that nature and protected areas cannot be purely abiotic. The IUCN was criticised by Brilha (2002) for its lack of geological programmes. Although it is true that most are biological and little reference is made to geoconservation in any of its documents or sites, one of the networks does involve a Caves and Karst Specialist Group (formerly Task Group) which has prepared and published guidelines for the management of these sensitive systems. The first chapter of the 800-page book *Managing Protected Areas: a global guide*, edited by Lockwood, Worboys and Kothari (2006) and conceived at the 5th IUCN World Parks Congress held in Durban, South Africa, in 2003, has a short section on the physical environment including geodiversity which it describes as 'important to society' (Worboys and Winkler, 2006), though most of the rest of the book is devoted to managing biological protected areas.

Probably because of the limited work on geoconservation within the IUCN, the Geological Society of Spain proposed a motion (CGR4.MOT055) at the 4th Session of the IUCN's World Conservation Congress in Barcelona, Spain in October 2008. It called on IUCN members and Commissions, especially the WCPA, 'to support the Secretariat in the design, organisation, hosting and funding of future sessions on Geodiversity and Geological Heritage to ensure that this mechanism will achieve the widest possible involvement of government, independent sector groups and international organisations around the world'. It also requested the Director General to 'convene a continuing series of meetings on Geodiversity and Geological Heritage in the regions in partnership with members and other organisations; and establish a Secretariat focal point to facilitate the organization of these meetings and to provide their continuity'. The motion was supported by the Congress, but has subsequently had little impact due to lack of funding. Further discussion took place at the 5th IUCN meeting in South Korea in 2012 where Motion M056 was passed (Diaz-Martinez, 2012) including the following key operating statements:

- to ensure that, when reference is made in the IUCN Programme 2013–2016 to nature in general, preference be given to inclusive terms like nature, natural diversity or natural heritage, so that geodiversity and geoheritage are not excluded;

- to promote and support local socioeconomic development initiatives, such as UNESCO Geoparks, based on the sustainable use of geoheritage, including the proper management of geoheritage in protected areas.

6.6 Geosites

A Global Indicative List of Geological Sites (GILGES) was established in the early 1990s by UNESCO, the IUCN and the International Union of Geological Sciences (IUGS). The list included hundreds of sites that were intended to be of 'first-class importance to global geology ... outstanding examples representing major stages of the Earth's history, significant ongoing geological processes in the development of landforms, such as volcanic eruption, erosion, sedimentation, etc., or significant geomorphic or physiographic features, for example, volcanoes, fault scarps or inselbergs' (Cowie and Wimbledon, 1994). Several problems arose from the establishment of this list. One was the range in size of the sites from huge national parks to metre-sized fossil localities (Cowie and Wimbledon, 1994). But the most serious problems came in judging the geological value of sites and in achieving consistency between countries (Cleal *et al.*, 2001).

Thus in 1995, the IUGS replaced GILGES with a more rigorous and comprehensive scheme known as Global Geosites and this was subsequently endorsed by UNESCO. The aim was to compile a global list, with supporting documentation, of the world's most important geological sites. The initial work was co-ordinated by the IUGS's Global Geosites Working Group (GGWG) and the list was stored as a computer database at the IUGS Secretariat in Trondheim, Norway. The aim was a 'bottom-up approach' with geoscientists in all countries being encouraged to compile their own registers (where they did not already exist), which can then be scrutinised by the wider geological community. The end result was 'not to search for token "best sites": it is to identify natural networks of sites that represent geodiversity' (Cleal *et al.*, 2001, p. 10). An example of geosite work in Kazakhstan was described by Nusipov, Fishman and Kazakowa (2001) and in Spain by Garcia-Cortés (2009). Although the global range and aims of the scheme were abandoned by the IUGS in 2006, there is still some multinational collaboration occurring in Europe co-ordinated by ProGEO, the European Association for the Conservation of the Geological Heritage (see Section 6.11).

6.7 Geomorphosites

This is an initiative of the International Association of Geomorphologists (IAG) which established a Geomorphosites Working Group at their 5th International Conference held in Tokyo in 2001. The aim is to define, assess, protect and promote geomorphosites as 'landforms of interest that appeared specifically valuable in terms of natural heritage' (Reynard and Coratza, 2007). Several journal special issues have been devoted to geomorphosites (e.g. Piacente and Coratza, 2005; Reynard and Coratza, 2007, 2011) and a book has been published (Reynard, Coratza and Regolini-Bissig, 2009). These all provide useful case studies of geomorphological conservation, though the emphasis tends to be on landforms with few examples of geomorphological process sites. Examples of the application of geomorphosite research include Garavaglia and Pelfini (2011) and Coratza *et al.* (2011).

6.8 GSSPs

The International Commission on Stratigraphy (ICS), a commission of the IUGS, has a programme to reach international agreement and definition of all the main stratigraphic boundaries within the geological timescale. The result will be a global network of over 100 Global Boundary Stratotype Sections and Points (GSSPs). The history, philosophy and application of the concept of GSSPs was reviewed by Walsh, Gradstein and Ogg (2004). The programme commenced in 1977 and is still ongoing with full documentation of ratified sites being published mainly in the IUGS journal *Episodes* (http://www.episodes.co.in/www/backissues/312/312.htm).

The requirements for GSSP status were outlined by Remane *et al.* (1996) and have subsequently been amended by the ICS. They include:

- stratigraphic completeness across the GSSP level;
- adequate thickness of section above and below;
- continuous sedimentation;
- absence of synsedimentary, tectonic or metamorphic disturbance;
- abundance and diversity of well-preserved fossils;
- support from magnetostratigraphy, chemostratigraphy and dating to increase the possibilities of global correlatability;
- accessibility, including logistics, national politics and property rights;
- provisions for conservation and protection.

This last point is discussed here.

Table 6.2 gives a full list of the 115 stratigraphic boundaries in the geological column. Twelve of these in the Precambrian are defined chronometrically (Plumb, 1991) and will remain so, and thus have no terrestrial rock record that may be lost or damaged. That leaves 103 GSSPs to be defined on the basis of lithostratigraphy, but this includes the Pleistocene/Holocene boundary in the North GRIP ice core in Greenland archived, in the University of Copenhagen, Denmark. Thus there are 102 GSSPs that are defined, or probably will be defined, on the basis of field outcrops. Of these, 64 had been ratified up to 1 January 2013 (Subcommission for Stratigraphic Information website, https://engineering.purdue.edu/Stratigraphy/gssp/index.php?parentid=all). Of the remaining 38, two have ratified temporary chronostratigraphic boundaries that will be replaced eventually by GSSPs. For the other 36, agreement has yet to be reached though in most cases there are shortlists of one or more candidate GSSPs. For example, there are three candidate GSSPs (in Poland, United States and Germany) for the base of the Coniacian. The most important sites (Series

Table 6.2 Summary of the GSSP table (from Subcommission for Stratigraphic Information; https://engineering.purdue.edu/Stratigraphy/gssp/index.php?parentid=all).

System	Series	Stage	Age (Ma)	Location	Status	Reference (vol./issue)
	Holocene	**Holocene**	**0.0118**	**North GRIP ice core, Greenland**	**Ratified 2008**	***Episodes* 31/2**
		Tarantian	0.126	Amsterdam-Terminal borehole, Netherlands	Accepted by ICS (2008); on hold by IUGS	*Episodes* 31/2
		Ionian	0.781	Awaited		
		Calabrian	1.806	Vrica, Italy	Ratified 1985	*Episodes* 8/2
Quaternary	**Pleistocene**	**Gelasian**	**2.588**	**Monte San Nicola, Sicily, Italy**	**Ratified 1996**	***Episodes* 21/2**
		Piacenzian	3.6	Punta Picolo, Sicily, Italy	Ratified 1997	*Episodes* 21/2
	Pliocene	**Zanclean**	**5.333**	**Eraclea Minoa, Sicily, Italy**	**Ratified 2000**	***Episodes* 23/3**
		Messinian	7.246	Oued Akrech, Morocco	Ratified 2000	*Episodes* 23/3
		Tortonian	11.63	Monte dei Corvi Beach, Italy	Ratified 2003	*Episodes* 28/1
		Serravallian	13.82	Fromm Ir-Rih Bay, Malta	Ratified 2007	*Episodes* 32/3
		Langhian	15.97	Awaited		
		Burdigalian	20.44	Awaited		
Neogene	**Miocene**	**Aqitanian**	**23.03**	**Lemme-Carrioso, Italy**	**Ratified 1996**	***Episodes* 20/1**
		Chattian	28.1	Awaited		
	Oligocene	**Rupelian**	**33.9**	**Massignano, Italy**	**Ratified 1992**	***Episodes* 16/3**
		Priabonian	37.8+/−0.5	Awaited		
		Bartonian	41.2+/−0.5	Awaited		
		Lutetian	47.8+/−0.2	Awaited	Ratified 2011	*Episodes* 34/4
	Eocene	**Ypresian**	**56**	**Dababiya, Egypt**	**Ratified 2003**	***Episodes* 30/4**

		Thanetian	59.2	Zumaia, Spain	Ratified 2008	*Episodes* 34/4
		Selandian	61.1	Zumaia, Spain	Ratified 2008	*Episodes* 34/4
Palaeogene	**Palaeocene**	**Danian**	**66+/−0.05**	**Oued Djerfane, Tunisia**	**Ratified 1991**	***Episodes* 29/4**
		Maastrichtian	72.1+/−0.2	Tercis les Bains, France	Ratified 2001	*Episodes* 24/4
		Campanian	83.6+/−0.2	Awaited		
		Santonian	86.3+/−0.5	Awaited		
		Coniacian	89.8+/−0.3	Awaited		
		Turonian	93.9+/−0.2	Pueblo, Colorado, USA	Ratified 2003	*Episodes* 28/2
	Upper	Cenomanian	100.5+/−0.4	Mount Risou, France	Ratified 2002	*Episodes* 27/1
		Albian	113+/−0.4	Awaited		
		Aptian	126.3+/−0.4	Awaited		
		Barremian	130.8+/−0.5	Awaited		
		Hauterivian	133.9+/−0.6	Awaited		
		Valanginian	139.4+/−0.7	Awaited		
Cretaceous	**Lower**	**Berriasian**	**145+/−0.8**	**Awaited**		
		Tithonian	152.1+/−0.9	Awaited		
		Kimmeridgian	157.3+/−1	Awaited		
	Upper	Oxfordian	163.5+/−1.1	Awaited		
		Callovian	166.1+/−1.2	Awaited		
		Bathonian	168.3+/−1.3	Ravin du Bes, France	Ratified 2008	*Episodes* 32/4
		Bajocian	170.3+/−1.4	Cabo Modego, Portugal	Ratified 1996	*Episodes* 20/1
	Middle	Aalenian	174.1+/−1	Fuentelsaz, Spain	Ratified 2000	*Episodes* 24/3
		Toarcian	182.7+/−0.7	Awaited		
		Pliensbachian	190.8+/−1	Robin Hood's Bay, UK	Ratified 2005	*Episodes* 29/2
		Sinemurian	199.3+/−0.3	East Quantoxhead, UK	Ratified 2000	*Episodes* 25/1
Jurassic	**Lower**	**Hettangian**	**201.3+/−0.2**	**Kuhjoch section, Tyrol, Austria**	**Ratified 2010**	

(*continued overleaf*)

Table 6.2 (*continued*)

System	Series	Stage	Age (Ma)	Location	Status	Reference (vol./issue)
		Rhaetian	209.5+/−1.5	Awaited		
		Norian	228.4	Awaited		
	Upper	Carnian	237+/−1	Prati di Stuores, Italy	Ratified 2008	*Albertiana* 36
		Ladinian	241.5+/−1	Bagolino, Italy	Ratified 2005	*Episodes* 28/4
	Middle	Anisian	247.1+/−0.2	Awaited		*Albertiana* 36
		Olenekian	250+/−0.5	Awaited		*Albertiana* 36
Triassic	**Lower**	**Induan**	**252.2+/−0.5**	**Meishan, China**	**Ratified 2001**	***Episodes* 24/2**
		Changhsingian	254.2+/−0.3	Meishan, China	Ratified 2005	*Episodes* 29/3
	Lopingian	Wuchiapingian	259.8+/−0.4	Penglaitan, China	Ratified 2004	*Episodes* 29/4
		Capitanian	265.1+/−0.4	Nipple Hill, Texas, USA	Ratified 2001	
		Wordian	268.8+/−0.5	Guadalupe Pass, Texas, USA	Ratified 2001	
	Guadalupian	Roadian	272.3+/−0.5	Stratotype Canyon, Texas, USA	Ratified 2001	
		Kungurian	279.3+/−0.6	Awaited		
		Artinskian	290.1+/−0.2	Awaited		
		Sakmarian	295.5+/−0.4	Awaited		
Permian	**Cisuralian**	**Asselian**	**298.9+/−0.2**	**Aidaralash Creek, Kazakhstan**	**Ratified 1996**	***Episodes* 21/1**

	Upper Pennsylvanian	Gzhelian	303.7+/−0.1	Awaited		
		Kasimovian	307+/−0.2	Awaited		
	Middle Pennsylvanian	Moscovian	315.2+/−0.2	Awaited		
	Lower Pennsylvanian	Bashkirian	323.2+/−0.4	Arrow Canyon, Nevada, USA	Ratified 1996	*Episodes* 22/4
	Upper Mississippian	Serpukhovian	330.9+/−0.3	Awaited		
	Middle Mississippian	Visean	346.7+/−0.4	Pengchong, China	Ratified 2008	
Carboniferous	**Lower Mississippian**	**Tournaisian**	**358.9+/−0.4**	**La Serre, France**	**Ratified 1990**	***Episodes* 14/4**
		Farmennian	372.2+/−1.6	Coumiac Quarry, France	Ratified 1993	*Episodes* 16/4
	Upper	Frasnian	382.7+/−1.6	Col du Puech de la Suque, France	Ratified 1986	*Episodes* 10/2
		Givetian	387.7+/−0.8	Jebel Mech Irdane, Morocco	Ratified 1994	*Episodes* 18/3
	Middle	Eifelian	393.3+/−1.2	Wetteldorf, Germany	Ratified 1985	*Episodes* 8/2
		Emsian	407.6+/−2.6	Zinzil'ban Gorge, Uzbekistan	Ratified 1995	*Episodes* 20/4
		Pragian	410.8+/−2.8	Velká Chuchle, Czech Republic	Ratified 1989	*Episodes 12/2*
Devonian	**Lower**	**Lochkovian**	**419.2+/−3.2**	**Klonk, Czech Republic**	**Ratified 1972**	***IUGS Series A*, 5**
	Pridoli		423+/−2.3	Reporyje, Czech Republic	Ratified 1984	*Episodes* 8/2
		Ludfordian	425.6+/−0.9	Ludlow, UK	Ratified 1980	*Episodes* 5/3
	Ludlow	Gorstian	427.4+/−0.5	Ludlow, UK	Ratified 1980	*Episodes* 5/3
		Homerian	430.5+/−0.7	Sheinton Brook, UK	Ratified 1980	*Episodes* 5/3
	Wenlock	Sheinwoodian	433.4+/−0.8	Hughley Brook, UK	Ratified 1980	*Episodes* 5/3
		Telychian	438.5+/−1.1	Cefn-cerig Road Section, UK	Ratified 1984	*Episodes* 8/2
		Aeronian	440.8+/−1.2	Trefawr Track Section, UK	Ratified 1984	*Episodes* 8/2
Silurian	**Llandovery**	**Rhuddanian**	**443.8+/−1.5**	**Dobb's Linn, UK**	**Ratified 1984**	***Episodes* 8/2**

(continued overleaf)

Table 6.2 *(continued)*

System	Series	Stage	Age (Ma)	Location	Status	Reference (vol./issue)
		Hirnantian	445.2+/−1.4	Wangjiawan North Section, China	Ratified 2006	*Episodes* 29/3
		Katian	453+/−0.7	Black Knob Ridge, Oklahoma, USA	Ratified 2006	*Episodes* 30/4
	Upper	Sandbian	458.4+/−0.9	Sularp Brook, Sweden	Ratified 2002	*Episodes* 23/2
		Darriwilian	467.3+/−1.1	Huangnitang Section, China	Ratified 1987	*Episodes* 20/3
	Middle	Dapingian	470+/−1.4	Huanghuachang Section, China	Ratified 2007	*Episodes* 28/2; 32/2
		Floian	477.7+/−1.4	Diabasbrottet, Sweden	Ratified 2002	*Episodes* 27/4
Ordovician	**Lower**	**Tremadocian**	**485.4+/−1.9**	**Green Point, Newf'land, Canada**	**Ratified 2000**	***Episodes* 24/1**
		10	489.5	Awaited		
		Jiangshanian	494	Duibian B Section, Zhejiang, China	Ratified 2011	
	Furongian	Paibian	497	Wuling Mts, China	Ratified 2003	*Lethaia* 37
		Guzhangian	500.5	Louyixi, China	Ratified 2008	*Episodes* 32/1
		Drmian	504.5	Drum Mts, Utah, USA	Ratified 2006	*Episodes* 30/2
	3	5	509	Awaited		
		4	514	Awaited		
	2	3	521	Awaited		
		2	529	Awaited		
Cambrian	**Terreneuvian**	**Fortunian**	**541+/−1**	**Fortune Head, Newfoundland, Canada**	**Ratified 1992**	***Episodes* 17/1 and 2**

Ediacaran	635	Enorama Creek, Australia	Ratified 1990	*Lethaia* 39
Cryogenian	850	Defined chronometrically at present. GSSP to follow	Ratified 1990	*Episodes* 14/2
Tonian	1000	Defined chronometrically	Ratified 1990	*Episodes* 14/2
Stenian	1200	Defined chronometrically	Ratified 1990	*Episodes* 14/2
Ectasian	1400	Defined chronometrically	Ratified 1990	*Episodes* 14/2
Calymmian	1600	Defined chronometrically	Ratified 1990	*Episodes* 14/2
Statherian	1800	Defined chronometrically	Ratified 1990	*Episodes* 14/2
Orosirian	2050	Defined chronometrically	Ratified 1990	*Episodes* 14/2
Rhyacian	2300	Defined chronometrically	Ratified 1990	*Episodes* 14/2
Siderian	2500	Defined chronometrically at present. GSSP to follow	Ratified 1990	*Episodes* 14/2
Era				
Neoarchean	2800	Defined chronometrically	Subcomm. decision (1996) not submitted to ICS	Infomally in *Episodes* 15/2
Mesoarchean	3200	Defined chronometrically	Subcomm. Decision 1996 not submitted to ICS	Infomally in *Episodes* 15/2
Palaeoarchean	3600	Defined chronometrically	Subcomm. Decision 1996 not submitted to ICS	Infomally in *Episodes* 15/2
Eoarchean	4000	Defined chronometrically	Subcomm. Decision 1996 not submitted to ICS	Infomally in *Episodes* 15/2
Eon				
Hadean	4560	Formation of planet. Informal term		

boundaries in the Cenozoic, System boundaries in the Mesozoic and Palaeozoic) are highlighted in Table 6.2.

Clearly, these sites are of crucial importance to international stratigraphy and ought to be retained for the future and thus need to be protected from loss or damage (though not from natural processes that retain exposure of the outcrops). While inclusion on the GSSP list clearly identifies the scientific importance of these sites it does not provide any legislative protection. This has to be the responsibility of the countries or regional/local authorities in which the sites are located. Some GSSPs do have protection through national legislative programmes. For example, Dobb's Linn in Scotland, the GSSP marking the Ordovician/Silurian boundary, is a Site of Special Scientific Interest (SSSI) with protection through the UK's Countryside and Rights of Way Act (2000). Similarly, the Precambrian/Cambrian boundary at Fortune Head, Newfoundland, Canada was designated as an Ecological Reserve in 1992. Also on Newfoundland, the GSSP marking the Cambrian/Ordovician boundary (Cooper, Nowlan and Williams, 2001) at Green Point (Figure 6.3), lies within Gros Morne National

Figure 6.3 Green Point, Newfoundland, Canada. The GSSP for the Cambrian/Ordovician boundary is about half way along the section. See plate section for colour version.

Park and is therefore protected by the Parks Canada legislation and the Gros Morne National Park Management Plan.

On the other hand most of the GSSPs in the developing world and even several in the developed world are currently unprotected or unpromoted. For example, the GSSP for the base of the Middle Jurassic Series and Aalenian Stage at Fuentelsaz, Guadalajara, Spain has no legal protection (Carcavilla *et al.*, 2009) other than the Spanish laws requiring permissions from the regional government before any palaeontological sampling can be carried out (Page, Meléndez and Henriques, 2008). Sites in the developing world may be particularly vulnerable. For example, there are two GSSPs in Morocco, neither of which has legal protection. The Giventian GSSP at Jebel Mech Irdane is said to be 'extremely isolated and unlikely to be threatened in any way' (Walliser *et al.*, 1995) but the Messinian Stage GSSP at Oued Akrech, near Rabat (Hilgen *et al.*, 2000) is in a road cutting and could be lost by, for example, road widening and/or slope grading. Wimbledon (2012) draws attention to the destruction of internationally important sections on the south coast of the Crimean peninsula in Ukraine where research to define the base of the Cretaceous had been taking place, the last System boundary still with no defined GSSP.

Remane *et al.* (1996), in their revised guidelines for the establishment of GSSPs, recommended that:

> When making a formal submission to ICS, the concerned Subcommission should try to obtain guarantees from the respective authority concerning ... permanent protection of the site. ... ICS should attempt to finalise, within 3 years after IUGS ratification, any remaining official steps for the protection of the site with the authorities of the country in which the GSSP is located. (p. 80)

There are two points to be made about these statements. First, several GSSPs were ratified prior to these 1996 recommendations for geoconservation. Secondly, the published reports on most GSSPs ratified since 1996 give little or no information about the geoconservation status of the sites. But as Page (2004) points out, the establishment of GSSPs is a conservation-driven activity in itself since the intention is to select key sites that will exist into the future as stratigraphic reference points. It follows from the above discussion that not enough attention has been given to the crucial need for conservation of GSSPs, but how should this be achieved?

The lack of effective protection for most GSSPs is a serious matter given their scientific importance to geology and the time and effort expended by the ICS and IUGS to identify, agree and ratify the sites since 1977. Without adequate protection GSSPs will remain vulnerable to activities that may damage or even destroy them (Page, 2004). While it must remain the responsibility of nation states to protect sites within their territories, international recognition of their standing and importance would help to bring the importance of the GSSP network to the attention of national governments and other authorities. However, the two existing international site programmes aimed at geoconservation are unsuitable to be applied to the GSSP network for the reasons outlined here.

One possibility that has been suggested is that the GSSP site network could be established as a serial World Heritage Site. Serial sites are those where a number of individual sites with a common theme are linked together as one serial World Heritage site (see Chapter 7). However, serial sites are normally within the same region, the same country or in adjacent countries. Given the number and widespread global distribution of GSSPs and the requirement for their management, a single serial GSSP World Heritage Site seems an unlikely and unworkable outcome. Another possibility is that groups of sites could be proposed as serial World Heritage Sites and this has indeed been suggested by Page *et al.* (2008) for the 11 Jurassic GSSPs. Not all of these have yet been agreed and ratified and there would again be the need for co-ordination between several countries. Page *et al.* (2008) point out most of these are in Europe and this project may yet become the ultimate goal for the International Subcommission on Jurassic Stratigraphy. However, not all the other Subcommissions are likely to be able to follow suit and thus this approach is unlikely to provide the full consistency and impact that global recognition for the whole network would bring.

The Global Geopark programme is certainly unsuitable, as geoparks normally comprise substantial territorial areas within which there may be several geosites (see Chapter 8), rather than the individual sites of limited extent represented in the GSSP network. As well as geoconservation, they are also concerned with promoting sustainable development and tourism (Henriques, 2004) and thus are not necessarily a suitable means of protecting sensitive sites.

Consequently, the best approach would be for the ICS, IUGS, IUCN and UNESCO to work together to establish a third geoconservation site network, namely a separate GSSP network to sit alongside the World Heritage and Global Geopark networks. This would have several benefits. For example, it would create a strong, unified and consistent network recognised by UNESCO that would publicise the importance of these sites to national governments and encourage them to provide legislative and/or other protective/management methods for the sites (Henriques, 2004). It would avoid individual Subcommissions of the ICS from the considerable effort involved in achieving World Heritage status for their group of sites. It would also assist UNESCO and the IUCN in their efforts to achieve a series of sites representative of the geological column (see Chapter 7). Here is a network of sites that has already been subject to rigorous selection processes and comprehensive international scrutiny by geological experts, and finally to ratification by the IUGS, an organisation that already advises UNESCO on applications for World Heritage and Global Geopark status. Thus there can be no question about the integrity of the network already established and there would be only very limited resource implications for UNESCO. But in recognising this third site network, UNESCO would achieve immense benefits in bringing international attention to the set of sites that define the fundamental basis of the geological history of the planet.

It is however, recognised that this proposal is unlikely to be quickly adopted by UNESCO which has still (in January 2013) to fully commit to adopting Global Geoparks as a UNESCO programme (see Chapter 8). Nonetheless, when and if that adoption takes place, the next step for international geological and geoconservation community should be to vigorously promote the GSSP network as a new UNESCO programme. In the meantime, pressure should be exerted by national geological communities supported if possible by the ICS, IUGS, IUCN and UNESCO to persuade governments or regional/local authorities to give legal protection to individual GSSPs within their territories.

6.9 PaleoParks

The PaleoParks initiative has been developed by the International Palaeontological Association (IPA) during several International Geological Congress (IGC) meetings since 1996. The aim is to identify and protect important, endangered palaeontological sites whether currently protected or not. Lipps (2009) outlines the goals and objectives of this programme and a number of case studies are described in his e-book (Lipps and Granier, 2009). For example Fedonkin *et al.* (2009) describe 'paleo-piracy' of Vendian (Ediacaran) fossils in the White Sea–Arkangel region of Russia and the measures being proposed to control these illegal activities. Goldstein (2009), on the other hand, describes the management of the fossil resources (a coral-stromatoporoid biostrome) at the Falls of Ohio State Park in the middle of a metropolitan area on the Indiana/Kentucky border, United States.

Nominated sites will be assessed by a committee of the IPA and local palaeontologists. Unprotected sites will be brought to the attention of the relevant authorities in each country, explaining that the sites have been identified as scientifically important and could be of educational, geotouristic and therefore economic value. The IPA will work with the authorities to protect the sites both formally and in practice given the potential impact of visitors.

6.10 The European dimension

There are as yet no specific European geoconservation directives or policies but some of the biological conservation directives are still useful. For example, the European Habitats Directive (1992) allows the designation of Special Areas of Conservation (SACs) and the Birds Directive (1979) provides for Special Protection Areas (SPAs). Together they form a Natura network of sites that may, coincidentally, contain geological or geomorphological features. Brancucci, Burlando and Marin (2002), for example, found that at least 25% of Natura sites in Liguria, Italy contain important geological and geomorphological interests. More importantly the conservation management of many of the sites will have to recognise the fundamental dependency of habitats on soils, geology, landforms and active processes. 'The lessons currently being learned in managing sites designated under the Habitats Directive and through the biodiversity process will be valuable in demonstrating the wider importance and relevance of Earth science understanding' (Gordon and Leys, 2001b, p. 9).

But there have been other European initiatives including the Environmental Impact Assessment Directive (see Chapter 11), Environmental Action Programmes and establishment of the European Environment Agency. The Bern

Convention of the Council of Europe was established to conserve European wildlife and habitats, and has led to a network of Areas of Special Conservation Interest (ASCIs) otherwise known as the Emerald Network. On 5 May 2004 the Committee of Ministers of the Council of Europe adopted Recommendation Rec(2004)3 (see Box 6.2), which also has a series of useful appendices examining the philosophy and practice of geological and geomorphological conservation, existing geoconservation programmes and proposals for management, education and training (https://wcd.coe.int/ViewDoc.jsp?id=740629andSite=COE).

Box 6.2 Council of Europe Rec(2004)3

Members states are recommended to:

- 'identify in their territories areas of special geological interest, the preservation and management of which may contribute to the protection and enrichment of national and European geological heritage; in this context, take into account existing organisations and current geological conservation programmes;
- 'develop national strategies and guidelines for the protection and management of areas of special geological interest embodying the principles of inventory development, site classification, database development, to ensure sustainable use of areas of geological interest through appropriate management;
- 'reinforce existing legal instruments or develop new ones, to protect areas of special geological interest and moveable items of geological heritage, making full use of existing international conventions;
- 'support information and education programmes to promote action in the field of geological heritage conservation;
- 'strengthen co-operation with international organisations, scientific institutions and NGOs in the field of geological heritage conservation;
- 'allocate adequate financial resources to support the initiatives proposed above;
- 'report to the Council of Europe on the implementation of this recommendation five years after its adoption, so that an assessment of its impact may be carried out'.

ProGEO is the European Association for the Conservation of the Geological Heritage and aims 'to promote the conservation of Europe's rich heritage of landscape, rock, fossil and mineral sites'. It organises conferences, publishes a newsletter (*ProGEO News*) and the journal *Geoheritage*, is compiling a list of European geosites and supports an integrated approach to nature conservation in Europe (Erikstad, 2008). It has recently adopted a Geoconservation Protocol and published a manual on geoconservation in Europe (Wimbledon and Smith-Meyer, 2012).

The European Water Framework Directive (WFD) was adopted in 2000. Its aim is to establish a new integrated approach to the protection, improvement and sustainable use of Europe's rivers, lakes, estuaries, coastal waters and groundwaters. The Directive requires member states to prepare River Basin Management Plans (RBMPs) and 'Programmes of Measures' (PMs) that must ensure that all surface waters are of 'good' status by 2015. This will only be adequately achieved by partnership working amongst a great many organisations responsible, for example, for agriculture, fisheries, flood management, conservation and land-use planning. Although mainly aimed at improving water quality and aquatic ecosystems, the WFD does refer to hydromorphological systems and should lead to physical restoration of rivers and fluvial processes.

Also relevant to geoconservation is the European Landscape Convention (2000) and its objectives are outlined in Chapter 10.

6.11 Other International agreements

Other international initiatives have been mainly concerned with protecting biodiversity, for example international agreements on trade of wild animals (CITES) or whaling. The Ramsar Convention for wetland conservation includes, coincidentally, many important geomorphological/ hydrological sites. There is, as yet, no international Convention on Geodiversity to match the Convention on Biological Diversity (CBD). There was active interest in promoting such a convention and international working more generally in the early 1990s (e.g. Knill, 1994; Creaser 1994b; Dixon 1996a) but this has not been sustained and perhaps now needs to be resurrected. Antarctica has its own set of treaties aimed at conserving its environment (Box 6.3).

Box 6.3 Antarctica

Antarctica is roughly one and a half times the size of the United States and 98% of it is covered by an ice-sheet up to 3700 m thick, containing about 90% of the Earth's fresh water. If melted, this would raise the sea-level by about 60 m. The ice-sheet reaches the sea in several places to form ice-shelves and much of the coast is locked by pack ice in winter. Some areas are ice-free, particularly around the Antarctic Peninsula and on high mountain peaks of the Transantarctic Mountains. These divide the continent and the ice-sheet into two parts, East and West Antarctica (Sugden, 1982).

Until recently, the main practical use of Antarctica has been for scientific research, but it has been suggested that substantial oil, gas and mineral reserves may be present on the outer continental shelf. On the Antarctic mainland, coal, oil, gas and iron ore have been found, and it has been estimated that there is sufficient iron ore in the Prince Charles Mountains to satisfy world demand for 200 years. Fortunately there are abundant supplies of iron ore elsewhere in the world that are considerably easier to exploit. The oil and gas reserves would also be difficult and expensive to exploit but as world oil starts to run out, the pressure may fall on these resources (Sugden, 1982).

There were many disputes over sovereignty and management of Antarctica prior to 1959 when 12 governments, seven with territorial claims (Argentina, Australia, Chile, France, New Zealand, Norway and the United Kingdom) and five others (Belgium, Japan, South Africa, United States and Union of Soviet Socialist Republics) signed the Antarctic Treaty. This did not resolve the sovereignty issue but rather was 'masterful in its vagueness' and 'has yet to be fully tested under conditions of economic pressure' (Buck, 1998, pp. 59–60). The Treaty governs the whole area south of 60°S and states that this area should be used only for peaceful purposes and the advancement of scientific knowledge. Article 1 of the Treaty explicitly prohibits any military activities and there have been various subsequent agreements dealing with environmental matters. The Treaty has now been signed by 44 countries representing over 80% of the world's population. Several further agreements have been adopted including a Protocol on Environmental Protection, which has recently banned mineral and oil exploration in Antarctica for at least 50 years (Buck, 1998), but there have been other conflicts over fishing rights and demands from non-treaty nations (Hansom and Gordon, 1998).

6.12 Conclusions

Geodiversity and geological heritage are a global resource for both geologists and the public. The many threats to this resource outlined in Chapter 5 mean that it is entirely appropriate and necessary that international recognition and efforts are given to geoconservation, and this chapter has outlined the important international organisations, initiatives and approaches involved so far. The next two chapters outline two of the most successful international programmes of relevance to geoconservation—World Heritage Sites and Global Geoparks.

7

World Heritage Sites

Little real progress will be possible until governments around the world start taking their national geological heritage seriously, and begin to submit more geological nominations to the World Heritage Committee.

Patrick Boylan (2008)

7.1 The World Heritage Convention

The World Heritage Convention (WHC), which is concerned with 'protecting the world's cultural and natural heritage', was adopted by the General Conference of UNESCO in 1972. To date, 190 countries have ratified the convention, making it an important international conservation instrument. It is not intended to provide protection for all important sites in the world, but only for a select list of the most outstanding international cultural and natural areas. An analysis by Badman and Bomhard (2008) showed that natural World Heritage Sites comprise just over 10% of the total area of all IUCN-classified protected areas, although without the Great Barrier Reef, Australia, which covers nearly 35 million hectares, the figure drops to 8%.

No formal limit is imposed either on the total number of sites included on the List or on the number of properties that can be submitted by any State. The key point is that sites should be of 'outstanding universal value' which means that they should have 'significance that is so exceptional as to transcend national boundaries and to be of common importance for present and future generations of all humanity. As such, the permanent protection of this heritage is of the highest importance to the international community as a whole'. This quote is taken from the extensive *Operational Guidelines for the Implementation of the World Heritage Convention* which, together with the List and description of sites, appears on the UNESCO website (www.unesco.org). The *Operational Guidelines* are updated periodically.

Geodiversity: Valuing and Conserving Abiotic Nature, Second Edition. Murray Gray.

7.2 Nomination and inscription of sites

An initial stage is for countries to develop and submit a 'Tentative List' of sites that they consider to be of 'outstanding universal value' and that they intend to consider nominating over a period of 5–10 years. Countries are encouraged to develop their Tentative List in collaboration with site managers, local and regional government, local communities, non-governmental organisations (NGOs) and other interested parties and partners. Nominated sites for World Heritage status must already be on the Tentative List of the country making the nomination. There is a Tentative List Submission form that requests the names of the sites, their location, a brief description and a justification of their outstanding universal value. Of the 190 countries that have ratified the Convention, 169 have submitted Tentative Lists containing a total of 1561 properties (sites) (November 2012).

Formal nominations to be inscribed on the List can be made by countries at any time by submitting a 'nomination file' which is typically an exhaustive set of documents including a nomination report together with maps, appendices and other supporting materials. The main nomination report must include the following:

- Identification of the site, including its boundaries and any buffer zone;
- Description of the site (or sites in the case of serial nominations);
- Justification for inscription, including the criteria (see Section 7.3);
- State of conservation and factors affecting the site, including condition and threats;
- Protection and management, including legislation, regulatory, contractual, planning and/or traditional methods and the way these operate. A management plan is also a requirement;
- Monitoring plans, including periodicity of inspection;
- Documentation to substantiate the nomination including maps, photographs, appendices, bibliography, etc.;
- Contact information of responsible authorities;
- Signature on behalf of the nominating country.

To be successful with a nomination there must be evidence of the full commitment of the nominating government in the form of relevant legislation, staffing, funding and management plans. It is also essential that local people should participate in the nomination process to make them feel a shared responsibility with the State for the maintenance of the site. Joint nominations are encouraged in cases where outstanding areas stretch across national boundaries. A series of sites (serial sites) in different geographical locations may be nominated as a single World Heritage Site provided that the sites are related as belonging, for example, to the same geomorphological formation, and provided that it is the series as such that is of outstanding universal value and not its components taken individually. The Australian Fossil Mammal Sites (Riversleigh/Naracoote) (1994) is an example of a twin locality site, the Dolomites (2008) in Italy comprise nine separate mountain groups and the Cornwall and West Devon Mining Landscape (2006) in the United Kingdom has 10 component localities.

Figure 7.1 The World Heritage Emblem on plaque at Yosemite National Park.

Following nomination, UNESCO invites the IUCN to independently evaluate nominated sites and, in the case of geoscience sites, the International Union of Geological Sciences (IUGS) often provides expertise to assist in assessing whether or not nominated sites satisfy the necessary criteria and conditions. Following a review of this expertise and a site visit, the IUCN makes a recommendation to the World Heritage Committee. The latter has representatives from 21 member countries on a rotating basis and meets once a year in different countries to decide which sites should be inscribed on the World Heritage List. On occasions it may defer a decision and seek further information from the nominating country.

Inscribed sites are entitled to use the World Heritage 'Emblem' (Figure 7.1) that symbolises the interdependence of cultural and natural sites. The central square is a form created by humans and the circle represents nature, the world and is also a symbol of protection.

7.3 Criteria for selection

Article 2 of the Convention states that 'natural heritage' comprises:

- 'natural features consisting of physical and biological formations, which are of outstanding universal value from the aesthetic or scientific point of view;
- geological and physiographical formations and precisely delineated areas which constitute the habitat of threatened species of animals and plants of outstanding universal value from the point of view of science or conservation;
- natural sites or precisely delineated natural areas of outstanding universal value from the point of view of science, conservation or natural beauty'.

In order to interpret the type of sites that might be suitable for inclusion on the World Heritage List, the *Operational Guidelines* list the 10 criteria, at least one of which should be met. Until 2004, World Heritage Sites were selected on the basis of six cultural and four natural criteria, but the revised Operational Guidelines have simplified these into a single list of 10 criteria. The following are the criteria relevant to geological/geomorphological sites:

(vii) contain superlative natural phenomena or areas of exceptional natural beauty and aesthetic importance;
(viii) be outstanding examples representing major stages of earth's history, including the record of life, significant ongoing geological processes in the development of landforms, or significant geomorphic or physiographic features;

Sites must also meet the conditions of integrity and must have an adequate protection and management system in place. Integrity is defined as a measure of the wholeness or intactness of the geoheritage and thus the site must have a statement of integrity addressing the extent to which the site:

(a) includes all elements necessary to express its outstanding universal value;
(b) is of adequate size to ensure the complete representation of the features and processes necessary to the convey the site's significance;
(c) suffers from adverse effects of development and/or neglect.

For criterion (vii), sites should include areas that are essential for maintaining the beauty of the site. For example, a waterfall site should include adjacent catchment and downstream areas that are integrally linked to the maintenance of the aesthetic qualities of the site.

For criterion (viii), the site should contain all or most of the key interrelated and interdependent elements in their natural relationship. For example, an 'ice age' area 'should include the snow field, the glacier itself and samples of cutting patterns, deposition and colonization (e.g. striations, moraines, pioneer stages of plant succession etc.); in the case of volcanoes, the magmatic series should be complete and all or most of the varieties of effusive rocks and types of eruptions be represented'. Table 7.1 lists all 82 sites inscribed under criterion (viii) as of November 2012.

Although World Heritage status by itself does not give any legal protection, all sites on the List must have conservation and management measures administered by national or regional authorities to ensure that the integrity of the site is maintained or enhanced in the future. This should include a management plan and adequate long-term legislative, regulatory and institutional structures to protect the site. Where necessary for proper conservation, a 'buffer zone' around a site where uses are restricted is encouraged in order to give an added layer of protection.

Table 7.1 The World Heritage Sites inscribed under the criterion viii of the Guidelines.

Name	Country	Date inscribed/ extended	Criteria (other than viii)	Size (ha)	Themes
Tassili n'Ajjer	Algeria	1982	i,iii,vii	72 00 000	Arid
Ischigualasto/Talampaya NPs	Argentina	2000		2 75 369	Fossil
Los Glaciares	Argentina	1981	vii	6 00 000	Glacial
Tasmanian Wilderness	Australia	1982, 1989	iii,iv,vi,vii,ix,x	14 07 513	Glacial
Macquarrie Island	Australia	1997	vii	4 50 000	Tectonic
Purnululu NP	Australia	2003	vii	2 39 723	Arid/Karst
Great Barrier Reef	Australia	1981	vii,ix,x		Reef
Shark Bay	Australia	1991	vii,ix,x	21 97 300	Fossil
Wet Tropics of Queensland	Australia	1988	vii,ix,x	8 94 420	Fossil
Heard & McDonald Islands	Australia	1997	ix	6 58 903	Volcanic
Gondwana Rainforests	Australia	1986, 1994	ix,x	3 70 000	Volcanic
Uluru-Kata Tjuta NP	Australia	1987, 1994	v,vi,vii	1 32 556	Arid
Willandra Lakes	Australia	1981	iii	2 40 000	Fossil/Lacustrine
Riversleigh/Naracoorte	Australia	1994	ix	10 300	Fossil
Fraser Island	Australia	1992	vii,ix	1 84 000	Coastal
Pirin NP	Bulgaria	1983, 2010	vii,ix	38 350	Glacial
Canadian Rocky Mountain Parks	Canada	1984, 1990	vii	23 06 884	Mountain/Glacial/ Fossil
Nahanni NP	Canada	1978	vii	4 76 560	Fluvial/Karst
Miguasha NP	Canada	1999		87	Fossil
Dinosaur Provincial Park	Canada	1979	vii	7493	Fossil
Gros Morne NP	Canada	1987	vii	1 80 500	Glacial/Tectonic
Joggins Fossil Cliffs	Canada	2008		689	Fossil
Kluane/Wrangell-St Elias, Glacier Bay/Tatshenshini-Alsek	Canada/USA	1979, 1992, 1994	vii,ix,x	98 39 121	Mountain/Glacial
Three Parallel Rivers of Yunnan Protected Areas	China	2003	vii,ix,x	17 00 000	Fluvial
China Danxia	China	2010	vii	82 151	Red-bed landforms

(*continued overleaf*)

Table 7.1 (*continued*)

Name	Country	Date inscribed/ extended	Criteria (other than viii)	Size (ha)	Themes
South China Karst	China	2007	vii	47 558	Karst
Chengjiang Fossil Site	China	2012		512	Fossil
Talamanca Range La Amistad Reserves and NP	Costa Rica/ Panama	1983, 1990	vii,ix,x	5 70 045	Mountain/Glacial
Plitvice Lakes NP	Croatia	1979, 2000	vii,ix	29 482	Fluvial/Lacustrine/Karst
Desembarco del Granma NP	Cuba	1999	vii	41 863	Coastal/Karst/Tectonic
Virunga NP	DR Congo	1979		8 00 000	Tectonic/Volcanic/Glacial
Ilulissat Icefjord	Denmark	2004		4 02 400	Glacial
Morne Trois Pitons NP	Dominica	1997		6957	Volcanic/Fluvial
Galapagos Islands	Ecuador	1978, 2001	vii,ix,x	14 066 514	Tectonic/Volcanic
Sangay NP	Ecuador	1983	vii,ix,x	2 71 925	Volcanic/Mountain
Wadi al Hitan	Egypt	2005		20 015	Fossil
High Coast/Kvarken Archipelago	Finland/Sweden	2000, 2006		1 94 400	Glacial
Gulf of Porto	France	1983	vii,x	11 800	Coastal/Volcanic
Pyrenees - Mont Perdu	France/Spain	1997, 1999	iii,iv,v,vii	30 639	Mountain/Glacial
Messel Pit Fossil Site	Germany	1995		42	Fossil
The Wadden Sea	Germany/ Netherlands	2009	ix,x	9 82 004	Coastal
Rio Platano Biosphere Reserve	Honduras	1982	vii,ix,x	3 50 000	Fluvial
Caves of Aggtelekl Karst & Slovak Karst	Hungary/ Slovakia	1995, 2000		56 651	Karst
Lorentz NP	Indonesia	1999	ix,x	23 50 000	Glacial/Coastal/Tectonic
Isole Eolie	Italy	2000		1216	Volcanic
The Dolomites	Italy	2009	vii	1 41 903	Mountain/Glacial
Monte San Georgio	Italy/Switzerland	2003, 2010		1089	Fossil
Lake Turkana NPs	Kenya	1997	x	1 61 485	Fossil/Lacustrine
Jeju Volcanic Island & Lava Tubes	Korea	2007	vii	9475	Volcanic
Gunung Mulu NP	Malaysia	2000	vii,ix,x	52 864	Karst
Durmitor NP	Montenegro	1980, 2005	vii,x	32 100	Glacial/Fluvial

Tongariro NP	New Zealand	1990, 1993	vi,vii	79 569	Volcanic
Te Wahipounamu	New Zealand	1990	vii,ix,x	26 00 000	Mountain/Glacial/Tectonic
West Norwegian Fjords	Norway	2005	vii	NK	Glacial/Fluvial/Coastal
Huascaran NP	Peru	1985	vii	3 40 000	Mountain/Glacial
Lake Baikal	Russia	1996	vii,ix,x	88 00 000	Lacustrine
Volcanoes of Kamchatka	Russia	1996, 2001	vii,ix,x	38 30 200	Volcanic/Glacial
Lena Pillars Nature Park	Russia	2012		12 72 150	Fluvial
Valee de Mai Nature Reserve	Seychelles	1983	vii,ix,x	20	
Skojan Caves	Slovenia	1986	vii	413	Karst
Vredfort Dome	South Africa	2005		30 000	Meteorite
Teide NP	Spain	2007	vii	18 990	Volcanic
Pitons Management Area	St Lucia	2004	vii	2909	Volcanic
Laponian Area	Sweden	1996	iii,v,vii,ix	9 40 000	Mountain/Glacial/Fluvial
Swiss Alps Jungfrau-Aletsch	Switzerland	2001		82 400	Glacial
Swiss Tectonic Area Sardona	Switzerland	2008		32 850	Tectonic/Glacial/Karst
Ngorongoro Conservation Area	Tanzania	1978		8 22 800	Volcanic
Giant's Causeway & Coast	UK	1986	vii	70	Volcanic/Coastal
Dorset & East Devon Coast	UK	2001		2550	Fossil/Coastal
Everglades NP	USA	1979	ix,x	5 92 920	Fluvial/Coastal
Yellowstone NP	USA	1978	vii,ix,x	8 98 349	Volcanic
Great Smoky Mountains NP	USA	1983	vii,ix,x	2 09 000	Mountain/Fluvial/Fossil
Mammoth Cave NP	USA	1981	vii,x	21 191	Karst
Papahanaumokuakea	USA	2010	iii,vi,ix,x	36 207 499	Volcanic/Coastal/Reef
Carlsbad Caverns NP	USA	1995	vii	18 926	Karst
Grand Canyon NP	USA	1979	vii,ix,x	4 93 077	Fluvial/Stratigraphic
Hawaii Volcanoes NP	USA	1987		92 934	Volcanic
Yosemite NP	USA	1984	vii	3 08 283	Mountain/Glacial
Canaima NP	Venezuela	1994	vii,ix,x	30 00 000	Fluvial
Ha Long Bay	Vietnam	1994, 2000	vii	1 50 000	Karst/Coastal
Phong Nha-Ke Bang NP	Vietnam	2003		85 754	Karst
Mosi-oa-Tunya/Victoria Falls	Zambia/Zimbabwe	1989	vii	6860	Fluvial

7.4 Endangered sites

States are expected to monitor the condition of sites on the List and to submit periodic reports on the condition of sites within their territory and any legislative, administrative or other actions they have taken to implement the Convention. In particular, reports are expected each time exceptional circumstances occur or work is undertaken which may have an effect on the state of conservation of a site on the list. Where a site on the List is threatened by serious and specific danger and major operations are necessary for its conservation, it may be placed on the List of World Heritage in Danger. Listed examples of where this may occur include:

- 'severe deterioration of the natural beauty or scientific value of the property, as by human settlement, construction of reservoirs which flood important parts of the property, industrial and agricultural development including use of pesticides and fertilizers, major public works, mining, pollution, logging... etc.';
- 'human encroachment on boundaries or in upstream areas which threaten the integrity of the property';
- 'a modification of the legal protective status of the area'.

Several sites have been placed on this list in the past including the Galapagos Islands, Ecuador (impact of tourism and immigration) and Mount Nimba Strict Nature Reserve, Guinea/Ivory Coast (iron-ore mining concession). The Everglades National Park in the United States is now back on the List in Danger due to the impacts of urbanisation.

Sites may be deleted from the list where they have deteriorated to the extent that they have lost the characteristics that led them to be included on the list, but this only occurs after several years of monitoring. So far only two sites have ever been deleted from the World Heritage List—the Arabian Oryx Sanctuary in Oman in 2007 (unilaterally reduced in size by 90%) and Dresden in Germany in 2009 (construction of a new four-lane bridge). The IUCN has recommended that the WHC should involve NGOs, academic institutions and local people in monitoring the conservation status of sites in order to make WHC processes more transparent and democratic (Feick and Draper, 2001).

7.5 Towards a 'representative, balanced and credible' list

As of November 2012, there are 962 World Heritage Sites in 157 countries, 745 of which are cultural sites, 188 are natural sites and 29 are mixed sites. The vast majority are located in developed regions of the world, particularly in Europe. This imbalance between cultural and natural sites and towards the developed world was of concern to UNESCO and in 1994 it launched a Global Strategy for a 'representative, balanced and credible' World Heritage List. The aim was, and is, to identify and fill the major gaps, thematic and spatial, in the List and this is being achieved in several ways. One has been to encourage more countries to ratify the Convention and develop Tentative Lists, and to request countries whose heritage is already well represented to slow down their rate of nominations or only nominate sites in under-represented thematic categories.

In relation to geological and geomorphological sites, the IUCN carried out a study of existing sites (Dingwall, Weighell and Badman, 2005). One analysis in this study built on a previous study of fossil sites (Wells, 1996) which recommended choosing sites 'that contain well-preserved fossil accumulations of high species diversity which in combination best document the story of community and environmental change through time' (p. 10). Wells concluded that the few fossil sites then on the List 'are not representative of the history of life on earth' (p. 39). Dingwall *et al.* reviewed the position as of 2005 by which time several more fossil sites had been inscribed. Nonetheless, their analysis (Table 7.2) demonstrated some significant gaps in the representation of the geological column, particularly the absence of sites in the Silurian and Cenozoic. Dingwall *et al.* (2005) also developed a list of 13 thematic areas 'as a broad conceptual framework for geological World Heritage'. Table 7.3 lists these geothemes and also the number of sites (in 2005) falling into each theme. It will be noted that some themes are much better represented than others. Particularly notable is the low number of stratigraphic sites (see also Boylan, 2008). One of the aims of Dingwall *et al.*'s (2005) work was to assist the World Heritage Committee and its advisors to identify possible gaps in coverage of the World Heritage List. In a sense therefore, we can say that UNESCO and the IUCN are attempting to make the list more representative in both a chronostratigraphic and thematic sense, i.e. the List is aiming to represent the world's outstanding geodiversity.

A thematic study of World Heritage Cave and Karst sites was carried out by Williams (2008). The stimulus for this was an IUCN evaluation report to the 31st World Heritage Convention in Christchurch in 2007, which concluded that karst systems (including caves) are relatively well represented on the List. 'Therefore in the interests of maintaining the credibility of the World Heritage List, IUCN considers that there is increasingly limited scope for recommending further karst nominations.' and that these should only be promoted where:

- 'There is a very clear basis for identifying major and distinctive features of outstanding universal value that has been verified by a thorough global comparative analysis';
- 'The basis for claiming outstanding universal value is a significant and distinctive feature of demonstrable and widespread significance, and not one of many narrow and specialized features that are exhibited within karst terrains'.

Williams' (2008) analysis identified poor representation of caves and karst in South America, Africa, Australasia and the South Pacific, North, Central and South Asia, and the Middle East. He also identified poor representation of arid, semi-arid and periglacial karsts, while karsts on evaporite rocks were totally unrepresented on the List. He also concluded (p. 10) that 'although karst is well represented in the humid zone, especially in the Northern Hemisphere, the sites

Table 7.2 Geological World Heritage Sites and the stratigraphic column (after Dingwall, Weighell and Badman, 2005).

Geological period	Biological event	World Heritage Site
Quaternary	Humans appear Ice Age	Naracoote (Australia)
Pliocene		
Miocene		Riversleigh (Australia)
Oligocene		
Eocene		Messel Pit (Germany)
Palaeocene	First primates	
Cretaceous	Extinction of dinosaurs Origin of flowering plants	Dinosaur Provincial Park (Canada)
Jurassic	Age of dinosaurs First birds	Jurassic Coast (UK)
Triassic	First mammals/dinosaurs	Monte San Georgio (Switzerland)
Permian		Grand Canyon (USA)
Carboniferous	First reptiles	Mammoth Cave (USA)
Devonian	First amphibians/forests	Miguasha (Canada)
Silurian	First land plants	
Ordovician	First fish/coral	Gros Morne (Canada)
Cambrian	First trilobites	Burgess Shale (Canada)
Precambrian	First algae/bacteria	

Table 7.3 The 'Geothemes' of Dingwall *et al.* (2005) and the number of sites falling into each theme.

Geotheme	No. of sites
Tectonic and structural features	3
Volcanoes/volcanic features	13
Mountain systems	11
Stratigraphic sites	2
Fossil sites	11
Fluvial/lacustrine systems and landscapes	10
Caves and karst	7
Coastal development	8
Reefs, atolls and oceanic islands	1
Glaciers and ice caps	6
Ice Ages	7
Arid and semi-arid landforms and landscapes	4
Meteorite impact	1

currently on the World Heritage List are not sufficient to represent karst values of the region of Europe from which karst took its name—the Dinaric Karst'. He therefore recommended that countries in this area 'consider a transnational serial nomination that would include a representative range of karst values and features of all scales above and below ground from the mountains to the sea. Such a property should represent the international type-site of karst by illustrating a full range of the region's karst features and by providing evidence for their evolution'. Elsewhere he recommended that countries where the geographical, morphoclimatic and lithological gaps in representation had been identified, should be encouraged to consider the potential for nominating karst sites. Again, a clear intention emerges to work towards a World Heritage List of karst sites that is more globally representative, i.e. at least in part, more representative of karst geodiversity.

A further thematic study, of World Heritage Volcanoes, was carried out by Wood (2009). The IUCN report to the 31st World Heritage Convention again provided the stimulus for the study as it advised that volcanic features are relatively well represented on the List, and thus suggested that there was a decreasing potential for further nominations of volcanic properties. It argued that these should be restricted to filling any gaps in the present global coverage of volcanic sites where there is 'a very clear basis for identifying major and distinctive features of outstanding universal value that has been verified by a thorough global comparative analysis'. Wood's study included all World Heritage Sites inscribed up to the end of 2008 as well as those on all Tentative Lists. He found that 57 inscribed properties contained some volcanic geology, with 27 of these containing at least one active volcano, while 40 volcanic properties were included in the Tentative Lists, 25 with one or more active volcanoes. Wood concluded that while the World Heritage List appears to possess good overall representation of volcanic features, deeper analysis revealed 'some gaps that might be filled by future nominations'. These included:

- some important features of basaltic volcanism, such as fissure volcanoes, sub-glacial volcanic edifices and continental flood basalts;
- some features or more silicic volcanism, including calderas and large ash or pumice flows (ignimbrites);
- some of the world's most iconic volcanoes. Wood particularly noted the absence of Mt Etna, Italy; Thera (Santorini), Greece; Mt Fuji, Japan (though this is on the Japanese Tentative List); Paracutin, Mexico; Mt Mayon, Philippines; Mt St Helens and Crater Lake, United States; Laki, Iceland; Mt Pelée, Martinique, and Tambora, Indonesia.

Thus again, here, we see an attempt to ensure that the global geodiversity of volcanic features is included on the World Heritage List.

7.6 Validity of inscription criteria

Some World Heritage Sites are clearly geologically important but lack a criteria (vii) or (viii) designation. For example, Sigiriya in Sri Lanka (sheer-sided volcanic neck) and Thingvellir in Iceland (astride the European/North American plate boundary) are listed only as cultural sites, but must surely qualify as having outstanding universal value under criterion (vii) at least. Surtsey, a new volcanic island in Iceland that erupted in 1963, is only listed as an ecological site important for monitoring colonisation. Several other cultural sites have geological content but the important point is that it is not sufficient that cultural sites have some geological interest: instead, the geology should also be of 'outstanding universal value'. Thus Petra in Jordan is only a cultural site because although carved in red sandstone, this is not regarded as outstanding geologically. Alexandrowicz, Urban and Miskiewicz (2009) have carried out a study of the 13 Polish World Heritage Sites currently classified only for their cultural importance. They concluded that two of these—the Wieliczka Salt Mine and the Park Muzakowski (containing a glaciotectonised push moraine)—would merit reassessment to include natural as well as cultural criteria during the next revision of the World Heritage List. There are also some hidden gems. The Canadian Rocky Mountain Parks World Heritage Site, for example, contains the famous Burgess Shale Cambrian fossil site in Yoho National Park (Figure 7.2) and Ujung Kulon National Park in Indonesia, contains the remnants of the Krakatoa volcano.

7.7 Case studies

7.7.1 Wadi Al-Hitan ('Whale Valley'), Egypt (2005)

Wadi Al-Hitan is located in Egypt's Western Desert, about 150 km south-west of Cairo and was inscribed on the World Heritage List in 2005. It is the most important site in the world in demonstrating the evolution of whales, particularly in portraying their transition from land animals to marine mammals. The number, concentration, quality, and accessibility of the fossils set in an attractive and protected landscape makes this site the most important in the world for demonstrating this phase of evolution and in full accord with the IUCN recommendations on the inscription of fossil sites. The fossils are from the Eocene Epoch, 40–37 million years ago, and belong to the earliest, now extinct suborder of whales: Archaeoceti. About 400 whale skeletons have been found in the valley, some from the genus *Basilosaurus*. The largest skeleton found is 21 m long with well-developed five-fingered flippers on the forelimbs and with fully functional hind legs, knee caps, ankles, feet and toes. Some of the skeletons are fully exposed on the surface while others are slowly being uncovered by surface erosion. This creates problems in managing the sites, particularly with increasing visitor numbers. The site lies within the Wadi El-Rayan Protected Area and visitors are allowed only on prearranged guided tours along prescribed trails. The site is regularly patrolled and monitored to discourage intruders who often arrive in four-wheel drive vehicles. Etches (2006) has produced an educational/interpretive plan for the protected area, including an expanded visitor centre, trails and interpretive panels. He draws attention to dangers of visitors removing whale teeth and bones as souvenirs and recommends no public access to the site without a guide. Several other vertebrate and invertebrate species have been found on the site including sharks, crocodiles, rays, turtles and crabs.

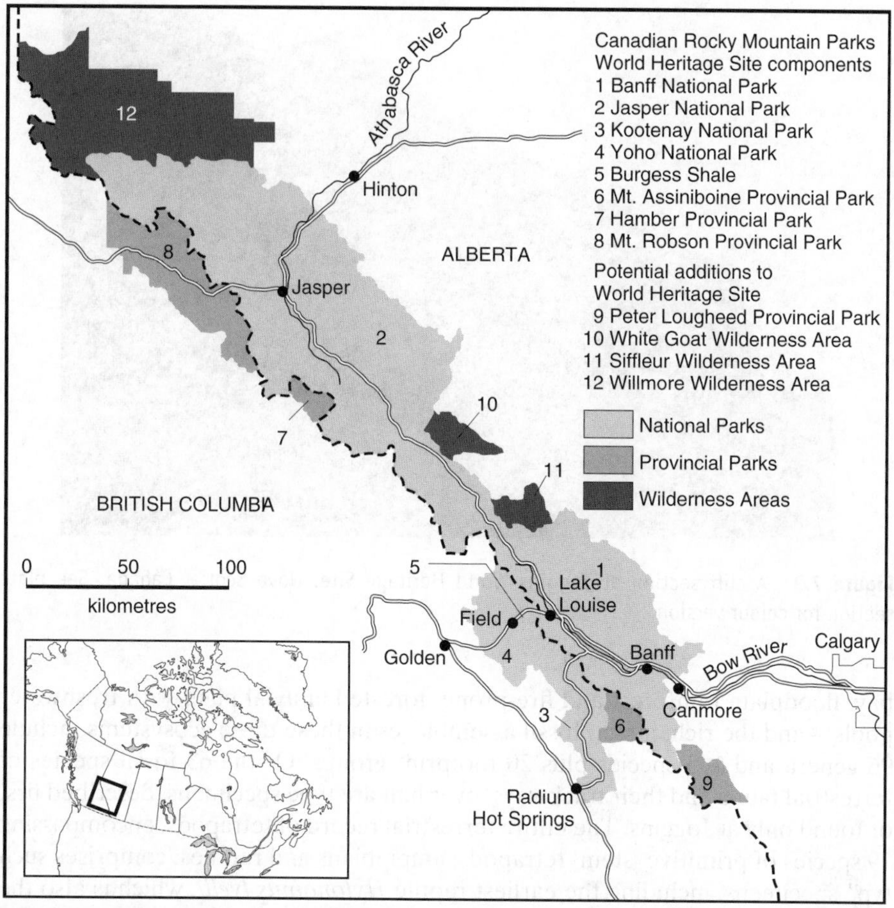

Figure 7.2 Rocky Mountain Parks World Heritage Site in Canada comprises a number of National and Provincial Parks. The Burgess Shale Fossil Site is included in the Site (after Feick and Draper (2001) by permission of Tourism Recreation and Research, www.trrworld.org).

7.7.2 Joggins Fossil Cliffs, Canada (2008)

This site, inscribed at the 32nd Convention in Quebec City in 2008, lies at the head of the macro-tidal Bay of Fundy in the province of Nova Scotia, eastern Canada. It is considered to represent the world's thickest (4442 m) and most complete record of Pennsylvanian strata (318–303 Ma) with a remarkable record of terrestrial life from this time, all magnificently exposed as steeply dipping strata in sea cliffs up to 30 m high and related shore platforms and reefs along 14.7 km of coastline (Figure 7.3). As such the site is representative of the Carboniferous Period or 'Coal Age'. The fossils include the remains of three tropical ecosystems—estuarine

Figure 7.3 A cliff section at Joggins World Heritage Site, Nova Scotia, Canada. See plate section for colour version.

bay, floodplain rainforest and fire-prone, forested alluvial plain with freshwater pools—and the rich, intact, fossil assemblages in these three ecosystems include 96 genera and 195 species plus 20 footprint groups. Of the 63 fossil species of terrestrial fauna and their trackways, over half are type specimens, described first or found only at Joggins. The entire terrestrial record of tetrapods, encompassing 19 species of primitive 'stem' tetrapods, amphibians and reptiles, comprises such type specimens, including the earliest reptile *Hylonomus lyelli*, which is also the oldest known representative of the amniotes. The site has been researched for almost 200 years including by Sir Charles Lyell, Sir William Dawson and Charles Darwin (Falcon-Lang, 2006). Lyell described it as 'the finest example in the world' of the Pennsylvanian 'Coal Age' and Darwin noted 68 horizons of upright trees occurring at the site. Joggins has played a vital historic role in the development of seminal geological and evolutionary principles, including subsidence of the earth's crust, the origin of coal, and the inherent nature of the fossil record as it informs evolutionary theory (Joggins Fossil Institute, 2007). The site has a range of legislative protection (Figure 13.5), and also has a new visitor centre, opened in 2008.

7.7.3 Grand Canyon National Park, United States (1979)

This National Park in Arizona, United States, would be recognised worldwide as one of the natural wonders of the world and is richly deserving of its World Heritage Site status, which it achieved in 1979. It now covers nearly half a million hectares and is dominated by the spectacular Grand Canyon, 550 m to 30 km wide, about 1500 m deep and 447 km long. It has formed during the past 6 million

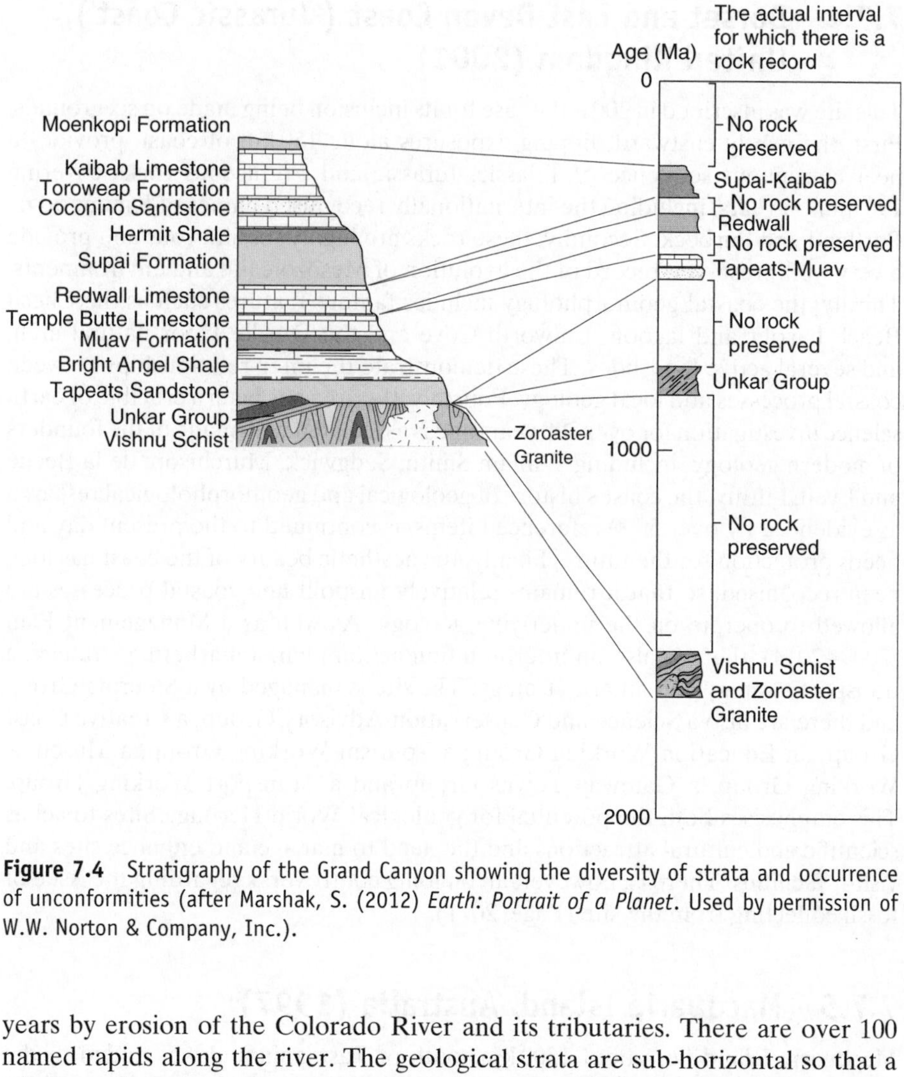

Figure 7.4 Stratigraphy of the Grand Canyon showing the diversity of strata and occurrence of unconformities (after Marshak, S. (2012) *Earth: Portrait of a Planet*. Used by permission of W.W. Norton & Company, Inc.).

years by erosion of the Colorado River and its tributaries. There are over 100 named rapids along the river. The geological strata are sub-horizontal so that a trip from the canyon rim to the river takes us through over 1000 million years of geological time from Triassic to Precambrian though there are many gaps in the record represented by unconformities (Figure 7.4). The earliest Precambrian strata, the Vishnu Metamorphic Complex, is barren, but the late Precambrian Bass Limestone, within the Unkar Group, has early plant fossils. The Palaeozoic strata contain both marine and terrestrial fossils demonstrating alternations of submergence and emergence.

An updated management plan was completed in 1995 and involved local citizens, local Indian tribes, and public and private agencies. The park's most serious management issue is that of tourism. The annual influx of over 5 million visitors is gradually degrading the park's natural resources by, for example, footpath erosion and development, though the park is zoned with over 90% managed as wilderness.

7.7.4 Dorset and East Devon Coast ('Jurassic Coast'), United Kingdom (2001)

This site was inscribed in 2001, the case for its inclusion being made on six grounds. First, the gently eastward dipping exposures along 150 km of coast, provides a near continuous sequence of Triassic, Jurassic and Cretaceous rocks covering 190 million years, including the internationally recognised names of Kimmeridge, Portland and Purbeck. Secondly, these rocks are highly fossiliferous and provide a very well-preserved record of the evolution of Mesozoic life and environments. Thirdly, the coastal geomorphology includes famous features such as the Chesil Beach barrier and lagoon, Lulworth Cove and the Durdle Door natural arch, and several active landslides. These demonstrate the inter-relationships between coastal processes and local geology. Fourthly, the area has been a crucible of earth science investigation for over 300 years and is associated with many of the founders of modern geology, including William Smith, Sedgwick, Murchison, de la Beche and Lyell. Fifthly, the coast's history of geological and geomorphological research as evidenced by over 5000 referenced items, is continued to the present day and needs protection for the future. Finally, the aesthetic beauty of the coast has long been recognised, so that it remains relatively unspoilt and coastal processes are allowed to operate on the underlying geology. As well as a Management Plan (2009–2014), there is also an interpretation action plan, a marketing strategy, a transport strategy and an arts strategy. The site is managed by a Steering Group and there are also a Science and Conservation Advisory Group, a Creative Coast Group, an Education Working Group, a Tourism Working Group, a Museums Working Group, a Gateway Towns Group and a Transport Working Group. This emphasises both the potential for geological World Heritage Sites to act as scientific and cultural attractions and the need to manage and enhance sites and visitor facilities. There is, however, an ongoing controversy regarding the issue of fossil collecting from the site (Page, 2011).

7.7.5 Macquarie Island, Australia (1997)

Macquarie Island lies about 1500 km south of Tasmania and was added to the World Heritage list in 1997 as an exposed section of a huge oceanic ridge extending upwards from a depth of 2.5 km. It started life between 30 and 11 million years ago as a small divergent margin ridge. The southern three-quarters of Macquarie Island is composed of fissure erupted basalts, including pillow lavas. Miocene abyssal oozes are found between some of the pillows.

About 10 million years ago the spreading stopped and instead reversed to produce an upward pressure and movement of abyssal rocks. The most important products of this upward movement occur in the north of the island where an ophiolite sequence of upper mantle rocks is exposed comprising sea-bed lavas, a sheeted dolerite complex, massive, layered gabbros and finally a mixed zone of peridotite believed to be from 6 km below the ocean floor. 'No drill hole has ever penetrated these depths and these exposures provide a rare opportunity for geologists to gain an understanding of rocks from the uppermost mantle' (Pemberton, 2001b).

Other features seen on the island include fault-controlled structures and frequent earthquakes, raised shorelines, landslides, solifluction lobes and patterned ground (Viney, 2001).

A series of conservation objectives has been established that will be implemented through the following practical policies:

- Potential adverse impacts on geodiversity and earth processes will be assessed when planning any development or action, including land rehabilitation and stabilisation;
- Management practices and development will avoid or otherwise minimise impacts on the integrity of sites of geoconservation significance;
- Scientific research will be conducted in a way which avoids impacts on geodiversity, sites of geoconservation significance or the aesthetics of significant exposures. Geoscientific research must be consistent with the values for which the area was nominated and ought to be justified in this context;
- The use of coring devices and other mechanical sampling devices for geoscientific research will not be permitted unless special permission is provided. Any approval will be strictly controlled and monitored. Similar conditions will apply to the use of explosives for geoscientific or management purposes;
- The impacts on geodiversity will be monitored including adherence to conditions identified in scientific collecting permits;
- Sites of geoconservation significance will be identified and mapped, and recommendations will be made on their future management (Pemberton, 2001b).

7.7.6 Messel Pit, Germany (1995)

The Messel Pit Fossil Site lies in a former shale quarry about 35 km south-east of Frankfurt in Germany. During the Eocene, subsidence along faults created a lake basin on the floor of which sediments and their contained fossils accumulated over a long period of time and to a thickness of circa 190 m. Brown coal and oil shale were mined here starting in 1859 and fossils were soon being unearthed by the miners, though generally they quickly dried out and crumbled. It was not until the twentieth century that the fossil finds began to attract national attention and scientific investigation. In the 1960s the 'transfer' technique was developed which allowed the fossils to be stabilised using resin. After commercial mining stopped in 1971, amateur, uncontrolled fossil collecting reached a peak, particularly when a plan to turn the quarry into a landfill site was announced. However, this was met with a campaign to save the site for future research and in 1987 the idea was quashed following a series of legal proceedings. The State of Hessen stepped in to buy the site in 1991 and then worked with others to have it inscribed on the World Heritage List, which was achieved in 1995.

The Messel Fossil Pit is regarded as the richest site in the world for understanding the palaeoenvironments of the Middle Eocene around 48 Ma, particularly the early evolution of mammals. The quality of preservation of the fossils is remarkable and includes fully articulated skeletons often with skin, hairs, feathers and stomach contents intact. The total number of fossils discovered runs into thousands and there is also a diversity of fossils with birds, reptiles, fish, insects and plant remains all discovered as well as the mammal fossils, altogether making this one of the most important fossil sites in the world. Its fame was cemented with the unveiling in 2009 of *Darwinius masillIae*, commonly known as Ida, a near-complete primate fossil collected by an amateur and revealed only 20 years later when an amnesty was declared on illegally collected fossils. Further important finds are expected as collecting continues under permit.

7.8 Conclusions

The World Heritage Convention has recognised 80 internationally outstanding geoscience sites and is attempting to create a World Heritage List more representative of the world's geodiversity, spatially, temporally and thematically. This is to be welcomed but there is still some way to go before achievement of the objective to have a 'representative, balanced and credible list'.

8

Global Geoparks

Geoparks are not just about rocks, they are about people.

Chris Woodley-Stewart, North Pennines Geopark, United Kingdom

8.1 History

The idea of a network of European Geoparks was initiated at the 30th International Geological Congress (IGC) in Beijing in 1996 (Zouros and McKeever, 2008). The triple aims were conserving geoheritage, improving public understanding of the geosciences, and promoting regional economic development. The European Geoparks Network (EGN) was established in 2000 by four founding members (Figure 8.1): Haute-Provence in France (ammonites, bird footprints, etc.), Lesvos in Greece (petrified forest), Maestrazgo in Spain (dinosaurs) and Vulkaneifel in Germany (volcanic craters and structures). The following year, the UNESCO Executive Board decided that it would not adopt a 'Geoparks Progamme' but was prepared to support ad hoc efforts with Member States as appropriate. Thus an agreement on co-operation was reached between the then Division of Earth Sciences of UNESCO and the European Geoparks Network. In the subsequent few years further geoparks joined the European network and at the same time a network of geoparks was being established in China (Xun and Milly, 2002). As a result, the Global Network of National Geoparks (GGN) was established in 2004 with 17 European Members and 8 from China. The First International Conference on Geoparks was held in Beijing, China in the same year. This has been followed by biennial Global Geopark conferences as follows:

- 2006 Belfast, Northern Ireland;
- 2008 Osnabruck, Germany;
- 2010 Langkawi Geopark, Malaysia;
- 2012 Mount Unzen Geopark, Japan;
- 2014 Stonehammer Geopark, Canada.

Geodiversity: Valuing and Conserving Abiotic Nature, Second Edition. Murray Gray.

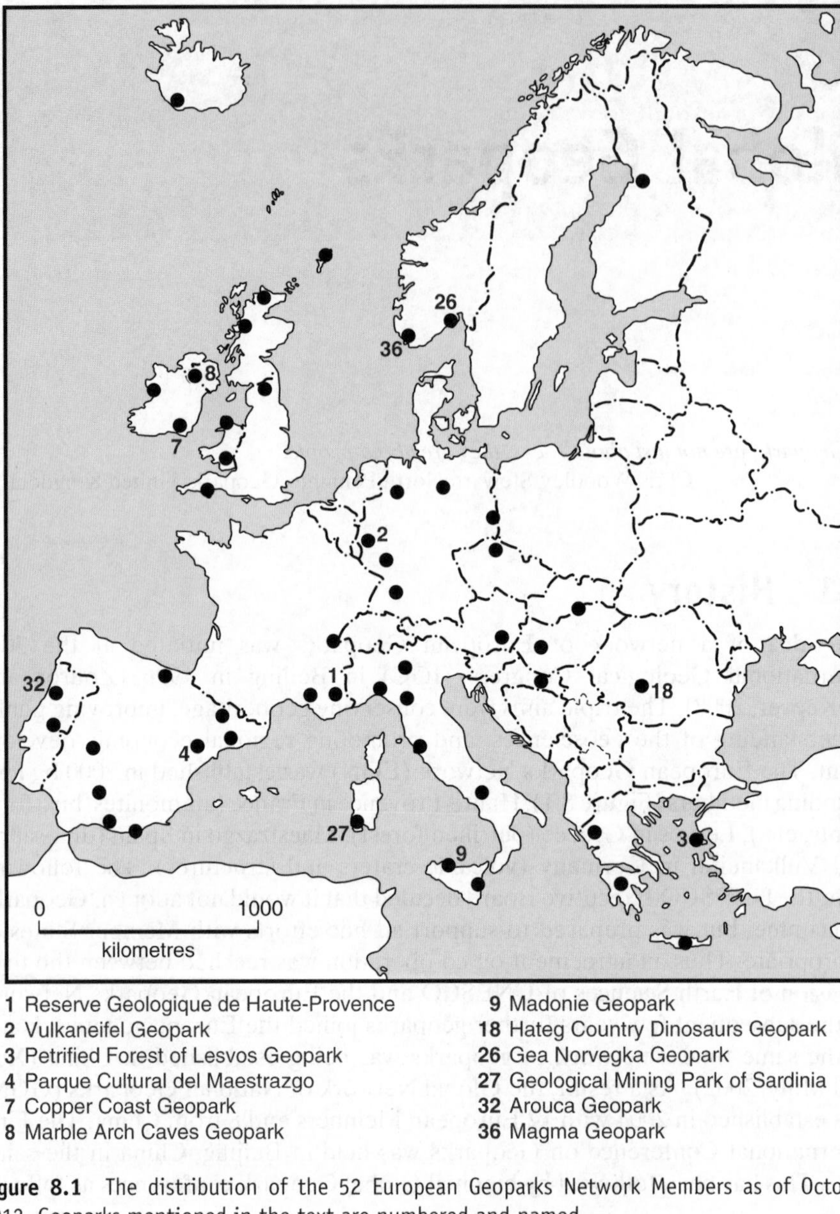

1 Reserve Geologique de Haute-Provence
2 Vulkaneifel Geopark
3 Petrified Forest of Lesvos Geopark
4 Parque Cultural del Maestrazgo
7 Copper Coast Geopark
8 Marble Arch Caves Geopark
9 Madonie Geopark
18 Hateg Country Dinosaurs Geopark
26 Gea Norvegka Geopark
27 Geological Mining Park of Sardinia
32 Arouca Geopark
36 Magma Geopark

Figure 8.1 The distribution of the 52 European Geoparks Network Members as of October 2012. Geoparks mentioned in the text are numbered and named.

In the Madonie declaration, signed at the EGN conference in Madonie Geopark, Italy (Figure 8.1) in October 2004, the EGN was recognised as the official branch of UNESCO GGN Network in Europe. In fact, the EGN has provided an important model for UNESCO in rolling out the GGN worldwide, which has happened quite rapidly. By October 2012 the GGN had 90 members in 27 countries, although most are still in Europe (52) and China (27). Japan has five Global Geoparks and, outside Europe, countries with single geoparks are

Brazil, Malaysia, Canada, Vietnam, South Korea and Indonesia. Many countries including Italy, Canada, Japan and China have National Geopark Development Committees, and there are also incipient Geopark Networks in other continents/regions including Latin America and Caribbean, Africa and Arabian and Asia-Pacific. Further rapid expansion of the network is anticipated with aspiring geoparks in many new countries (Patzak and Missotten, 2010). There can be no doubt that Geoparks are a tremendous success story for the geoscience and geoconservation communities.

UNESCO has played an important role in supporting the Global Geoparks initiative mainly because of the interdisciplinary and international nature of the co-operation in studying Earth systems while also sustaining local communities. However, as stated above, the GGN is not as yet a full UNESCO programme, and instead the official title is the 'Global Network of National Geoparks, assisted by UNESCO'. This assistance includes policy advice, visibility, global attention and a label of excellence. UNESCO also provides a GGN Secretariat, a website, publications and organises evaluation of applications to join the GGN network. It is to be hoped that Global Geoparks will be accepted as a full UNESCO Programme in the near future, and the signs are encouraging. At the 36th UNESCO General Conference in Paris in November 2011, Resolution 36C/14 was adopted that includes exploring the idea of transforming the Geoparks initiative into a formal UNESCO Geoparks Programme.

8.2 Principles

According to Patzak and Eder (1998) and Eder (1999), the original aim of geoparks was to enhance the value of nationally important geological sites while creating economic development, employment and geotourism as part of an integrated programme. These have remained the guiding principles behind geoparks which now have three clearly defined aims:

- conservation of geoheritage;
- geological education for the wider public;
- sustainable socio-economic development mainly through geotourism.

Most importantly, the aim is to allow local communities to take ownership of their geological and other heritage by protecting it, promoting it and, by doing so, gaining some sustainable economic benefit from it. McKeever (2010) stressed that a geopark is not:

- an area of outstanding geological heritage alone;
- a small, single site of geological interest;
- a fenced off area just for scientists;
- a geological theme park;
- an area with no local community involvement;
- an area with no sustainable development strategy;
- a formal designation with legal obligations or restrictions.

On the other hand, a geopark is:

- an area with a particular geological heritage of international significance;
- an area with a sustainable strategy or plan involving local communities;
- an area that uses its geological heritage as its primary promotional tool but also promotes other aspects of its natural and cultural heritage such as archaeology, history, biodiversity, gastronomy, crafts and cultural traditions;
- an area that attempts to re-establish the link between humanity and our planet;
- an area that respects human rights.

The four founding members of the EGN signed a Charter in Lesvos, Greece, on 5 June 2000 and all new EGN members are obliged to accept this Charter (see Box 8.1).

Box 8.1 The EGN Charter

1. A European Geopark is a territory which includes a particular geological heritage and a sustainable territorial development strategy supported by a European programme to promote development. It must have clearly defined boundaries and sufficient surface area for true territorial economic development. A European Geopark must comprise a certain number of geological sites of particular importance in terms of their scientific quality, rarity, aesthetic appeal or educational value. The majority of sites present on the territory must be part of the geological heritage, but their interest may also be archaeological, ecological, historical or cultural.
2. The sites in a European Geopark must be linked in a network and benefit from protection and management measures. The European Geopark must be managed by a clearly defined structure able to enforce protection, enhancement and sustainable development policies within its territory. No loss or destruction, directly or via sale, of the geological values of a European Geopark may be tolerated. In this respect European Geoparks are managed within the framework established by the Global Geoparks Network Charter (see later).
3. A European Geopark has an active role in the economic development of its territory through enhancement of a general image linked to the geological heritage and the development of Geotourism. A European Geopark has direct impact on the territory by influencing its inhabitants' living conditions and environment. The objective is to enable the inhabitants to re-appropriate the values of the territory's heritage and actively participate in the territory's cultural revitalization as a whole.
4. A European Geopark develops, experiments and enhances methods for preserving the geological heritage.
5. A European Geopark has also to support education on the environment, training and development of scientific research in the various disciplines of the Earth Sciences, enhancement of the natural environment and sustainable development policies.

6. A European Geopark must work within the European Geopark Network to further the network's construction and cohesion. It must work with local enterprises to promote and support the creation of new by-products linked with the geological heritage in a spirit of complementarity with the other European Geoparks Network members.

The GGN has a longer set of Operational Guidelines agreed in 2004 and subsequently amended. For example, there is a requirement to have a thorough management plan, partnership working, and appropriate publicity and promotion of the geopark designation. In relation to the collection and/or sale of geological materials from a geopark, the following statement is made,

> A Geopark must respect local and national laws relating to the protection of geological heritage. In order to be seen to be impartial in its management of the geological heritage, its managing body must not participate directly in the sale of geological objects within the Geopark (no matter from where they are) and should actively discourage unsustainable trade in geological materials as a whole, including shortsighted selling of Earth heritage, minerals and fossils. Where clearly justified as a responsible activity and as part of delivering the most effective and sustainable means of site management, it may permit sustainable collecting of geological materials for scientific and educational purposes from naturally renewable sites within the Geopark. Trade of geological materials based on such a system may be tolerated in exceptional circumstances, provided it is clearly and publicly explained, justified and monitored as the best option for the Geopark in relation to local circumstances. Such circumstances will be subject to debate and approval by the GGN/EGN on a case by case basis.

Local involvement is seen as essential with nomination coming from local communities and local authorities committed to developing and implementing a management plan that promotes and protects the geodiversity of the local landscape and meets local economic needs.

According to McKeever (2010), territories must demonstrate at the time of their application for membership of the GGN that they are already a *de facto* geopark, i.e. operating as a geopark in all but name. Thus the pre-requisites are:

- significant geological heritage;
- geoconservation activities;
- sustainable tourism activities;
- educational activities;
- community involvement;
- strong management structure;
- secure financial basis.

Applications for GGN membership may be made to the Geoparks Secretariat in Paris between 1 October and 1 December in each year, and are restricted to two applications per country per year except for countries new to the network when three applications are permitted in the first year. The application dossier will comprise a maximum of 50 pages and must precisely follow a set format with topics that include definition of the territory, scientific description, arguments for nomination, general information on the territory and letters of support. The application's self-evaluation is also submitted, which includes questions on geodiversity, including:

- How many geological periods are represented in your area?
- How many clearly defined rock types are represented in your area?
- How many distinct geological or geomorphological features are present within your area?

After initial checking of applications there is a field visit by independent geopark experts and a recommendation to a Geoparks Bureau who will make the decision about admission to the GGN. It can take up to a year to complete the process. Applicants are now required to keep the UNESCO National Commissions and relevant Ministries informed of their proposed applications. Thus what began as very much a bottom up, local community movement, has over the years become more top-down and bureaucratic. It is also fairly expensive to maintain membership and, perhaps because of this, several national geoparks have developed outside the GGN (see below).

GGN status lasts for four years after which there is a re-evaluation process that assesses:

- progress in geoconservation;
- progress in education;
- progress in sustainable tourism;
- progress in community involvement;
- progress in economic development;
- involvement in the GGN/EGN;
- continuing financial security;
- visibility within the community.

Following this, members can then be given a:

- 'green card' if it is operating fully and strongly as an EGN member;
- 'yellow card' if there is some identified problem. Membership is renewed for two years to allow the issue(s) to be addressed. There is then a further re-evaluation process which can either result in a 'green card' and a four year renewal if everything is satisfactory, or:
- 'red card' if the original problem remains, and membership of the EGN is lost.

Geoparks are expected to form partnerships with community and tourism organisations in their areas, including schools, tour operators, accommodation providers, activity operators, restaurants and agritourism producers. They are also expected to work/network with other geoparks using e-mail, and many geoparks exchange staff and/or exhibitions.

8.3 The European Geopark Network (EGN)

As stated earlier, the EGN acts as the GGN in Europe. As of October 2012 there were 52 geoparks in the EGN and their distribution is shown in Figure 8.1. The EGN has a structure comprising an Advisory Committee (12 members including representatives of UNESCO, IUGS and IUCN) that gives advice on the development and expansion of the Network, and a Co-ordination Committee (comprising 2 members from each geopark) that makes decisions about the operation and management of the Network. The Co-ordination Committee elects an EGN Co-ordinator and Vice-Co-ordinator to liaise with other international organisations and prepare the agenda of meetings in partnership with the meeting host. The Co-ordination Committee meets at least twice per year in different geoparks to discuss the Network's progress, share best practice and experiences, and make decisions on new applications for European Geopark (and thus GGN) status and can also consider extensions to geopark boundaries. These meetings help to develop 'brand identity', geotourism marketing strategies and geoconservation approaches.

The EGN holds an annual conference and in 2010 at Lesvos Petrified Forest Geopark in Greece it celebrated the 10th anniversary of the founding of the Network. As well as meetings and conferences, a Geopark Fair is often held during the annual conference (Figure 8.2). This allows geoparks to publicise their initiatives and activities. There is also an annual geoparks week or fortnight of special activities, exhibitions, walks, talks, workshops, childrens' projects, music, food, etc., for the public. Intensive training courses are also run by the EGN on topics covering geoconservation, geotourism, geopark management and so on.

Figure 8.2 Part of the Geopark Fair at the European Geoparks Network Conference in Naturtejo Geopark, Portugal in 2009.

An EGN magazine is published regularly giving news of events, conferences and activities in several of the geoparks in the Network. Most EGN members run schools' programmes, guided walks, museum exhibitions and other initiatives. Descriptions of the work of many of the European Geoparks can be found in the attractive book edited by Zouros (2008). In it, Zouros and McKeever (2008) give several examples of Geopark partnership work supported by EU funding. Ramsay *et al.* (2010a, 2010b, 2010c, 2010d) review the educational, regional development, research and geotourism activities and successes of the first 10 years of the EGN.

One of the aims of geoparks is to promote sustainable socio-economic development, and although there has been little formal research to establish the impact of geopark status in achieving this there are certainly some notable examples of local employment. For example, Marble Arch Caves Geopark in Northern Ireland (Figure 8.1) in 2010 had 8 full-time permanent staff, 35–40 seasonal staff and 4 others giving a proportion of their time to geopark work (K. Lemon, pers. comm.). In 2008, it became the world's first transnational geopark as a result of including parts of County Cavan in the Republic of Ireland. Haute-Provence Geopark in France is even more successful, attracting over 100 000 visitors per year and employing several hundred staff in the tourist season, but by far the most successful geoparks are in China (Zhao *et al.*, 2012). Shilin (Stone Forest) Geopark, for example, attracts over 2 million visitors per year and employs thousands of staff. But as Zouros (2012) points out, this is only part of the geopark story because other job opportunities are created in the local area in hotels, guest houses, restaurants, food and drink producers, local artisan businesses and other tourist enterprises. 'In this way Geopark visitors experience not only the rich natural heritage of the area ... but also the culture, tradition and local production of the region'.

8.4 Other 'geoparks'

Several areas describe themselves as 'geoparks' without being members of the EGN or GGN, and some countries have national geopark networks all or some of whose members are not part of the EGN or GGN. Germany, for example, has a number of national geoparks, including the 'Ruhr Area National GeoPark', as well as EGN/GGN members (Wrede and Mügge-Bartolovic, 2012) and Japan has a similar geopark structure. Switzerland is establishing its own independent network of geoparks but has also agreed a procedure for Swiss geoparks to apply for EGN/GGN membership (Reynard, 2012). Taiwan is not permitted to be a member of the UN and therefore any geoparks that are established there cannot become part of the UNESCO-assisted GGN. This includes Yehliu Geopark on the northern coast of the island covering only 24 ha of coastal erosion features including 'candle rocks' and the 'Queen's Head'. As of September 2012, China had over 140 national geoparks, but only 27 of which were GGN members. In other places, aspiring GGN members that are not yet accepted into the network use the 'geopark' label prematurely, and some areas that have left the network continue to use the name, though not the logo. Finally, some areas just call

Figure 3.12 Buttes at Monument Valley, USA.

Figure 3.15 Incised meanders of the San Juan River, USA.

Geodiversity: Valuing and Conserving Abiotic Nature, Second Edition. Murray Gray.

(a)

(b)

Figure 3.18 Diversity of relict coastal erosional features: (a) underct cliffline and (b) natural arch, Scotland.

Figure 4.17 Geotourism at the Grand Canyon.

Figure 4.22 The 'White House' in Canyon de Chelly, Arizona.

Figure 6.3 Green Point, Newfoundland, Canada. The GSSP for the Cambrian/Ordovician boundary is about half way along the section.

Figure 7.3 A cliff section at Joggins World Heritage Site, Nova Scotia, Canada.

Figure 8.7 Tall standing petrified tree at Lesvos Petrified Forest Geopark during the GGN's 10th Anniversary Conference field trip, 2010.

Figure 9.9 Speleothems from Carlsbad Cavern National Park.

Figure 9.25 Folded carbonaceous rocks with siliceous intercalations at Ag Pavlos, Crete, Greece (Photo by permission of Elpida Athanassouli).

Figure 9.29 Part of Bungle Bungle National Park in Western Australia showing the typical horizontal stripes (see text for explanation). (Photo by permission of Grant Dixon Photography.)

Figure 9.31 Mitre Peak (1692 m), Milford Sound, Fiordland National Park, New Zealand. The fjord landscapes and diverse geological features of this National Park are also protected through World Heritage Area status. (Photo by permission of Graeme Butterfield.)

Figure 10.2 Rock outcrops and mesomorphology often contribute to landscape character. An example from the granite landscapes of Peneda-Geres National Park, Portugal showing the physical, biological and cultural layers of landscape.

Figure 11.2 The cottages under threat and the Chalk cliffs at Birling Gap, East Sussex, England.

Figure 13.4 'The Wave' on the Utah/Arizona border United States.

themselves 'geoparks' without any intention to apply for official membership. All this must be very confusing for the public and troubling to the EGN and GGN since they are working hard to develop a quality Geopark brand and there is no quality control on these unofficial geoparks.

8.5 Geoparks and geodiversity

Even a casual surfing of the EGN website (www.europeangeoparks.org) reveals that the network demonstrates the great diversity of European geology from the mining heritage of the Copper Coast Geopark in Ireland and the Mining Park of Sardinia in Italy to the fossiferous parks such as Haute-Provence in France or Hateg Country Dinosaur Geopark in Romania to the igneous rocks of Magma Geopark in Norway (Figure 8.1). Some European Geoparks are mainly based around a single type of geological feature, for example Lesvos Petrified Forest Geopark in Greece or Vulkaneifel Geopark in Germany, but even these can demonstrate diversity. For example, the latter's website states that 'the concentration and variety of maar-craters give the Vulkaneifel an outstanding position among the worldwide volcanic regions'. But the large majority of European Geoparks try to attract visitors by promoting their geodiversity. The clearest example of this is the Gea Norvegical Geopark in Norway (see Box 8.2).

Box 8.2 Gea Norvegica Geopark, Norway

This geopark's application dossier for nomination as a European Geopark (Dahlgren, 2006) carries on its front cover the description 'Unique geodiversity in an old rifted continent' and inside we find the statement that:

> Unique for the Gea Niorvegica Geopark is the extreme geodiversity present. During the very long time-span of geological evolution, a great variety of rock types and geological deposits formed through a wealth of processes in widely different geological environments. ... Textbook examples can be found in almost any field of geology, e.g.:
>
> - Deep orogenic environments
> - Rift environments
> - Ductile deformation: folds and mylonites
> - Faults and fractures
> - Sediments of highly different origins
> - Glacial deposits and glacial features
> - Volcanics, plutons, dykes, sills, laccoliths, pegmatites
> - Metamorphic rocks in orogens and contact aureoles
> - Metasomatic rocks and hydrothermal vein deposits
> - Weathering and soils
> - A wide range of fossils and minerals
> - Many different geological resources (gravel, minerals, rocks, oil)

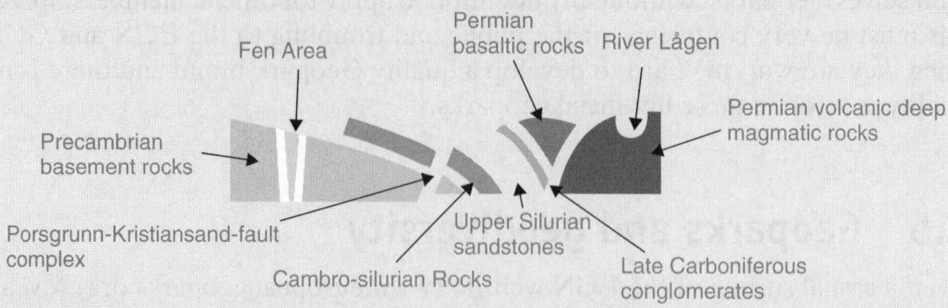

Figure 8.3 The Gea Norvegica Geopark logo explained in terms of the structural elements of the area. (Reproduced with permission from the Gea Norvegica European and Global Geopark, Norway).

The simplified geological structure of the geopark forms the basis for the attractive logo (Figure 8.3). Specific examples of geodiversity include:

- the 40 types of rhomb porphyry identified in or near the geopark (Figure 8.4);
- the more than 150 different minerals identified in the Langesundfjord area of the geopark, over 50 of them on the one small island of Løven;
- the Rognstranda area where well-developed Precambrian banded gneiss (Figure 3.3) has been sculpted by glacial abrasion, and where a Precambrian–Cambrian unconformity is seen;
- the Fen Carbonate Complex around Ulefoss;
- the Larvik plutonic complex with its famous larvikites;
- Younger Dryas end moraines and Lateglacial sea-levels.

Figure 8.4 Examples of the diversity of rhomb porphyry on display in Gea Norvegica Geopark.

Other European geoparks emphasising their geodiversity on their websites include:

- North West Highlands Geopark, United Kingdom—'The geopark contains some of the most important and diverse geological and geomorphological features in Britain';
- Cabo de Gata-Nijar Geopark, Spain—'its geological diversity derives from the predominance of volcanic substrata and recent coastal deposits';
- Terra Vita Geopark, Germany—'by means of an enormous variety of different sedimentary rocks, which are accessible in natural or artificial openings, this huge area can be investigated, reconstructed and explained to visitors';
- Berstrasse-Odenwald Geopark, Germany—'the region between the Rhine, Main and Neckar exposes not just a great variety of magmatic and sedimentary rocks, but also the tracks of two tectonic events';
- Copper Coast Geopark, Ireland—'The Copper Coast is an outdoor geology museum with a geological heritage that reflects the variety of environments under which the area has evolved over the last 460 million years'.

Several geoparks in other parts of the world also advertise their geodiversity. For example, according to the website of Stonehammer Geopark, in New Brunswick, Canada, it displays 'a billion years of a relatively continuous geologic history. ... Geological stories include plate tectonics, mountain building, volcanism, earthquakes, glaciation, sedimentology, mineralogy, geological time, weathering, evolution of invertebrates, vertebrates and plants, geomorphology and the history of geology'. Similarly, Itoigawa Geopark in Japan has 'an extraordinary variety of geosites' including rock diversity, age diversity, topographic diversity and a 'deep relation between Man and the Earth'. Jehu Island Geopark in South Korea has 'well-preserved diverse volcanic landforms'.

8.6 Other geopark case studies

A book series, *Geoparks of the World*, is being published, the first of which covers Danxiashan Global Geopark in China (Peng and Eder, 2012). A selection of other Global Geoparks is outlined here.

8.6.1 Langkawi Geopark, Malaysia (2007)

Langkawi Geopark comprises a 99-island archipelago off the north-west coast of Peninsular Malaysia, close to the Thailand border. In addition to its rich geodiversity and biodiversity, the geopark is also a land of many legends, myths, cultures and traditions and therefore is a good example of the multidisciplinarity of most geoparks. On the island there is a Geopark School (whose pupils are called 'Geoparkians') and a 'Geopark Hotel'.

The Geopark is managed by the Langkawi Development Authority (LADA) established in 1990 to promote tourism to the islands and now incorporating a Geopark Division. A Memorandum of Understanding was signed between LADA and the University Kebangsaan Malaysia and this has been very significant in developing the geopark initiative through the input of the university geologists. There is also active involvement of Kedah State in which Langkawi is located, and the Forestry Department of Peninsular Malaysia.

The geology of Langkawi comprises four sedimentary rock formations, one granitic formation and a series of Quaternary deposits, mainly alluvium.

- The oldest formation is the Cambriam Machunching Formation, mainly sandstones with some conglomerate and shale forming the Machuching Mountains in the north-west of the main island, Pulau Langkawi and designated as the Machuching Geoforest Park. Visitors can now access the mountain ridge via an impressive cable car system (Figure 8.5).
- The Setul Formation comprises a series of limestone and detrital beds formed from the Ordovician to the Middle Devonian and found mainly in the east of the archipelago. Most of the outcrop in the east of Pulau Langkawi is designated as the Kilim Geoforest Park, which can be visited by boat from Tanjang Rhu. The landscape here comprises a conical karst and visitors can explore Bat Cave with its impressive speleothems. On the coastline there are many limestone sea-notches, caves, arches and stacks.
- The Late Devonian to Early Permian Singa Formation comprises dark coloured shale and siltstone found in central and south-west Palau Langkawi and offshore islands.
- The Middle to Late Permian Chuping Formation comprises limestones found mainly on the western coast of Pulau Dayang Bunting which is the third Geoforest Park in the archipelago and home of the legend of the Lake of the Pregnant Maiden.

Figure 8.5 The upper Machuching cable car station in Langkawi Geopark, Malaysia.

- The granitic rocks are found in stocks and bosses throughout the archipelago.
- Quaternary alluvium is mainly found in the western valleys and coast, among the limestone valleys of Kilim Geoforest Park and on the coast at Kuah, the islands' capital.

Leman *et al.* (2007) describe the geodiversity of Langkawi including the rock, mineral, fossil, sedimentary structure, deformation structure, geomorphic and landscape diversity found in the archipelago and they describe the several geological monuments and protected geosites.

8.6.2 Hexigten Geopark, China (2005)

This Chinese Geopark lies in Inner Mongolia, 400 km north of Beijing. Its 1750 km^2 area comprises eight diverse park areas:

- Arshihaty granite forest—is a 'forest' of stone columns created by frost shattering and wind erosion.
- Qingshan granite potholes—this is a 2 km^2 area of over 300 potholes up to a few metres in diameter, eroded by glacial meltwater.
- Dalai Nur volcanic landscape—lies on the north-west shore of the Dalai Nur lake. The volcanic features present include a basalt plateau and the plugs of extinct volcanoes which formed islands in the lake when it stood at a higher level.
- Reshuitang thermal springs—is a mineral hot spring bath area in the eastern part of the geopark.
- Huanggangliang nunatak—at over 2000 m, Mount Hanggangliang is the highest peak in the Greater Khingan mountain range. A glacial trimline exists at 1500–1700 m above which the peak is extensively periglaciated demonstrating that it remained as a nunatak during the last glaciation.
- Pingdingshan scenic cirque group—this group of cirques in the south-east of the geopark provides further evidence of the glacial history of northern China.
- Xilamulun River canyon—the Xar Moron River flows north-east through this impressive canyon which was the original homeland of the Khitan people.
- Hunshandake sand sea—this is one of the four largest sand sea in China covering an area of 53 000 km^2. Since the 1960s the fraction of Hundshandake covered by sand dunes increased from 2% to 33% due to rapid desertification as a result of overgrazing. Only the part of the sand sea on the south shore of Dalai Nur lies within the geopark.

8.6.3 Arouca Geopark, Portugal (2009)

This is a relatively new and small geopark of 328 km^2 coinciding with the Municipality of Arouca, about 40 km south-east of the city of Porto and 30 km inland from the Atlantic Ocean. It is an area of mountains and valleys strongly influenced by the Hercynian orogeny. The geopark is managed by the Arouca

Geopark Association (AGA), a non-profit-making association of private law created by public deed in 2008. The AGA is composed of (i) a General Assembly with 24 members (seven from public organisations and 17 private members), (ii) a Board of Directors with five members (two public and three private and whose chairman is the President of the Municipality), and (iii) a small Council of Auditors. In addition, the science policy and activities of the AGA are supported by a Scientific Advisory Board.

Geologically, the geopark comprises five main elements:

- a basement of metasedimentary rocks of the Schist–Greywacke Complex;
- an early Palaeozoic sequence of Ordovician and Silurian quartzites and schists;
- important Carboniferous clastic outcrops;
- six main Variscan granitic stocks, bosses and dykes, some of which are linked to exploited mineralizations;
- Quaternary river and slope deposits.

Research on the geodiversity of the geopark has identified 41 geosites, the most important of which are:

- the Canelas Quarry and Geological Interpretative Centre. This quarry is in Middle Ordovician slate and produces roofing slates and other building materials. This is an internationally important site with a museum developed by the quarry operator and opened to the public in 2006 which houses palaeontological exhibits including trilobites, bivalves, gastropods cephalopods, crinoids, ostracods, graptolites and ichnofossils. However, the most spectacular fossils are some of the largest trilobites found worldwide (up to 90 cm) and trilobite clusters (Gutiérrez-Marco *et al.*, 2009).
- Pedras Parideiras (rocks delivering stones) is a small granitic body in the village of Castanheira. The granite contains rare biotite nodules up to 12 cm in diameter with a quartz/feldspar nucleus. Weathering of the biotite outer layer releases the nodules from the surrounding rock and this had led to a popular myth related to female fertility.
- 'Maize Bread' rocks of Junqueiro. Here two giant granitic boulders present polygonal weathering that is likened to the surface cracking of maize bread (Figure 8.6).
- Regoufe Wolfram and Tin Mines. These mines opened in 1915 but reached their peak production during the Second World War. The mines were abandoned shortly afterwards but visitors can still see many remains of the old mining activities.

8.6.4 Lesvos Petrified Forest Geopark, Greece (2000)

The Greek island of Lesvos lies in the north-east Aegean Sea near the Turkish coast (Figure 8.1). On its western tip lies the impressive Lesvos Petrified Forest Geopark. It was a founder member of the EGN and has been designated and protected as a Greek Natural Monument (Presidential Decree 443/85). The

Figure 8.6 A 'maize bread' rock in Arouca Geopark, Portugal.

Figure 8.7 Tall standing petrified tree at Lesvos Petrified Forest Geopark during the EGN's 10th Anniversary Conference field trip, 2010. See plate section for colour version.

outstanding feature of the Geopark is the large number of petrified trees, both standing and lying. But it is the standing trunks, one of which is over 7 m tall (Figure 8.7) while another is 13.7 m in circumference and many of which have their root systems intact and fully developed, that give the Geopark international importance. A diverse range of fossilised trees is represented in the Geopark, including sequoia, pine, cypress, palm, oak, plane, laurel, cinnamon, alder and poplar. As well as the large number of tree trunks and roots, fruits, leaves and seeds have also been recovered. The petrified forest was created during intense volcanic activity in the northern Aegean area about 20 million years ago, but excavations reveal several different periods of forest development separated by ash and pyroclastic layers, indicating that volcanic activity was spasmodic allowing forest to develop between eruptions.

The Natural History Museum of Lesvos Petrified Forest located in the west coast village of Sigri has excellent exhibitions of petrified wood, leaves etc., of the evolution of plant life, of volcanic activity and of the volcanic history of the Aegean. The Museum also coordinates research, educational and geotouristic activities in the Geopark. It receives about 90 000 visitors annually

8.6.5 Hong Kong Geopark, China (2011)

Hong Kong is usually thought of as an international financial and business centre, a city of skyscrapers and neon lights. But surrounding the city is some remarkable geology that has been recognised recently through the establishment of a Chinese National Geopark in 2009 and GGN membership in 2011. The most remarkable feature is undoubtedly the columnar jointing in the acidic volcanic rocks on the islands of south-east Hong Kong (Figure 8.8), believed now to be part of a super-caldera eruption during the Cretaceous. The geopark also displays spectacular coastal landforms (e.g. cliffs, caves, natural arches, stacks, shore platforms, beaches), tectonic features (e.g. folds, faults, tilted layers), and sedimentary formations from Devonian to Palaeogene (e.g. breccia, conglomerate, sandstone,

Figure 8.8 Columnar jointing in Hong Kong Geopark.

siltstone, mudstone and shale some of which are fossiliferous). The geopark is split into two regions (the Sai Kung Volcanic Rock Region and the Northeast New Territories Sedimentary Rock Region), each of which has four Geo-Areas. The Geopark also contains many cultural features including traditional fishing villages and temples as well as biodiversity. The village of Lai Chi Wo is a traditional Hakka village enclosed by fung shui walls and woods to retain wealth and keep out evil spirits.

8.6.6 Jeju Island Geopark, South Korea (2010)

Jeju Island lies off the south coast of the Korean peninsula and essentially consists of a single shield volcano called Hallasan, rising to 1950 m and one of Korea's three sacred mountains. Satellite cones occur on the volcano flanks and there are a number of offshore islands, remnants of past eruptive cones or lava flows. The island measures circa 74 × 32 km and contains nine geosites (including one geocluster). Seven of these geosites are designated as Korean National Natural Monuments. The geosite cluster is at the summit of Hallasan and comprises several crater lakes, a trachyte dome, several scoria cones, two lava tube caves and other features. The area was designated as a natural monument in 1966, a national park in 1970, a UNESCO Biosphere Reserve in 2002 and a UNESCO World Heritage Site in 2007. The other geosites include a lava dome, two tuff rings, a tuff cone, a columnar basalt site, a waterfall, the world's longest lava tube (13 km) and a section showing vocaniclastic sediments interbedded with fossiliferous, non-volcanic sediments from the Pliocene.

8.7 Conclusions

Hayward (2009) comments that given the number of existing categories of protected area in New Zealand, he sees 'no need for the introduction of a new category, such as Geoparks'. This is not a view that is widely shared because the Geopark story is one of remarkable success in drawing public attention to the place of geology within the wider environment and local culture. From the formation of the idea in the late 1990s to the first four European Geoparks in 2000, to the creation of the Global Geoparks Network in 2004, to today's 90 members, the development of the initiative has been a remarkable phenomenon and one that has been extraordinarily good for geoscience and geotourism. Those responsible for this success and those managing these geoparks, should be very proud. This is not to say that there are not concerns, for example as expressed by Hose and Vasiljevic (2012, p. 38,) that the focus of the geopark movement has become more on geotourism and economic development and less on geoconservation and geological research. There are also areas that promote themselves as geoparks outside the formal EGN/GGN system. Nonetheless, the UNESCO supported Global Geopark brand is only likely to expand and flourish in the future because of the perceived benefits that this status brings.

9 National Geoconservation

National Parks: America's 'best idea'

James Bryce, British Ambassador to the United States, 1912

9.1 Introduction

The most recent IUCN and UNEP-WCMC figures give the total nationally protected areas as covering almost 25 million km^2. Most of this (over 16 million km^2) is on the continents, representing about 11% of the Earth's land surface. About 7.5 million km^2 comprise marine protected areas, an estimated 2.3% of the world's oceans. Both the number and area of protected sites increased rapidly during the twentieth century (Figure 9.1). However, Mather and Chapman (1995) asked us to view such figures with caution. For example, designation means little if there is no respect for the principle and no system of enforcement. This is the problem of the so-called 'paper parks'. Secondly, it may be easier to designate remote locations that are not under threat and where the need for conservation is lower, than areas where threats from mining, logging or agriculture are greater and where conservation measures are less welcomed. Several countries are working to increase the number and area of marine/coastal protected zones.

One of the main types of protected area is national parks. The American concept of national parks was formally adopted by the IUCN in 1969 and national parks that conform to it have been established all over the world. Most countries now define part of their territory with this name. Today, more than 100 nations contain some 1200 national parks or equivalent preserves, though the aims differ from country to country. In some places cross-border parks have been established such as the Glacier/Waterton International Peace Park and World Heritage Site on the United States/Canada border (Figure 9.2).

In the sections that follow, some details of national geological site protection systems are described. I am only too well aware of the gaps and deficiencies in this information, particularly for developing nations. For more information on specific countries, government and geological websites can often provide detailed information.

Geodiversity: Valuing and Conserving Abiotic Nature, Second Edition. Murray Gray.

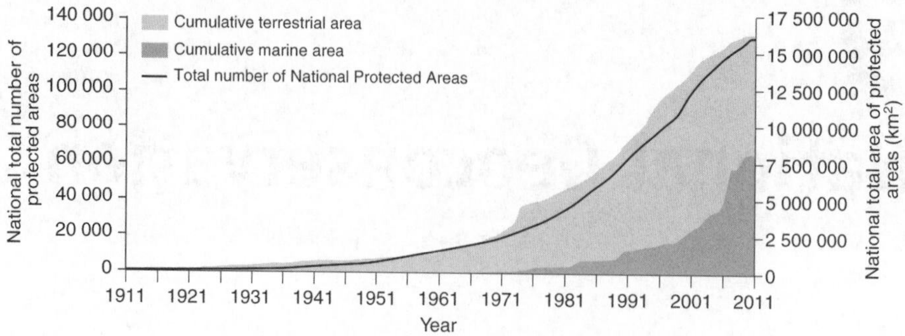

Figure 9.1 Growth in the number and area of nationally protected areas (modified by permission of UNEP-WCMC. Source: IUCN and UNEP-WCMC (2012) The World Database on Protected Areas (WDPA), February 2012. UNEP-WCMC, Cambridge, UK; www.wdpa.org/Statistics.aspx).

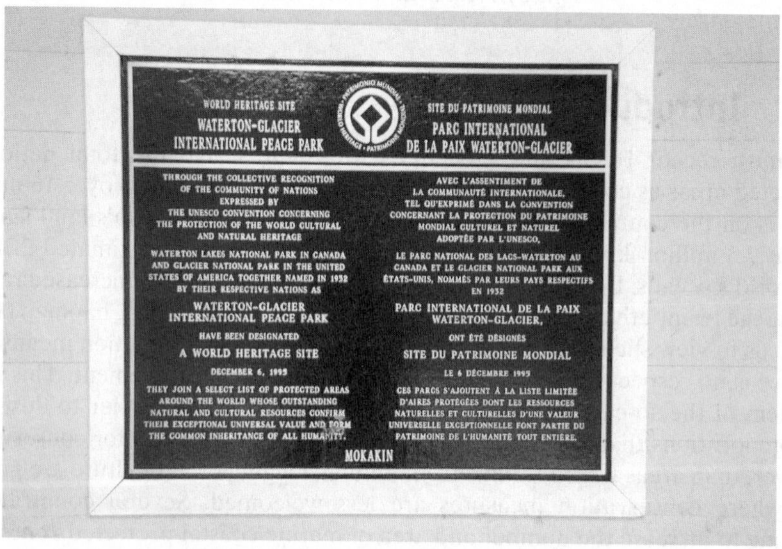

Figure 9.2 Plaque commemorating the inauguration of the Waterton/Glacier Peace Park and World Heritage Site on the Canada/United States border.

9.2 United States

Miller (1998, p. 388) claims that, 'No other nation on Earth has set aside such a large portion of its land for the public's enjoyment and use'. Over one-third (35%) of the country's land is managed by the federal government. However, of this 35%, 73% is in Alaska and 22% in the western states, leaving only 5% in the central, eastern and southern states. Many of these areas have been established primarily for their wildlife resources rather than their geodiversity. Nonetheless, the protection of land for wildlife usually also protects the contained geology and geomorphology. Examples include the national forests, national wildlife or marine refuges, national wilderness preserves and rangeland areas. National Resource lands are mainly rangelands, administered by the Bureau of Land Management with the emphasis being on providing a secure domestic supply of energy and strategically important non-energy minerals.

9.2.1 The National Park System

Thomas and Warren (2008) describe the early history of geoconservation in the United States. An important step was taken in 1933 when an Executive Order transferred 63 national monuments and military sites from the Forest Service and War Department to the National Parks Service. The General Authorities Act (1970) brought all areas administered by the National Park Service into one National Park System. Areas included in the system are 'cumulative expressions of a single national heritage; that, individually and collectively, these areas derive increased national dignity and recognition of their superb environmental quality through their inclusion jointly with each other in one national park system preserved and managed for the benefit and inspiration of all people of the United States'. These were major steps in improving the integration of America's national system, which includes historical and cultural as well as scenic and scientific areas and sites. Thus as well as National Parks, the units include National Monuments, National Rivers, National Seashores and several other categories. However, there are still anomalies in the system in that the Grand Staircase—Escalante National Monument in Utah is administered not by the National Park Service but by the Bureau of Land Management.

Today, the US National Park System comprises over 390 units (of all types) covering about 35 million hectares in 49 States (none in Delaware), the District of Columbia and overseas dependencies. Some cover huge tracts of land. The largest is the Wrangell-St Elias National Park and Preserve in Alaska and Canada at over 5 million hectares, which is over 16% of the entire system. The system attracts nearly 300 million visitors per year and is run by 20 000 full-time employees.

Anyone can propose an addition to the system, but to be considered nationally significant it must meet all four of the following standards:

- it is an outstanding example of a particular type of resource;
- it possesses exceptional value or quality in illustrating or interpreting the natural or cultural themes of the nation's heritage;
- it offers superlative opportunities for recreation, for public use and enjoyment, or for scientific study;
- it retains a high degree of integrity as a true, accurate, and relatively unspoilt example of the resource.

Three other criteria are important. First, it must not represent a feature already adequately represented in the system. Adequacy of representation is determined on a case-by-case basis taking account of character, quality, quantity, and so on. Secondly it must be feasible in terms of land ownership, acquisition costs, access, threats to the resource, and staff or development requirements.

Thirdly, additions to the National Park System will not usually be recommended if another arrangement can provide adequate protection and opportunity for public enjoyment. This might include management by state or local governments, Indian tribes, the private sector, or other federal agency, or might include the designation of federal lands as wilderness, areas of critical environmental concern, national conservation areas, national recreation areas, marine or estuarine sanctuaries, and national wildlife refuges. Requests for boundary changes may also be made in order to address operational problems or include resources that are critical to the park's purposes.

The National Parks website (www.nps.gov) gives a list of 10 examples of natural areas that might meet the criteria, six of which refer explicitly to geological or geomorphological features:

- 'an outstanding site that illustrates the characteristics of a landform or biotic area that is still widespread';
- 'a landform or biotic area that has always been extremely uncommon in the region or nation;
- 'a site that possesses exceptional diversity of ecological components ... or geological features (landforms, observable manifestations of geological processes);
- 'a site that contains rare or unusually abundant fossil deposits;
- 'an area that has outstanding scenic qualities such as dramatic topographic features, unusual contrasts in landforms or vegetation, spectacular vistas, or other special landscape features;
- 'a site that is an invaluable ecological or geological benchmark due to an extensive and long-term record of research and scientific discovery'.

This represents an impressive awareness of the geological/ geomorphological heritage of the United States, with landscape, landforms, processes and fossils all mentioned at least once. However, there is no mention of rocks, minerals or soils and in no sense is the system representative of the full geodiversity of the United States. Rather it conserves the most spectacular scenic areas, many geomorphological curiosities and some important fossil sites.

Areas are usually added to the National Park System by an act of Congress following an investigation by the National Parks Service, a positive recommendation to add the area and a congressional hearing. Congress decides on the park boundaries and which of several titles to apply to the area. The National Park Service also considers changes to park boundaries and makes recommendations. Some issues related to park boundaries are outlined in Box 9.1.

Box 9.1 Some boundary issues at US National Parks

Delimiting the right boundaries for protected areas is a key consideration. In a number of US National Parks, the boundaries have been, or should be, expanded to protect further georesources or in response to increased understanding of geological processes.

- A significant example is Petrified Forest National Park in Arizona which was authorized to be increased in area by 130% in 2004 (Figure 9.3). This means that it now includes much of the area where the wood-bearing Triassic Chinle Formation outcrops and from which significant quantities of wood were being removed and sold commercially (Figure 9.4).

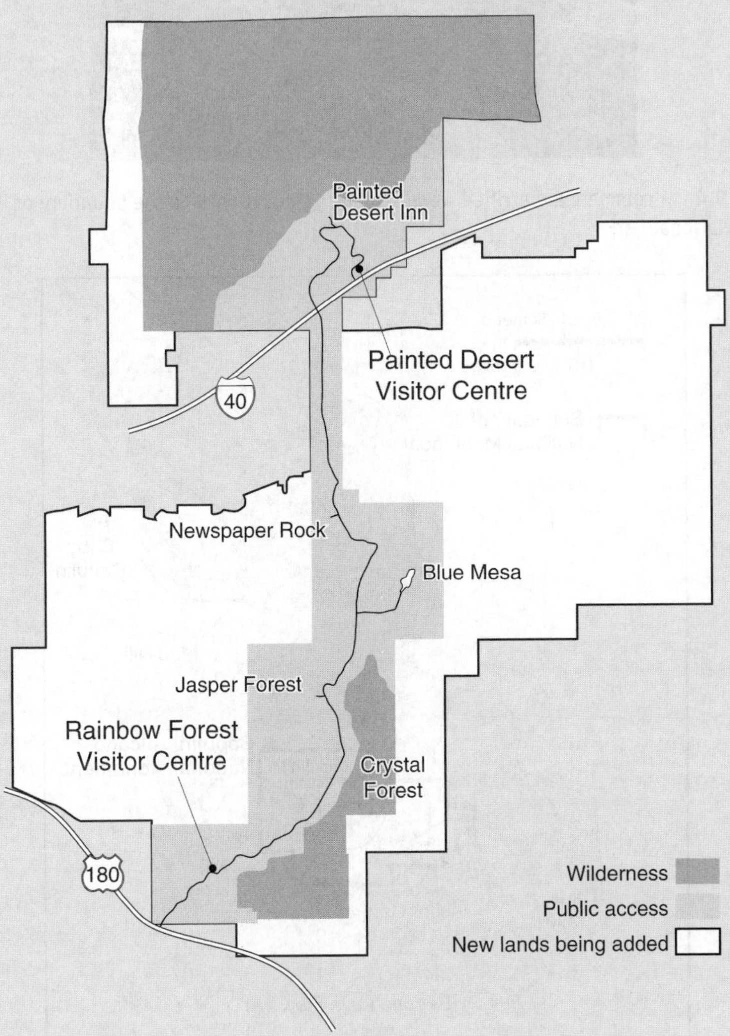

Figure 9.3 Map showing the pre- and post-2004 boundaries of Petrified Forest National Park.

- At Great Sand Dunes National Park in Colorado, the boundary was expanded to include areas of mountain forest from which rivers were returning sand blown there by winds from the sand dune area. This is a

Figure 9.4 A commercial petrified wood storage yard just outside the boundary of Petrified Forest National Park.

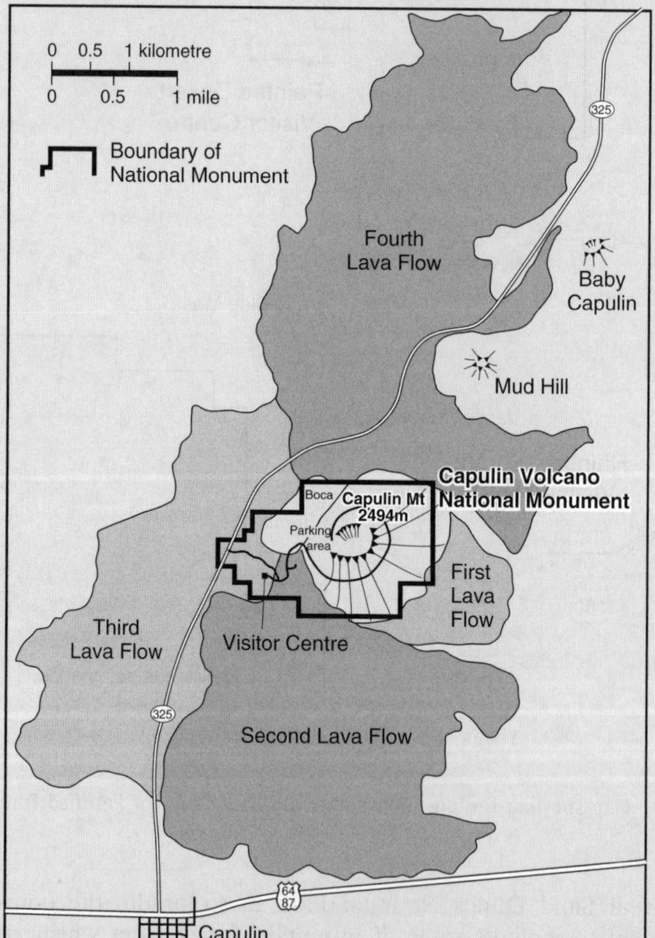

Figure 9.5 Map showing that the boundary of Capulin Volcano National Monument includes only the main cone and 'Boca' and excludes the lava flows and secondary cones.

good example of where increased understanding of physical systems led to a realisation of the need to extend park boundaries. Extension proved possible because the mountain forests were in the ownership of another federal agency—the National Forest Service.

- Extensions to boundaries may be less straightforward where private landowners are involved. At Capulin Volcano National Monument in New Mexico, the site boundary includes only the cone and a small adjacent area known as the Boca. It excludes the related lava flows, some of which display excellent pressure ridges areas, as these are in private ownership (Figure 9.5).
- The boundary of Sunset Crater National Monument in Arizona includes a single cone together with the Bonita Lava Flow. But the cone is part of a fissure eruption that occurred circa 1000 AD. It would be beneficial to include further volcanic features along the fissure within the site boundary including some in the ownership of the National Forest Service where quad-biking is permitted but is crushing the cinder outcrops.

Figure 9.6 shows the units in the National Park System in some of the western states and lists the main unit types, though these titles have not always been applied consistently. The title 'National Park' has traditionally been reserved for the most spectacular natural areas with a wide variety of features (e.g. Figure 9.7), where hunting, mining, and other consumptive activities such as grazing are generally prohibited. The National Park Service Organic Act (1916) gives general protection to national park units after their adoption. However, there are often pre-existing private mineral rights, though these can if necessary be dealt with by land acquisition (as long as suitable compensation is paid) or regulation under the Mining in the Parks Act (1976).

Many of the units in the National Parks system have been established wholly or largely for their geological or geomorphological interest. Box 9.2 describes some of the sites and legislation relating to speleological sites, and Box 9.3 deals with palaeontological parks and monuments. But several other units are well known for their geological/geomorphological interests. These include canyons (e.g. Bryce Canyon National Park), volcanoes (e.g. Hawaiian Volcanoes National Park, Hawaii; Crater Lake National Park, Oregon), geysers and hot springs (e.g. Yellowstone National Park, Wyoming; Hot Springs National Park, Arkansas), rivers (e.g. Buffalo Natural River, Arkansas; Natural Bridges, National Monument, Utah), glaciers and glacial geomorphology (e.g. Glacier Bay National Park, Alaska; Ice Age National Scenic Reserve and Trail, Wisconsin) and sand dunes (Great Sand Dunes National Monument, Colorado; White Sands National Monument, New Mexico). Lillie (2005) describes the geology of over 100 national parks and uses these to illustrate plate tectonic processes, while Sprinkel, Chidsey and Anderson (2000) describe the geology of national park units in Utah as well as 10 state parks (see later).

The National Parks and Conservation Association in the United States has been concerned for many years about human impacts on the national parks and encourages citizens to become involved in park protection battles. Writing

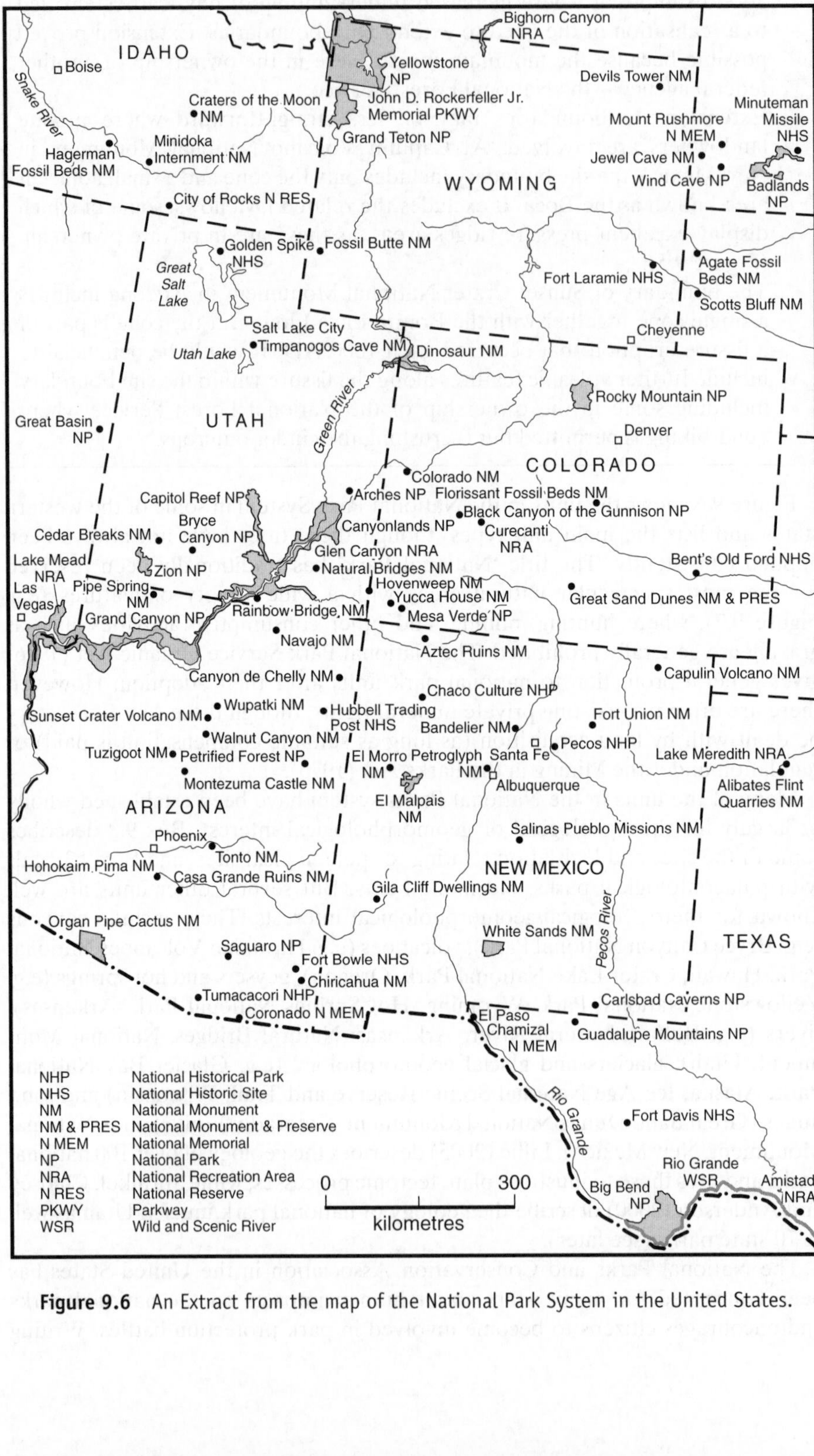

Figure 9.6 An Extract from the map of the National Park System in the United States.

Figure 9.7 Eastern entrance to the spectacular landscape of Glacier National Park, Montana, United States.

in 1994, the organisation argued that 'A tug of war exists between those who see parks as potential revenue sources, and those who share the view of the National Park Service's original enabling legislation, which states that parks must be preserved unimpaired for future generations'. Nowhere has this tug of war been more fiercely fought than over the Hetch Hetchy Valley in Yosemite National Park, California. Controversially dammed in 1913 to provide a supply of water to San Francisco, it has been the subject of campaigns and debate to restore it to its original condition ever since (Righter, 2005; Simpson, 2005). For example, Simpson argues (p. 318) that 'The O'Shaughnessy Dam and Hetch Hetchy Reservoir were a mistake. ... The project would never be approved today, and it should not have been approved in 1913, even recognizing the complicated history and context of the time', and (p. 325) 'It's time to undo the past and make a profound statement about our present priorities and place on the planet—not because it's cost effective ... but because it is the right thing to do' (see Figure 9.8).

Smaller scale effects may also have an impact. The Devils Tower in Wyoming became the first US National Monument in 1906, having been climbed first on 4 July 1893 with the assistance of a wooden ladder. Since then there have been thousands of bolt and piton-aided ascents which have caused significant damage and defacement to the Tower (see Section 5.10). As a result, a *Climbing Management Plan* has been introduced. The Tower, which is composed of phonolite porphyry, is about 180 m high and 250 m in diameter with a relatively flat top and columnar-jointed sides (Figure 4.19). It was featured in the film *Close Encounters of the Third Kind*, and partly as a result of that publicity it attracts over 450 000 visitors per year, 5000 of whom have come specifically to climb the pinnacle.

Figure 9.8 The O'Shaughnessy Dam and Hetch Hetchy Reservoir.

Box 9.2 Caves and Karst

Caves and karst features, including over 3900 caves, occur in over 100 units of the National Park system. These include Mammoth Cave in Kentucky, claimed to be the longest cave in the world with over 500 km of mapped passages in a complex system of caves and tunnels on multiple levels. The caves contain many impressive speleothems and the surrounding area demonstrates impressive karst topography.

Similar characteristics occur at the Carlsbad Caverns National Park in New Mexico (Figure 9.9) where the caves display an impressive array of speleothems including stalactites, stalagmites, columns (one of which, the Monarch, is one of the world's tallest at 25 m high), flowstone and other features. Unfortunately, many of the cave's smaller and more delicate formations have been damaged over the years by careless visitors. A count in 1993 found 32,000 damaged speleothems. The Carlsbad Caverns now exclude visitors from most of the network, run ranger led guides into some caves and have regulations which include the following conservation measures:

- 'Touching cave formations is prohibited. Formations are easily broken and the oil from your skin permanently discolours the rock';
- 'Smoking, or any use of tobacco, is not permitted. Eating and drinking are not permitted except in the Underground Lunchroom';
- 'Throwing coins, food or other objects in cave pools is forbidden. Foreign objects ruin the natural appearance of the pools and are difficult to remove.

Also, the chemical reaction between foreign objects, the water, and the rock can leave permanent stains';

- 'Photography is permitted when you tour on your own, but not when you accompany a ranger-guided tour. Photographers should not rest their tripods or other camera equipment on formations or step off the trails'.

Figure 9.9 Speleothems from Carlsbad Cavern National Park. See plate section for colour version.

A landmark decision was the passage of the Lechuguilla Cave Protection Act (1993) named after a cave in the Carlsbad Caverns National Park that was threatened by oil and gas exploration on adjacent land. The Act states that the cave has 'internationally significant scientific, environmental and other values' and should be protected against adverse effects of mineral exploration, tourism and development. The Act withdraws all federal lands inside the boundaries of a protected cave area from all forms of mineral and geothermal leasing. The protected area was established by an expert panel of geologists and speleologists.

Other measures have been taken to document and manage US karst and caves. In 1988, the US Congress created a major impetus for cave conservation and management by passing the Federal Cave Resources Protection Act (1988). The purposes of the Act are:

- 'to secure, protect, and preserve significant caves on Federal lands for the perpetual use, enjoyment and benefit of all people'; and

- 'to foster increased cooperation and exchange of information between governmental authorities and those who utilize caves located on Federal lands for scientific, education, or recreational purposes'.

The Act directs the Departments of the Interior and Agriculture to record all significant caves on federal lands, to provide management of these resources and to disseminate information on them. The Act made it an offence for any person who knowingly 'destroys, disturbs, defaces, mars, alters, removes or harms any significant cave' or sells, barters or exchanges any cave resource, but licences can be issued for research purposes. This was followed up in 1998 by the passing of the National Cave and Karst Research Institute Act (1998) to establish a research institute within the Carlsbad Caverns National Park.

Box 9.3 Palaeontology

There is a considerable diversity in the fossil record in the National Park System. Petrified trees, leaves, wood, pollen, shells, bones, teeth, eggshells, tracks, burrows and coprolites have been found in more than 145 National Park System units. However, the palaeontological resources of the Parks is still being fully assessed (Santucci, Kenworthy and Kerbo, 2001). The number of palaeontologists employed in the State Park System is very low and in many parks that have significant palaeontological resources there are no geological staff. The fossils discovered range from the Precambrian stromatolites in Glacier National Park (Montana) to the Quaternary mammal bones found throughout the Alaskan parks. Some sites in the National Park system have been established solely or mainly for their palaeontological interest, mostly in the western states:

- Agate Fossil Beds National Monument, Nebraska, is a large fossilised Miocene waterhole where hundreds of skeletons have been discovered, but where the majority are believed to be preserved *in situ*. Among the extinct mammal fossils discovered are *Menoceras, Morpus, Dinohyus* and *Stenomylus*.
- Dinosaur National Monument, Colorado and Utah, where many dinosaur and other fossils have been found over the past 100 years. The rocks within the Monument range from Precambrian to Quaternary and many of them are fossiliferous, but the Upper Jurassic Morrison Formation, a fluviatile sandstone, is one of the most prolific dinosaur-bearing units in the world. The dinosaurs include *Apatosaurus* (*Brontosaurus*), *Diplodocus, Stegosaurus* and *Allosaurus*. One wall of the Dinosaur Quarry building is formed from a tilted sandstone face exhibiting many dinosaur bones (Figure 13.2).
- Florissant Fossil Beds National Monument, Colorado where a great diversity of fossil plants and insects are exceptionally well preserved in Eocene lake mudstone. They are poorly exposed today, but since the late 1800s

over 60 000 fossils have been collected for museums, universities and private collections all over the world. A number of petrified tree stumps are also exposed (Figure 9.10).

Figure 9.10 Petrified Tree stumps in Florissant National Monument.

- Fossil Butte National Monument, Wyoming where 100 m of the Eocene limestones, mudstones and volcanic ash containing some of the most perfectly preserved fossil fish and other aquatic animals and plants are exposed.
- Hagerman Fossil Beds National Monument, Idaho, where over 140 vertebrate and invertebrate species have been discovered in Pliocene river sands and lake clays. Eight of the species have been found nowhere else and 44 were discovered here first, including the Hagerman horse (*Equus simplicidens*).
- John Day Fossil Beds National Monument, Oregon, where 40 million years of Cenozoic rocks (Eocene-Pliocene) contain some of the world's most diverse and rich plant and animal fossil beds, including hundreds of Eocene plant species in the Clarno nutbeds and over 100 Miocene mammals groups in the John Day Formation.
- Petrified Forest National Park, Arizona, where Late Triassic fluviatile sedimentary rocks of the Chinle Formation contain silicified wood from trees such as *Araucarioxylon, Woodworthia* and *Schilderia.* These were first discovered in the 1850s but by 1900 removal of the 'petrified wood' led to calls to preserve the remaining major sites. Collecting within the park is prohibited but commercial dealers collect from the same deposits

outside the park. Consequently Congress authorised extension of the park boundaries in 2004 (see Box 9.1).

Fossil collecting throughout the National Parks System is prohibited by the Antiquities Act (1906) and the National Park Service Organic Act (1916), referred to above, which prevent removal of 'objects of antiquity' from federal lands, though again research permits can be issued. However, a large number of theft and vandalism incidents affecting fossil sites are recorded each year and this has increased following staff training (Santucci, 1999).

9.2.2 National Natural Landmarks Program

This programme was established by the Secretary of the Interior in 1962 under the Historic Sites Act (1935) and is administered by the National Parks Service. A National Natural Landmark (NNL) is a nationally significant natural area as designated by the Secretary of the Interior. To be nationally significant 'a site must be one of the best examples of a type of biotic community or geologic feature in its physiographic province'. Examples include rock exposures, landforms and fossils. It is a goal of the program to identify, recognize and encourage the protection of sites containing the best remaining examples of geological and geomorphological components of the nation's landscape (www.nature.nps.gov/nnl). About 30% of NNLs are entirely privately owned, but they are normally only designated with the agreement of the landowners. In return for their goodwill in protecting the integrity of the feature they receive a certificate and, sometimes, a plaque. As of January 2013 there were 594 designated NNLs and some examples of geological NNLs are given in Box 9.4.

Further sites can be added following nomination by groups or individuals, often as a result of designation in state natural area programmes (see later). Whether the site is added to the programme will depend on the primary criteria of illustrativeness and condition of the specific feature, and secondary criteria of rarity, diversity and values for science and education. For example, in 1972 it was suggested that 'Lakes and Ponds' should include examples of the following: large deep lakes, large shallow lakes, lakes of complex shape, crater lakes, kettle lakes and potholes, oxbow lakes, dune lakes, sphagnum-bog lakes, lakes fed by thermal streams, tundra lakes and ponds, swamps and marshy areas, sinkhole lakes, unusually productive lakes and lakes of high productivity and high clarity.

Evaluations of sites are carried out by scientists, and landowners are notified. If accepted by the National Parks Service, the sites are placed in the Federal Register for a public comment period. If confirmed by the National Parks Service, they are designated by the Secretary of the Interior and listed on the National Register of Natural Landmarks.

NNL designation does not dictate activities that can occur within the site, but the designation has been used by individuals and organisations to draw attention to development threats (Gibbons and McDonald, 2001). Additionally, the Secretary is required to provide an annual report to Congress, prepared by the National Parks Service, on damaged or threatened NNLs. If sites are deemed to have lost the values that originally led to their designation they can be removed from the list.

The scheme has the potential to create a network of sites representative of the scientific geodiversity of the United States, but at the present time it is a long way from doing so, and there seems little prospect of this is the near future. There is also a need for better protection of those sites already designated.

Box 9.4 Examples of geological National Natural Landmarks

The following are examples of a few of the National Natural Landmarks that have been designated for their geological interest.

- *Dinosaur Trackway, Connecticut* was discovered in 1966 when a bulldozer operator discovered a slab of rock at Rocky Hill showing tracks of a three-toed creature. Originally believed to be made by the bird *Eubrontes*, it has subsequently been identified as a bipedal animal with sharp claws and a stride length of over a metre. Comparison with dinosaur bones in Arizona suggests that the tracks may have been made by *Dilophosaurus*, a carnivore about 8 m long and 2.5 m tall, though there is uncertainty about this identification.
- *Diamond Head, Hawaii* is a prominent peak on the rim of an extinct volcano in the south-east of Oahu. It was designated as an NNL to protect its slopes from commercial development at Waikiki Beach.
- *La Brea Tar Pits, California* are situated in downtown Los Angeles, and contain one of the richest, best-preserved and heavily researched assemblages of Quaternary vertebrates in the world. About 60 species of mammal have been found, including extinct species of native horse, camel, mammoth, mastadon, longhorned bison and sabre-toothed cat. Thousands of individuals of one sabre-toothed cat genus, *Smilodon*, have been found at La Brea, and it has been adopted as California's State fossil. Over 135 species of bird have also been found, including vultures, condors, eagles and giant, extinct stork-like birds known as teratorns. Numerous mollusc, insect and plant species have also been discovered and the fossils have been dated to between 40 000 and 8000 yr BP. Many are on display at the George C. Page Museum adjacent to the tar pits.
- *Enchanted Rock, Texas*, is a large, pink granite, exfoliation dome, representing the weathered top of a Precambrian batholith, just over 1000 million years old. It rises 125 m above general ground level.
- *Hickory Run Boulder Field, Pennsylvania*, lies within the Hickory Run State Park (see later) and comprises a large, periglacial blockfield. It covers a flat area measuring circa 120 m by 500 m and some of the blocks are 7–8 m long (Figure 9.11).

Figure 9.11 The periglacial blockfield at Hickory Run National Natural Landmark, Pennsylvania, United States.

- *Barringer Meteor Crater, Arizona* is one of the most spectacular impact craters in the world. It is about 1200 m in diameter and 170 m deep and is surrounded by a rampart rising circa 45 m above the surrounding plain. The meteor, circa 50 m in diameter and weighing several hundred thousand tonnes, impacted about 50 000 years ago. The site is privately owned by descendents of Daniel Barringer who first identified the meteor impact origin of the landform, but it is open to the public with an excellent visitor centre.

9.2.3 National Wild and Scenic Rivers System

The Wild and Scenic Rivers Act (1968) allows rivers and stretches of rivers that have outstanding scenic, recreational, geological, wildlife, historical or cultural or other similar values to be 'preserved in free-flowing condition'. The legislation states that the goal is to preserve the character of the river, and generally prohibits widening, straightening, dredging, filling, damming of the rivers or mining in their vicinity. There are three classes:

- wild—free of impoundments, generally inaccessible except by trail, watersheds or shorelines essentially primitive and waters unpolluted, 'vestiges of primitive America';

- scenic—free of impoundments, watersheds or shorelines largely primitive, shorelines largely undeveloped, accessible in places by roads;
- recreational—readily accessible by road or railroad, may have some development along their shorelines, may have undergone some impoundment or diversion in the past.

The lengths of US rivers now designated under the Act is increasing and by 2008, the 40th anniversary of the Act, over 20 000 km had been designated on 203 rivers in 39 States, though much of this is in the north-western states and Alaska. Oregon, for example, has 48 designated rivers including the Klamath River. Other examples include Michigan's AuSable and Pere Marquette rivers, Connecticut's Farmington River, Louisiana's Saline Bayou, West Virginia's Bluestone River and Massachusetts' Concord, Sudbury and Assebet rivers (www.rivers.gov/rivers).

9.2.4 State Parks

All states in the United States designate State Parks that are intended to be areas for public recreational activities, but many also contain features of scientific interest, including important geological sites, a situation that often leads to conflicting pressures. This is particularly true where state parks are located near urban areas. Some examples of geologically oriented State Parks are outlined in Box 9.5.

Box 9.5 Examples of State Parks in the United States with geological associations

- *Falls of the Ohio State Park* lies on the banks and bed of the Ohio River in Clarksville, Indiana. Between August and October the low water levels reveal extensive exposures of Silurian—Middle Devonian limestone forming part of a patch reef system. The fossils here include corals, sponges, bryozoans, trilobites, molluscs, brachiopods, crinoids and blastoids. Fossil collecting is prohibited within the park, but there is an interpretive centre with gallery of exhibits and video presentation, while the web site allows a virtual tour of the park (Goldstein, 2009: www.fallsoftheohio.org/).
- *Sun Lakes State Park* in Washington was formerly known as Dry Falls because of the impressive 120 m high, 5 km long, dried-up waterfall that operated during the Quaternary Ice Age. Peak discharge is estimated to have been about 10 times greater than the present Niagara Falls.
- *Archbold Pothole State Park* in Pennsylvania boasts what it claims as the World's largest glacial pothole measuring about 11 m deep and 12 m in diameter. It is excavated through sandstone, shale and anthracite coal at its base. In fact it was discovered in 1884 by a coal miner extending a mine shaft, prior to which it had been filled with hundreds of tonnes of rounded

stones and other debris. It is believed that it formed as a plunge pool at the foot of a glacial moulin.

- *Goosenecks State Park* in Utah is interesting as it comprises simply a viewpoint, car park, campsite and toilet block. But from it, the spectacular incised meanders of the San Juan River can seen (Figure 3.15).

9.2.5 State Natural Areas

Most states also designate Scientific and/or Natural Areas. These are specifically cited for their ecological and geological interests, though the ecological sites are predominant. They are usually open to the public for nature observation and education, but are not parks or recreational areas. Consequently there are generally no or only limited facilities for visitors. Most states, e.g. Maine, Texas and Colorado, call these localities 'Natural Areas' and have passed relevant legislation (e.g. the Maine Natural Areas Program Legislation (1993), but in Minnesota they are 'Scientific and Natural Areas'. Some examples from Minnesota are outlined in Box 9.6

Box 9.6 Scientific and Natural Areas in Minnesota, United States

There are over 150 Scientific and Natural Areas (SNAs) in Minnesota but it is estimated that 500 are needed to adequately protect significant features. The large majority are ecological sites, but the need to protect landforms, rock outcrops and fossil remains is also recognised. The programme, begun in 1969, has the goal of ensuring that no single rare feature is lost from any region of the state. The following are some examples of geological sites and these are very well-documented on the websites, all of which show the site boundaries, access roads and parking areas (see Figure 9.12).

- Agassiz Dunes SNA, is a dune field associated with Glacial Lake Agassiz, dammed by an ice-sheet located over the site of the existing Great Lakes. The Sand River flowed into the lake at this locality and formed a delta 12 000–9000 years ago. Subsequently the sand was blown into dunes, and some have been reactivated to give secondary dunes and modern blow outs.
- Gneiss Outcrops SNA (Figure 9.12) are Precambrian rocks dated to circa 3.6 million years ago making them some of the oldest known rocks on earth. Other outcrops along the Minnesota River have been mined and developed. The granitic gneiss is banded with light pinks and red most common, with grey to black hornblende-pyroxene or garnet-biotite gneiss being less common. At the nearby Blue Devil Valley SNA (Figure 9.12) skinks can be seen sunning themselves on the granite outcrops thus demonstrating an important geodiversity/biodiversity link.

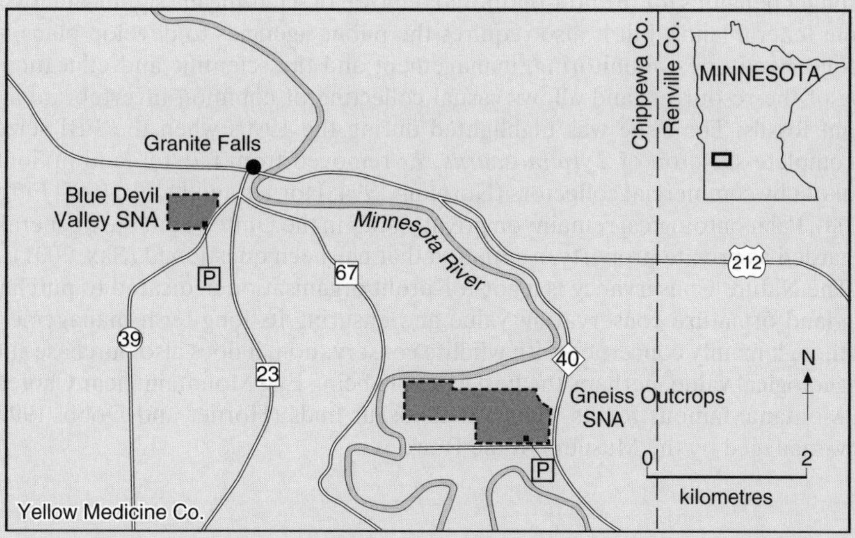

Figure 9.12 An illustration of the website information for Scientific and Natural Areas in Minnesota (redrawn from www.dnr.state.mn.us/snas/index.html and published with permission, © Department of Natural Resources, State of Minnesota).

- Grey Cloud Dunes SNA, comprises two terraces of the Mississippi River. Sand dunes 3–6 m high occur on both terraces, some with impressive crescentic shapes.
- Ripley Esker SNA, is a short section of a circa 10 km long, steep-sided, meandering esker.
- Yellow Bank Hills SNA, is a small area of kame mounds, ridges and kettle holes composed of sand and gravel.

9.2.6 Other legislation and protected areas

In 1987, a major review of fossil collecting in the United States was published (Raup *et al.*, 1987) partly in response to differences in policy between States. It made strong recommendations setting out the need for uniformity in national policy. It recommended that fossil collecting for research purposes be encouraged, but that commercial collecting should be done through permit regulated by palaeontologists. It also recommended a public education programme for collectors, amateur researchers and landowners. However, it engendered much heated argument and was not pursued within government. It was followed in 1992 by a new Bill (known as the Baucas Bill, after Senator Max Baucus) proposing a Vertebrate Palaeontology Resources Protection Act.

However, this too was not passed and instead the Palaeontological Resources Preservation Act was included in Title VI, Subtitle D of the Omnibus Public Land Management Act (2009) with similar aims to the Baucus Bill. It introduces stronger penalties for the non-permitted removal of scientifically significant fossils from federal lands, but it also requires the public agencies to develop plans for public inventories, monitoring, management and the scientific and educational use of the resources, and allows casual collecting of common invertebrate and plant fossils, The issue was highlighted during the 1990s when the FBI seized a complete skeleton of *Tyrannosaurus rex* removed from federal land in South Dakota by commercial collectors (Norman, 1994; Horner and Dobb, 1997; Fiffer, 2000). Palaeontological remains on private lands in the United States are generally regarded as private property, a situation that has been questioned (Sax, 2001).

The Nature Conservancy is a not-for-profit organisation dedicated to purchasing land of nature conservation value and ensuring its long-term management. Although mainly concerned with wildlife conservation, it does also purchase sites of geological value, perhaps the best example being Egg Mountain, near Choteau in Montana, famous for its Maiasaur dinosaur finds (Horner and Dobb, 1997), now managed by the Museum of the Rockies.

9.3 Canada

9.3.1 National Parks

Canada also has a National Park system established under the National Parks Act and revised by the Canada National Parks Act (2000). As of January 2013, there are 44 National Parks administered by Parks Canada. Many of these contain important geological and geomorphological sites or landscapes and some have been specifically designated for their geological interest, including the Gros Morne National Park in Newfoundland (Box 9.7). Four National Parks, including Yoho National Park and its famous Burgess Shale sites, together with three Provincial Parks make up the Rocky Mountains World Heritage Site (see Figure 7.2).

Box 9.7 Gros Morne National Park, Newfoundland, Canada

This National Park and World Heritage Site has been established mainly for its geological interest. The Park was the site of a constructive plate margin 600 million years ago with thick basalt sequences visible at Western Brook Pond. From 570 to 420 million years ago the area was an ocean and sedimentary strata containing a wealth of early Palaeozoic fossils were formed. However, the plate divergence was reversed about 460 million years ago, producing some unusual stratigraphic relationships. In particular the juxtaposition of oceanic lithosphere and mantle ophiolites overlying continental crust is demonstrated in The Tablelands area in the south of the park where the rocks ('serpentine barrens') contain heavy metals that are toxic to most plant life though some specialised arctic and alpine plants are still able to grow. Other geological points of interest include the global stratotype of the Cambrian–Ordovician boundary (see Section 6.8 and Figure 6.3), 40 important fossil sites including graptolites and trilobites, a massive rock sag at Bonne Bay. However, the main exposures of the famous Cow Head Breccia with its limestone boulders over 2 m in diameter, the result of underwater landslides (Figure 9.13), lie just outside the Park boundary. Geomorphological features include several fjords and ribbon lakes cut into the Gros Morne Plateau, sea cliffs and intertidal platforms.

Figure 9.13 The Cow Head Breccia, Newfoundland.

The National Parks legislation and guidance makes it clear that the primary objectives in the national parks are related to 'ecological integrity' and 'ecosystem management'. Although 'ecological integrity' is defined in the Act as including 'abiotic elements' and although this is clearly understood by many park managers, there is no specific reference to geology, geomorphology or pedology in the Act or related guidance. This is clearly a weakness in Canadian geoconservation.

The long-term aim of Parks Canada, started in the 1970s, is to establish at least one National Park in each of the established 39 Natural Regions in the country so that the system as a whole is representative of the natural diversity of the country's landscapes. Currently 25 of the 39 Natural Regions have at least one National Park and the Rocky Mountains Natural Region has five (Jasper, Banff, Yoho, Kootenay and Waterton Lakes). The Natural Regions on the Canadian Shield are those most poorly represented by National Parks network.

All National Parks are required under the Canada National Parks Act to have a strategic Management Plan, which establishes a vision for the following 15 years or more and is updated every five years following public consultation and approval by the Minister. Some of these include sections on geology and landforms outlining objectives and key actions for the plan period. Box 9.8 gives the relevant extract from the Management Plan for Yoho National Park (2010).

Box 9.8 Extracts from the Yoho National Park Management Plan (2010)

'The Burgess Shale fossils are of international significance, consisting of more than 150 species of remarkably preserved marine animals from the middle Cambrian period, 505 million years ago. It is recognised as one of the most significant fossil sites in the world, and was designated as a World Heritage Site in 1980. Protection and presentation of the sites and fossil collections is an international obligation'.

'The Burgess Shale rocks and fossils are vulnerable to theft and vandalism (graffiti) and to abrasion caused by rocks shifting under people's feet ... Visitor access to the Walcott Quarry and Mount Stephen Trilobite Beds is permitted only through guided hikes'.

Objectives:

- The fossils of the Burgess Shale sites within the park are protected from vandalism and theft;
- Visitors have unparalleled opportunities to 'Step into the Wild' and experience the Walcott Quarry and Mount Stephen Trilobite Beds with qualified guides;
- Reluctant travellers or visitors unable to visit the sites have opportunities to learn about the Burgess Shale through off-site interpretation and digital media;
- Peer-reviewed scientific research continues to increase understanding of Burgess Shale faunas, their geological context, and the nature of exquisite fossil preservation, and is incorporated in Parks Canada educational efforts.

Key actions:

- Maintain restricted and closed areas around Walcott Quarry and Mount Stephen Trilobite Beds, and evaluate other sites to determine if additional closures are required to protect fossil resources;
- Maintain guided access to the two main Burgess Shale sites, providing visitors with outstanding opportunities to visit and learn about an iconic palaeontological site;
- Monitor visitor satisfaction with guided hikes to the Walcott Quarry and Mount Stephen Trilobite Beds;

- Maintain association with the Royal Ontario Museum, and develop links with other global research institutions and individuals, as appropriate;
- Encourage peer reviewed scientific research on Burgess Shale sites, particularly those that have been less studied;
- Permit fossil excavation on Fossil Ridge and Mount Stephen as part of legitimate scientific research only if it can be demonstrated that the research questions cannot be answered using existing fossil collections, and if research is supported by peer reviewers.

All Canadian National Parks are also zoned into five categories plus 'environmentally sensitive sites', and some geological/geomorphological interests are further protected by these zones. For example, in Jasper National Park the Surprise Valley is designated as a 'Special Preservation' Zone 1, being part of the Maligne karst system. This valley is drained entirely underground through limestone of the Upper Devonian Palliser Formation, and is associated with one of the largest underground river systems in North America. It also contains deep sinkholes in glacier drift, sink lakes, and some of the finest rillenkarren in North America. No new access will be provided to this area. The same is true of Rabbitkettle Hot Springs in Nahanni National Park because of the physical delicacy of the tufa mounds.

9.3.2 National Marine Conservation Areas

Canada has the longest coastline of any country in the world—over 243 000 km of coastline on three oceans plus 9500 km along the Great Lakes. National Marine Conservation Areas (NMCAs) in Canada are being established under the Marine Conservation Areas Act (2002) in order to protect this coastal heritage from threats such as dumping, undersea mining and oil and gas exploration. They include the seabed and the overlying water column, species within these and may also include coastal wetlands, estuaries, islands and adjacent terrestrial zones. The aim of Parks Canada is to establish a network to represent each of the 29 marine regions of the country and also the Canadian Great Lakes coastline. Geology (particularly geomorphology) is included in the list of criteria being used to select potential representative marine areas.

As of January 2013, there are four NMCAs: Fathom Five National Marine Park in Georgian Bay in Lake Huron in Ontario; Lake Superior NMCA also in Ontario; Saguenay–St Lawrence Marine Park in Quebec, established by special complementary federal and provincial legislation allowing for co-operative management with the Province of Quebec (Saguenay–St Lawrence Marine Park Act, 1997); and the Gwaii Haanas NMCA Reserve on the Pacific Coast, which is in the planning stages.

9.3.3 National Landmarks

In the 1970s and 1980s it was intended to establish a network of National Landmarks in Canada, but only one has ever been established—Pingo National Landmark near Tuktoyaktuk in the Northwest Territories. This is an important geomorphological site as it protects 8 of the area's 1350 pingos, a quarter of the world's pingos. The 16 km^2 National Landmark area includes the Ibyuik pingo the second highest pingo in the world at 49 m.

9.3.4 Heritage Rivers

A particular designation of 'heritage rivers' was introduced in 1986 when 110 km of the French River became the first of several Canadian Heritage Rivers, though habitat protection, cultural heritage and recreational value can be as, or more, important than geomorphology in their recognition (McKenzie, 1994). In fact the approach taken is a holistic one involving cultural, biological and geological sciences. Nonetheless, it is clear that geological processes have been and are responsible for developing these spectacular rivers, and the criteria used to decide whether a waterway can be designated include several that are geological/geomorphological. For example, a river can be designated if:

- it is an outstanding example of river environments as they are affected by major stages and processes in the Earth's evolutionary history;
- it is an outstanding representation of significant ongoing fluvial, geomorphological and biological processes;
- it contains along its course, habitats with rare or outstanding examples of natural phenomena, formations or features, or areas of exceptional natural beauty.

Before designation takes place public consultation procedures and management guidelines are drawn up, which always exclude mining as an allowable activity. After designation, monitoring of river quality, habitats, etc., takes place. As of January 2013, 37 heritage rivers have been designated and five more nominated (www.chrs.ca).

9.3.5 Provincial designations

The Canadian Provinces have considerable autonomy in designating Provincial Parks, Nature Reserves, Natural Areas, etc., and a variety of landowner agreements and private stewardship initiatives also exist (Davidson *et al.*, 2001). Almost all are owned and managed by the provinces though some are managed jointly with local indigenous groups. For example, Nisga'a Memorial Lava Bed Provincial Park in British Columbia is managed jointly by BC Parks and the Nisga'a Tribal Council. Attempts have been made to get all provinces to list their protected areas under the IUCN categories but with limited success. Hence a range of protected areas exist that make inter-province comparison difficult.

The absence of any geological designation category in the provinces has led to some interesting treatments of geological sites. For example, in Newfoundland the global stratotype of the Precambrian–Cambrian boundary has had to be designated as an Ecological Reserve! Three provinces are discussed here—Ontario, Alberta and Quebec.

Ontario Ontario's provincial park system underwent a rapid expansion during the twentieth century. A major review of Provincial Park Policy took place in the 1970s based on the principle that the system of protected areas should represent the diversity of biological, geological and historical features that comprise the province's landscape. This led to the designation of over 150 new parks, many at least partly on the basis of Precambrian, Palaeozoic and Quaternary stratigraphy and geomorphology. At the same time about 300 new geological Areas of Natural and Scientific Interest (ANSIs) on private land were identified including key type sections, reference sections and type areas. These are managed or overseen by the Ontario Ministry of Natural Resources (OMNR).

Private landowners are eligible for tax relief through the Conservation Land Tax Program if they agree to protect the geological interest of an ANSI on their property. Furthermore, under the Ontario Heritage Act, conservation easements can be established, the first being an Ordovician type section and fossil locality near Owen Sound. 'In return for the loss of certain development rights, the landowners were provided with an income tax rebate for the full value of this gift to the province' (Davidson *et al.*, 2001, p. 230).

A further review of Ontario's parks and protected areas took place in 1996 and the Provincial Parks and Conservation Reserves Act (2006) brought further changes, so that Ontario now has more than 620 protected areas covering nearly 9% of the province's area. The Act streamlined legislation and increased protection for provincial parks and conservation reserves. It also requires the Minister of Natural Resources to publicly report on the health of parks and protected areas at least every five years. According to the official website (www.ontarioparks.com), 'Provincial parks protect a representative sample of Ontario's earth science features'.

There are six categories of provincial park (Wilderness Parks, Nature Reserves, Cultural Heritage Parks, Natural Environment Parks, Waterway Parks and Recreation Parks), each of which is zoned with some or all of the following categories: Natural Environment Zones (NE), Development Zones (D), Wilderness Zones (WI), Nature Reserve Zones (NR), Historical Zones (HI) and Access Zones (A). The matrix of existing Park classes and Zone types is shown in Figure 9.14.

Alberta Alberta has eight existing categories of protected areas designated under the Provincial Parks Act (1980), Wilmore Wilderness Park Act (1980) and Wilderness Areas, Ecological Reserves and Natural Areas Act (1989). In addition some geomorphological sites are protected under the Historical Resources Act (1980). The categories range from Wilderness Areas, where the only human activity allowed is foot access and which are therefore more highly protected than National Parks, to Recreation Areas, where outdoor recreation is promoted and encouraged.

Park class	Zone type					
	NE	D	WI	NR	HI	A
Wilderness			●	●	●	●
Nature Reserve				●	●	●
Historical	●	●		●	●	●
Natural environment	●	●	●	●	●	●
Waterway	●	●	●	●	●	●
Recreation	●	●		●	●	●

NE Natural Environment Zones
D Development Zones
WI Wilderness Zones
NR Nature Reserve Zones
HI Historical Zones
A Access Zones

Figure 9.14 Ontario's Provincial Park classes and zone types (© Queen's Printer for Ontario, 1992. Reproduced with permission).

A major review in the late 1990s brought the protected areas together and expanded them under Alberta's 'Special Places' programme. Because of overlaps and inconsistencies, the Alberta government has been reviewing the legislation with the intention of consolidating, updating and improving it. As of September 2011, there were 478 protected areas in the Alberta system covering 27.7 km^2. The largest category is the Recreational Areas, though many of these are being downgraded or transferred to local government. The management policies in different categories of the network are shown in Figure 9.15. There is little active enforcement in the protected areas, but among the strict management measures in existence is public exclusion from the Plateau Mountain Ecological Reserve cave to prevent ice crystal melting.

Several protected areas are specifically designated for their geological/geomorphological interest or the geohabitats they provide for wildlife. Examples include sand dunes (Athabasca Dunes Ecological Reserve; Richardson River Dunes Wildland Provincial Park), glacial features (Marguerite River Wildland Provincial Park), periglacial features (Plateau Mountain Ecological Reserve), river canyons and coulees (Brazeau Canyon Wildland Provincial Park; Red Rock Coulee Natural Area), karst landscapes (Whitemud Falls Wildland Provincial Park and Ecological Reserve), erratics (Okotoks Erratic Historic Site) and fossil sites (Dinosaur Provincial Park, see Box 9.9). The most recent additions to the system include Glenbow Ranch Provincial Park (2008) on the Bow River between Calgary and Cochrane that contains an abandoned sandstone quarry, source of early twentieth century building stone in Edmonton and Calgary, and an 800 ha extension to Writing On Stone Provincial Park (2011) with impressive hoodoos, sandstone cliffs and a number of petroglyph sites.

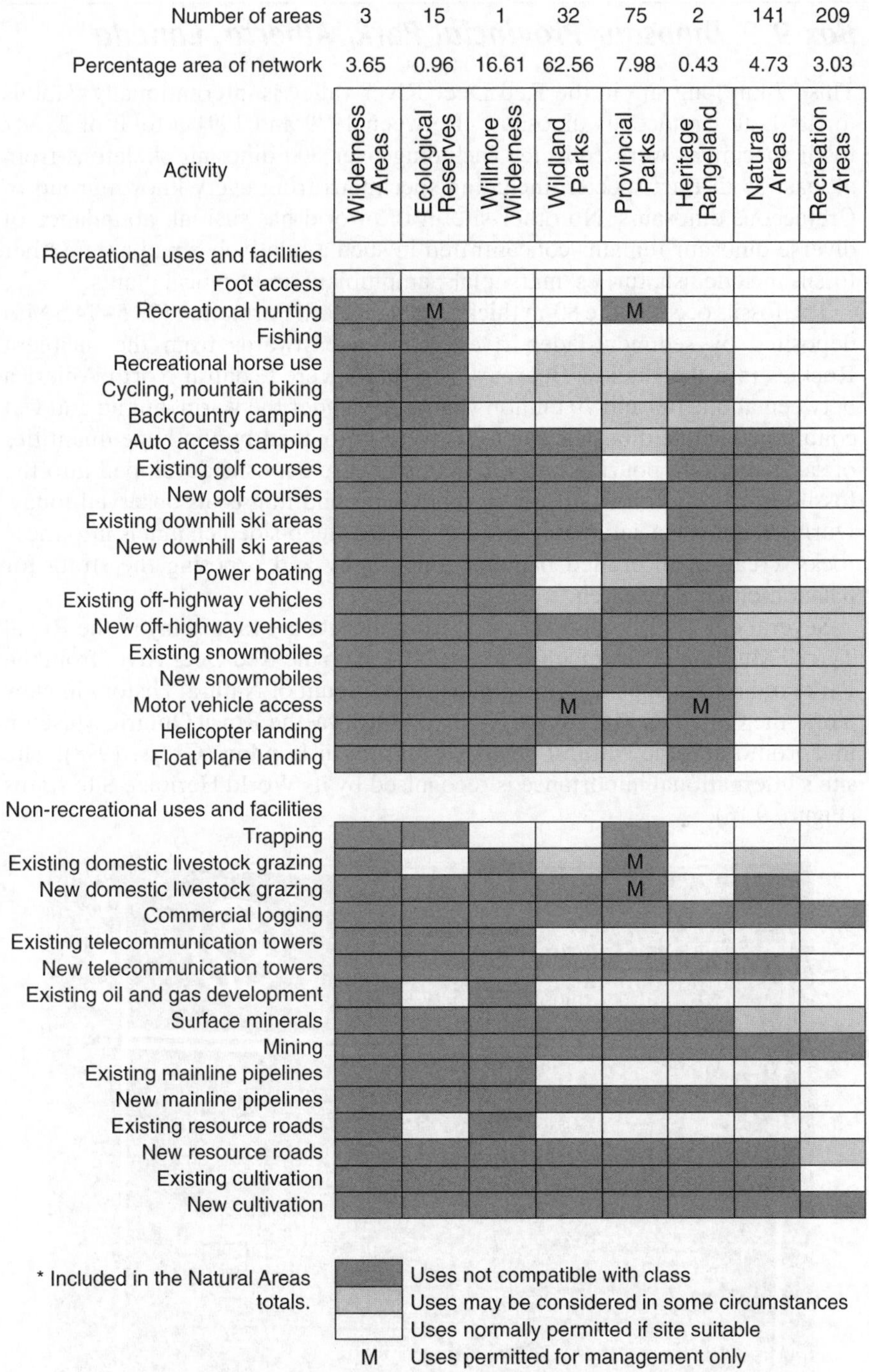

Number of areas
3
15
1
32
75
2
141
209
Percentage area of network
3.65
0.96
16.61
62.56
7.98
0.43
4.73
3.03
Activity
Wilderness Areas
Ecological Reserves
Willmore Wilderness
Wildland Parks
Provincial Parks
Heritage Rangeland
Natural Areas
Recreation Areas
Recreational uses and facilities
Foot access
Recreational hunting
Fishing
Recreational horse use
Cycling, mountain biking
Backcountry camping
Auto access camping
Existing golf courses
New golf courses
Existing downhill ski areas
New downhill ski areas
Power boating
Existing off-highway vehicles
New off-highway vehicles
Existing snowmobiles
New snowmobiles
Motor vehicle access
Helicopter landing
Float plane landing
Non-recreational uses and facilities
Trapping
Existing domestic livestock grazing
New domestic livestock grazing
Commercial logging
Existing telecommunication towers
New telecommunication towers
Existing oil and gas development
Surface minerals
Mining
Existing mainline pipelines
New mainline pipelines
Existing resource roads
New resource roads
Existing cultivation
New cultivation
M
* Included in the Natural Areas totals.
Uses not compatible with class
Uses may be considered in some circumstances
Uses normally permitted if site suitable
M Uses permitted for management only

Figure 9.15 Management policies within Alberta's protected areas. (Reproduced with permission of Archie Landals, Government of Alberta.)

Box 9.9 Dinosaur Provincial Park, Alberta, Canada

This 27-km long site in the Red Deer River valley is internationally famous for its Late Cretaceous dinosaurs. Between 1979 and 1991 a total of 23 347 fossil specimens were collected including over 300 dinosaur skeletons from at least 35 distinct species, including specimens from every known group of Cretaceous dinosaurs. No other site in the world has such an abundance of diverse dinosaur remains concentrated in such a relatively small area. Other fossils include fish, turtles, marsupials, amphibians and tropical plants.

The fossils occur in the 80-m thick Dinosaur Park Formation (76.5–74.5 Ma) deposited by sediment-laden rivers flowing eastwards from the incipient Rockies into the shallow Bearpaw Sea that covered central North America between about 100 and 70 million years ago. Under the warm, humid and wet conditions, whole dinosaur carcasses were entombed by the huge quantities of sand and mud pouring into the system These deposits developed into the fossiliferous claystones, siltstones, sandstones and thin coals observed today. During the last glaciation, meltwaters eroded steep-sided channels into these rocks, creating an eroded badland topography and exposing the strata for palaeontological research.

Several dinosaur fossils excavated from the site are on display in the Royal Tyrrell Museum in Drumheller, about 200 km up the Red Deer River from the Park, and others appear in the American Museum of Natural History in New York, the Canadian Museum of Nature in Ottawa, the Royal Ontario Museum in Toronto and the Natural History Museum in London (Gross, 1998). The site's international importance is recognised by its World Heritage Site status (Figure 9.16).

Figure 9.16 Plaque commeorating Dinosaur Provincial Park as a World Heritage Site, Alberta, Canada.

Quebec Quebec has about 1100 protected sites within 17 different judicial or administrative designations. Nonetheless they cover only 2.8% of the land area of the province. To meet international obligations Quebec has agreed to increase its protected areas to 8% of the land area and has recognised that one existing deficiency is in protecting its geodiversity. A consultation exercise was carried out in 2002–2003 by a working group with the aim of establishing a network of protected geosites in the province (Government of Quebec, 2002; P. Verpaelst, pers. comm.). In 2005, Quebec added provisions to its Mining Act to protect outstanding geological sites of educational, scientific or conservation interest and which deserve protection because they are threatened, rare or vulnerable. Guidelines for managing the sites were published in 2010 (Ferron *et al.*, 2010) and by December 2011 about 60 sites were being evaluated for designation under the scheme.

9.3.6 Other legislation

Control of fossil collecting in Canada is the responsibility of the Provinces through the passing of appropriate Acts of Government. The first such Act (*The Historical Objects Protection Act* (1970) was passed by Nova Scotia. All Provinces now have such Acts to regulate fossil collection that generally involves the issuing of research permits. Alberta's *Historical Resources Act* (1980) prohibits all excavation other than by permit holders, but Norman (1994) detects inconsistency from province to province in the degree to which individual fossil groups are protected and in the rigour of law enforcement. There is also a Federal Law—the Federal Cultural Property Export and Import Act (1975)—that controls the export of fossils from Canada.

9.4 United Kingdom

'Britain can be regarded as having more geological diversity than any other comparable area on Earth' (Trewin, 2001, p. 59) and perhaps as a result has well-developed, if complex, systems for geoconservation. As stated in Section 6.2, there were many early efforts at nature conservation in general and earth science conservation in particular in the United Kingdom, but major steps were taken in the late 1940s (Prosser, 2008). One was the establishment by Royal Charter of the Nature Conservancy as the UK government's nature conservation agency covering England, Scotland and Wales (but not Northern Ireland) and the other was the passing of the National Parks and Access to the Countryside Act (1949). These measures empowered the Nature Conservancy to establish National Nature Reserves and Sites of Special Scientific Interest on the basis of their flora, fauna, geology or 'physiography' (geomorphology). The impetus was a series of government committee reports (Prosser, 2013a) including that of the Wildlife Conservation Special Committee (Huxley, 1947), and it is indeed fortunate that the Committee included both a geologist (A.E. Trueman) and a geomorphologist (J.A. Steers). The Nature Conservancy was later retitled the Nature Conservancy Council, but under the Environmental Protection Act (1990) it was divided into three country-based agencies—English Nature (EN), Scottish Natural Heritage (SNH) and the Countryside Council for Wales (CCW). For Scotland and Wales this was more than a simple change of name since the organisations were merged with the Countryside Commission, which had responsibility for the cultural landscape. In England, a similar merger was achieved under the Natural Environment and Rural Communities Act (2006) with this new organisation being renamed Natural England (NE). In order to achieve some consistency across the United Kingdom and advise on international nature conservation policy, the Joint Nature Conservation Committee (JNCC) was established under the Environment Protection Act (1990). Regrettably it ceased all its in-house geoconservation work in 2011. Further changes have recently taken place in Wales through the merger of the Countryside Council for Wales with both the Environmental Agency and Forestry Commission in Wales to form Natural Resources Wales (NRW).

In the United Kingdom a large number of site designations are used and some believe that there is an urgent need for simplification of the 'existing maelstrom of countryside designations' (Adams, 1996, p. 148). Up-to-date summaries of geoconservation in the United Kingdom are provided by Cleal (2012) and Prosser (2013a).

9.4.1 National Parks

British national parks are very different to those of the United States and Canada in that they are 'neither national (for they are not nationally owned) nor parks (for they include large areas of farmland and private ground to which the public has no right or even custom of access) ... '. This 'peculiarly British' national park system would be recognised internationally not as 'national parks' but as 'protected landscapes' (IUCN Category V, Table 6.1). 'What the British call "national parks" are intended not so much to conserve uninhabited wilderness as to protect inhabited landscapes where the land should be managed for a multiplicity of purposes—conserving their character, promoting their enjoyment and supporting human life in many diverse ways' (MacEwen and MacEwen, 1987, pp. 3–4).

There are now 15 National Parks in the United Kingdom (Figure 9.17) including the Norfolk and Suffolk Broads, an area of rivers and flooded peat diggings in Eastern England which has had a status equivalent to a National Park for over 20 years but not yet the name. The most recent additions to the network are Loch Lomond and the Trossachs (2002) and Cairngorms (2003) in Scotland, and The New Forest (2005) and South Downs (2010) in England.

It is usually only coincidental if important geology, landforms or soils are included within these areas, though their scenic beauty and often ancient geology inevitably mean that landscape and geoscientific interests often coincide. Thus many of the British national parks, such as the Lake District, Snowdonia, Dartmoor, Cairngorms and others, although designated for their landscape beauty, all contain important geological and/or geomorphological interests. All National

Figure 9.17 Location of Britain's National Parks. Note that the Broads have a status equivalent to a National Park but not yet the name.

Park Authorities are required to produce regularly updated Management Plans and many refer to geology and geomorphology. For example, the Lake District National Park Management Plan (2004, Policy NC4) seeks to 'promote sustainable management of the Lake District's geology and geomorphology by ensuring that irreplaceable features or exposures are not degraded, and a full range of representative features remains accessible', while the Loch Lomond and the Trossachs National Park Management Plan (2007) contains a policy (G1) which states that 'Geodiversity that contributes to the Park's special qualities will be safeguarded and enhanced. This will be informed by the development of a Geodiversity Action Plan ... ' (see Section 12.5).

9.4.2 Nature Reserves

The Royal Charter establishing the Nature Conservancy empowered it to establish National Nature Reserves (NNRs), which are areas preserved primarily to maintain and enhance their scientific status and research potential. The first geological NNR, designated in 1952, was the Piltdown Skull Site in East Sussex, later exposed as a hoax and revoked a few years later (Prosser, 2009). NNRs are generally owned by the State and managed by one of the national nature conservation agencies. As of January 2013 there were 13 NNRs designated primarily for their geological and geomorphological interests (Prosser, 2013b). Examples include the Wren's Nest NNR in Dudley, West Midlands (Figure 9.18) designated for its ripple beds containing Silurian fossils including trilobites, Cwm Idwal NNR in Snowdonia, Wales which contains an important series of moraines mentioned by Charles Darwin, and the Portrush sill in Northern Ireland where there was an important historical debate between the Neptunists and Vulcanists over the origin of igneous rock. Scotland has a number of important geological NNRs including the Isle of Rum, an important Tertiary igneous centre, Glen Roy with its famous ice-dammed lake shorelines known as the Parallel Roads of Glen Roy, Staffa with its remarkable basalt columns and sea caves including Fingal's Cave, and Knockan Crag where the nineteenth century geologists Peach and Horne discovered the principle of thrusting (Moine Thrust) bringing older rocks over younger ones.

Local authorities also have the power to designate Local Nature Reserves (LNRs) and there are over 700 of these in England, about 70 of which have a geodiversity interest and 14 of these are primarily geological (Prosser, 2013b). Marine Nature Reserves are discussed in Section 9.4.6.

Figure 9.18 Wren's Nest, National Nature Reserve, Dudley, England.

9.4.3 Sites of Special Scientific Interest (SSSIs)

The National Parks and Access to the Countryside Act and subsequently the Wildlife and Countryside Act (1981, as amended 1985) also required the Nature Conservancy to designate sites worthy of protection purely based on their scientific value. Both biological and geological Sites of Special Scientific Interest (SSSIs) were to be established, but this status did not give a site direct protection. Rather it simply ensured that local authorities were notified of the sites and that a full consultation process took place if developments affecting SSSIs were proposed. The 1981 Act enables 'potentially damaging operations' (PDOs) to be identified for each site (see Section 5.1). The Town and Country Planning Acts (1947) and (1990) ensured that local authorities consulted the Nature Conservancy and its successors on relevant planning applications and allowed them to adopt appropriate planning policies (see Chapter 11). In Northern Ireland, Areas of Special Scientific Interest (ASSIs) give similar protection. Most SSSIs and ASSIs are on private land.

The situation was strengthened significantly in England and Wales by the Countryside and Rights of Way Act (2000) (usually known as the CROW Act). This puts the emphasis on positive site management and partnership between the conservation agencies and landowners and occupiers, rather than paying them not to carry out operations that could damage sites (Prosser and Hughes, 2001). However, if agreement cannot be reached, the Act allows action to be taken to prevent site deterioration through neglect or deliberate damage. The Act also makes it an offence for anyone, including third party fossil or mineral collectors or vandals, to knowingly or recklessly damage an SSSI. The Act also gives the agencies powers to introduce bylaws on SSSIs to further protect them from third party damage, and strengthens their right to enter private land to investigate offences and monitor the condition of SSSIs (Prosser and Hughes, 2001). The Act gives right of access to several categories of open land and therefore is helpful in giving access to geological exposures in mountain and moorland areas. In Scotland, similar provisions are included in the Nature Conservation (Scotland) Act (2004).

Up until 1977, SSSIs were selected on the basis of available information and consultation with the country's geoscientists, but this resulted in very inconsistent site networks (Gordon, 1994a). A major review, the Geological Conservation Review (GCR), took place between 1977 and 1990 to establish a systematic list of important geological and geomorphological sites (Nature Conservancy Council, 1990; Ellis *et al.* 1996; Ellis, 2008, 2011). The site series was intended to reflect 'the range and diversity of Great Britain's Earth heritage' (Ellis *et al.*, 1996, p. 45) defined as:

1. sites of importance to the international community of Earth scientists, e.g. time interval or boundary stratotypes, type localities for biozones, internationally significant type localities for particular rock types, mineral or fossil species, historically important type localities or features;
2. sites that are scientifically important because they contain exceptional features;

3. sites that are representative of an Earth science feature, event or process which is fundamental to Britain's Earth history.

This third category of sites has been described as the 'key component of the GCR' in that it ensures the selection of a network of representative sites (Ellis, 2011, p. 357). In practical terms, a number of operational criteria were employed (Ellis *et al.*, 1996) so that preference is given to sites:

- where there is a minimum of duplication of interest between sites;
- where it is possible to conserve any proposed site in a practical sense;
- that are least vulnerable to potential threat;
- that are more accessible;
- that show an extended, or relatively complete record of the feature of interest;
- that have a long history of detailed research study and re-interpretation;
- that have potential for future study;
- that have played a significant part in the development of the Earth sciences, including former reference sites, sites where particular geological phenomena were first recognised, and sites that led to the development of new theories or concepts.

'Application of these criteria ensures that sites chosen for a particular network in the GCR have the greatest collective scientific value and can be conserved' (Ellis *et al.*, 1996, p. 75). Attempts were made to minimise the number of sites designated so that only those that are necessary to characterise 'the network' were selected. Ellis *et al.* (1996, p. 76) define 'the network' as those sites that 'demonstrate the current understanding of the range of Earth science features in Britain … '. In other words, it is a network of sites that represents the geodiversity of the country (see Section 13.1). The size of each site was also kept to a minimum, though there are exceptions in upland areas with a range of features.

About 3000 GCR sites were identified, but because of overlaps (e.g. a fossil site within a coastal geomorphology site), about 2010 SSSIs have been designated by the three UK nature conservation agencies (Prosser, 2013b). This represents almost 30% of the United Kingdom's 6500 SSSIs as of July 2012, the remainder being biological. GCR site descriptions are being published in a set of 45 volumes, each made up of blocks covering a particular geological period, rock or landform type and/or part of the country. The first volume on the *Quaternary of Wales* (Campbell and Bowen, 1989) was published in 1989 and an introductory volume explaining the aims, processes and products of the review was published in 1996 (Ellis *et al.*, 1996). As of January 2013, 39 volumes had been published in several phases with different publishers, and the series is due to be completed in the next few years as special volumes of the journal *Proceedings of the Geologists' Association*. Gordon and Leys (2001b, p. 7) refer to it as 'the most comprehensive review undertaken by any country'. In Northern Ireland, the equivalent of the GCR has identified about 330 sites, and 121 have been designated as earth science ASSIs (Prosser, 2013b). Box 9.10 gives some examples of important earth science SSSIs in the United Kingdom.

Box 9.10 Examples of United Kingdom SSSIs

The following are some examples of the circa 2010 earth science SSSIs currently being designated in the United Kingdom:

- Siccar Point, Berwickshire is known as Hutton's unconformity. Upturned Llandovery, Silurian greywackes and shales overlain by sub-horizontal Devonian sandstones and breccias, led James Hutton in 1788 to recognise the principle of the angular unconformity the essentially cyclical nature of erosion and deposition, and the great age of the Earth. As such it is one of the most important geological sites in the world (Figure 9.19).

Figure 9.19 Hutton's unconformity at Siccar Point SSSI, Scotland.

- Rhynie Chert, Aberdeenshire, is internationally famous as containing some of the world's oldest terrestrial plant fossils (Devonian), as well the earliest known wingless insect (*Rhyniella*) and a very fine micro-arthropod fauna, including mites, springtails and a small aquatic shrimp-like creature (*Lepidocaris*). The site is a silicified peat produced when the site was flooded by silica-rich hot water. This killed and preserved the fauna before the tissue decayed and so preserved a complete three-dimensional ecosystem (Ellis *et al.*,1996; Rice, Trewin and Anderson, 2002). The structure of the xylem, stomata and sporangia with spores in place can be observed.
- Blakeney Esker, Norfolk, is England's best-developed esker (Figure 9.20). Gravel quarrying has revealed its internal structure and has demonstrated that the gravels infill a series of channels cut into the underlying marly drift, indicating that the esker was formed subglacially (Gray, 1997b). However,

gravel quarrying also destroyed large parts of the esker prior to designation (Figure 9.20).

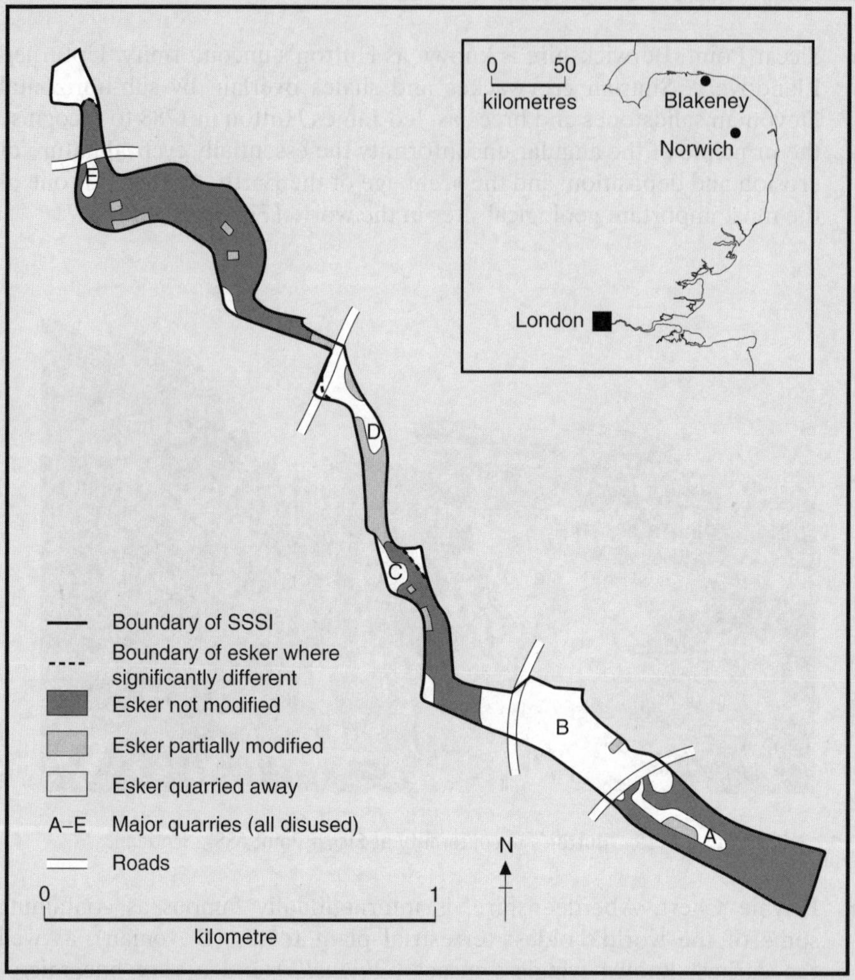

Figure 9.20 Map of the Wiveton Downs SSSI (Blakeney Esker), England showing quarried areas.

- River Feshie, in the Scottish Highlands is one of the UK's few braided, gravel-bed rivers. It regularly moves its course and is building a fan at its confluence with the river Spey. It is therefore an important fluvial geomorphology site (Werritty and Brazier, 1994).
- Dundry Main Road South Quarry SSSI and Barns Batch Spinney SSSI in Somerset are two of the UK sites with important historical connections (Prosser and King, 1999). It was here in the 1790s that William Smith,

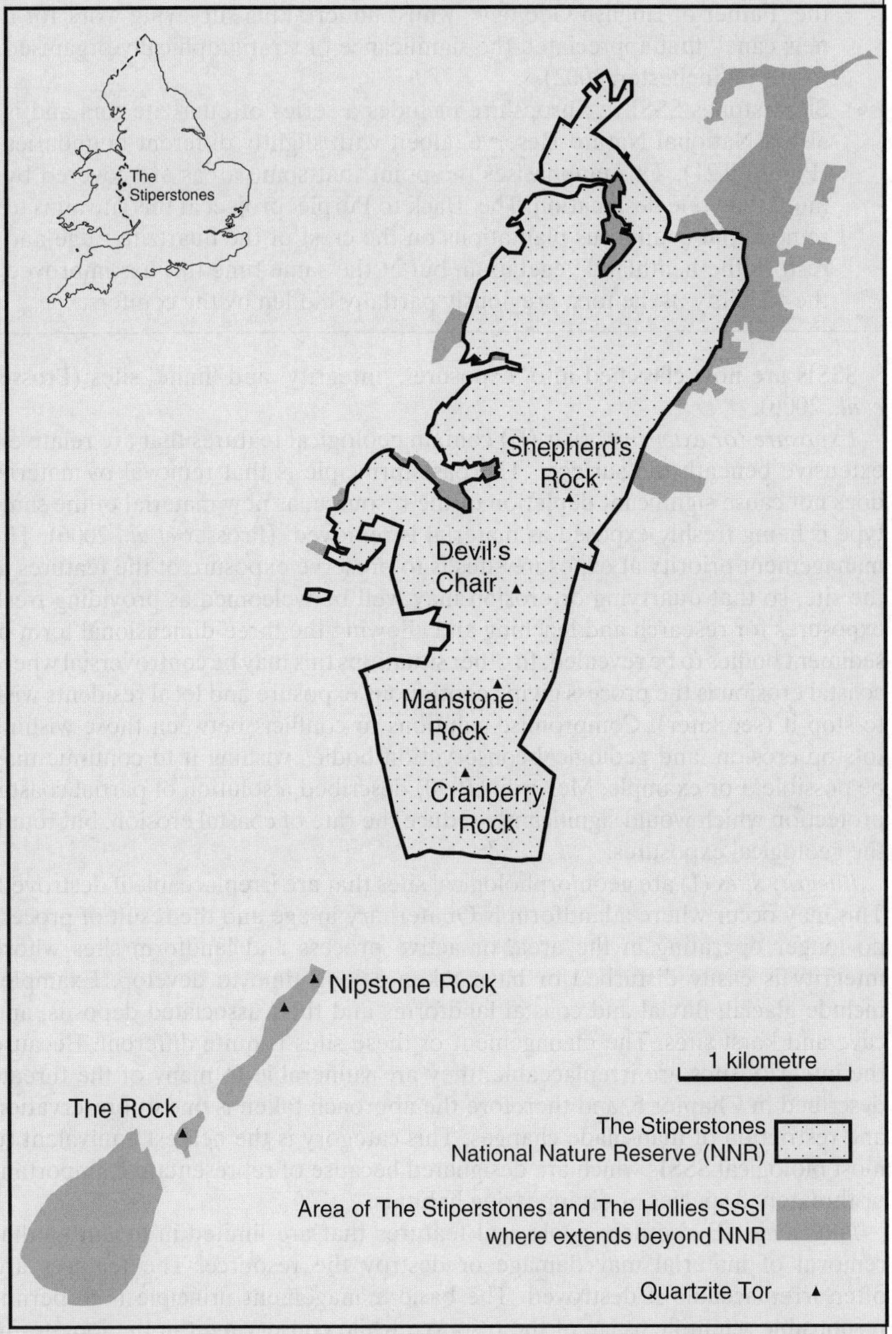

Figure 9.21 Boundaries of the Stiperstones National Nature Reserve and Site of Special Scientific Interest. (Reproduced with permission of Natural England.)

the 'Father of English Geology', whilst undertaking surveying work for a new canal, first appreciated the significance of stratigraphically organised fossils (Winchester, 2002).

- Stiperstones SSSI in Shropshire includes a series of quartzite tors and is also a National Nature Reserve, albeit with slightly different boundaries (Figure 9.21). This emphasises the point that some areas are covered by more than one designation. The 'Back to Purple' project at this site aims to remove the coniferous plantations on the crest of the quartzite ridge and restore the heathland vegetation, but at the same time this has improved the visibility of the tors, previously partially hidden by the conifers.

SSSIs are now classified into 'exposure', 'integrity' and 'finite' sites (Prosser *et al.*, 2006):

Exposure (or extensive) sites (E) contain geological features that are relatively extensive beneath the surface. 'The basic principle is that removal of material does not cause significant depletion of the resource, as new material of the same type is being freshly exposed as material is removed' (Prosser *et al.*, 2006). The management priority at exposure sites is to preserve exposure of the features at the site, so that quarrying or erosion may well be welcomed as providing fresh exposures for research and teaching and allowing the three-dimensional form of sediment bodies to be revealed. In other situations this may be controversial where coastal erosion is the process giving continuous exposure and local residents wish to stop it (see later). Compromise solutions to conflicts between those wishing to stop erosion, and geological conservation bodies wishing it to continue may be possible. For example, McKirdy (1990) described a solution of partial coastal protection which would significantly reduce the rate of coastal erosion, but retain the geological exposures.

Integrity sites (I) are geomorphological sites that are irreplaceable if destroyed. This may occur where a landform is Quaternary in age and the result of process no longer operating in the area, or active process and landform sites whose integrity is easily disturbed or have taken a long time to develop. Examples include glacial, fluvial and coastal landforms and their associated deposits, and cave and karst sites. The management of these sites is quite different. Because the integrity sites are irreplaceable, they are vulnerable to many of the threats described in Chapter 5, and therefore the approach taken is one of preservation and restriction of man-made changes. This category is the nearest equivalent to most biological SSSIs which are designated because of representative, important or threatened species or disappearing habitats.

Finite sites (F) 'contain geological features that are limited in extent so that removal of material may damage or destroy the resource. The features are often irreplaceable if destroyed. The basic management principle is to permit responsible scientific usage of the resource while conserving it in the long-term. Hence, it is often necessary to implement controls over removal of material' (Prosser *et al.*, 2006).

Prosser *et al.* (2006) give a classification of exposure, integrity and finite sites (see Table 9.1) and in their Table 2.2 they show the main potential threats

Table 9.1 Classification of UK Earth Science Sites (after Prosser, Murphy and Larwood, 2006).

Category	Type of site	Site code
Exposure	Active quarries and pits	EA
	Disused quarries and pits	ED
	Coastal Cliffs and foreshore	EC
	River and stream sections	EW
	Inland outcrops	EO
	Exposure underground mines and tunnels	EU
	Extensive buried interest	EB
	Road, rail and canal cuttings	ER
Integrity	Static (fossil) geomorphology	IS
	Active process geomorphology	IA
	Caves	IC
	Karst	IK
Finite	Finite mineral, fossil or other geological	FM
	Mine dumps	FD
	Finite underground mines and tunnels	FU
	Finite buried interest	FB

and issues, the conservation techniques and then provide case studies for each category in the classification.

Some site documentation is usually required before a site is designated. Normally this will outline the site's location, importance and the potential threats to it. The 45 volumes of the GCR series give details of all the sites in terms of 'Highlights', 'Introduction', 'Description', 'Interpretation' and 'Conclusions'. At a more detailed level, Site Documentation Reports exist for some GCR sites. In the case of Scotland the primary purpose of this series of reports is to describe and explain the scientific interest of each of the GCR sites in plain English. As well as the scientific interest of the site and annotated bibliography, the reports describe the current condition of the site as a baseline for site condition monitoring programmes, geomorphological sensitivity, potential threats, site management and potential usage. The Moss of Achnacree and Achnaba SSSI report runs to 85 pages including 4 maps and over 40 photographs of the site (Bentley, 1996).

Ellis *et al.* (1996) believe that all sites need a 'conservation plan' giving information about the threats to the site, how it would deteriorate naturally without intervention, and what action would be desirable, or even essential, to maintain the features(s) of interest. 'This will lead naturally to a consideration of what site-specific and practical Earth heritage conservation measures will be needed to ensure that the features of special interest are not obscured, destroyed or damaged and also to indicate the recommended frequency of monitoring' (Ellis *et al.*, 1996, pp. 89–90). This type of approach was developed in Scotland where Werritty, Duck and Kirkbride (1998) developed a 'Land Attributes' system which refers to the characteristics of an earth science site which should be taken into account for monitoring and conservation management (see also Kirkbride *et al.*, 2001).

The Environmental Protection Act (1990) requires 'common standards throughout Great Britain for the monitoring of nature conservation' and the above approach is an example of the search for best practice in this regard. The JNCC published the first Common Standards Monitoring Report in 2006 (JNCC, 2006) in three sections covering Geology, Species and Habitats. The report demonstrates that the geological SSSIs are generally in a much better condition than the species and habitat ones, with 86.4% of sites being in favourable condition. However, 7.5% of mineral sites are listed as destroyed or part destroyed, by far the highest figure of any site type.

The UK government set a target of having 95% of SSSIs in favourable condition by 2010. In consequence, English Nature/Natural England implemented an enhancement programme termed 'Face Lift' aimed at clearing vegetation, talus or overburden from sites to re-expose the geological interest of natural or quarry sections (Murphy, 2002). Some of the work was carried out by local amateur groups, for example through the Ludlow series near Ludlow (Oliver and Allbutt, 2000).

Even with all this protection and management in place, SSSIs can still be lost. This is particularly true where permission for a particular use pre-dates SSSI designation. An example of this is Webster's Clay Pit in Coventry, England where landfilling in the 1990s has totally buried an important Westphalian site (Prosser, 2003; see Section 5.3). Similarly, Brighton and Hove Council in 2002 granted itself planning consent to stabilise the cliff face SSSI at Black Rock, Brighton, obscuring the Quaternary deposits and mammalian fossil assemblage (Bennett, 2003). Box 9.11 describes an important legal case concerning the redesignation of an SSSI in Suffolk where several houses on a clifftop are facing erosion by the sea. Coastal erosion is rarely recognised as a beneficial aspect of nature in creating iconic coastal scenery and scientifically important exposures.

Box 9.11 Pakefield to Easton Bavents SSSI, Suffolk

At Easton Bavents, Suffolk, this SSSI comprises an actively eroding cliffline in which Quaternary sediments containing important microtine vole fossils in particular, are exposed. This is the type site for the Baventian Stage of UK Early Quaternary stratigraphy. The actively eroding cliffline was threatening a number of properties near the cliff edge and between 2003 and 2005 one of the residents arranged for several thousand tonnes of waste soil to be deposited on the beach in front of the cliffs forming a circa 20 m wide terrace, about 1 km long (Figure 9.22). Although the cliffs had been designated as an SSSI in 1989, erosion over this time had taken the cliffline outside the designated boundary. In order to prevent further damage to the SSSI, in 2008 Natural England decided to redesignate the current cliffline at Easton Bavents together with 225 m of land inland. The residents appealed against this redesignation through Judicial Review on various grounds including whether erosion could be considered as conservation, and the principle of designating land for future study rather than its current scientific interest. The High Court Judge dismissed these arguments, in particular believing that 'Conservation ... may also involve allowing natural processes to take their course'. However, the Judge concluded that the Judicial Review should succeed, but only on ecological grounds. Consequently he refused to support the redesignation of the site. In 2009, Natural England was successful in getting this judgement overturned in the Court of Appeal where the Judge supported the geological arguments of the High Court. In particular, Lord Justice Mummery concluded that 'one would have thought that allowing natural processes to take their course, and not preventing or impeding them by artificial means from doing so, would be a well recognised conservation technique in the field of nature conservation'. An attempt by residents to take the case to the Supreme Court and European Courts was refused in February 2010 (Prosser, 2011).

Figure 9.22 Terrace of material deposited against the cliffs at Easton Bavents SSSI. (Photo by permission of Mike Page.)

9.4.4 Local Geological Sites (LGSs)

Up until 2006, several terms were used to describe locally important nature conservation sites. These included County Wildlife Sites (CWSs), Sites of Importance for Nature Conservation (SINCs), Local Nature Reserves (LNRs) and Regionally Important Geological/Geomorphological Sites (RIGS).

The last category was introduced in the late 1980s to meet the need for more local involvement in earth science conservation in the same way that many local wildlife groups identify local wildlife sites. The aim was to set up a country-wide network of sites, established and managed locally by volunteer groups of mainly amateur geologists. This idea proved to be very successful with about 50 groups now operating in the United Kingdom, covering most counties of England, with some groups also established in Wales and Scotland. Britain is very fortunate and unique in having an army of amateur geologists organised into numerous local geological societies and voluntary groups who, since the 1990s, have contributed hugely to geoconservation efforts in the country, mainly through identifying and establishing local geological site networks and developing Local Geodiversity Action Plans (Section 12.5) (Burek, 2008a, 2008b; Whiteley and Browne, 2013).

RIGS are not designated or protected by law, but they are increasingly given some protection by local authority planning policies (see Chapter 11). A national body (UKRIGS, since renamed GeoConservationUK) was established in 1999 and has a development strategy aimed at expanding the RIGS network, strengthening work quality and partnerships and securing significant funding for the work. It produces newsletters, promotes good practice and holds conferences.

In 2006 a report (DEFRA, 2006) proposed a rationalisation of these local conservation terms and recommended the use of the term 'Local Site' which may be subdivided into 'Local Wildlife Site' or 'Local Geological Site'. The benefits of the latter include:

- providing protection and management for locally important geological sites;
- increasing public awareness, understanding and enjoyment of geology/geomorphology;
- forming a focus for partnerships between local authorities, conservation organisations and local people.

Despite this call for more consistency in the use of names, in England the term 'Local Geological Sites' is used, but in Scotland they are called 'Local Geodiversity Sites' and Wales still uses the term 'Regionally Important Geodiversity Sites'. In recent years, several former RIGS groups have changed their names and status to reflect the fact that their aims are broader than just recording sites. Some of these groups have come together under the umbrella organisation known as the 'Geology Trusts' which include the counties of Herefordshire and Worcestershire, Gloucestershire, Monmouthshire, Oxfordshire, Shropshire, Warwickshire, Wiltshire and Bedfordshire. In total, some 4400 local geological sites have been established (Prosser, 2013b).

In Wales, a complete audit of RIGS has been produced (Kendall, 2012). For each site there is documentation giving basic details such as name, location, type and date registered, followed by a 'Statement of Interest', 'Geological setting/context' and 'References'. Finally there are sections covering such information as accessibility and safety, ownership and planning control, current and potential usage, site condition, potential threats and management options. This style of documentation is used in Wales to provide consistency of recording of RIGS sites and represents the best of existing practice (Burek, 2008b).

9.4.5 Limestone Pavement Orders

Limestone pavements constitute 'a unique and finite part of Britain's physical landscape' (Bennett and Doyle, 1997, p. 106). They 'are limestone outcrops which have been stripped of any pre-existing soil or other cover by some scouring mechanism, generally but not exclusively, glacial scour' (Goldie, 1994) and subsequently subject to joint weathering and solution producing a variety of surface morphologies (Figure 9.23) including clints (abraded and weathered surface blocks) and grykes (weathered joints). In Europe they occur predominantly in the Carboniferous limestones of Upland Britain and Ireland. Serious loss of these pavements by unscrupulous operators has occurred in Yorkshire and Cumbria in Britain to satisfy the demand for weathered limestone in garden rockeries and walls where their weathered characteristics are valued by horticulturalists. 'Clints have been sold at garden centres to purchasers who are probably ignorant of their beauty and interest *in situ*. . . . Pavements thus damaged have a much altered geomorphology, a messy, ugly, broken surface with much loose debris and rough remnant clint tops, lacking attractive runnelling, and with a much depleted, or even totally destroyed, flora' (Goldie, 1994, p. 216).

Because of this growing impact in the 1960s and 1970s, Limestone Pavement Orders (LPOs) were introduced under Section 34 of the Wildlife and Countryside Act (1981, as amended 1985). This is one of the few pieces of legislation to protect a specific landform type and make it a criminal offence to damage the landforms so designated. About 100 orders have been made, and all significant limestone pavements in England are now protected. Though some illegal removal still continues, the strengthening of legislation in England meant increased pressure on the Irish pavements. Limestone pavements are also an important habitat

Figure 9.23 Limestone Pavement at Malham Cove, Yorkshire Dales National Park.

(see Box 4.1) and are specifically identified in Annex 1 of the European Habitats Directive as a natural habitat of community interest whose conservation requires the designation as Special Areas of Conservation (SACs) (Goldie, 1994).

9.4.6 Marine Protected Areas

Marine Nature Reserves (MNRs) may be designated under *the* Wildlife and Countryside Act (1981), but as of January 2013, there are only three—at Lundy Island, Skomer and Strangford Lough—though others have been proposed. More importantly, the Marine and Coastal Access Act (2009) has created a new type of Marine Protected Area (MPA) called a Marine Conservation Zone (MCZ). The intention is that these will protect important marine ecological, geological and geomorphological areas in English inshore waters and offshore waters around England, Wales and Northern Ireland. English waters are divided into four areas—North Sea, south-east seas, south-west seas and Irish Sea—and an active Marine Conservation Zone Project led by Natural England and the JNCC and involving a range of stakeholder groups was completed in 2011 identifying and proposing MCZs in each of these areas (Burek *et al.*, 2013).

Although mainly concerned with protecting marine wildlife and habitats, the relevant guidelines for selecting MCZs (Ashworth, Stoker and Aish, 2010) gave scope for geological and geomorphological sites to be included:

- First, the 32 coastal GCR sites that have a significant intertidal or subtidal portion and have not yet been designated as SSSIs should be considered for MCZ designation.
- Secondly, when identifying broad-scale habitat MCZs, consideration should be given to including important geological and geomorphological features (Ashworth *et al.*, 2010). Research by Brooks, Roberts and Kenyon (2009) identified 6500 subtidal geological and geomorphological seabed features in UK waters classified into five types—glacial process features, marine process features, mass movement features, features indicating past changes in sea-level, and geological process features. The features were then assessed for their conservation importance including aspects such as rarity, exceptionality and sensitivity (Brooks *et al.*, 2009).

Recommendations made in 2011 were for 127 MCZs across all 4 sea regions, but only 19 of these were geological or geomorphological features. According to Burek *et al.* (2013) this reflects the biological emphasis of the process, the range of stakeholder groups, each with their own interests, and the lack of understanding of the importance and vulnerability of marine geoscience features. In December 2012 the government accepted only 31 of these areas covering 10 000 km^2 but few geological/ geomorphological sites have survived. One exception is Haig Fras (a 45-km long granitic submarine ridge in the Southern Celtic Sea).

In Scotland, the Marine (Scotland) Act (2010) together with the Marine Scotland government directorate is allowing similar progress to be made on

Scottish MPAs, while Wales, Northern Ireland and the Isle of Man are expected to follow suit. It should be noted that Marine Protected Areas in the United Kingdom also include European marine sites (SACs and SPAs), the marine components of SSSIs and marine Ramsar sites.

9.4.7 Other legislation and policies

Geology, and particularly geomorphology, may also be recognised in some of the other designation types. For example, hill or coastal landforms and processes are often a major part of the recognition of the 41 Areas of Outstanding Natural Beauty (AONBs) and over 40 Heritage Coasts in England and Wales, the coastline of North Norfolk for example being recognised under both categories (Funnell, 1994; Adams, 1996). Together these two categories cover over 12% of the land area of England and Wales and the emphasis within them is on landscape conservation. Under the Countryside and Rights of Way Act, local authorities are obliged to produce management plans for AONBs so there is an opportunity for these plans to include geoconservation objectives and promote the continued use of local stone for ongoing quality developments.

Twelve Nature Improvement Areas were selected in England in 2012 and government funding provided for specific projects within them. A number of these have a geological element, Birmingham and the Black Country being perhaps the most geologically orientated example.

Scotland has had a chequered history of debate regarding other protected areas (Cullingford and Nadin, 1994). It does have about 40 National Scenic Areas (NSAs), which are equivalent to AONBs and which cover over 1 million hectares, including Ben Nevis and Glen Coe. It also has a few regional parks (including the Pentland Hills near Edinburgh) and about 35 country parks. These two categories are primarily intended to provide recreational opportunities, but they allow for conservation work and management agreements with private landowners.

A number of conservation organisations own land in the United Kingdom and manage it for conservation purposes. One example already mentioned is the National Trust and National Trust for Scotland that together own about 350 000 ha of land or over 1% of the British land surface. Ownership includes many important mountain, moorland and coastal landscapes, including half the coastline of Cornwall and over1000 km of coast in total. Their *Enterprise Neptune* project has been responsible for purchasing stretches of coastline over the past 50 years. All these areas are managed to protect natural landscapes and their biodiversity, but greater attention is being given to geoconservation following the publication of a National Trust Geological Policy in 2007. Much important geoheritage is included in National Trust land holdings, e.g. large parts of the glaciated landscapes and Ordovician/Silurian volcanic sequences of Snowdonia National Park in Wales.

The UK Wildlife Trusts also own and manage reserves, and although their primary aim is wildlife conservation, some also take an active interest in geodiversity and geoconservation. An example is the Scottish Wildlife Trust (www.swt.org.uk)

which has a policy document on geodiversity (Scottish Wildlife Trust, 2002). In it the Trust:

- 'recognises Geodiversity as an essential component of our natural heritage';
- 'believes that land management practices should recognise conservation of geodiversity as a major aim and attribute high value and importance to this';
- will promote education about Geodiversity by raising awareness through interpretation on appropriate reserves and clubs for young geologists;
- will promote the conservation of Geodiversity through its works on its reserves and its support for RIGS (see earlier).

At least 17 of the Trust's current reserves have major geological or geomorphological interests within them and it recognises the need to 'highlight this geodiversity interest' and incorporate it in management planning.

The Wildlife and Countryside Act (1981) provides a framework to regulate fossil collecting at designated SSSIs or NNRs, In addition, the Nature Conservation (Scotland) Act (2004) included provision for Scottish Natural Heritage to prepare a Scottish Fossil Code. This was completed in 2008 following wide consultation and includes the following main elements:

- Seek permission—you are acting within the law if you obtain permission to extract, collect and retain fossils;
- Access responsibly—Consult the Scottish Outdoor Access Code prior to accessing land. Be aware that there are restrictions on access and collecting at some locations protected by statute;
- Collect responsibly—Exercise restraint in the amount collected and the equipment used. Be careful not to damage fossils and the fossil resource. Record details of both the location and the rocks from which fossils are collected;
- Seek advice—If you find an exceptional or unusual fossil do not try to extract it; but seek advice from an expert. Also seek help to identify fossils or dispose of an old collection;
- Label and look after—Collected specimens should be labelled and taken good care of;
- Donate—If you are considering donating a fossil or collection choose an accredited museum, or one local to the collection area.

Other collecting codes are used in the United Kingdom including a general code promoted by the Geologists' Association, and also a number of local codes for particular areas or sites, e.g. West Dorset Coast (Edmonds, 2001). The Geologists' Association has also published a Code of Conduct on Rock Coring.

9.5 Republic of Ireland

Ireland is an interesting example since, apart the geological importance of its six National Parks, including The Burren with its important karst geomorphology and limestone pavements (Drew, 2001), 'It would be fair to say that there has been no significant tradition of geological conservation (in Ireland) until relatively recently' (Parkes and Morris, 2001, p. 79). As a consequence a major review was undertaken and the Irish Geological Heritage Programme (IGH) instituted. It is instructive to outline the way in which a country has recently initiated a geoconservation programme more or less from scratch.

The IGH Programme is run as a partnership between the Geological Survey of Ireland (GSI), which undertakes scientific appraisal and site selection, and the National Parks and Wildlife Service (NPWS) of the Department of Environment, Heritage and Local Government (DEHLG) that will carry out the statutory designation of Natural Heritage Areas (NHAs) under the Wildlife (Amendment) Act (2000) and their management. The aim is to identify and document the wealth of geological heritage in the Republic of Ireland, protect it against the ever-increasing threats (including through the Irish planning system), and promote its value to landowners and the public. However, Parkes and Morris (2001) predicted a slow process of designation probably as themed blocks.

In fact, 16 themes have been identified and these are listed in Table 9.2. For each theme, expert panels were established whose primary purpose was to 'ensure inclusiveness of all acknowledged expertise and maximum scientific rigour in the selection process, as any legal challenge to the validity of a site will be based and assessed on scientific robustness' (Parkes and Morris, 2001, p. 82). Each panel has then worked with the GSI to refine the final choice of sites, following which consultants were appointed to undertake desk studies and site reports. Field visits to photograph sites and establish site boundaries were the final stage, with particular attention paid to personal contact with landowners. Parkes (2008) and Gatley and Parkes (2012) have given progress reports.

Table 9.2 Irish Geological Heritage Programme Themes (after Parkes and Morris, 2001).

IGH1	Karst
IGH2	Precambrian to Devonian Palaeontology
IGH3	Carboniferous to Pliocene Palaeontology
IGH4	Cambrian—Silurian
IGH5	Precambrian
IGH6	Mineralogy
IGH7	Quaternary
IGH8	Lower Carboniferous
IGH9	Upper Carboniferous and Permian
IGH10	Devonian
IGH11	Igneous intrusions
IGH12	Mesozoic and Cenozoic
IGH13	Coastal Geomorphology
IGH14	Fluvial and Lacustrine Geomorphology
IGH15	Economic Geology
IGH16	Groundwater

Of 1100 geosites on the IGH Indicative List, about 60% had been provisionally selected for recommendation as NHAs by 2012, though none had been formally designated due to resource constraints. According to Parkes and Morris (2001, p. 82) 'Each theme is intended to provide a national network of NHA sites and will include all components of the theme's scientific interest'. In other words, what the system is intended to establish is a representative selection of Ireland's geodiversity, but unique, exceptional and internationally important sites are also included. They also emphasise that only a minimum number of sites will be selected with minimum duplication of interest, but they recognise the potential vulnerability of this minimalist approach. The following preferences were also used in site selection:

- sites with an assemblage of characteristics or features or a range of interests;
- sites with the most complete, undamaged or 'natural' record;
- sites with a history of research or future research potential;
- sites that have yielded results with a wider significance, e.g. . archaeological, historical or ecological interest.

Box 9.12 gives an example of an important palaeontological site.

In addition to the NHAs, a network of non-statutory County Geological Sites (CGS) is being established (similar to RIGS in the United Kingdom) with some level of protection achieved through the Irish land-use planning system. Concern over damage to limestone pavements at The Burren has led to the introduction of the Burren Code which discourages visitors from removing limestone or building cairns and dolmens from shattered limestone or field wall stones.

Box 9.12 The Valencia Island Tetrapod Trackway

A mid Devonian tetrapod trackway comprising about 200 prints was discovered on Valencia Island, County Kerry in 1993. This makes it the most extensive and possibly the oldest of only seven comparable sites in the world, having been dated to at least 385 Ma. It represents a major evolutionary step from aquatic vertebrates to air breathing amphibians and is therefore worthy of conservation. It is however threatened by unscrupulous collectors and visitor pressures and was in urgent need of protection. Since it occurs on slabs only 10–30 cm thick it could be removed relatively easily in sections for museum curation, but there was also a desire to conserve it *in situ* in order to retain its field context.

After negotiation, the site has been purchased by the Irish government which manages it in conjunction with the GSI and local Valencia Heritage Society. The site is already within an NHA designated on ecological grounds, and day-to-day vigilance is being provided by the local population who have been made aware of the site interest and threats. Access to the site is being improved together with overlook points and interpretation panels (Parkes and Morris, 2001; Parkes, 2001).

9.6 The rest of Europe

Many counties in continental Europe have a long tradition of geoconservation work. This section attempts a summary of current legislation, programmes and projects, greatly aided by the recent publication of an edited book on *Geoheritage in Europe and its Conservation* (Wimbledon and Smith-Meyer, 2012), which describes geoconservation activities in all the major countries of Europe. Only a summary is possible here and readers are directed to Wimbledon and Smith-Meyer (2012) for the details.

9.6.1 Northern Europe

Norway's first Nature Conservation Act was passed in 1910, and in 1919 the first site to be protected partly for geology, was Tofteholmen island in Oslofjord. In 1923 an erratic block in southern Norway was designated and in 1931, all limestone caves in Rana/Nordland area were protected. At the time only 5–8 caves were known but this is now increased to 20–30 (Erikstad, 2012). 'This protection is a good example of geological "species protection" ... ' (Johansson, 2000, p. 46). New nature protection laws were passed in 1954 and 1970 and the latter allowed national parks, landscape protection areas, nature reserves and natural objects to be protected. Thus during the 1970s systematic nature conservation planning was initiated, concentrating on national and regionally important areas and on features that were perceived to be in danger of becoming extinct, for example mires, wetlands and seabird colonies. Large national parks were established and include important geological and geomorphological features within such famous areas as Jotunheimen, Jostedalsbreen and Hardangeridda National Parks.

Later, work began on a regional basis on Quaternary geology, palaeontology and mineralogy. As a result, regional protection plans have been implemented for Quaternary geology of the Finnmarks and Hedermarks districts (40 sites), sedimentary rock art in the Oslo area (64 sites) and mineral localities in Sør-Norge (16 sites). Between 167 and 214 geological sites are protected in Norway (Erikstad, 2012) but there are about 900 sites of Quaternary interest registered in the national database. In addition, there is a Norwegian system of protected rivers with 387 watercourses currently included (Huse, 1987; Smith-Meyer, 1994; Erikstad, 2012). In 2009 an important Nature Diversity Act was passed that specifically includes geodiversity, biodiversity and landscape diversity. Today the focus of work is on landscape scale protection including the establishment of Landscape Protected Areas and the expansion of the current network of 35 National Parks (excluding Svalbard). These cover nearly 10% of the country and the plan is to establish new, and extend existing, National Parks. They already include most of the country's glaciers (Erikstad, 2012). The polar archipelago of Svalbard has separate legislation (e.g. Svalbard Environmental Act, 2002). About 65% of Svalbard's land area is protected including seven national parks and the large Nordaustlandet reserve which covers over 19 000 km^2.

In Iceland, the Ministry for the Environment, established in 1990, has responsibility for nature conservation, but the Environment Agency has a role in monitoring and supervising nature conservation sites, providing public information, and commenting on proposals for major developments. There are, however, three large protected areas not managed by the Environment Agency: the Vatnajökull National Park and Briedafjordur Conservation Area which are both directly managed by the Ministry, and the Thingvellir National Park managed by a committee under the auspices of parliament (Ásbjörnsdóttir, Einarsson and Jónasson, 2012). The Nature Conservation Act (1999) increased the possibility of protecting geodiversity as it included volcanic craters, rootless vents, lava fields, waterfalls, hot springs as well as sinter and travertine deposits. 'Unfortunately, this part of the act has proved useless since it has never been taken seriously by the authorities'. This is partly due to lack of professional geological capacity in the relevant organisations (Ásbjörnsdóttir *et al.*, 2012). In Iceland, geological formations of outstanding beauty or scientific interest can be designated as natural monuments including waterfalls, volcanoes, hot springs, caves, rock pillars and beds containing fossils or rare minerals. The Act stipulates that there must be a buffer zone around each monument to allow it to be appreciated. More than 35 geological natural monuments have been designated, including the Laki fissure cones (Figure 9.24), though many important geological sites are also included within other categories. Past efforts at systematic site selection in Iceland have had limited success but new work is underway to identify the geodiversity of Iceland and classify its geoheritage (Ásbjörnsdóttir *et al.*, 2012).

In Denmark in 1984, the Ministry of the Environment published a map of over 200 Areas of National Geological Interest with the intention that regional and local authorities would protect them and take account of them in their planning processes (Holm, 2012). This has subsequently been expanded into a digital map of 400 areas of international, national and regional geological interest supported by web-based descriptions and information on each site. The Danish National Committee of Geology has nominated 38 of the areas as Geosites but none has legal protection. However, some geological sites are protected through their inclusion in areas protected for biodiversity, etc. (Holm, 2012). A survey of Denmark's coastline has been undertaken with the aim of identifying areas of geological, geomorphological or coastal dynamic interest (Johansson, 2000).

In 1999 Sweden's environmental laws were transferred into the Environmental Code whose aim is to promote sustainable development and this has brought broader aims than nature conservation. Geological sites may be indirectly protected by general regulations in the Code such as those protecting shorelines and watercourses. National Parks, Nature Reserves and Natural Monuments can be designated under the Code. Sweden has 29 National Parks almost 90% of which are in the Swedish Mountain Area. Lundqvist *et al.* (2012, p. 356) comment that 'Today, a large part of the geological heritage of Sweden is not rightfully recognised.... the consideration of important but small geological sites is often neglected'. Despite the plea by Johansson (2000) for a geological perspective in nature conservation, this report 'has regrettably not met any substantial response in either policies or practical measures in Sweden' (Lundqvist *et al.*, 2012, p. 352).

Figure 9.24 Part of the Laki fissure eruption Natural Monument in Iceland showing some of the cinder cones.

Finland has 35 National Parks some of which include important geological or geomorphological features For example, Päijänne National Park includes some of Finland's best esker systems. In fact Finland has a National Esker Conservation Programme (1984) in response to damage caused by decades of aggregate extraction. The main criteria for the selection of esker sections are representativeness, diversity and rarity. Currently there are 159 eskers protected covering 97 000 ha, about 6% of the total esker area (Kananoja *et al.*, 2012). A National Mire Protection Programme was established in 1979. There are a few hundred geological natural monuments in Finland the majority being large erratics, with examples of potholes, gorges, cliffs and caves also being legally protected. Finland has also developed management guidelines applying to all its protected areas (Metsähallitus, 2000). One of the main principles (p. 15) is 'not to interfere with natural processes without good reasons related to nature conservation'. Mining is prohibited within protected areas but traditional gold panning is permitted by license. In the past 20 years several inventories of geological sites have been created including an assessment of scenically valuable rocky outcrops that has so far recognised over 1000 nationally valuable sites. A second rock outcrop project focussing on smaller geologically and geomorphologically important bedrock sites was launched in 1989 and recognised 160 internationally or nationally important sites. A national survey of moraines project (MORMI) was carried out between 1991 and 2007 and established about 600 internationally or nationally valuable sites. The TUURA project (2005–2011) assessed wind-blown sand formations and beaches. Most of these projects were carried out by the Ministry of the Environment and Geological Survey of Finland. In addition there is a database of 1000 caves in Finland that could be the basis for a cave protection programme (Kananoja *et al.*, 2012).

Estonia's *Nature Conservation Act* (2004) protects 18% of its land area, including about 500 geosites, 300 of which are erratic boulders. Estonia also has five National Parks, the first of which—Lahemaa National Park (1971)—comprises interesting geological structures, stratotypes, landscapes and erratics (Raudsep, 2012). Geology is also protected inside Nature Conservation Areas which include the Kaali Meteorite Craters formed on the island of Saaremaa about 3000 years ago. The Act also conserves 150 Landscape Conservation Areas, 344 Special Conservation Areas and 1198 individual natural objects, including the erratics mentioned above. Raudsep (2012, p. 113) comments that 'sometimes there are discussions about the necessity and reasonableness of a small country having so many protected areas and small objects. Yet, it seems clear that damage or destruction of unique geological or other sites would be harmful not only to Estonian nature ... but to all Baltic Sea countries'.

In April 2001, the Latvian Cabinet approved the protection of 206 geological and geomorphological nature monuments in Latvia including 88 cliffs or rock outcrops, 32 caves, 7 waterfalls, 34 erratic boulders, 21 springs, 8 Quaternary formations and a Devonian fish site. However, there are probably more geosites within other designated areas (Pavils *et al.*, 2012). A similar system exists in Lithuania with 170 geological (erratics, rock outcrops, sinkholes), 35 geomorphological (eskers, kettles, ravines, hills), 36 hydrogeological (lakes, springs) and 23 hydrographical (rivers, oxbows) sites, making a total of 264 protected geosites in the country in May 2010 (Satkunas, Lincius and Mikulenus, 2012).

9.6.2 Eastern Europe

Many eastern European countries have a long history of geoconservation and detailed site networks. This is certainly true of the Czech Republic, Slovakia, Hungary and Poland. In the Czech Republic the most important legislation is Act 114/1992 Coll. that enables landscapes, caves, karst, fossils and minerals to be protected as well as the establishment of National Parks and Protected Landscape Areas. The following categories exist (with the number of sites where geology is the main reason for protection in brackets (Budil *et al.*, 2012): National Nature Reserves (31), National Nature Monuments (54), Nature Reserves (120), Nature Monuments (287). The Czech Geological Survey maintains a GIS database of over 2500 sites of which 1440 have legal protection. The records include geological characteristics, degree of conservation, objectives of conservation and conflicts of interest (www.geology.cz). Many sites also have a photographic record. According to Budil *et al.* (2012, p. 99) 'management is variable and is dependent on the activity of the authorities involved'.

Not surprisingly, Slovakia has followed a similar history of geoconservation but has a new Nature and Landscape Protection Act (2002) protecting National Natural Landmarks, Natural Landmarks, Nature Reserves and National Nature Reserves. Slovakia's extensive limestone outcrop means that caves and karst have special recognition in law, including cave protection zones on the surface. Waterfalls, fossils and mineral sites also have legal protection. From 2008 to 2011 a database of geological sites was established and by November 2011 contained details of 479 sites in 11 categories (Liscak and Nagy, 2012). However, Liscak (2012, p. 320) argues that 'there is still a lack of a sound geoconservation policy with a well-defined strategy, and a clear and transparent procedure based on fundamental geological information for the protected and managed geosites'.

In Hungary, sites protected mainly for their geological/geomorphological value include four National Parks (including the World Heritage listed Aggtelek National Park), one Landscape Protection Reserve (with geology contributing to the designation of 16 others) and 15 Nature Conservation Areas (and 15 joint geological/biological sites). In addition, 34 geological sections of key Devonian–Triassic stratigraphy have been designated as one Natural Monument. Over 4000 caves are protected, 146 of which are 'strictly protected' due to their size and/or significance. Other protected natural assets include crystals/aggregates of 11 mineral species (including azurite, cuprite, malachite, native copper and opal) if they are of exceptional size (Bolner-Takács, Cserny and Horváth, 2012).

Poland had developed the protection of earth science features by the 1930s, but the Nature Protection Acts (1991 and 2004) reviewed the existing series of protected areas developed piecemeal over many years and introduced some new ones. There are now 10 categories of protected area in Poland with Inanimate Nature Reserves, Inanimate Nature Monuments (including hundreds of erratic boulders) and Documentary Sites having been specially created for geoconservation (Alexandrowicz, 2012). However, as in most countries, many important geosites occur within categories protected for other reasons.

This is certainly true in Russia where 'only a few important geological features and landforms are protected expressly because of their geoheritage values ... plenty of important and even unique geosites are threatened' (Lapo and Vdovets, 2012, p. 298). Similarly in Ukraine 'only a small number of geosites has presently any proper protection if they are within State Reserves ... ' (Gritsenko, Rudenko and Stetsyuk, 2012, p. 388). See also Wimbledon's (2012) comments on the lack of geosite protection and proper planning systems at an important site in Ukraine.

In contrast, since the1990s Belarus has been undertaking systematic inventories and protection of geosites . By 2009, 22% of the country had been surveyed, with 221 zakazniks (sanctuaries), 231 natural monuments, 109 landforms, 330 erratics and 42 outcrops having been identified. In 1985, in the grounds of the Institute of Geological Sciences on the outskirts of Minsk, a remarkable 'Park of Stones' was created by assembling over 2000 glacial erratics from all over Belarus (Vinokurov and Goldenkov, 2012).

9.6.3 South-Eastern Europe

In Romania, natural monuments (including geological sites) have been established over the past 60 years during which time about 200 geological sites and more than 250 geomorphological, speleological and landscape areas have been protected as nationally important, including 25 considered to be of international importance (Andrasanu and Grigorescu, 2012). But 'the management of these areas is practically non-existent ... due to the fragmentation of responsibility for identifying, designating and managing protected areas between central (government) and local authorities' (Grigorescu, 1994, p. 468). However, 'with increasing public interest and organisational involvement, the real management of geosites is beginning' (Andrasanu and Grigorescu, 2012, p. 286). In Bulgaria, over 360 geosites are protected, but in many cases these reflect 'the folk history of the landscape' through features such as 'upright rocks' and 'rock mushrooms' and 'do not have scientific value' (Todorov and Nakov, 2012, p. 77).

In Greece, 'There is no national recognition or systematic governmental registration of protected geosites as yet ...' (Theodosiou, 2012, p. 150), but an inventory and database of over 1000 geosites has been established (e.g. Figure 9.25). A number of the 'protected monuments of nature are specifically designated for their palaeontological or geomorphological interests. 'Unfortunately, since 2011 due to the international and European crisis, things have deteriorated in every domain, geoconservation included (Theodosiou, 2012, p. 147). The first national study of geological heritage conservation in Albania was carried out by

Figure 9.25 Folded carbonaceous rocks with siliceous intercalations at Ag Pavlos, Crete, Greece. (Photo by permission of Elpida Athanassouli.) See plate section for colour version.

the country's Geological Survey in 1999 and led to an inventory of some 350 sites most of which are protected as natural monuments (Serjani, 2012).

In Former Yugoslav Republic of Macedonia, 23 geosites and 2 caves have legal protection as natural monuments while others are currently proposed. However, none of these 'is under real protection. Those facts mean that at any identified geosites, including those already protected, collecting minerals and fossils may occur, as well as export from the country' (Klincharov and Petkovska, 2012, p. 231). In Serbia, about 100 geosites are protected or incorporated in National Parks, Nature Reserves or Natural Monuments, but an inventory of around 550 geoheritage sites reflecting the geodiversity of Serbia has been established (Maran, 2012). Important loess/palaeosol sequences occur in the Vojvodina region (Vasiljevic *et al.*, 2010). However, 'Most of the geologically significant objects are not protected and those that are protected ... are not made accessible to the public' and 'are not receiving the treatment they deserve' (Mijoviç, 2012, p. 306). Bosnia and Herzegovina have little current geoconservation activity (Sijaric, 2012) but in Croatia, new Nature Protection Acts (2003 and 2005, amended 2008) have allowed special treatment of geological heritage. As a result there are now 56 protected geosites in Croatia, 52 as Nature Monuments and 4 as Special Nature Reserves, this being 12% of all protected areas in the country. In addition the 8 National Parks and 11 Nature Parks in Croatia 'all abound in geologically important sites' (Marjanac, 2012, p. 87). These include Plitvice Lakes National Park, a World Heritage Site since 1979. 'The legal basis is finally favourable for conservation of the geological heritage, but physical protection is questionable' (Marjanac, 2012, p. 90). Finally, Slovenia has about 30 geological, 115 geomorphological and 102 hydrological protected Natural Monuments designated under the Nature Conservation Act (1999). The mercury mining area of Idrija is particularly interesting and is a World Heritage Site and aspiring Geopark (Kavcic, 2012). However, in Slovenia, 'the protection of biodiversity is increasingly equated with nature protection ...' (Hlad, 2012, p. 326).

9.6.4 South-Western Europe

There has been an upsurge of interest in geoconservation in Italy, partly through the passing of the outline law on protected areas (L.394/91). This includes in its provisions scope to protect geological and geomorphological features of national or international significance due to their 'natural, scientific, aesthetic, cultural and recreational value ...' (Gisotti and Burlando, 1998, p. 11). The categories of protected areas include parks, reserves and natural monuments. More recently, the Italian Geological Survey has been engaged in an Italian Geosites Project involving the compilation of a database in GIS format (D'Andrea *et al.*, 2004; Brancucci *et al.*, 2012). After a break, the project was restarted in 2009.

Alcalá and Morales (1994), Melendez and Soria (1994) and Alcalá (1999) describe the effectiveness of two national laws passed during the 1980s in protecting Spain's geological heritage. First, the *Law of Conservation of Natural Spaces and Wild Flora and Fauna* (1989) established four categories of protected natural area: National Parks, Natural Reserves, Natural Monuments, Protected

Landscapes. National Parks are designated by the Spanish Parliament and managed by both the State Administration and the Autonomous Communities in which they lie. Responsibility for designating and managing the other categories of sites lies locally. New research on geosites by the Geological Society of Spain and Geological Survey of Spain led to the inclusion of geodiversity conservation within the new Natural Heritage and Biodiversity Act (2007). It defines the concept of geodiversity (see Table 1.1) and establishes that the Ministry of Environment, collaborating with regional governments and scientific institutions will maintain an Inventory of Sites of Geological Interest representative of Spanish geodiversity (Garcia-Cortés, Gallego and Carcavilla, 2012). Natural monuments include geological or palaeontological sites designated for their special interest due to the unique importance of their scientific, cultural or scenic value. The Law of Historical Heritage (1985) gives protection to sites of cultural interest including geological and palaeontological sites related to the history of mankind, generally treating them as subordinate to archaeological sites. These are the responsibility of the Autonomous Communities and some of the latter have also developed their own heritage laws. Box 9.13 gives a case study from La Palma in the Canary Islands.

Box 9.13 La Palma, Canary Islands, Spain

La Palma is at the western edge of the Canary Island chain and is the most volcanically active. Despite its relatively small size (45 × 28 km) its highest point is at 2426 m. Twenty protected areas have been designated on the island and these are shown in Figure 9.26. Some, like P-2 and P-4, have been designated primarily for their wildlife interest, and others, like P-8 and P-10 for a combination of landscape and wildlife. However, several are of primarily geological interest and these are briefly described here.

- P-1, Caldera of Taburiente National Park was designated in 1954 as Spain's fourth National Park. It is a huge basin 9 km in diameter and circa 1500 m deep with a sharp horse-shoe rim at over 2000 m above sea-level. In 1825 the German geologist Leopold von Buch applied the Spanish term 'caldera' (a large, deep pot or caldron) to this feature that he believed to be a volcanic crater. Nowadays it is believed to be the result of large-scale volcanically-induced landslides and the erosive action of 400 000 years of intensive rainfall.
- P-5, Cumbre Vieja Natural Park. This is the largest protected area on the island and comprises the volcanic spine of the southern half of the island. It contains five historically active volcanoes and many inactive cones. It has recently been predicted that further volcanic activity could result in instability of the western side of the area resulting in a landslide into the Atlantic of 500 km^3 of rock. The resultant mega-tsunami would inundate coasts from Brazil to Canada with waves up to 50 m high.
- P-6, Mountain of Azufre Natural Monument is an impressive volcanic cone on the eastern coast.

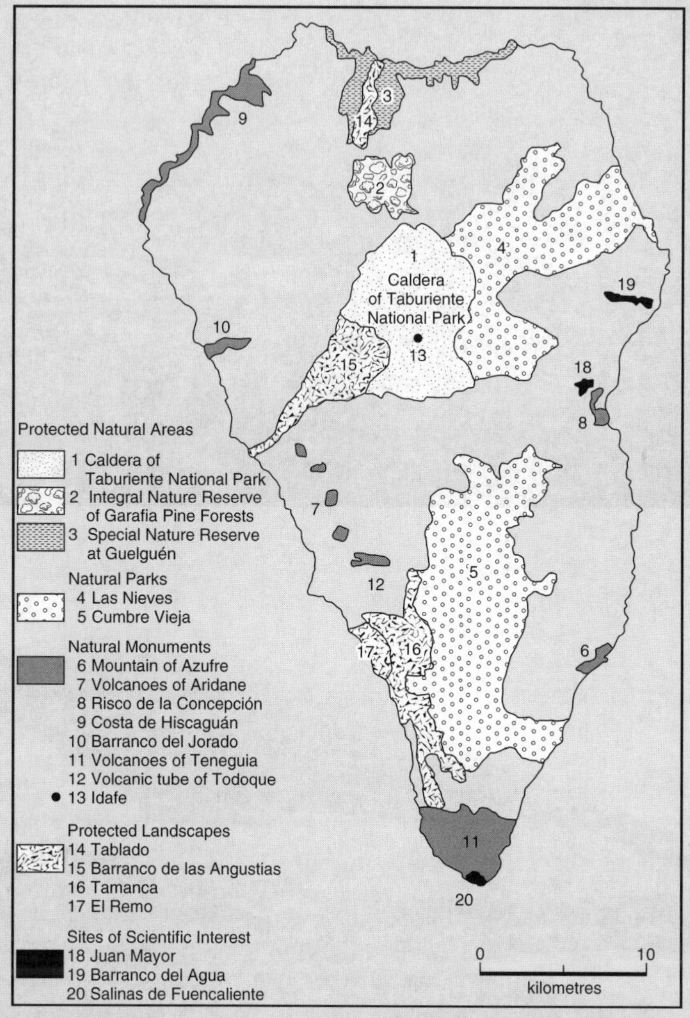

Figure 9.26 The 20 protected areas on the Spanish Canary Island of La Palma (after Santos, 2000).

- P-7, Volcanoes of Aridane Natural Monument comprises four separate volcanic cones in close proximity. Banana cultivation is having an impact on their setting (Figure 9.27) and Santos (2000, p. 129) comments that 'some of them are affected by building work and aggregate extraction'.
- P-8, Risco de la Conceptión Natural Area is an igneous intrusion on the outskirts of the island's capital Santa Cruz.
- P-11, Volcanoes of Teneguia Natural Monument. This protected area at the southern tip of the island contains several volcanic cones, including the most recent eruption point of Teneguia (1971) and the impressive cone of San Antonio (1677).

(a)

(b)

Figure 9.27 Banana cultivation in plastic greenhouses encroaching on one of the Volcanoes of Aridane Natural Monument, La Palma: (a) the general setting of the volcanic cone and surrounding greenhouses and agriculture, and (b) a greenhouse partially excavated into the cone.

- P-12, Volcanic Tube of Todoque Natural Monument. This area contains a lava tube 500m long. It was unearthed by a mechanical digger and lies in an area of ropy lava formed during the San Juan lava flow (1949). Santos (2000, p. 132) comments that 'All of the best ropy lava fields on La Palma were from the 1949 eruption of the San Juan volcano, but most of them have been savagely destroyed, with only small isolated areas remaining'.

- P-13, Idafe Natural Monument is an upstanding dyke remnant within the Taburiente caldera and is believed to have been sacred to the indigenous population.

Brilha (2005, 2012) provides a summary of geoconservation in Portugal. There are 32 protected areas in the country: 1 National Park (Peneda-Geres, Figure 10.2), 13 Natural Parks, 9 Natural Reserves, 2 Protected Landscapes and 7 Natural Monuments (Brilha, 2012). All 7 Natural Monuments are geological, 5 of them on the basis of dinosaur footprints, 1 for Jurassic stratigraphy (Cabo Modego which is also a GSSP), plus the gap in the Ordovician quartzite ridge at Portas de Ródao. Research has been carried out in 2 Natural Parks in northern Portugal to inventory and interpret the geological heritage (e.g. Pereira *et al.*, 2007). The autonomous regions of Madeira and Azores have their own legislation and site categories (e.g. Figure 9.28). Over 270 lava tube caves are included in the Azorean Speleological Inventory (Costa *et al.*, 2008).

9.6.5 West/Central Europe

France has the benefit of having 12 National Geological Reserves. These include a number of stratotypes and palaeontological sites, and they also include the European/ Global Geoparks of Haute-Provence and Luberon. Several hundred important geological sites also occur in other protected areas and an inventory of these has been created (Guiomar and Pages, 2012). In Belgium and the Netherlands, only a few geological sites have legal protection (Jacobs, 2012; van den Ancker and Jungerius, 2012).

'Germany has a very long tradition in the protection of geosites, which we call *geotopes* ... ' Röhling, Schmidt-Thomé and Goth (2012, p. 133). However, 'Due to the differing regional competences and procedures within and between the German Federal states, reliable and comparable data on the total number and significance of geosites are not available (Röhling *et al.*, 2012, p. 136). A project to select the German geotopes of international and national significance was initiated in 2003 and resulted in 77 sites being chosen.

Geotopes are also identified in Austria, but again nature conservation is the responsibility of the 9 federal states. Approximately 42% of Austria is protected under at least one of 14 different categories of site protection, so that many geosites are automatically protected. As a result, 'The diverse geological and geomorphological heritage of Austria is ... to a very high extent protected ... ' (Hofmann and Schönlaub, 2012). In 1928, Austria passed one of the first laws dedicated especially to the protection of a geological feature—caves (Trimmel, 1994). Criteria for establishing protected caves include scientific value, for historical research, palaeontology, geological structures, sediments, etc. The law also made it possible to create buffer zones around cave entrances. Over 10 000 caves

Figure 9.28 Signage for the Natural Regional Monument Furnas do Enxofre, Terceira, Azores, Portugal.

are now documented and the protection system has been regionalised, though this has led to inconsistencies in enforcement (Trimmel, 1994).

Switzerland also has an inventory of geosites first compiled by the Swiss Academy of Sciences (SCNAT) in 1999 and updated as the Inventory of Swiss Geosites in 2006–2012. However, it has no legal status as it does not fall within the framework of the Nature Protection Act (1966) (Reynard, 2012; Stürm, 2012). Nature protection is the responsibility of the Swiss cantons and as a result, geoconservation varies substantially from one canton to another. Stürm (1994) expressed concern that existing procedures are too weakly implemented to protect important landscapes and deposits in the face of a booming demand for sand and gravel aggregates. Stürm's concern relates to the moraine landscape near Zurich. Although the landscape described by Stürm (1994) has been added to the *Federal Register of Landscapes of National Importance*, the problem is that planning authorities do not always designate such landscape areas in their land-use plans. In such circumstances, the inventory 'will remain limited and vague' (Stürm, 1994, p. 27).

9.7 Australia

Reviews of the history of geoconservation in Australia have been published by Brocx (2008) and Joyce (2010). Brocx concludes (p. 116) that although significant progress has been made in identifying geoheritage sites, 'there still are large deficiencies and gaps in the conceptualisation of the issues and in the process of their preservation and management'. But we begin with Australia's National Parks.

9.7.1 National Parks

Australia's first National Park was established in 1879 on the southern margin of Sydney. Subsequently the number has grown gradually and Australia now has over 500 National Parks established and managed by the States and Territories under various National Parks and Wildlife Acts or similar passed in the 1970s. In addition, six National Parks are administered by the Australian Government, including Uluru-Kata Tjuta National Park (Figure 4.24). The Australian National Parks cover over 25 million ha or 3.42% of the land area. The main aim has been to conserve scenery, wilderness or biological communities, and rarely have geological and geomorphological sites been included. But, an example of a national park designated primarily for its geology/geomorphology is Purnululu National Park in the northern part of Western Australia, also known as Bungle Bungle National Park. It comprises 770 km^2 of horizontally striped sandstone domes about 100 m high separated by a labyrinth of gorges. The striping is the result of cyanobacteria being able to grow on layers with higher clay content but not on those that dry out rapidly, giving the striking, alternating orange and grey layer effect (Figure 9.29). It became a national park in 1987 and a World Heritage Site in 2003.

An example of an Australian National Park primarily designated for its biodiversity is Fraser Island, Queensland. It comprises tall rainforest growing on sand dunes, but the latter are important in their own right (see Box 9.14). Fortunately, Queensland's Nature Conservation Act (1992) introduced dynamic processes and landforms as aspects to be considered in selecting subsequent national parks.

Box 9.14 Fraser Island, Queensland

Fraser Island is a National Park and World Heritage Site lying off the south-east coast of Queensland, Australia. Its primary conservation interest is its globally unique ecosystem of rainforest types, but at 122 km long and 5–25 km wide it is claimed to be the largest sand island in the world and the sand extends 30–60 m below present sea-level. The dunes have been dated as a complete sequence from before the last interglacial (120 000–140 000 years BP) through to the early Holocene (less than 10,000 BP). The sand derives from granites, sandstones and metamorphic rocks in river catchments to the south and from the seafloor. The hydrology of the sand mass is also important with an extensive lens shaped aquifers within the porous sand mass and perched aquifers above less permeable organically-bound sands. These also underlie the circa 40 dune lakes, some of which contain long sediment records stretching back 300 000 years.

9.7.2 Conservation reserves and sites

In addition to national parks, there are over 2700 other conservation reserves in Australia covering 3.6% of the land surface and designated by States and territories and with a variety of names, including conservation parks, nature reserves, State parks, wilderness parks and Aboriginal areas. However, geology is a secondary player in these reserves and indeed some prominent geological environments have been deliberately excluded because of their potential economic value.

In 2004, the Australian Heritage Council was established to replace the Australian Heritage Commission (AHC) as the body responsible for advising the government on heritage matters, including the National Heritage List (NHL) established under the Environment Protection and Biodiversity Conservation Act (1999). This is a list of items and areas worthy of protection for future generations, highlighting the 'major stories of Australia—the evolution of the land, the qualities of its people and the diversity of its culture' (Joyce, 2010, p. 50). About 20 geoscience sites are included on the List, including the Ediacara Fossil Site in the Flinders Range, South Australia and the Dinosaur Stampede National Monument in Queensland.

Figure 9.29 Part of Bungle Bungle National Park in Western Australia showing the typical horizontal stripes (see text for explanation). (Photo by permission of Grant Dixon Photography.) See plate section for colour version.

Many more geological and geomorphological sites are included on the now superseded *Register of National Estate* (RNE). This was intended to be a 'representative list of the places that demonstrate the main stages and processes of Australia's geological history' as well as 'rare or outstanding natural phenomena, formations, features, including landscapes and seascapes' (Joyce, 2010, p. 45). A total of 691 geological and geomorphological sites are included on the Register (Joyce, 2010), including the Franklin River, Maria Island and Mole Creek karst in Tasmania. However, since 2012 the RNE has been maintained only as a non-statutory, publicly available archive, the sites themselves having been transferred to state, territory or local heritage registers.

The Geological Society of Australia (GSA) has carried out much work to identify 'those features of special scientific or educational value which form the essential basis of geological education, research and reference' and thus are worthy of protection and preservation (Joyce, 1997, 2010). Initially GSA subcommittees in each state developed their own methods for selecting sites of geoheritage significance but later a more systematic approach was used. Joyce (1995, 1999) described the methodology used in reviewing sites to be added to the RNE. The methodology is summarised by the acronym IDEM (Identification, Documentation, Evaluation, Management)—and to record site information, a new form called LCAN (Location, Classification, Assessment, National Estate Criteria) was used. The aim was to draw up a consistent classification of geological features according to physical type (section, quarry, landform, etc.), geological type (palaeontological, igneous, geomorphic, etc.), status (local to international) and use (research, education, reference). The intention was to 'provide an impetus to the development of Australia's geological heritage, and to the formal listing of further features on the Register of the National Estate' (Joyce, 1999).

The Geological Society of Australia's conservation policy has been criticised by Tasmanian workers. Pemberton (2001a), for example, is critical of the GSA approach since 'sites can only be significant in terms of scientific or educational use values and appears to emphasise geological features. This narrow interpretation of significance may be appropriate for the GSA but it only encompasses part of the aims of conserving geodiversity … and does not recognise that relevant expertise may reside with non-GSA scientists'. Pemberton (2001a) also believes that 'Current process sites have been seriously under-represented in Australia inventories of significant sites'. Similarly, Sharples (2002a) believes that 'historically most geoconservation work in Australia has been focussed on a 'geological heritage' approach, in which geodiversity … was seen as being important mainly for its value to scientific research and education'. Because this approach does not address issues of intrinsic values and ecological sustainability, it 'has largely been ignored as a minor issue in nature conservation programs because of its perceived lack of relevance to central issues of land management' (Sharples, 2002a). C. Sharples (pers. comm.) sees geoconservation as an almost essential and integral part of nature conservation, instead of it being a rather separate and almost unrelated issue, which is how the traditional 'geological heritage' approach tends to be presented in Australia.

9.7.3 Marine protected areas

Australia has recently become the world leader in the conservation of the marine environment through a systematic review and expansion of its marine reserve network. On 16 November 2012, the government announced the addition of 33 reserves covering 2.3 million km^2 to the network bringing the total to 60 reserves totalling 3.1 million km^2. As well as the Great Barrier Reef Marine Park, the network now includes a large area to the northeast known as the Coral Sea that includes other iconic reefs including Osprey Reef, Marion Reef, Bougainville Reef, Vema Reef and Shark Reef as marine national parks. In southwest Australia, the network includes the Perth Canyon, an underwater area bigger than the Grand Canyon, and the Diamantina Fracture Zone, a large underwater mountain chain. In addition to these Commonwealth marine reserves a number of other marine protected areas exist in Australia including marine and coastal parks.

9.7.4 State activity

The Australian states and territories have a large measure of autonomy and their jurisdiction takes precedence over the national government in conservation areas. The Geological Society of Australia and its Sub-committees have achieved a great deal in building up site inventories for individual states/territories or parts of states/territories. Joyce (1999) also discusses the different approaches to site identification. In some states sites are chosen as being representative of a particular geological type while in others they may be selected as being outstanding at state, national or international levels.

South Australia has created 'fossil reserves' that are controlled by the South Australian Museum, which actively manages access and collecting. The important Ediacara Fossil Reserve about 700 km north of Adelaide, is one example where a range of Precambrian soft-bodied animals including jellyfish was discovered in the 1950s, and is the earliest known animal assemblage in the world. Unfortunately, although collecting can only be done with permission, enforcement has not been adequate and the original site is virtually destroyed (Swart, 1994). Swart believes that enforcement and funding are major issues in geological conservation in South Australia, but notes that voluntary and community effort together with public education are probably the key to effective geological conservation.

Joyce (1999) described work to identify geological sites in Victoria, and Rosengren (1994) described Victoria's Late Cenozoic volcanic province consisting of 354 volcanic eruption points. He carefully described these features and classified them according to morphology and scientific significance. He expressed concern about the impact of quarrying on this geological and geomorphological heritage (see Section 5.2), but noted that in Victoria, the government department responsible recommends that '... the geological significance of a feature should be considered when assessing quarry applications' and there should be '... investigations ... to locate alternative sources of scoria and tuff where existing pits are found to be compromising significant volcanic features ...' (Guerin, 1992).

But the most impressive geoconservation work in Australia has undoubtedly been undertaken in Tasmania by the Department of Primary Industries, Water and Environment (DPIWE), the Parks and Wildlife Service, the Forest Practices Unit, Forestry Tasmania and private consultants (e.g. Kiernan, 1991, 1995, 1996, 1997a; Sharples 1993, 1995, 1997, 2002a, 2002b, 2011; Eberhard, 1994, 1996; Dixon, 1995, 1996a, 1996b; Houshold *et al.*, 1997; Pemberton, 2001a; Jerie, Houshold and Peters, 2001; Houshold and Sharples, 2008). Sharples (2002a) and Houshold and Sharples (2008) have summarised the philosophy and practice of the approach taken in Tasmania, which sees geoconservation as an essential and integral part of nature conservation and land management. Conservation of both traditional 'static' heritage sites and ongoing geomorphic and soil processes are recognised, the latter distinguishing the Tasmanian work from most of the rest of Australia. In addition the Regional Forest Agreement (Land Classification) Act (1998) made 'conserving geological diversity' a statutory management objective in all categories of conservation reserves in the state, while the Mineral Resources Development Act (1995) protects speleothems from uncontrolled destruction by collectors.

The Tasmanian Geoconservation Database (TGD) is a computerised database (with on-line access) of over 1500 Tasmanian geoconservation sites (e.g. the Maria Island Fossil Beds and the Pedra Branco phosphatic flowstone; see also Figure 9.30). The main aim is to provide information on the significance of sites for land managers and land management organisations. The database is actively managed by a 'Reference Group' established in 1999 consisting of government department representatives, academics, industry representatives and others, and includes geologists, geomorphologists and soil scientists.

Figure 9.30 Cambrian pillow lavas, City of Melbourne Bay, King Island, Tasmania. The lack of deformation is unusual and as a result they have been listed on the Tasmanian Geoconservation Database as a site of international significance. (Photo by permission of Michael Pemberton.)

The 2003 *State of Environment Report* for Tasmania found that of the 833 geological sites on the database at that time, 0.4% had been destroyed, 9.7% had some level of degradation and 24.2% were either endangered, threatened or potentially threatened. Of the 762 geomorphological sites, 20% had some level of degradation while 27.6% were either endangered threatened or potentially threatened. The figures for the 69 soil sites were much higher with 31.9% having suffered some degradation and 43.4% being endangered, threatened or potentially threatened.

A number of Codes of Practice also exist in Tasmania. For example, the Tasmanian *Forest Practices Code* (1993) makes provision for the protection of significant landforms by management prescription or special reservation (Kiernan, 1996). The Tasmanian Parks and Wildlife Service (TPWS) has issued a *Reserve Management Code of Practice* to guide all activities in protected areas within the State. This contains procedures for assessing and approving new activities and developments in protected areas and has a specific section on 'Geodiversity' (Box 9.15).

Box 9.15 Tasmanian Reserve Management Code of Practice (2003)

This Code of Practice has a section on Geodiversity that includes the following measures:

- relevant information on sites or areas of significance, or potential significance, should be recorded when planning an activity;
- consult the Tasmanian Geoconservation Database and other sources for the conservation significance and sensitivity;
- where an activity is approved at a site of geoconservation significance, measures should be implemented to avoid or minimise adverse impacts to geological, landform and soil features and processes. Specialist advice should be sought to determine these measures;
- permits and authorities required under the relevant acts and regulations should be obtained;
- threatening processes or activities should be addressed where they are likely to adversely affect a site of geoconservation significance or natural processes relevant to the integrity of a site of geoconservation significance;
- where rehabilitation activities are undertaken, these should, as far as practicable, restore relevant natural features and processes in accordance with approved guidance;
- modification of coastal landforms should be avoided except where there is a threat to infrastructure and the measures are consistent with *Tasmanian State Coastal Policy;*
- retention of parts or all of artificial exposures (e.g. road cuttings, quarries, etc.) will be considered where this can contribute to maintaining the values of sites of geoconservation significance;
- sites of geoconservation significance that are publicised or promoted should be managed to protect the values from threats arising from increased

visitation. Consideration should be given to controlling public access where this is likely to result in unacceptable impacts on site values;
- caves will be designated as Restricted Access Caves where unrestricted access is likely to result in unacceptable impacts to the cave environment.

Concern over the impact of coring in Tasmania has led to the adoption of a more careful approach to assessing research proposals and the restoration of cored sites (M. Pemberton, pers. comm.). For example:

- Proposals for coring or drilling significant outcrops (or locations in caves) are carefully considered and all other options explored, e.g. use of hand specimens;
- When coring is necessary, they are taken from overhangs or similar inconspicuous locations;
- Once a core is taken, the bottom of the hole should be filled with cement and the hole plugged with the cut-off core end;
- Old drill holes can be plugged with dyed cement or preferably with rock cores taken from loose rock samples or rock dust processed from these samples and mixed with cement or resin.

Another useful concept introduced in Tasmania is the conservation status (Dixon *et al.*, 1997). Five broad categories have been proposed:

- Secure—sites, processes or systems whose geoconservation values are not degraded, and are likely to retain their integrity, because the values are robust or their protection is provided for under existing management arrangements;
- Potentially threatened—sites, processes or systems whose values are not being actively degraded, but which are sensitive and whose protection is not specifically provided for under existing management arrangements;
- Threatened—sites, processes or systems whose values have been or are subject to degrading processes, although the values remain largely intact at the present time;
- Endangered—sites, processes or systems whose values have been or are subject to degrading processes that have had significant impact on values;
- Destroyed—sites, processes or systems whose values have been lost due to degrading processes.

Sharples (2002a) also suggests the development of performance indicators or targets, including data coverage indicators, condition and conservation status of sites, and indicators of the integrity of natural processes upon which significant sites and assemblages depend. A number of initiatives in this regard are discussed in Chapter 10.

9.7.5 Other legislation and protected areas

Another important geoconservation measure in Australia is the Protection of Movable Cultural Heritage Act (1986). Under the Act fossils, minerals and meteorites are included in a 'Control List' of movable objects that require a permit for export. Until 1983 this applied only to fossils with a value to museums of over $A1000, but this stipulation was removed in 1993. However, it is still the case that minerals valued at less than $A10 000 do not require an export permit (Creaser, 1994a). Applications for permits are assessed by expert examiners, but the government minister makes the decision whether to issue a permit. The maximum penalty for exporting without a permit is $A100 000 or five years' imprisonment, or both. Creaser (1994a) recounts examples of illegal fossil exporting from Australia.

Other organisations in Australia own their own reserves or lands in which conservation, including geoconservation, is practised. Each state and territory has its own National Trust and these own and protect important landscapes including coastal and upland areas. The Australian Council of National Trusts (ACNT) was formed in 1965 to represent Trust interests at the federal level and increasingly coordinates the work of the constituent Trusts. Another example is the Sydney Harbour Federation Trust, which produced a Comprehensive Plan in 2003. This recognises that 'Geodiversity has contributed much to the character of Sydney Harbour'. In protecting the geodiversity of its sites the Trust undertook to adopt the conservation principles, processes and practices contained in the Australian Heritage Charter (Australian Heritage Commission, 2002) and 'identify, protect, conserve and interpret significant sites such as volcanic dykes, Pleistocene sand dunes and laterites etc'. Indigenous people have voluntarily declared 24 Indigenous Protected Areas on their lands covering 200 000 km^2.

9.8 New Zealand

There are about 17 000 protected areas in New Zealand covering more that 8.6 million km^2 or about 32% of the country's land area. The first national park was established in 1894 covering the active volcanoes at Tongoriro. The 13 National Parks, consolidated under the *National Parks Acts* (1952 and 1980), have been established to protect landscapes and biological diversity but they contain some of the country's best earth science features, including fjords, waterfalls, glaciers and volcanoes (Figure 9.31). Kahurangi National Park includes most of the pre-Permian Palaeozoic rocks in New Zealand, including Trilobite Rock with New Zealand's oldest fossils. It is a tourist attraction but one that was damaged by indiscriminate collecting prior to its legal protection.

There are around 60 different types of protected areas in New Zealand but those most associated with geoconservation include Scenic Reserves (e.g. Moeraki Boulder scenic reserve protecting large spherical septarian concretions) and Scientific Reserves (e.g. the Curio Bay Scientific Reserve protecting a shore platform exposure of a Jurassic petrified forest; Figure 9.32).

In 1983, the Geological Society of New Zealand (GSNZ) began compiling a list of New Zealand's important and diverse geoheritage and there are now over 3400 sites on the Geopreservation Inventory (Hayward, 2009). The inventory was compiled by relevant experts rather than by systematic field surveys. In terms of significance, 291 have an international ranking, 1357 a national ranking and the

Figure 9.31 Mitre Peak (1692 m), Milford Sound, Fiordland National Park, New Zealand. The fjord landscapes and diverse geological features of this National Park are also protected through World Heritage Area status. (Photo by permission of Graeme Butterfield.) See plate section for colour version.

remainder are regionally important. Each site has also been given a vulnerability rating from 1 (= highly vulnerable to human modification) to 4 (= sites that could be improved by human activity), and 5 (= sites that have been destroyed). Their current protected status has also been documented. This information is maintained by the GSNZ with a copy held by the government Department of Conservation. The aim is to ensure protection and management of these sites for future research, education and public access (Buckeridge, 1994). The sites have been classified into 15 themes (e.g. landform, igneous geology, fossil, mineral, etc.) and sorted by region. These regional listings have been distributed to all land management agencies and conservation groups in the country (Hayward, 2009). Kenny and Hayward (1993) believed that 'The overriding objective of earth science conservation in New Zealand should be to ensure the survival of the best representative examples of the broad diversity of geological features, landforms, soil sites and active physical processes'. In effect, this is a call to have a site network that recognises and protects the geodiversity of New Zealand.

The Environment Act (1986) and the Conservation Act (1987) both recognise the intrinsic value of natural features and systems, although geoheritage and geoconservation are not specifically mentioned in the legislation. The Conservation Act (1987) and Resource Management Act (1991) require various authorities in New Zealand to prepare Conservation Management Strategies (CMSs) and Regional Coastal Plans some of these, including the Auckland CMS, have geoconservation objectives.

Figure 9.32 Petrified tree trunks of fossil Jurassic forest exposed on a marine platform, Curio Bay Scientific Reserve, New Zealand. (Photo by permission of Graeme Butterfield.)

In 1987 legislation was passed to halt further drilling for geothermal steam at Rotorua, following the failure of several geysers (see Section 5.4). The situation has now stabilised with a slight increase in bore pressures. Other aspects of New Zealand's geodiversity are protected by the Queen Elizabeth II National Trust Act (1977), which can protect by covenant privately owned areas of open space, usually in perpetuity, without jeopardising the rights of ownership. Two important fossil sites that benefit from this are Pyramid Valley swamp with its numerous intact moa skeletons, and Mangahouanga Valley with its large riverbed concretions containing Late Cretaceous dinosaur bones (Hayward, 2009). Section 11.1 deals with conservation and the planning system in New Zealand, while Table 9.3 outlines the quantitative selection of geoconservation sites in the Auckland area. New Zealand has no overall ban on the collection or export of fossils, though legislation does control export of type and iconic specimens under the Protected Objects Act (2006) (Hayward, 2009).

Table 9.3 The criteria and grading scheme for scoring the value of earth science sites in Auckland, New Zealand (simplified from Hayward, 2009).

Criterion	Grades	Points
A. Geological significance	None	0
	Local	4
	City-wide	8
	Regional	16
	National	32
	International	64
B. Rarity	Common locally or regionally	0
	Rare locally	4
	Rare in City or unique locally	8
	Rare regionally or unique in City	16
	Rare in NZ or unique regionally	32
	Rare globally or unique in NZ	64
C. Scientific potential	Poor	0
	Fair	2
	Good	4
	Excellent	8
D. Representativeness	Atypical	0
	Fair: not obviously representative	2
	Good: displays good and typical characteristics	4
	Excellent: displays exemplary characteristics	8
E. Diversity within a feature	One	0
	Several (2–4)	2
	Many (5–9)	4
	Numerous (>10)	8
F. Group significance		0 to 16
G. Visual contribution		0 to 16
H. Setting		0 to 8
I. Intactness		−8 to 0
J. Fragility		−32 to 0
K. Education value		0 to 16
L. Community value		0 to 16
M. History of geology value		0 to 16

9.9 The rest of the world

In many countries in the rest of the world geoconservation has barely begun and it is likely that significant losses of geoheritage are occurring. For example, in Central Eurasia comprising the newly independent states of Armenia, Azerbaijan, Georgia, Kazahkstan, Kyrgyzstan, Tajikistan and Turkmenistan covering 5 million km^2, many irreversible examples of damage can be cited (I. Fishman, pers. comm.). Examples of national site inventories include Kazakhstan (Nusipov, Fishman and Kazakowa, 2001: Fishman and Kazakova, 2012) and Iran where Amrikazemi (2010) has compiled an impressive photographic atlas of geosites (e.g. see Figure 3.2). In a number of other countries, including India, China, Taiwan, Brazil and South Africa, some attempts have been made to establish formal geoconservation systems.

India has over 100 National Parks, almost exclusively to protect wildlife or important wildlife habitats. It has also designated about 16 geological 'heritage sites' under the *National Monuments Act* (1974) (Prasad, 1994).

As well as 27 Global Geoparks and 140 national geoparks (see Chapter 8), China has over 200 National Parks, some of which are geologically-based World Heritage Sites. Geological interest ranks highly in the criteria used to select such areas (Jiang, 1994). China also has a law on 'preservation of cultural relics' (1982) that treats vertebrate and human fossils in an identical way to cultural objects. Enforcement has been strengthened recently to prevent illegal collecting of vertebrate fossils, some of which are used in traditional Chinese medicine.

Hong Kong, prior to its reunification with China, had a well-developed system of country parks, special areas, and Sites of Special Scientific Interest to protect a small territory faced with huge development pressures. The *Country Parks Ordinance* (1976) allowed 21 country parks to be established to protect wildlife, landscape and cultural sites, but also to encourage recreation and tourism. There is no reference to geological or geomorphological conservation in the Ordinance but activities such as mining are prevented in the parks. 'Special areas' can be designated inside or outside country parks and several of the 14 sites are of particular geological or landscape value. SSSI designation 'neither confers statutory power on the government to enforce preservation nor implies any restriction upon owners. ... Designation as an SSSI is intended to ensure that due consideration is given to protecting the site when considering development proposals that may affect it' (Workman, 1994, p. 295). Of the circa 50 SSSIs in Hong Kong, several have been designated primarily for their geological or geomorphological interest. There is currently a very active *Association for Geoconservation* in Hong Kong which helped to spearhead the efforts to create the Hong Kong Geopark (Section 8.6.5).

Taiwan has a system of national parks, nature preserves and scenic areas and a Cultural Heritage Preservation Law that give some protection against unsuitable land development projects (Wang, Sheu and Tang, 1994; Wang, Lee and King, 1999). Several national parks and three nature reserves were chosen for their geological or geomorphological interest (e.g. columnar basalt, mud volcanoes).

Yushan National Park, in the centre of the island, contains over 30 peaks above 3000 m as well as many canyons, cliffs, faults and folds. A five-year Earth heritage conservation strategy was initiated in 1994 with the aim of expanding the site database, increasing the protection of sites and improving public awareness (Wang *et al.*, 1999).

Brazil has over 60 National Parks including Iguacu National Park and World Heritage Site. As well as Araripe Global Geopark, several other GGN applications are being made. There is a national initiative co-ordinated by SIGEP (Brazilian Commission of Geological and Palaeobiological sites) aimed at approving a national list of the most important geosites from the many nominated by individual geoscientists. Two volumes of almost 100 site descriptions are already published with a third in preparation. Currently, there is considerable interest in the topics of geoconservation, geotourism and geoparks in Brazil (e.g. do Nascimento, Ruchkys and Mantesso-Neto, 2008) and considerable progress is expected in the next few years. Martill (2011) has argued that the laws protecting fossils in Brazil are draconian and ought to be scrapped.

About 6% of South Africa is covered by state protected areas managed by South African National Parks (SANParks). Like most of Africa's National Parks, they are primarily intended to protect wildlife, but many of the 18 National Parks contain important elements of geoheritage. For example, Golden Gates Highlands National Park is named after the colour of the sandstone cliffs, especially the sunlit Brandwag cliff, and the network also includes Table Mountain National Park. The SANParks website gives geology details for each of the Parks (www.sanparks.org). A particular issue in these parks has been the land claims of indigenous peoples and some settlement rights within parks are now starting to be restored, e.g. the Nama at Richersfield National Park and Makuleke at Kruger National Park (Magome and Murombedzi, 2003). Another important initiative is the establishment of trans-frontier conservation areas (TFCAs) and parks. These are conceived as relatively large areas of land including one or more protected areas that straddle boundaries between two or more countries. Although they are intended and allow protection and restoration of large-scale ecosystems, including traditional migration routes, they also protect many important abiotic elements (land form, soils, active processes, etc.). Figure 9.33 shows the location of existing or proposed TFCAs in southern Africa. The Great Limpopo Transfrontier Park was formally established in 2002. There is no systematic approach to the identification and conservation of geosites in South Africa although this is the ultimate aim, but there are some excellent local projects including the West Coast Fossil Park (Haarhoff and Prosser, 2007). South Africa has rigorous controls on collecting and export of fossils, largely due to earlier exploitation of its rich palaeontological heritage, including the remains of hominid fossils from the Plio-Pleistocene sites in the northern Transvaal and the removal of fossil reptiles from the Permo-Triassic of the Great Karoo Basin by collectors and international expeditions. As a result, the National Monuments Act (1969) protects all palaeontological as well as archaeological and historical sites.

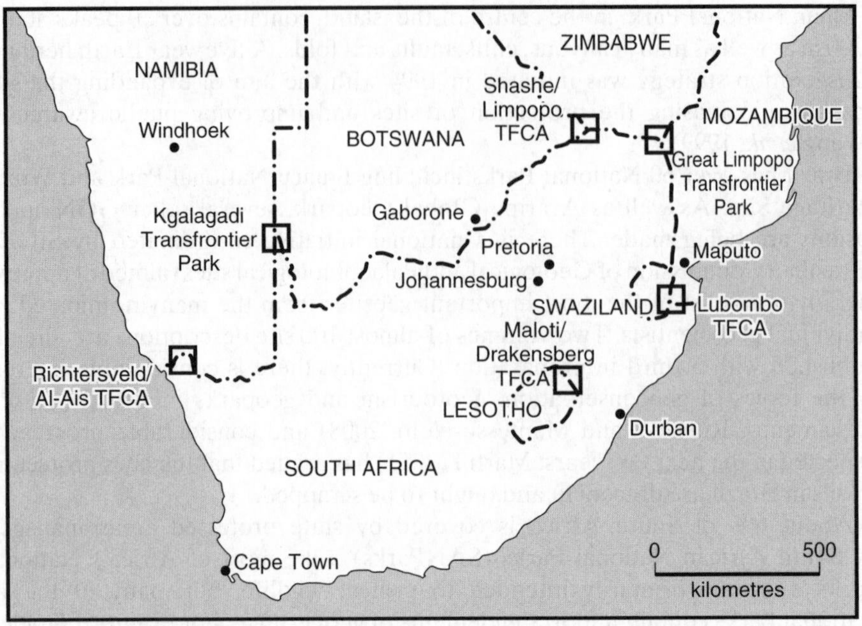

Figure 9.33 Existing and proposed Trans-Frontier Conservation Areas & Parks in Southern Africa (modified after Magome & Murombedzi (2003) Sharing South Africa's national parks: community land and conservation in a democratic South Africa. In Adams, W.M. & Mulligan, M. (eds) Decolonizing Nature. Earthscan, London, 108-134, by permission of William Adams).

9.10 Conclusions on protected area geoconservation

A number of issues emerge for this limited review in Chapters 6–9 of international and national geoconservation systems:

- *Diversity of systems.* The first obvious point to make is that systems of protected areas and nature conservation legislation vary greatly around the world. It is clear that biodiversity dominates nature conservation efforts in every country, but there are important geoconservation efforts being made in the UK, Australia, New Zealand, United States, Canada, Europe and some countries elsewhere in the world. The GCR programme in Britain is an outstanding effort though it is based only on scientific values. However much more needs to be done even in these countries to raise awareness of the value and threats to geodiversity and the need for conservation efforts. In the United States, for example, there are many geological sites including world famous palaeontological sites, stratotypes and unique mineral localities that have no legal protection. The GSSP network needs international geoconservation recognition. The Geoparks movement has been an outstanding success and is stimulating thinking about geoconservation in many developing countries where few efforts have been made in the past. Some countries have special designations for geological sites (e.g. 'Geological Monuments'), but others argue that this makes geodiversity look like a special case rather than being a standard objective of nature conservation.

- *Site selection approaches.* Those countries that have geoconservation site networks have different means of selecting sites. Some have used literature reviews to identify sites or panels of experts to reach consensus judgements about which sites should be included and which not. However, this may result in inventories of research undertaken, rather than systematic assessments of sites and features to be protected. Ideally systematic national surveys or inventories of the geodiversity resource should be established as the basis for selecting sites, and an example of this approach is in Spain where the Spanish Geological Survey has led the work (Carcavilla et al., 2009).
- *Site selection criteria.* A large number of criteria have been used to select sites for designation. Erikstad (1994) and Gordon (1994) have both compiled lists of criteria used in early surveys, including research value, rarity, vulnerability and representativeness. By 'representativeness' authors must mean that sites should represent the country's geodiversity. There is, however, also the issue of the level of detail. It may not be sufficient to conserve a representative granite, drumlin or vertisol, since these themselves display a considerable geodiversity. There is considerable overlap between many of these criteria and furthermore, site standing may be boosted where more than one criterion is present.

  Some authors try to select sites on the basis of significance and attempt to make the process more objective by defining sets of criteria and quantifying these in order to make judgements about which sites should be selected. Examples of this type of work include Pereira, Pereira and Caetano-Alves, (2007) in Portugal, Solarska and Jary (2010) in SW Poland, Bruschi, Cendrero and Albertos (2011) in northern Spain, Coratza *et al.* (2011) in Malta, and Fassoulas *et al.* (2012) in Crete. For example, Solarska and Jary (2010) used accessibility (scored 1–5), state of preservation (1–5), scientific worth (2–10) and education significance (2–10) to score sites and then summed the scores allowing 17 of the 38 original potential sites to be selected for geoconservation. However, very often the criteria used to select sites are not adequately explained (de Lima *et al.*, 2010) and the methods are at least partly subjective. A detailed quantitative assessment scheme has been developed by Auckland City Council in New Zealand involving 13 criteria (Table 9.3). A score of at least 40 points is needed before a site is considered for scheduling (Hayward, 2009).
- *Duplication or replication?* In selecting representative samples some countries, including the UK, try to avoid duplicating features by selecting only one example of different types of feature. While this is understandable in terms of trying to limit the escalation of sites, it does leave the type of feature vulnerable if the site selected becomes degraded. Replication provides some security against the complete loss of a particular aspect of geodiversity. Sharples (2002a) argued that the fewer examples of a type of feature exist, the more important it is to protect a higher proportion of the examples of the

type in order 'to guard against the possibility of all examples being degraded or destroyed for unforeseen reasons'. He also believes that the case for replication is stronger in the case of features that are highly sensitive, support biodiversity or are poorly understood. Replication also allows nations to have representatives of particular types of feature in different parts of the country, thus supporting local (or regional) geodiversity.

- *Values.* It is also clear that sites can have value in more than one way. As outlined in Chapter 4, sites may be of intrinsic, cultural, aesthetic, economic, functional, research/educational value etc. It is important to identify which of these values is being associated with sites. As noted above, in Britain statutory conservation and designations such as SSSIs are heavily science based whereas the Local Site network may have wider aesthetic, educational or other values. There is an issue here related to the fact that the public finds it difficult to understand and therefore fully support a very scientific and academically led approach. For this reason in some countries, such as Norway, educational and public orientated criteria are integrated with scientific criteria (Daly, Erikstad and Stevens, 1994), and in others, such as the United States, there are integrated systems of natural and cultural protected areas (National Parks System). This is also true of the World Heritage List. A related point is made by Boulton (2001) with reference to the selection of sites on scientific grounds. He points out (p. 51) that 'The definition of a Site of Special Scientific Interest, for example, is not a scientific judgement, but a societal judgement about the value of science set against the value of the site for other purposes'.
- *Scope.* Many systems, including the GCR programme in Britain and geological heritage programme in Australia, are biased towards static sites, particularly geologically-based ones, rather than dynamic processes. This has meant that geosites are often regarded a 'something of an oddity' (Pemberton, 2001a) divorced from mainstream nature conservation rather than being integrated into it. The New Zealand Geopreservation Inventory is one of the few to include soil sites.
- *Significance.* Sites may be graded for significance according to whether they are of international, national, regional/state or local importance.
- *Boundaries.* Drawing appropriate boundaries of protected areas is very important. Several examples from US National Parks/Monuments are discussed in Box 9.1 and in Box 9.11 we saw that the boundary of a Site of Special Scientific Interest had not taken into account the dynamics of coastal erosion so that the present portion of the cliff sections lay outside the boundary and had to be redesignated to include 225 m inland.
- *Access.* Dinis *et al.* (2010) have drawn attention to the fact that in Portugal access to private land is allowed in relation to concession contracts to explore for mineral resources, but not for scientific research. Access has been refused to a crucial area for research on the Lower Cretaceous stratigraphy of central Portugal including well-preserved early angiosperm assemblages. They argue that legal measures should be put in place to balance private rights and the public interest, particularly ensuring access to private property by official researchers. In the United Kingdom, the law allows government conservation

agencies access to designated SSSIs on private land in order to monitor site condition.

- *Site management and enforcement.* The management regimes for protected areas vary greatly. In some cases reliance is placed simply on the designation of the protected area and related legislation, which may include fines for violations. But laws are only as good as their enforcement and this is variable. Academic geoscientists are often very comfortable with site selection but less good at supporting practical site management or trying to influence decision-makers or a public with other agendas and priorities. There is also the issue of low penalties for law infringement. Pemberton (2001a) quotes a case in Tasmania where a person was found guilty of removing quartz crystals of Devonian age from a cave, but received only a $50 fine. 'That equates to a fine of about a cent for every million years. Interesting also to note that this material was stolen from the Tasmanian Wilderness World Heritage Area, an area with the highest possible level of conservation status' (Pemberton, 2001a). Likewise, Horner and Dobb (1997, p. 242) referring to vertebrate fossils in the United States believe that 'Compared to what a commercial collector stands to earn from the sale of such fossils, the fines are negligible, encouraging collectors to treat them as one of the costs of doing business'. The management regime employed should reflect many of the above decisions, for example the value and significance of sites, but many protected areas have no or little management and often suffer damage as a result. At Lake St Clair in Tasmania, Kiernan (1996) argued that the establishment of a national park had not served to protect the glacial geoheritage. Committing the energy and resources to enforcing legislation is usually as challenging as getting the legislation enacted.
- *Fortuitous geoconservation.* Much geodiversity conservation is being achieved in areas protected for other purposes, e.g. biodiversity, scenic value, etc.
- *International agreements.* There have been several calls for greater international consistency in geoconservation practice (e.g. Creaser, 1994b). Stürm (1994) argued that an International Convention would be useful in promoting and supporting the implementation of the geotope concept within national planning systems. Possibilities might include an international Convention on Geodiversity (ICG) to match its influential biodiversity equivalent or a European Directive on Geoconservation to match the Natura 2000 Birds and Habitat Directives.

  Norman (1994) pointed out that fossils are a national and finite resource, but questioned their status, ownership and rights of sale on the open market. Commercial sale can remove them from the public domain into private collections or abroad. He believed that regulations should be in place to prevent contentious issues, (e.g. relating to who owns fossils, whether there is a right to sell, export and import controls) and that there should be consistency within and between nations. Specifically he recommended that there should be an international forum for considering transnational issues of ownership and trade of fossils. Norman believed that this approach is essential to avoid individual countries developing policies *ad hoc* in response to individual issues or local political initiatives.

Part IV

Geoconservation: the 'Wider Landscape' Approach

10

Geoconservation in the 'Wider Landscape'

We need to expand our geoconservation thinking from a site-based to a more integrated landscape-scale approach.

Colin Prosser *et al.* (2011)

The protected area and legislative approaches outlined in Chapter 6, Chapter 7, Chapter 8 and Chapter 9 play an absolutely crucial role in global efforts at conserving geodiversity and providing legal support. But there has been an increasing recognition that while they are unquestionably necessary, they are not, by themselves, sufficient to enable the full and sustainable management of the world's geodiversity.

This chapter and the next two attempt to explain a very wide range of approaches and initiatives that have been developed in the past 20 years or so and are still being developed. They can be considered as examples of good practice for others to borrow or follow. Of course, many traditional approaches to land-use management have attempted to apply sustainable land management practices, for example promoting soil conservation, preventing sand dune blow-outs and protecting karst systems. But some of the new approaches and techniques described in these three chapters can lead to a more holistic approach to geoconservation, thus ensuring a sound future for the physical or abiotic resources of the planet. Together these approaches represent a new, sustainable approach to the management of the physical environment. The threats are many and will require constant vigilance if we are to leave future generations with a geoheritage that is not significantly depleted and degraded.

In this chapter, it is first argued that we must go beyond the designation and management of protected areas to assess, value and protect geodiversity in the wider landscape and indeed in cities. Secondly, we must learn to work with nature rather than seeking to dominate or subjugate it. This will involve recognising

Geodiversity: Valuing and Conserving Abiotic Nature, Second Edition. Murray Gray.

local diversity and distinctiveness, the principles of geomorphological process dynamics and the issues of landscape restoration and landform design.

10.1 The need for a 'wider landscape' approach

What has become known as the 'wider landscape' or 'protecting beyond the protected' approach began in relation to protecting fauna, which is dynamic and cannot identify when it is leaving the protection of a designated area. In addition, in many countries natural areas had become fragmented with the result that animal and plant populations had become isolated from each other, islands in a sea of deteriorating natural environments. Thus the idea of 'wildlife corridors' or 'greenways' linking protected areas emerged which could allow wildlife to move along these corridors from one protected area to the next. In some countries, whole ecological networks of protected areas linked by these corridors were created, at least on paper (Jongman and Pungetti, 2004). One country that has developed this idea is the Netherlands, which has established a National Ecological Network (Hootsmans and Kampf, 2004). Subsequently, the concept of biodiversity has extended nature conservation philosophy to the whole landscape, including urban areas, identifying the need to protect habitats and species wherever they occur.

It is clear that the same thinking can be applied to abiotic nature, since rocks, landforms, processes, soils, etc., occur everywhere and are vulnerable to many threats. In the same way that fauna moves away from protected areas, so geomorphological processes are dynamic and difficult to conserve by the protected area approach. For example, protection of an underground cave system or a lake is problematic if the rivers flowing into them are polluted. The whole river catchment area needs to be managed sustainably in order to protect the cave system or lake in the long term. Similarly, natural coastal processes often operate on a large scale, and interference with one part of a coastal cell may produce undesirable consequences for other parts of the coastline. In other cases the protected area approach was difficult to apply to important landscapes covering large areas.

The issue can be widened still further as neatly described in the following quote from Mather and Chapman (1995). Although they are mainly directed at biological conservation, they are equally applicable to geoconservation:

> Another problem is that a park is a discrete area, with a boundary, and therefore there is a danger that the impression is given that effective conservation can be achieved solely in such areas, and that it can be ignored elsewhere. It is now increasingly being recognised that conservation cannot successfully focus solely on protected areas (or on protected species), and that it needs to be applied to all aspects of environmental resources. The corollary is that the scientific management of protected areas and species cannot, in itself, be successful if it is divorced from the wider use of environmental resources. (Mather and Chapman, 1995, p. 130)

This in turn reflects McNeely's (1988, 1989) view that conservation has to be integrated into the wider framework of the management of environmental resources, and not viewed simply as a separate sector. He argued that conservation is too important to leave to scientists and far more of a social challenge than a biological one. Adams (1996, p. 116) makes similar points:

> A conservation strategy based on protected areas can make a holistic approach to conservation more difficult, encouraging the view that conservation is a sector or a land use. ... Protected areas tend to be treated as separate entities from surrounding land, soaking up available conservation resources. Protected area boundaries are often arbitrary lines on maps ... irrelevant to natural processes and environmental problems'.

In addition there is the danger that protected areas will lead government and public to 'begin to think that they have a free license to do whatever they want outside of its boundaries' (Brussard, Murphy and Noss, 1992, p. 158). Adams' (1996, p. 118) conclusion is that 'The traditional concern for individual protected areas must be transformed to a concern for whole landscapes'. Myers (2002, p. 54) believes that the need for a wider perspective is proved by the increasing impact of humans on protected areas: 'setting aside a park in the overcrowded world of the early twenty-first century is like building a sandcastle on the seashore at a time when the tide is coming in deeper, stronger and faster than ever'.

Sharples (2002a) recognizes that landforms often need to be seen as part of broader assemblages. 'Thus, whilst a glacial cirque produced by past glacial processes may be of some interest in isolation, its full significance only becomes apparent when it is considered in the context of an assemblage of related features'. Without the evidence provided by the assemblage 'the information which any one isolated feature can give us about past processes is limited'. In other words, one of the things we need to do in geoconservation is to consider more than individual sections and landforms and instead think about how to protect three-dimensional geological structures, landform assemblages and system functions.

Many aspects of management of the geoheritage will be discussed in this chapter, but the aim is first to explore the issue of how geodiversity in the wider landscape can be assessed. Of course, in some fields of earth science the recording of the information has a long history, but this is not the purpose of this chapter. Rather, the focus here is on the less well- known assessment systems of the wider georesource intended for conservation management purposes. It is also important to note that much has been written on other aspects of geoconservation, including the curation of specimens (rock, mineral, fossil) (e.g. Collins, 1995), conservation of paper records and photographs (e.g. Ellis, 1993), and conservation of building stones (e.g. Ashurst and Dimes, 1990; Siegesmund, Vollbrecht and Weiss, 2001), but these are beyond the scope of this book.

10.2 The physical landscape layer

Before moving on to give examples of the application of a 'wider landscape' approach within geoconservation and land management, it is worth outlining the contribution that the abiotic layer makes to landscape (Gray, 2009). It is useful to think of landscape as comprising three primary layers (Figure 10.1). At the base is a physical layer, comprising the rocks, sediments and soils, the landscape topography and the physical processes operating on the landscape. Above this is the biological layer comprising flora and fauna, wildlife habitats and ecosystems. Finally there is a cultural layer involving land-use, buildings and infrastructure, all of which have evolved through time and therefore include historical land-uses. This layer also includes the human experience of landscape through the senses of sight, sound, etc., and through memories and associations of landscape.

Any one of these three primary layers may dominate a landscape. The cultural layer dominates most obviously in towns and cities or in rural areas where agricultural land-uses dominate. The biological layer dominates in many non-urban and non-agricultural areas, particularly where there is no obvious topography, for example. the tropical rain forests of the Amazon or Congo basins. The physical layer can dominate in natural areas where vegetation is scant or absent, e.g. glacial, mountain, desert, cliffs, rocky coasts, braided river channels and volcanic areas.

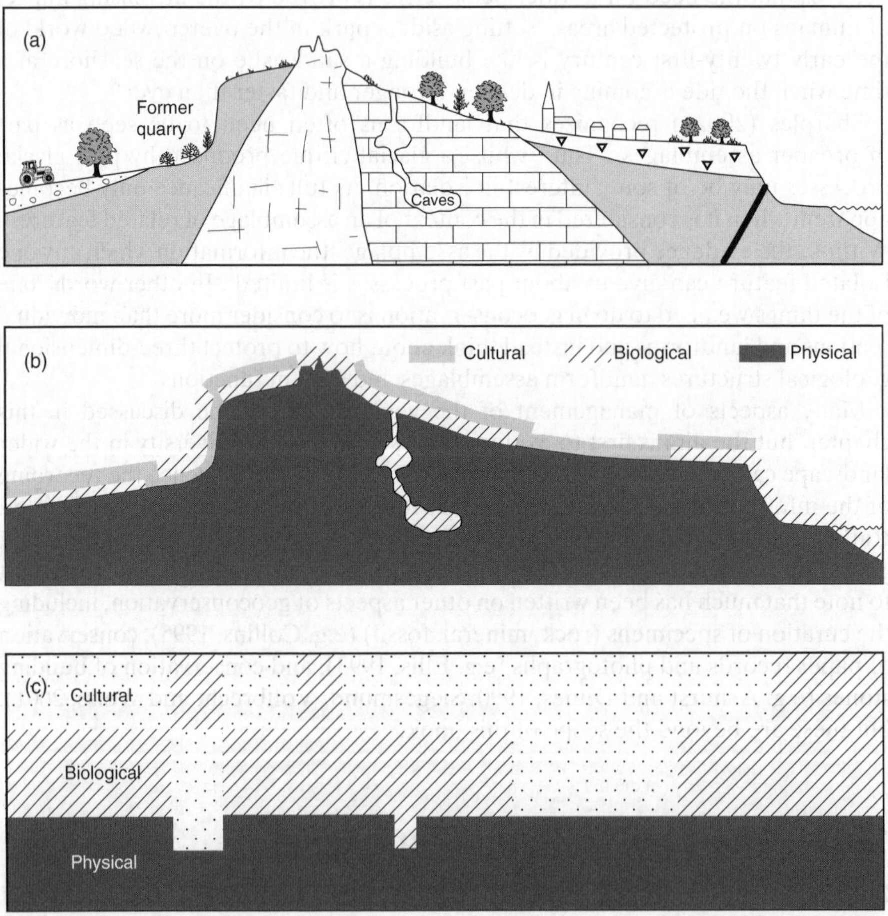

Figure 10.1 The three layers of landscape: (a) as a real landscape, (b) as an initial representation of the layer types, and (c) as a further abstraction into layers (after Gray, 2009).

The layers are clearly interrelated and may intermingle. For example, soil is a complex mixture of weathered regolith, organic matter and a living soil ecology. Wildlife may recolonise anthropogenic landscapes. Archaeological artefacts may be found within river gravels or interbedded with volcanic ash, and cities are predominantly constructed from physical materials. But despite this intermingling, the three layers are still clearly distinguishable and they combine to form modern landscapes.

Rocks and sediments are ubiquitous elements of the landscape but are not always ubiquitously visible due to being obscured by soil, vegetation, buildings, etc. As outlined earlier, they are therefore most obvious in areas where these obscuring materials are sparse or absent. In these areas, the geodiversity of rock type, colour and structure becomes apparent. For example, contrast the white Chalk cliffs of Dover, England, with the vertically bedded sandstones of Uluru, Australia, with the horizontally bedded sedimentary rocks of the Grand Canyon, United States, with the grey granite landscapes of the Sierra Nevada, United States. As well as the obvious differences between these, the mesomorphology of rock outcrops provides character to landscapes (Figure 10.2). But even where rocks are buried, they often have a major influence on topography through structural controls or variable rock resistance to erosion.

Like rocks and sediments, land form or topography is ubiquitous in the landscape, and can even be detected, albeit in modified form, within towns and cities. In many places specific landform types can be identified and named but the term 'land form' rather than 'landforms' is preferred here since most landscapes cannot be described in terms of standard landform categories, yet they have a physical form that remains essentially natural over large parts of the world.

Figure 10.2 Rock outcrops and mesomorphology often contribute to landscape character. An example from the granite landscapes of Peneda-Geres National Park, Portugal showing the physical, biological and cultural layers of landscape. See plate section for colour version.

Land form helps to give the landscape its local distinctiveness and provides landmarks that become familiar to, and are valued by, local people.

Other physical components of the landscape layer include physical processes (including volcanic and tectonic, glacial and periglacial, fluvial and slope, coastal and aeolian processes and agents) and soils (whose major contribution to landscape is through their influence on vegetation). There are also temporary physical manifestations in the landscape, including, snow, sea-ice, sunshine, rain, cloud cover and other weather effects (Gray, 2009).

10.3 Geoconservation initiatives in 'the wider landscape'

10.3.1 European Landscape Convention

Reflecting the desire for a wider landscape approach to conservation, the European Landscape Convention, promoted by the Council of Europe, was adopted on 19 July 2000 and signed by 18 European countries on 20 October 2000. It is the first international instrument devoted exclusively to the protection, management and planning of the landscape in its entirety and the aim is to organise European co-operation on landscape issues. It requires public authorities to adopt policies and measures at local, regional, national and international level for protecting, managing and restoration of landscapes throughout Europe. It covers all landscapes, 'both outstanding and ordinary', that determine the quality of people's living environment. It includes rural areas, villages and towns, urban and peri-urban landscapes, coasts and inland areas. According to the Convention, 'landscape means an area, as perceived by people, whose character is the result of the action and interaction of natural and/or human factors'.

Article 5 of the Convention requires signatories to:

- recognise landscapes in law as an essential part of people's surroundings, an expression of diversity of their shared cultural and natural heritage, and a foundation of their identity;
- establish and implement landscape policies aimed at landscape protection, management and planning;
- establish procedures for the participation of the general public, local and regional authorities, and other parties with an interest in landscape policies;
- integrate landscape into its regional and town planning policies and other policies relevant to landscape.

Article 6 requires signatory nations to increase awareness of landscape through education and training, identification and analysis of the landscapes within their own territories, and recognition of the forces and pressures transforming them.

As of November 2012, the Convention has 40 signatories but there are some notable absentee countries including Germany, Iceland and Russia. Natural England has produced an ELC Action Plan (Natural England, 2009) giving examples of projects successful and underway. The English National Park Authorities

Association has also published a leaflet giving examples of landscape protection, planning and management (Charlton, undated).

10.3.2 National Character Areas (NCAs) in the United Kingdom

In the United Kingdom, starting in the 1990s, the sufficiency of the site-based approach to nature conservation was reassessed. First, it was argued that although designated sites represent the best of the country's biodiversity and geodiversity, they cannot by themselves preserve all that is valuable in the natural environment. Preserving only the best sites, or a representative series of sites with no duplication, was seen as a rather elitist approach. There are many sites and areas outside the boundaries of protected areas that will be almost as valuable and many ordinary landscapes that are valued by local communities as part of their familiar and cultural environment. It was argued that there are strong arguments for greater recognition and respect for the distinctiveness of regional and local geodiversity. As Jarman (1994, p. 41) remarked, 'Ordinary landscapes can be rewarding too'. Secondly, a site-based approach had led to a highly fragmented system of conservation and did not promote the conservation interests of the areas between sites or the wider integrity of sites in their geological/geomorophological setting.

Following the devolution of the nature conservation function in the United Kingdom in 1991 (see Section 9.4), English Nature undertook a 'fundamental review of its approach to wildlife and geological conservation, to analyse the strengths and weaknesses of previous activities, and determine how best to take conservation forward in the 1990s' (Duff, 1994, p. 122). This review established the need not only to improve the management of designated sites, but also identified the importance of seeing nature conservation in a wider context, both geographically and in terms of countryside management. A consultation paper, issued in 1993 (English Nature, 1993), argued that English Nature must 'take an integrated approach to the whole environment and not simply view sites and habitats in isolation'.

The solution proposed was to base future conservation strategy around a system of 'Natural Areas' covering the whole country. English Nature therefore divided England into 120 of these areas (97 terrestrial and 23 maritime) on the basis of similarities in landscape, landform and other natural characteristics. 'Natural Areas are tracts of land unified by their underlying geology, landforms and soils, displaying characteristic natural vegetation types and wildlife species, and supporting broadly similar land uses and settlement patterns' (Duff, 1994, p. 121). The advantages of using these Natural Areas as the basis for developing national and local conservation strategies were seen as:

- offering 'a more effective framework for the planning and achievement of nature conservation objectives than do administrative boundaries' (English Nature, 1998);
- providing a framework for appraisal and evaluation to describe what is important and why, and for setting objectives to protect our characteristic biodiversity and geological heritage;

- enabling English Nature 'to look at the resource in an integrated way, and not just to focus on the special sites' (Duff, 1994). In turn this will achieve a less fragmented approach to nature conservation;
- increasing the integration of geological and biological conservation;
- acting as a framework for involving local communities in valuing the distinctiveness of their local areas and developing their sense of place in relation to nature conservation objectives.

The Natural Areas approach was anticipated to become the fundamental basis on which future nature conservation strategies in England would be founded.

The Natural Areas approach was subsequently modified into Joint Character Areas with the inclusion of cultural landscapes and these were subsequently relaunched as National Character Areas (NCAs, see Figure 10.3). They are now intended to be the way in which Natural England describes the whole natural environment/landscape—geology, geomorphology, soil, biodiversity, historic environment, cultural landscape, and so on. NCAs will be used to set the context for local objective setting and decision-making and are aimed at being fundamental to Natural England's future integrated philosophy and on-the-ground implementation. A complete set of NCA profiles, including some information on geology, geomorphology and soils, is due to be published by the end of 2013.

Similar work in Scotland divided the country into 21 Natural Heritage areas in very similar ways to Natural Areas in England (Mitchell, 2001; Scottish Natural Heritage, 2002). For each of the 21 areas, Scottish Natural Heritage prepared a 'Local Prospectus' which described the natural heritage for the area and the processes that have produced it, discussed the changes taking place, including human impacts, and presented a series of key goals and specific actions which aimed to close the gap between the vision and current trends over a 25-year period (Scottish Natural Heritage, 2002). The accompanying CD-ROM had detailed descriptions of the geology, palaeontology, geomorphology, soils and landscapes of both Scotland as a whole and each of the 21 areas. There are also discussions on existing trends and pressures and the state of the georesource. Altogether this adds up to a very significant effort to describe the geoheritage of Scotland and is a milestone that other countries could well follow.

The Countryside Council for Wales' LANDMAP Project is more closely related to a landscape character assessment approach and this is described here.

10.3.3 Landscape Character Assessment (LCA) in the United Kingdom

Landscape Character Assessment (LCA) came to prominence in the United Kingdom in the 1990s (Countryside Commission, 1993a). It is a technique that is closely related to the National Character Area approach initiative described earlier, but often at a finer scale. Landscape character is defined as 'a distinct and recognisable pattern of elements that occur consistently in a particular type of landscape. Particular combinations of geology, landform, soils, vegetation, land use, field patterns and human settlement create character' (Swanwick and Land

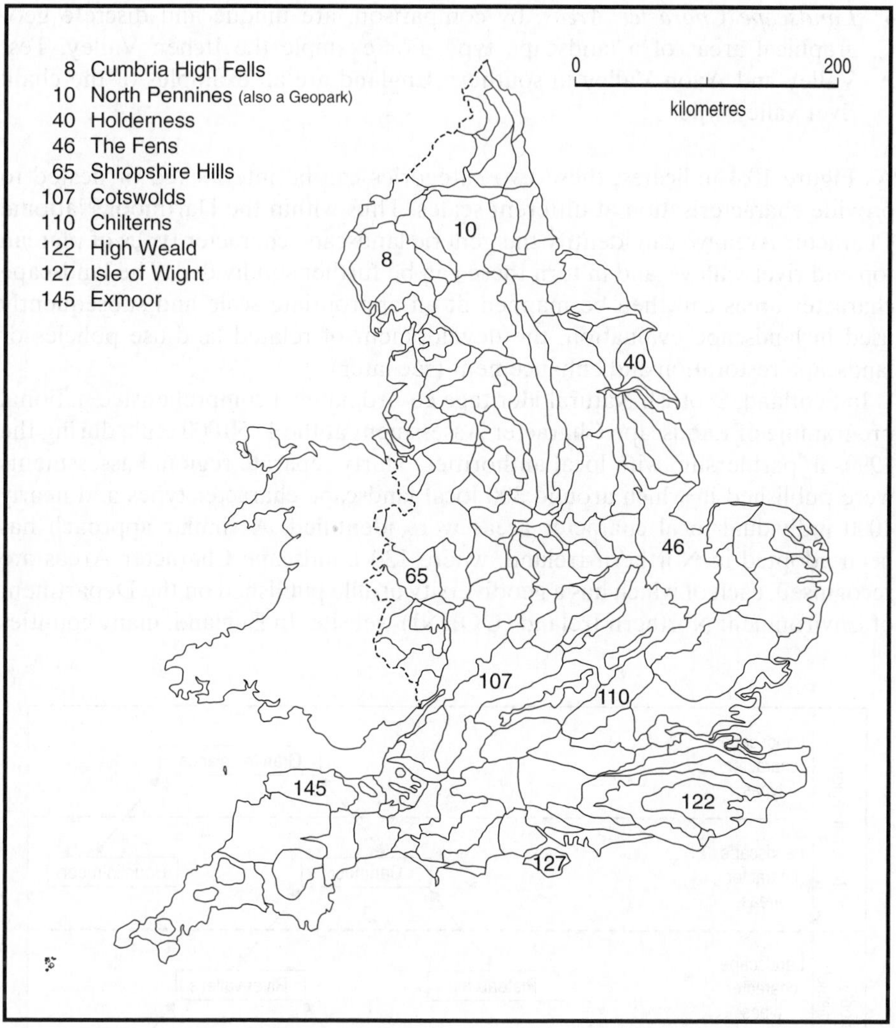

Figure 10.3 National Character Areas in England with some examples numbered and named. (© Crown copyright and database right 2012. All rights reserved. Natural England Licence No. 100022021.)

Use Consultants, 2002, p. 9). In this discussion, the emphasis will be placed on the physical elements of the landscape (geology, geomorphology, soils) within the context of the wider landscape character.

Characterisation is the name given to the practical steps involved in identifying two categories of landscape character:

- *Landscape Character Types*, which are generic in nature and which share many common combinations of geology, landform, drainage, vegetation and human influences. For example, chalk river valleys or rocky moorlands are landscape character types.

- *Landscape Character Areas*, by comparison, are unique and discrete geographical areas of a landscape type. For example the Itchen Valley, Test Valley and Avon Valley in southern England are all examples of the chalk river valley type.

As Figure 10.4 indicates, these two categories can be interlocked or nested to provide characterisation at different scales. Thus within the Dartmoor National Character Area we can identify the generic landscape character types of plateau top and river valleys, and in turn these can be further subdivided. The landscape character areas can then be mapped at an appropriate scale and subsequently used in landscape evaluation, the development of related land-use policies or landscape restoration and enhancement (see later).

In Scotland, Scottish Natural Heritage co-ordinated a comprehensive national programme of Landscape Character Assessment at the 1 : 50 000 scale during the 1990s in partnership with local authorities. Thirty separate regional assessments were published in which around 300 local landscape character types and nearly 4000 individual local character areas were identified. A similar approach has been adopted in Northern Ireland, where 130 Landscape Character Areas are recognised, each of which has a geodiversity profile published on the Department of Environment Northern Ireland (DOENI) website. In England, many counties

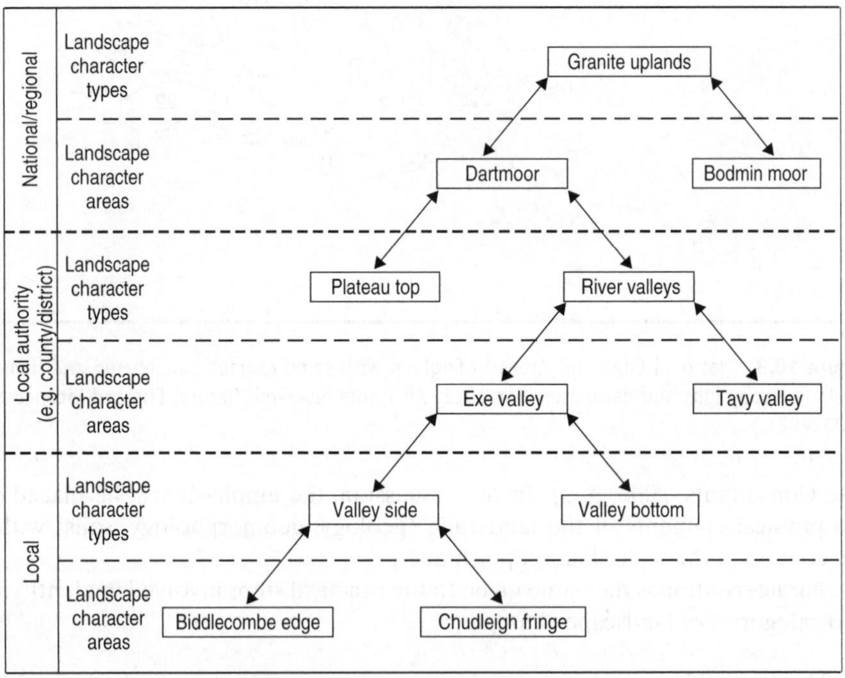

Figure 10.4 The relationship between Landscape Character Types and Landscape Character Areas (after Swanwick, C. and Land Use Consultants (2002) Landscape Charcter Assessment Guidance for England and Scotland).

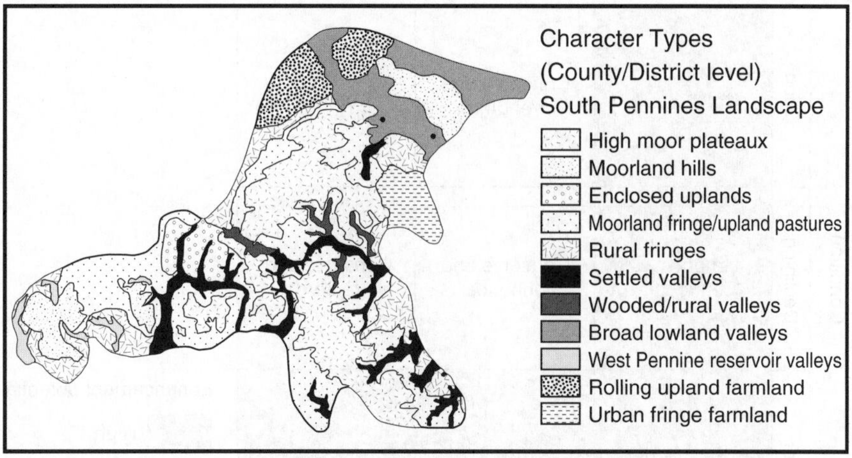

Figure 10.5 A landscape character map for part of the South Pennines, England (after Swanwick, C. and Land Use Consultants (2002) *Landscape Character Assessment Guidance for England and Scotland*. © The Countryside Agency & Scottish Natural Heritage).

and districts have also undertaken landscape character assessments for their areas (e.g. Figure 10.5) but there is some inconsistency in approach due to the fact that different consultants have worked to different briefs, and the coverage is somewhat patchy (Swanwick and Land Use Consultants, 2002).

The landscape characterisation stage may be an end in itself, in which case the landscape character map and descriptions of landscape types are value-free statements of the current appearance of the landscape. In turn, they can be used to raise awareness of the distinctiveness and diversity of the landscape and to encourage appreciation of the differences between places. However the landscape character process will not normally be an end in itself. Rather, the process will go beyond neutral statements of landscape character, to provide an assessment of the character that can inform particular decisions. 'The focus should be on ensuring that land use change or development proposals are planned and designed to achieve an appropriate relationship (and most often a 'fit') with their surroundings, and wherever possible contribute to enhancement of the landscape ...' (Swanwick and Land Use Consultants, 2002, p. 52).

Several approaches have been taken to making judgements about landscape character, though they are often used in combination (Swanwick and Land Use Consultants, 1999). The approaches include:

- *key landscape characteristics and guidelines*: these are the factors that are particularly important in creating landscape character. It follows that if the distinctive character of the landscape is to be maintained and enhanced, then the key characteristics must be protected from adverse change.
- *landscape quality and strategies*: quality here refers to the physical state of repair of the landscape or parts of it, and suggests future landscape strategies. For example, if the typical character of the landscape is very apparent and the

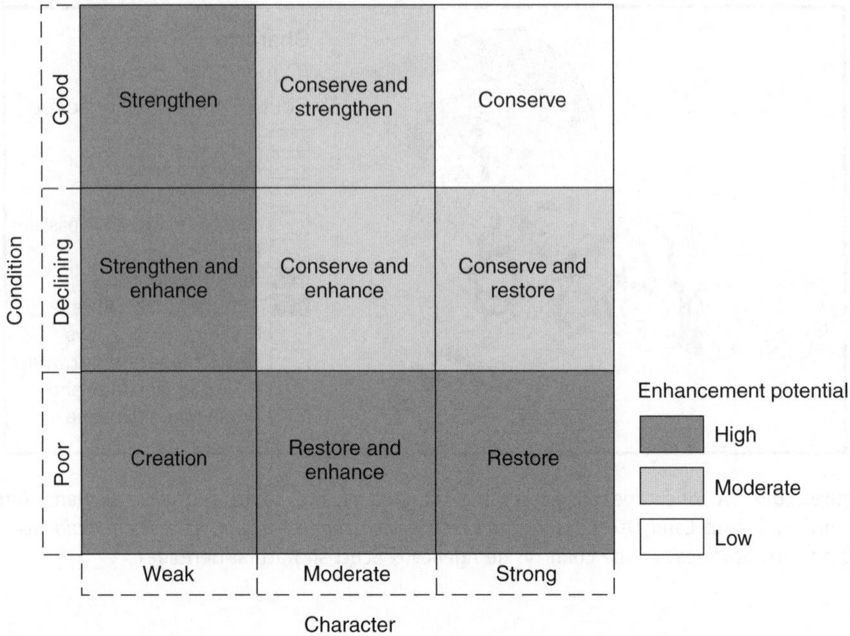

Figure 10.6 Strategies for managing landscape character (after Warnock, S. and Brown, N. (1998) A vision for the countryside. *Landscape Design*, 269, 22–26. Reproduced with permission of Landscape Design Trust).

features within it are in good repair, then the strategy might reasonably be to conserve its character. Conversely, if the typical characteristics are weakly defined or in poor condition, then the strategy may be to strengthen or restore its character. This is illustrated in Figure 10.6 (Warnock and Brown, 1998).

- *landscape sensitivity and capacity*: sensitivity concerns the degree to which a particular landscape character type or area can accommodate change without unacceptable adverse consequences for its character . Capacity is similar and deals with the amount of change of a particular type that a landscape can accept without detrimental effects on landscape character. 'Robustness' is a term used to describe both concepts. A robust landscape will have low sensitivity and a high capacity to accept change. For example, undulating landscapes may be of lower sensitivity, higher capacity and generally more robust in accepting landscape change, than flat and featureless landscapes. These are familiar concepts for geomorphologists but perhaps not for most landscape assessors.

10.3.4 Geological landscapes and geodiversity characterisation

In Wales, a national programme called LANDMAP has been developed in a similar way to landscape character assessment but based on a GIS system and with a series of five overlain layers mapping different landscape variables including a 'Geological Landscape' layer (Page, 2008). These geological landscapes are examined at several scales or levels:

Level 1: General landscape character
Level 2: Large-scale terrain or topography
Level 3: Medium-scale typifying terrain or topography
Level 4: Small-scale landform.

Figure 10.7 shows Radnorshire mapped at Level 3 (Page, 2005). The system allows full descriptions of the geological and topographical character of each area to be included in the database along with other geological information including information on quarries and mines, SSSIs and a bibliography of key sources. The scheme is an excellent example of how physical landscapes can be mapped, described and applied to integrated and sustainable landscape management.

Geodiversity characterisation is similar to the Welsh LANDMAP approach in that it uses a GIS system and tries to establish the contribution that geodiversity makes to landscape. The GIS-based mapping of geology, topography, soils,

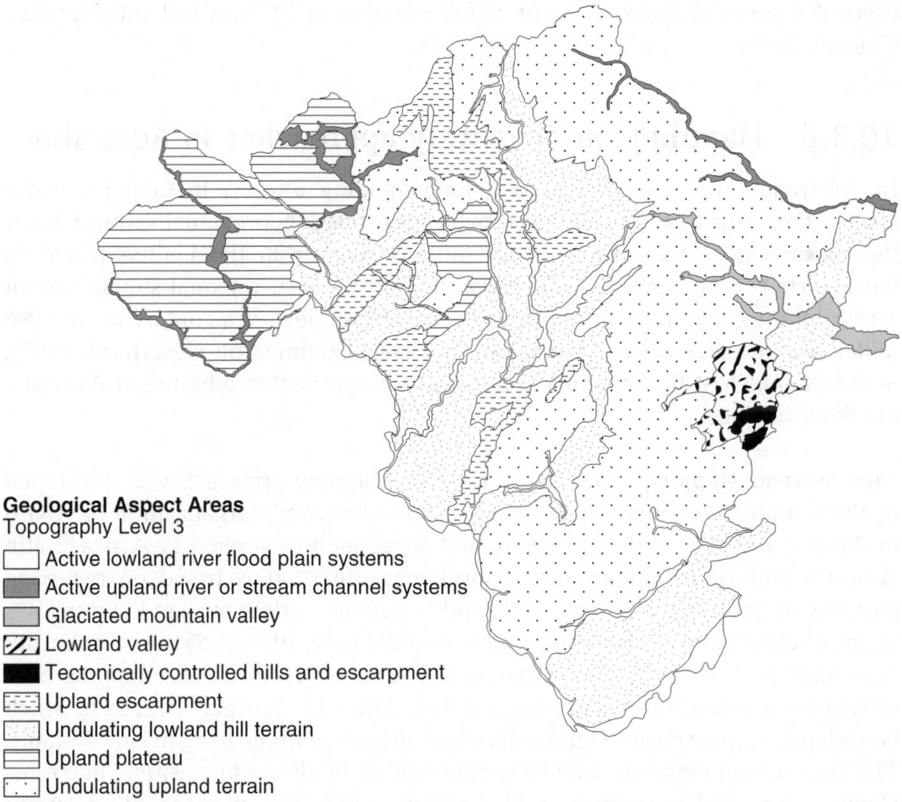

Figure 10.7 The geological landscape of Radnorshire mapped at Level 3 (after Page (2005) Radnorshire LANDMAP—Geological Landscapes Aspect Layer: Technical Report. Reproduced with permission of Natural Resource Wales, © Crown Copyright and database right 2013. Ordnance Survey 100019741).

hydrology, mineral working and the historic environment is used to define Geodiversity Character Areas and their component Geodiversity Character Zones. An example of this work has been undertaken on the Tendring peninsula of Essex, England where Essex County Council (2009) has published a map of 18 Geodiversity Character Areas and a related report.

10.3.5 Landscape Character Assessment elsewhere in Europe

According to the Irish Heritage Council (2006), nearly all Western European countries have or are implementing a landscape character assessment system. In the Irish Republic, trial projects were undertaken in Counties Clare and Leitrim (Heritage Council, 2002) and by 2006, 19 of the 29 counties had completed LCAs of some sort. However, because of inconsistencies and deficiencies, the Heritage Council recommended that a national programme of assessment should be initiated in collaboration with the Department of Environment, Heritage and Local Government. This should be in accordance with the recommendations of a national *Baseline Audit and Evaluation* of LCA in Ireland (Heritage Council, 2006).

10.3.6 Mapping and inventory approaches in Australia

In Australia there has also been a long-standing interest in taking a wider view of the natural land resource (sometimes termed 'georegionalisation'). Over the years, some of the same issues outlined earlier in the United Kingdom have started to be discussed. For example, the topic of regional subdivision of Australia for geoconservation purposes was the subject of a conference in 1996 under the auspices of the Australian Heritage Commission (Eberhard, 1997). Some examples of the mapping and inventory approaches adopted in Australia are described here.

Land Systems Mapping The Land Systems Mapping approach was developed by the Commonwealth Scientific and Industrial Research Organisation (CSIRO) in Australia in the 1950s and 1960s, and subsequently applied in Africa, Latin America and Asia (Cooke and Doornkamp, 1990). It is based on recurring patterns of geology, landform, soils and vegetation; 'thus the land system is a scientific classification of country' (Stewart and Perry, 1953, p. 55). As Cooke and Doornkamp (1990, p. 21) appreciated, the simplest criterion for mapping land system areas is land form. The system is based on a land form hierarchy in which land elements form parts of land units which in turn combine to form land systems. The land system maps are usually supplemented by three-dimensional diagrams (Figure 10.8) and descriptive tables (Table 10.1). The purpose of the work was to provide a basis for assessing land resources and particularly agricultural potential, rather than for conservation objectives (King, 1987). Nonetheless, it is a conceptual forerunner of the modern georegionalisation approach.

Table 10.1 Description of the Units of the Napperby Land System, Australia (after Perry, 1962).

Napperby Land System (2600 km^2)
Geology: massive granite and gneiss, some schist, Precambrian age
Geomorphology: Granite hills up to 150 m high and plains with branching shallow valleys, less extensive rugged ridges with relied up to 15 m, and a dense rectangular pattern of narrow, steep-sided valleys.

Unit	Area	Landform	Soil	Vegetation
1.	Large	Granite hills: tors and domes up to 150 m high; bare rock summits and rectilinear boulder covered hill slopes, 40–60%, with minor gullies; short colluvial aprons, 5–10%	Outcrop with pockets of shallow, gritty or stony soils	Sparse shrubs and low trees over sparse forbs and grasses, *Triodia spicata*, or *Plectachne pungens* (spinifex)
2.	Medium	Close-set gneiss ridges and quartz reefs: up to 15 m high; short rocky slopes, 10–35%; narrow intervening valleys		
3.	Medium	Interfluves: up to 7 m high and 0.8 km wide; flattish or convex crests, and concave marginal slopes attaining 2%	Mainly red earths, locally red clayey sands and texture-contrast soils, stony soils near hills	Sparse low trees over short grasses and forbs or *Eragrostis eriopoda* (Woollybutt)
4.	Medium	Erosional plains: up to 1.6 km in extent, slopes generally less than 1%		
5.	Small	Drainage floors: 180—365 m wide, longitudinal gradients about 1 in 200	Mainly texture-contrast soils, locally alluvial soils and red earths	*Eremophila* spp.- *Hakea leucoptera* over short grasses and forbs; minor *Kochia aphylla* (cotton-bush)
6.	Small	Alluvial fans: ill-defined distributory drainage; gradients above 1 in 200	Alluvial brown sands and red clayey sands	Sparse low trees over short grasses and forbs or *Aristida browniana* (kerosene grass)
7.	Small	Rounded drainage heads: up to 180 m wide and 1.5 m deep on the flanks of Unit 3	Red earths	Dense *A. aneura* (mulga) over short grasses and forbs
8.	Very small	Channels: up to 45 m wide and 1.5 m deep and braiding locally	Bed-loads, mainly coarse grit	*E. camaldulensis* (red gum), *A. estrophiolata* (ironwood) over *Chloris acicularis* (curly windmill grass)

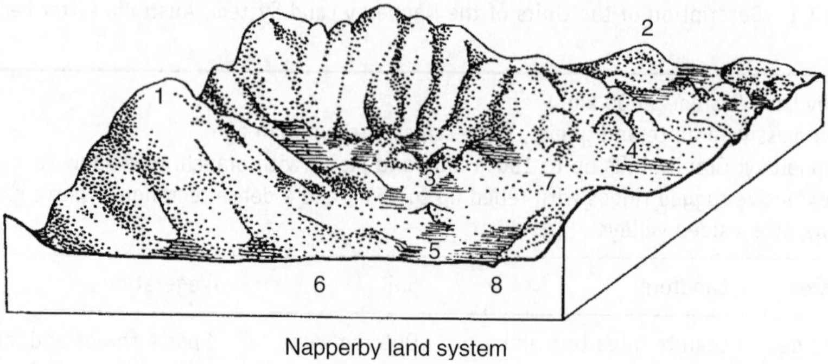

Figure 10.8 An example of a three-dimensional model of a land system in Australia: Napperby land system near Coniston, Northern Territories (copyright © CSIRO. Reproduced from Perry, R.A. (1962) General report on lands of the Alice Springs Area 1956–7. Land Research Series 6, CSIRO, Australia, with the kind permission of CSIRO).

Landform inventories Conservation objectives are very much the theme of the systematic inventory approach more recently developed in Australia, particularly in Tasmania. Sharples (2002a) describes three scales of geodiversity inventory:

- reconnaissance inventories, which are based largely on existing information and which have been prepared for most public land (but little private land) in Tasmania (e.g. Sharples, 1997);
- systematic and thematic inventories, which are comprehensive comparative assessments of all features and systems in a particular region or of a given theme over a larger area (e.g. Rosengren's, 1994, work on the Western Victoria lava province). A related approach is the classification-based approach in which a systematic classification of a particular aspect of geodiversity is established, an inventory of all known occurrences of each class is compiled and finally the best representative examples of each element are identified. This approach has been completed for karst, glacial and coastal features (Kiernan, 1995, 1996, 1997a). Kiernan (1996) developed a glacial 'landform communities' classification system containing 20 recognised types (Table 10.2). Kiernan argues (p. 195) that the adoption of this approach 'would facilitate the advancement of geoconservation from the traditional focus on geological monuments to a proper integration with management to protect environmental diversity'.

At the georegional scale, Houshold *et al.* (1997) and Jerie, Houshold and Peters (2001) describe an approach to producing an inventory of stream geodiversity in Tasmania. This is specifically being done because many stream types are not included in the current reserve system and there is a need for a more strategic approach to stream conservation in Tasmania. Jerie, Houshold and Peters (2001) are developing a GIS based Environmental Domain Analysis (EDA) which will result in a map of Tasmania showing stream regions where similar controls on stream development have acted through time. This

Table 10.2 Glacial landform communities (after Kiernan, 1996).

Glacial erosion communities	
1.	Minimally eroded ice dome communities
2.	Areally scoured ice dome landform communities
3.	Linearly eroded outley glacier landform communities
4.	Alpine valley glacier landform communities
5.	Cirque glacier landform communities
6.	Composite erosional landform communities
Glacial deposition communities	
7.	Fluted drift landform communities
8.	Active ice depositional landform communities
9.	Transition depositional landform communities
10.	Disintegration moraine landform communities
11.	End moraine landform communities
Glaciofluvial communities	
12.	Subglacial erosion landform communities
13.	Subglacial deposition landform communities
14.	Lateral margin communities
15.	Terminal/proglacial communities
Other landform communities	
16.	Glaciolacustrine landform communities
17.	Glacioaeolian landform communities
18.	Fjord glaciomarine landform communities
19.	Open water glaciomarine landform communities
20.	Other composite communities, e.g. glaciokarst

will help to improve stream conservation and management, for example by identifying representative elements of geodiversity in a scientifically rigorous way.

- detailed inventories, which are descriptions of particularly significant or sensitive systems at a level adequate to make specific management prescriptions for those particular systems' (Sharples, 2002a). Examples include the detailed inventories of the Exit Cave, Mole Creek and Junee-Florentine karsts in Tasmania.

Figure 10.9 shows how georegions might be identified by overlaying the significant physical factors. An example of this type of work is Dixon and Duhig's (1996) mapping of Pleistocene glacial georegions in Tasmania.

Sensitivity Zoning and Management This is another approach being developed in Tasmania that raises the possibility of mapping spatial variations in sensitivity, particularly where process–response understanding is well developed. 'The ability

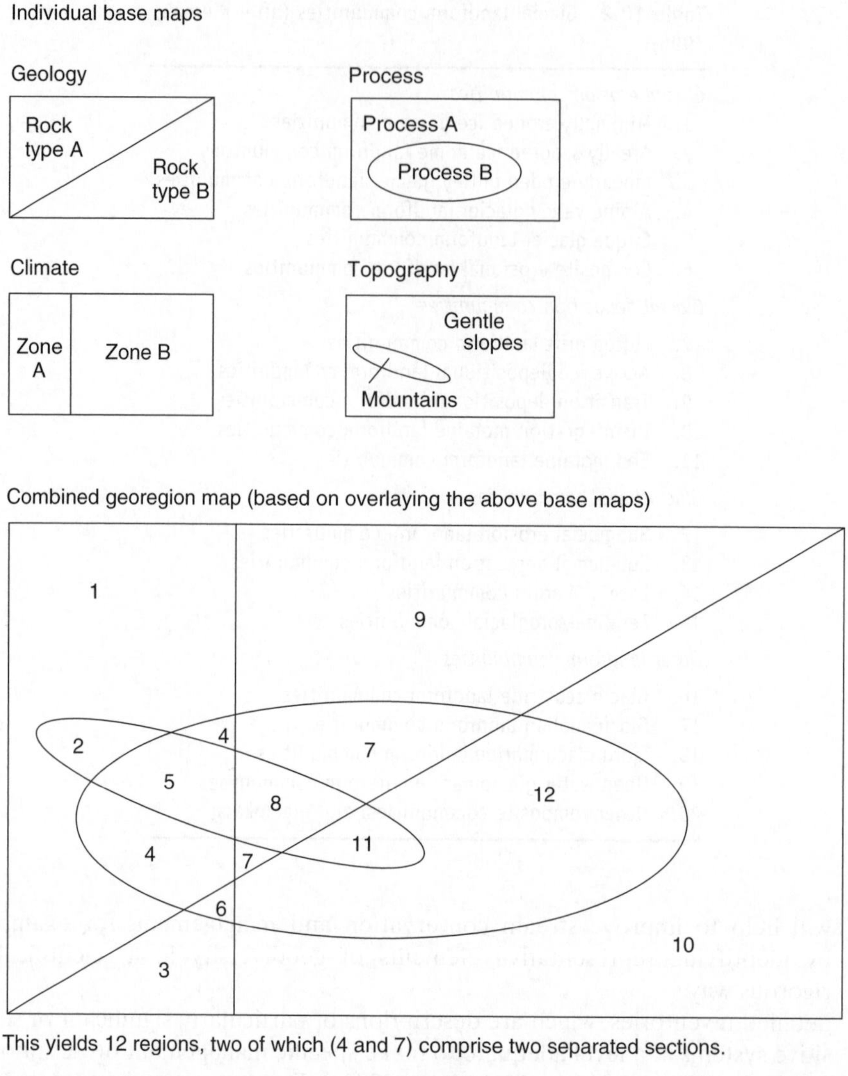

Figure 10.9 An illustration of how a georegional map can be constructed from individual base maps (after Sharples, C. (2002a) *Concepts and Principles of Geoconservation*. Tasmanian Parks & Wildlife Service at http://www.dpiw.tas.gov.au/inter.nsf/Attachments/SJON-57W3YM/$FILE/geoconservation.pdf. Reproduced with permission of Environment Australia).

to zone regions in this fashion provides an important planning tool which can be used to minimise conflict between conservation and development values at an early stage of planning' (Sharples, 2002a). An example of this approach is Eberhard's (1994, 1996) project to zone the Junee-Florentine Karst System into High, Medium and Low Sensitivity Zones. A large part of this karst system lies within the State forest, and this zoning system has the potential to reduce future conflict in the area by diverting forestry operations to less sensitive locations.

10.3.7 Potential geoconservation mapping in Peru

A related approach to identifying geoconservation needs in the wider landscape is described by Seijmonsbergen *et al.* (2010) in the Las Lagunas area of northern Peru. This is a formerly glaciated area lying between 3750 and 4600m above sea-level. Having produced a geological/geomorphological map of the area, they then created maps for four potential factors relevant to geoconservation potential. These four factors were scientific relevance and frequency of occurrence (primary factors) and disturbance and environmental vulnerability (secondary factors). Weighted scores were then allocated to each of the factors and the four maps were then combined into a final potential geoconservation map of the area with three categories—low, normal and high degree of significance for geoconservation (Figure 10.10). The importance of this study is that it provides a methodology and example of a spatially continuous assessment of geoconservation potential rather than the usual site based approach.

10.3.8 Natural Regions and Subregions in Canada

As outlined in Section 9.3, Parks Canada has divided the country into 39 Natural Regions and is using this scheme to try to ensure that Canada's National Park system represents the diversity of the country's natural landscapes. Some provinces have taken this further and divided the Natural Regions within their provinces into Natural Sub-regions and are using the resulting spatial systems

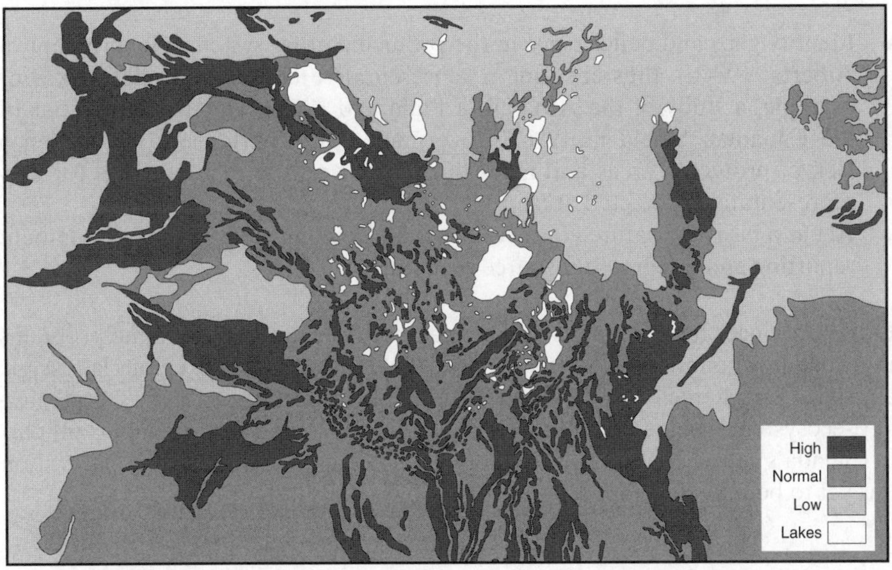

Figure 10.10 Geoconservation mapping of the Las Lagunas area of Peru (after Seijmonsbergen *et al.* (2010) A potential geoconservation map of the Las Lagunas area, Northern Peru, using GIS and remote sensing techniques. *Environmental Conservation*, 37, 107–115. Reproduced with permission of Cambridge University Press).

as a framework for managing the wider georesource. The approach is therefore similar to those being developed in the United Kingdom (see earlier).

One province that has developing this approach is Alberta where the six Natural Regions represented in the province have been further mapped into 20 Subregions (Figure 10.11) based on an amalgamation of two pre-existing schemes (Achuff, 1994). For example, the Rocky Mountain Natural Region in Alberta is divided into Alpine, Subalpine and Montane Subregions while the small section of Canadian Shield in NE Alberta is divided into the Athabasca Plain (underlain by Precambrian Athabasca Sandstone) and Kazan Upland (underlain by Precambrian igneous and metamorphic rocks) Subregions.

A further three levels of subdivision, referred to as Natural History Themes, have then been applied.

- Level I Theme—a broad landscape type within a Subregion. This is regarded as an important level for conservation and 20 Level 1 Themes have been applied across the 20 Subregions, resulting in 167 Level 1 Themes (not all themes are present in each Subregion);
- Level II Theme—a broad habitat/vegetation type within a Level 1 Theme;
- Level III Theme—a specific geological feature, plant community or species within a Level II Theme.

Together with the Regions and Subregions, these three Theme Levels make up a five-level classification system (Achuff, 1994; Government of Alberta, 1994a). Figure 10.12 gives an example. This has been used in Alberta to:

- Identify gaps and deficiencies in the protected areas system (Government of Alberta, 1994b), thus ensuring a representative conservation network. For example, a study of the Athabaska Plain and Kazan Upland Subregions of the Canadian Shield in 1996 (Government of Alberta, 1996) indicated a lack of protected areas and made suggestions on suitable locations for new representative protected areas.
- Act as a basis for nature conservation management, state of the environment reporting and integrated resource planning.

Other provinces have adopted similar schemes though there are some problems in reconciling boundary differences. Nevertheless, the Natural Regions landscape classification scheme is now being used for federal reporting on forestry strategies, agri-ecosystem schemes and biodiversity targets. It is therefore an important part of Canada's nature conservation strategy even if its application to geoconservation has yet to be fully developed.

10.3.9 Landscape physiographic units map of Italy

The Italian landscape physiographic units map is part of an ambitious project (Carta della Natura) introduced by national law (L.394/91) to provide a series of maps depicting the natural environment of the country, identifying the natural

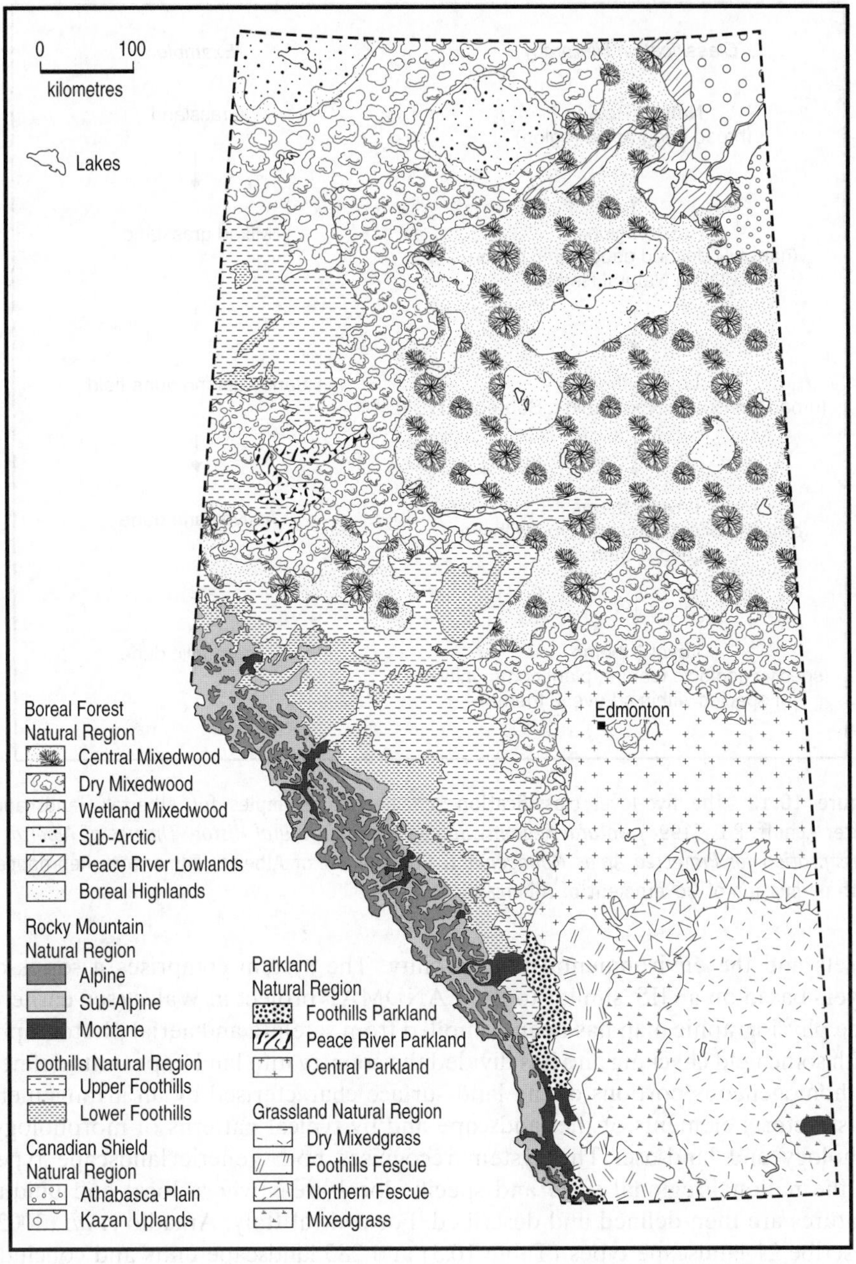

Figure 10.11 Natural Regions and Subregions represented in Alberta, Canada (after Government of Alberta (1994a) A Framework for Alberta's Special Places: Natural Regions, Report 1, Government of Alberta, Edmonton. Reproduced with permission of Government of Alberta).

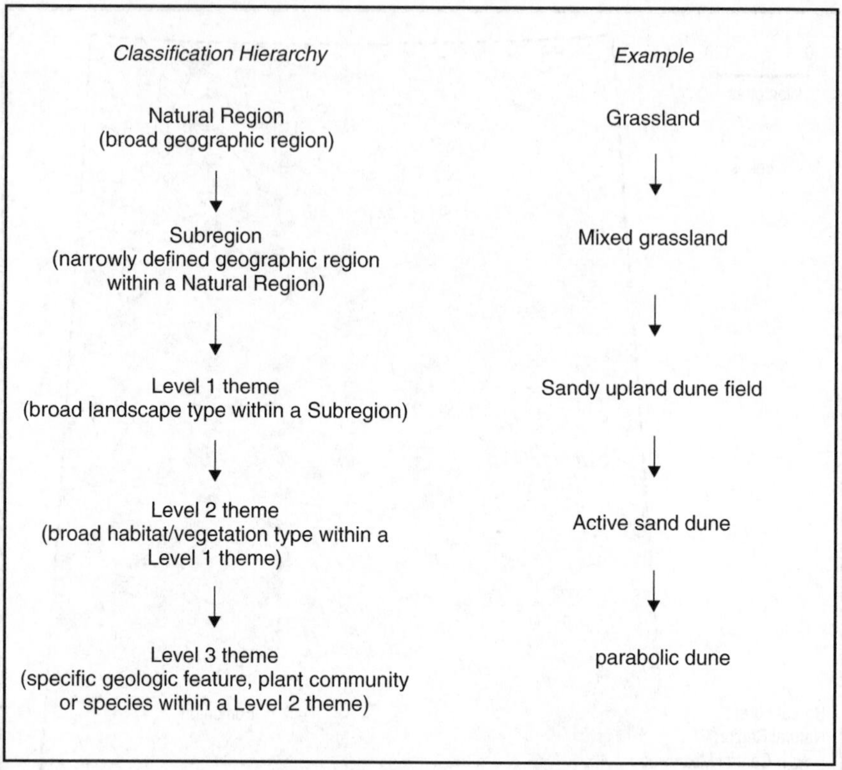

Figure 10.12 The five-level classification scheme and examples for Alberta's landscapes (after Achuff, P.L. (1994) *Natural Regions, Subregions and Natural History Themes of Alberta: A classification for Protected Areas Management*. Government of Alberta, Edmonton. Reproduced with permission of Government of Alberta).

assets and the environmental vulnerability. The system comprises a series of layers based on a GIS, similar to the LANDMAP project in Wales (see earlier). The physiographic map has been compiled from satellite and aerial photographs with some field checking and has divided the country into landscape units defined as homogeneous portions of the land surface characterised by an arrangement of structural elements of the landscape and by typical patterns of morphology, lithology and land-use. The system recognises both generic landscape types at the national/regional scale and specific local areas whose local and unique features are then defined and described. For central Italy, Amadio *et al.* (2002) describe 21 landscape types (Table 10.3) and 285 landscape units and conclude that 'geomorphology is the most important tool for the study of landscape at the regional hierarchical level to which the vegetation and the functional relationships are strictly related, to be defined' (p. 281).

The approach is similar to several of the other georegionalisation approaches described earlier, particularly Landscape Character Assessment. The system is organised in a GIS and will be used as a planning tool in the field of environmental quality and assessment (Amadio *et al.*, 2002; Lugeri *et al.*, 2012).

Table 10.3 Landscape Types of central Italy (after Amadio *et al.*, 2002).

	Landscape type	Landscape general structure
PC	Coastal plain	Flat or sub-flat area bordered by low coast
PA	Valley floor plain	Flat or sub-flat area inside a river valley
PP	Open plain	Flat, sub-flat or undulate, wide area with variable geometry
VI	Inner valley	More or less wide valley bordered by mountains on both sides
CI	Intra-montane basin	Closed and depressed area, surrounded by relief
TC	Carbonate plateau	Flat rocky area bordered by low limestone escarpments
CV	Volcanic hills and plateaux	Conical or tabular hills of volcanic origin
CA	Clay hills	Hills with smoothed to flat tops and graded slopes, mainly clay
CT	Siliciclastic hills	Wide hills of the forechain, mainly siliciclastic
CC	Carbonate rock hills	Hills of the chain and forechain, mainly carbonate
CE	Heterogeneous hilly landscape	Hilly landscape of lithologic and morphologic diversity
PI	Areas with isolated hummocks	Group of isolated hills separated by winding valleys
CP	Terrigenous hills with rocky ridges	Hills and ridges of the chain and forechain, rising abruptly
RC	Isolated coastal relief	Isolated rocky relief, surrounded by coastal plain
RI	Isolated rocky relief	Lower area of rocky relief bordered by abrupt breaks of slope
MV	Volcanic mountains	Mainly cone shaped mountains of volcanic origin
MT	Siliciclastic mountains	Mainly siliciclastic mountains often structured in ridges
MC	Carbonate rock mountains	Mainly carbonate mountains structured in ridges and massifs
MX	Crystalline massif	Group of mountains with well-defined steep crests and valleys
IS	Small islands	Islands of less than 500 km^2 with strongly defined coastlines
AM	Metropolitan areas	Built landscapes of urban texture

10.4 Georestoration

Given sufficient time, what is seen today as industrial dereliction can come to be valued as industrial archaeology. In fact a number of mineral waste dumps are now protected as archaeological monuments (Gray and Jarman, 2003). Furthermore, pits, quarries and other disturbed land often include very clear and important exposures of rock and sediments that need to be conserved for their scientific and educational values (Prosser, Murphy and Larwood, 2006). There is also a great interest in mining museums as the links between geological resources and their role in local social and cultural histories (Richards, 1996), and many are attracting significant tourist interest. Examples include the Geomining Historical and Natural Park of Sardinia, Italy (Arisci *et al.*, 2004), the Kennicott Copper Mine in Alaska, USA and the Great Orme Copper Mine in North Wales.

However, landscape restoration is becoming increasingly important as human society comes to appreciate the need for environmental improvement and reuse of derelict sites for new building. In this section I shall concentrate on the restoration of disused pits and quarries, river restoration, coastal restoration and managed realignment. Thus georestoration must consider restoration of both form and process.

There has, however, been considerable debate about landscape restoration. First, there has been discussion about the use of terms like 'restoration' and 'nature' and about the extent to which a true return to natural conditions can be achieved. Secondly, there has been concern that long-term restoration proposals for new sites can be used to justify major short- and medium-term environmental impacts (e.g. Katz, 1992; Eden, Tunstall and Tapsell, 1999), particularly where what is being proposed is not a restoration to pre-disturbance conditions. Thirdly, it is important to recognise that restoration is not about creating static places, but about allowing natural environmental dynamics to operate so that the landscapes continue to evolve. In Adams words (1996, p. 169–170) 'what we are doing is facilitating nature, and not making it. ... We must allow nature space to be itself, to function, to build and tear down'. And fourthly, increasingly the need for integrated rehabilitation is being stressed, where geo- and bio-conservation and management issues are considered together (see Chapter 14).

10.4.1 Quarry and pit restoration

Once pits and quarries have outlived their usefulness, many are simply abandoned. Thousands of abandoned mineral sites can be found within the US National Parks System and thousands more occur outside the Parks. Although some are important geological or wildlife sites, others are unsightly, may be environmentally polluting and often pose safety hazards for the public. Therefore attempts are often made to restore them. But Gregory (2000, p. 264) asks 'exactly how are the contours configured and the landscape recreated? Is advice sought from a geomorphologist or a physical geographer with an understanding of the landscape appropriate to that area, or is it undertaken by someone without such training?'

If the pits are below the water table, they may be quickly colonised by wildlife and/or water sports enthusiasts. If the pits or quarries are above the water table they can become important in preserving the geological interest of the quarry if the important faces can be left exposed, made safe and accessible, maintained and kept clear of colonising vegetation (Prosser *et al.*, 2006). Low-level pit restoration by grading and returning to agriculture or woodland will result in loss of the geological exposure and may create incongruous landforms.

An alternative approach is to remodel the pit and quarry as an authentic landform while retaining the geological interest of the quarry. An innovative attempt at this is described by Gunn (1993) and Gagen, Gunn and Bailey (1993), who used restoration blasting techniques in limestone quarries in Derbyshire, England to replicate the form of the local dry valleys. Given time, the older abandoned blackpowder blasted quarries would develop in this way, but the scale and methods used to excavate modern quarries means that they will continue to intrude upon the natural landscape for many centuries (Gunn, 1993, p. 196). Restoration blasting and seeding allows the process to be accelerated so that the appearance of a mature and attractive valley can be developed very quickly. The typical dry valleys (dales) of the area contain rock buttresses, rock headwalls, scree slopes and debris flows, and by careful formulation of a drilling and blasting pattern, a predictable suite of these landforms can be created from the abandoned quarry walls. 'The construction of these rock landforms together with their subsequent revegetation will enable quarried rock faces to be more easily harmonised with the surrounding unexcavated landscape' (Gagen, Gunn and Bailey, 1993, p. 25). The restoration programme can also be designed to allow access to features of scientific interest high on the quarry walls, but would be less appropriate where all the available exposure is regarded as important or where an integrity site is involved (see Section 9.4.3).

Nonetheless this technique represents a promising initiative, which is being adopted elsewhere. For example, a similar scheme has been promoted at Coniston in the Lake District National Park, England. Quarrying of green slate has occurred at Bursting Stone Quarry for many years, but it occupies a prominent position half way up the east flank of the hill called Coniston Old Man where the slate waste tips are artificially terraced. A planning application was submitted in 1997 which proposed regrading the slate waste to more natural profiles, seeding the slopes created and blasting the quarry face to create rock buttress features (Stephens Stephenson, 1998).

This is also the approach being used in the United States where the National Parks Service has an *Abandoned Mineral Lands Program* which focuses on re-establishing landscapes and environments that mimic the surrounding undisturbed lands. Volunteer labour is often used in this work, which applies not only to the mines, but also to the access roads. Good examples of successful restoration of this type occur at Redwood National Park and Joshua Tree National Park, California. At Lingerbay, in the Scottish Hebrides, the restoration aimed to create a flooded glacial corrie (Box 10.1).

Box 10.1 Lingerbay Superquarry, Scotland

In March 1991, Redlands Aggregates Ltd applied for permission to develop a large coastal superquarry in the Precambrian anorthosite outcrop at Lingerbay on the Isle of Harris in the Scottish Outer Hebrides. The plan was to remove 550 million tonnes of anorthosite over a 60-year period for use as general aggregate and armourstone removed by ship to south-east England, continental Europe and perhaps America (Owens and Cowell, 1996; McIntosh, 2001). A large proportion of Mount Roineabhal would have been removed (459 ha) and a substantial sea loch created (McKirdy, 1993; Bayfield, 2001). The restoration scheme aimed to restore the quarry basin progressively as a flooded coastal corrie. However, objectors questioned whether the resultant landform could be given a natural appearance in view of the difficulty of mimicking natural corrie backwalls and the need to comply with slope stability criteria (Owens and Cowell, 1996). The main concerns of Scottish Natural Heritage and local people were over the loss of a valued local landscape (McIntosh, 2001;

Warren, 2002). The applicants' case was that these impacts had to be balanced against the economic benefits in terms of local employment.

The application was 'called in' by the Secretary of State for Scotland in 1994 and an 85-day public inquiry was held in 1995. The Inquiry Inspector concluded that the proposals would 'completely change the landscape characteristics of Lingerbay by changing the scale and character of the coastline and its hinterland. ... The quarry would create an area of massive disturbance', but her overall conclusion was that there was a justified need for the aggregate which would make an essential contribution to national prosperity and was therefore in the national interest. However, this was not accepted by the environment minister of the newly devolved Scottish government who refused the application in November 2000, nine years after the application was made, on the grounds of landscape impact.

An alternative means of restoration of pits and quarries above the water table is by filling them with waste (after suitable lining systems have been installed), capping with suitable sediments and soil, and returning them to an appropriate afteruse (e.g. grassland, scrub, woodland, recreation). Carefully done, and given time, this can return the landscape to a similar or enhanced condition compared with the pre-excavation landscape (Box 10.2). However, if the quarry or pit contains important rock or sediment geological exposures, full landfilling will result in loss of the exposure. According to Bennett (1994), 31% of all Quaternary Sites of Special Scientific Interest in England are located in disused pits and quarries. In the case of large quarries it may be possible to engineer the site to leave a conservation face (see Figure 10.13). This becomes impractical and uneconomic in the case of small quarries unless the important strata are towards the top of a quarry face.

Box 10.2 Restoration of coal mining areas in the Lower Rhine, Germany

The largest mining area in Germany is found in the Lower Rhine area around Aachen, Cologne and Monchengladbach where lignite seams between 10 and 100 m thick are quarried in open pits up to 300 m deep. Mining has occurred here since the eighteenth century. The pits currently cover an area of about 90 km^2 with a further 150 km^2 having been previously worked. Most of this has now been restored by waste infilling and rehabilitation to forestry, agriculture, horticulture or industrial uses but the changes to the landscape, soils and hydrology of the area have been significant. Nonetheless, the older restored areas south of Cologne have matured into attractive lake and woodland landscapes used for recreational purposes. In the more recently quarried areas, the overburden is very carefully used to provide different soil types and different land uses. For example thick mixtures of gravels, sand and loess are used on slopes where forestry is planned, but loess and loess loam are spread on flatter ground meaning that extensive areas of land have been returned to agriculture (Aust and Sustrac, 1992).

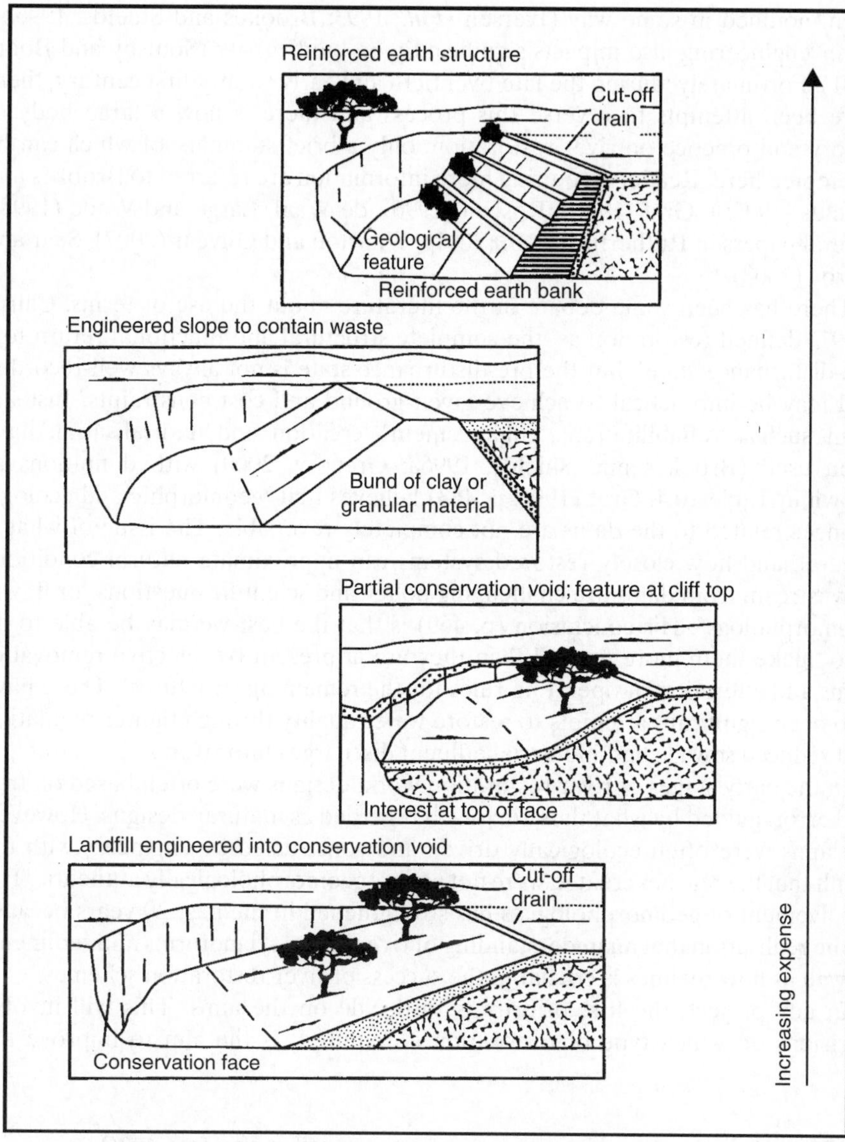

Figure 10.13 Some restoration schemes for waste disposal in quarries while retaining the geological interest of a face (from Bennett, M.R. & Doyle, P. (1997) Environmental Geology. Reproduced with permission of John Wiley & Sons).

10.4.2 River restoration

In his foreward to Brookes and Shields' (1996a) book on *River Channel Restoration*, Al Gore (1996, p. xiii) commented that 'the degradation of riverine systems over the entire planet is dismaying. ... In the "Lower Forty-Eight", in the United States, the Yellowstone River remains the only medium to large river which is unimpounded and does not suffer the impacts of regulated flows'. It is estimated that 98% of Danish streams and 96% of those in lowland England have been modified in some way (Iversen *et al.*, 1993; Brookes and Shields, 1996b). River engineering also impacts significantly on biodiversity (Soulsby and Boon, 2001). Fortunately, during the late twentieth and early twenty-first century, there have been attempts to reverse this process and there is now a large body of theory and practice on river restoration, only a brief summary of which can be attempted here. Readers requiring more information are referred to Brookes and Shields (1996a), Graf (1996), Rosgen (1996), de Waal, Large and Wade (1998), several papers in Rutherfurd *et al.* (2001), Wharton and Gilvear (2007), Sear and Darby (2008).

There has been some debate in the literature about the use of terms. Cairns (1991) defined restoration as 'the complete structural and functional return to a pre-disturbance state', but the pre-disturbance state is not always well-recorded and may be impractical to achieve given ground and cost constraints. Instead, terms such as 'rehabilitation', 'enhancement', 'creation' and 'naturalisation' have been used (Brookes and Shields, 1996a; Gregory, 2000) with definitions as shown in Table 10.4. Graf (1996, p. 443) believes that 'geomorphic and ecologic changes related to the dams are not completely reversible. The issue of what is natural, and how closely restored systems can approximate natural conditions downstream from dams are challenging policy and scientific questions for fluvial geomorphology'. His conclusion (p. 469) is that the best we may be able to do is to 'make them more natural than they are at present by selective removal of dams and alteration of operating rules for the remaining structures'. There have also been significant attempts to restore water quality through tighter regulation and reduced sewage and industrial effluent discharges into rivers.

In the early stages of river restoration work, designs were often based on trial and error, guided by what those involved regarded as 'natural' designs. However, the aims were often ecologically driven and constructed by engineers with the result that the reaches created were not always geomorphologically authentic. The involvement of geomorphologists has strengthened in the past 20 years because of the realisation that an understanding of river process, landforms and sediments as well as flow regimes is crucial to the success of river restoration schemes.

In any project, the first priority is to decide on the aims. This will involve decisions on which type of restoration is required. Is the aim to improve the

Table 10.4 Some terms used in river restoration (modified after Sear, 1994).

Recovery	The act of restoration of a river to an improved/former condition
Re-establishment	To make a river secure in a former condition
Enhancement	Any improvement of a structural or functional attribute
Rehabilitation	Partial return to a pre-disturbance structure or function
Reinstatement	To restore a river to a former condition
Restoration	The act of restoring a river to a former or original structural or functional condition
Creation	Development of a morphological and/or ecological resource that did not previously exist
Naturalisation	To return a system to a condition sustained by natural processes

aesthetics of the river, improve habitats, allow public access and recreation, etc.? Is intervention necessary or should the river be allowed to recover naturally? The next steps are to collect geomorphological and other data on the river, which will include historical information, hydrological data, bed sedimentology, catchment area processes, land-use data, land ownership, etc. The final steps include drawing up and evaluating restoration options and choosing a final design.

Among the issues to be resolved at the design stage are:

- design of the river planform, which will depend on functional constraints, floodplain constraints, etc. It will also be necessary to decide what level of future channel instability will be acceptable and how additional unwanted instability should be constrained (Gippel *et al.*, 2001);
- design of long profile (pools, riffles, bars, etc. See for example Wilkinson, Keller and Rutherfurd, 2001; Outhet *et al.*, 2001);
- design of cross-sectional channel shapes through the reach, which will depend on planform design, substrate and bank materials, desired low-flow width, etc.;

Over the past 20 years a more integrated approach has been taken, not only in reconnecting rivers to their floodplain systems (Brookes, Baker and Redmond, 1996; Richards, Brasington and Hughes, 2002) and viewing them in the context of their catchment areas (Kondolf and Downs, 1996; Abernethy and Wansborough, 2001), but also in linking gemorphological, ecological, water-quality, recreational and other aims into integrated river management strategies (Brookes and Shields, 1996c) that cover longer timescales (Brierley and Fryirs, 2005). Thus Lake (2001) sees three basic ways of stream restoration—restoration of particular reaches, restoration of longitudinal and lateral connectivity, and restoration of drainage basins involving co-ordinated activities at the stream catchment level. Hart and Poff (2002) and Pizzuto (2002) discuss the 'emerging science of dam removal'. There has also been an increasing interest in project appraisal carried out at various stages from conception through to several years of post-completion monitoring (Hulbert, Wharton and Copas, 2009).

Box 10.3 describes probably the world's most ambitious river restoration project on the Kissimmee River in Florida, United States. Powell (2002) describes attempts to restore more natural flows and sediment transport regimes on the Colorado River, which has been drastically altered since the construction of the Glen Canyon Dam. In the United Kingdom, the River Restoration Centre (RRC; formerly Project) is an independent, non-profit organisation whose aim is to encourage and co-ordinate good practice in river restoration (www.therrc.co.uk) and Figure 10.14 is an example of a UK scheme supported by them. They have published a useful *Manual of River Restoration Techniques* with examples from 15 projects (RRC, 2002). The London Rivers Action Plan is another UK example of restoring Thames' tributaries with straightened and/or channelised reaches and bringing aesthetic, recreational, biodiversity and geodiversity benefits in creating more naturally functioning rivers.

Box 10.3 River Restoration on the Kissimmee River, Florida, United States

The Kissimmee River lies in central Florida and drains into Lake Okeechobee. It has a 7800 km^2 drainage basin, and a 90 km long floodplain varying in width between 1.5 and 3 km (Toth, 1996). Prior to 1962 the river meandered for 166 km over this floodplain which was a rich wetland ecosystem supporting over 300 fish and wildlife species including resident and overwintering waterfowl and wading birds. 'The diversity and persistence of these biological resources were linked to dynamic river and floodplain habitat characteristics provided by basin hydrology and channel geomorphology' (Toth, 1996, p. 369).

However, the river was totally channelised between 1962 and 1971 to provide drained farmland for the expanding agricultural economy of central Florida. The major features of the scheme were a 9-m deep rectilinear canal cut through the floodplain and divided into five level reaches by water-control dams and separated from the floodplain by levees. The 26 headwater lakes were connected by canals and regulated as flood storage reservoirs. The scheme was successful in draining two-thirds of the floodplain but at the cost of a dramatic loss of habitat and wildlife. This impact soon led to calls from local communities and environmental groups for the river to be restored and in the 1980s and early 1990s a number of plans and feasibility studies were carried out, including a demonstration project between 1984 and 1989 on a 19-km reach (Pool B). In this scheme, water from the canal was diverted back through sections of the original channel leading to improved river aesthetics and wildlife habitats.

Lessons learnt from these studies led to a state-federal partnership plan for the dechannelisation of the river, including backfilling of 35 km of canal with original spoil which had been spread on the floodplain, removal of two dams and associated levees and re-excavation of 14 km of former river channel. Restoration work began in 1999 and is nearing completion in 2013 at a cost of over $400 million at 1997 prices. Flood protection is maintained in the residential stretches and peak discharges are reduced by increase flood storage capacities in the headwater lakes (Toth, 1996).

In some countries, legislation has aided river restoration. This is certainly the case in Denmark where the Watercourses Act (1982) restricts maintenance practices in order to safeguard local stream ecology. It also incorporated special provisions for stream restoration and its funding, and has led to several major restoration projects (Madsen, 1995). As in other cases, geomorphology has benefited from predominantly ecologically-driven measures.

10.4.3 Coastal restoration and managed realignment

In many parts of the world coastal erosion is resulting in loss of land and property. An understandable reaction to this is to try to prevent it occurring by erecting sea-defences, which may take the form of sea-walls, revetments, groynes, embankments, etc. As well as preventing coastal erosion, these constructions

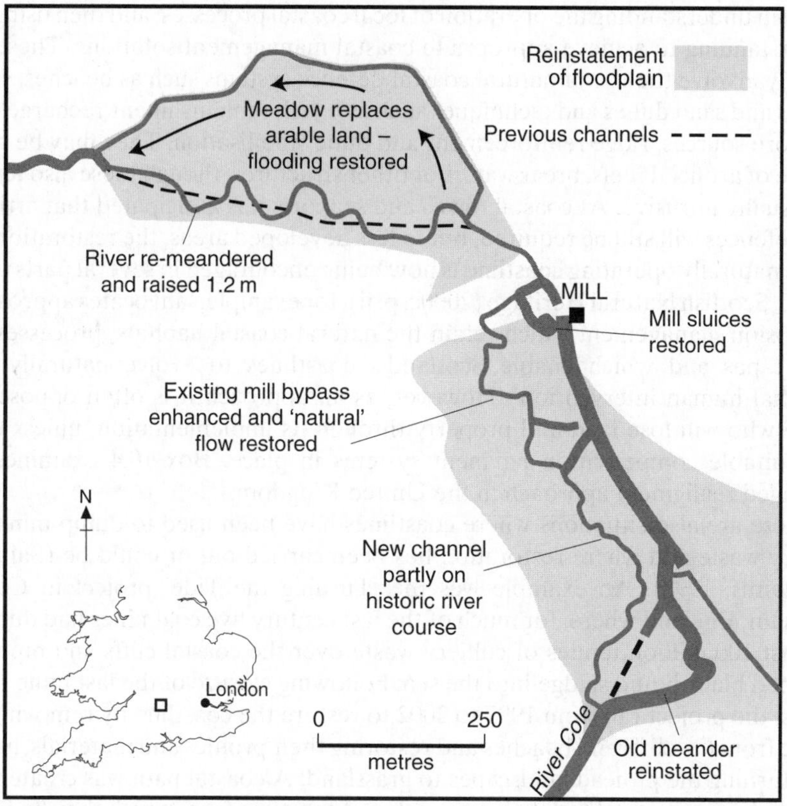

Figure 10.14 The river restoration scheme for the River Cole, England. (The River Restoration Centre (2012). Reproduced with permission).

may also prevent coastal flooding and allow marshland inside the defences to be drained and cultivated. However, this engineering approach has been questioned on a number of grounds over the past 20 years or so (Hooke, 1998, 1999; French and Reed, 2001).

First, sea-defences are expensive to install and maintain. Wave attack usually means that, even with regular maintenance, a sea-wall will have only a limited life before a replacement is needed, often to a larger design. Secondly, preventing coastal erosion and longshore drift through the construction of sea-walls and groynes, starves the coastline of sediment in a down drift direction and may accelerate coastal erosion there. In other words, human intervention is preventing the operation of natural coastal processes and disrupting natural systems. The evidence is that the traditional methods of coastal defence are not just ineffective, but actually exacerbate erosion problems. Thirdly, sea-defences such as concrete sea-walls or giant armour blocks often obscure coastal geological exposures and are not a very aesthetically pleasing addition to the coastline (see Section 5.5). In other words, they inevitably reduce natural coastal geodiversity.

The alternative approach that has been adopted in many areas over the past 20 years or more is what is often referred to as a soft engineering approach. The principle of this approach involves working with, rather than against, nature, firstly in understanding the operation of local coastal processes, and then using this understanding to achieve appropriate coastal management solutions. These will usually involve the use of natural coastal defence systems such as beaches, storm ridges and sand dunes and techniques such as beach replenishment/recharge from offshore sources, ridge reinforcement and dune stabilisation. They may be aided by use of artificial reefs, breakwaters or other structures, though these also tend to be visually intrusive. At coastal towns and seaports it is anticipated that artificial sea defences will still be required, but in less developed areas, the restoration of a more naturally operating coastline is now being encouraged in several parts of the world. Scottish Natural Heritage (2000c, p. ii), for example, 'advocates approaches to erosion management which retain the natural coastal habitats, processes and landscapes and which enable Scotland's coastlines to evolve naturally with minimal human intervention'. However, as an approach it is often opposed by those who will lose land and property through its implementation, unless there are suitable compensation payment systems in place. Box 10.4 examines the managed realigment approach in the United Kingdom.

There are also situations where coastlines have been used to dump mine and quarry waste and where restoration has been carried out or could be (Saiu and McManus, 1998). An example was the 'Turning the Tide' project in County Durham, England where, for much of the last century five coalmines had dumped at least 100 million tonnes of colliery waste over the coastal cliffs and pipelines pumped black liquid sludge into the sea. Following closure of the last mine in the 1990s, the project ran from 1997 to 2002 to restore the coastline by removing the waste from the cliffs and beaches and restoring their profiles and materials, as well as returning the pithead landscapes to grassland. A coastal path was created and the public encouraged to return to a restored coastline from which they had been excluded for so long. In 2001 Durham's coast was given Heritage Coast Status and in 2003 the Durham Heritage Trust was launched with the aim of achieving 'the integrated management of Durham Heritage Coast managed by and for local communities, protecting and enhancing the natural and cultural integrity of the area while developing and meeting the area's social and economic needs'. One of the Trust's projects involved creating a footpath alongside which are stones engraved with the names of different types of coal found in the Easington mine and a timeline of the area's mining history. In 2011 the Durham Heritage Coast won a Council of Europe Landscape Award for its coastal restoration achievements.

Box 10.4 Managed realignment and Shoreline Management Plans in the United Kingdom

Shoreline Management Plans were first drawn up in the 1990s for the coastline of England and Wales according to government guidelines (MAFF, 1995), and were also prepared for parts of the Scottish coast (Hansom, Crick and John, 2000). A second generation of plans covering the whole 6000 km of coastline of England and Wales is currently underway. Each Plan is based on a set of sediment cells or sub-cells, and each presents a strategy for coastal defence for that stretch of coast for the short, medium and long term (up to 100 years), taking into account coastal processes, human influences and other relevant factors. Given the complex administrative and legislative framework of coastal management responsibilities, the aim is to improve the integration of coastal management both within individual coastal areas and between them, thus bringing a more strategic approach (Hansom *et al.*, 2000).

A Shoreline Management Plan is 'a document which sets out a strategy for coastal defence of a specified length of coast, taking account of natural coastal processes and human and other environmental influences and needs' (MAFF, 1995). They should be heavily geomorphology-based, though geomorphologists are not always involved in the work (Hooke, 1999). The end product is usually a map of a coastal cell showing zones where particular shoreline management strategies are recommended. 'The choice of strategy is between four management options: hold the line, retreat the line, advance the line, or do nothing' (Hooke, 1999, p. 380–381). Variations on these themes may include installing limited defence schemes that slow down coastal erosion in order to retain important coastal exposures but prolong the lifetime of properties threatened by coastal erosion (McKirdy, 1990; Barton, 1998; Brampton, 1998).

Many of the estuaries and marshlands of south-east England have been enclosed by embankments and drained over the past two centuries to provide flood defence and additional grazing or arable land. This has resulted in coastal squeeze, which occurs when rising relative sea-levels raise the low water mark while the high water mark is held in place by the embankment. The width of the intertidal zone is therefore reduced with significant loss of geodiversity and biodiversity and buffering protection afforded to the embankment. These then become open to attack, with resultant need for expensive maintenance. Managed realignment involves setting back the line of actively maintained sea defences to a new line inland of the original and encouraging the operation of natural intertidal processes (salt marsh or mudflat) in the area between the old and new defences. The original front defence line can either be allowed to degrade naturally or may be wholly or partially removed.

An experimental site at Tollesbury in Essex was established in 1995 when the old sea wall was breached allowing the sea to flood low-lying agricultural farmland for the first time in over 150 years. Since then the site has been studied to discover the biotic and abiotic changes that are occurring as a direct result of seawater inundation and the effect on ebb and flow rates within the existing salt marsh creek systems. The conclusion was that managed realignment has the potential to alleviate some of the problems of rising sea-level in this area and produces benefits for biodiversity. In 2002 two full-scale schemes were established in England, allowing arable land to return to salt marsh at Freiston in Lincolnshire and at Abbott's Hall in Essex (Figure 10.15).

Figure 10.15 One of the breaches in the sea-defence embankment at Abbott's Hall, Essex, England.

10.4.4 Other land restoration projects

Community efforts at environmental restoration are increasingly important. One successful example is Landcare in Australia which now involves over 3000 groups and which has spawned other initiatives such as Coastcare and Rivercare. Eberhard and Houshold (2001) describe community efforts to restore degraded karstic features in the Mole Creek karst area of Tasmania, Australia. This includes removing sediment and rubbish from cave floors and entrances and cleaning sediment from delicate calcite formations.

In Hampshire, England, the construction of a motorway cutting (M3) through chalk downland (Twyford Down), enabled the closure and restoration of 1.5 km of the route of the old A33 at St Catherine's Hill. The old road was broken up and chalk spoil from the cutting was used to recontour the over-steepened hillside and recreate the pre-A33 slope (Eden, Tunstall and Tapsell, 1999). It was recognized, however, that this was not a return to an entirely natural state. Similarly, the National Trust of Scotland has restored high-altitude vehicle tracks on Beinn a' Bhúird in the Cairngorms.

A number of projects have involved restoration of drained bogs and wetlands by blocking channels and ditches to raise water levels and to restore the bog to a more natural condition. The aim is also to contribute to carbon storage efforts by preventing the drying out of upland bogs. Projects of this type have been undertaken in, for example, eastern Canada (Rochefort and Campeau, 1997), England (Johnson, 1997) and Northern Ireland (Gunn, 1995).

Koster (2009) describes projects to restore active drift sand areas in the Netherlands that have been stabilised by vegetation in the past 200 years, and Hildreth-Walker and Werker (2006) describe principles and practice of cave conservation and restoration.

Finally, old research core holes or climbers bolt holes can be restored by plugging them with dyed cement or with rock cores or rock dust taken from loose stones of the same rock type. MacFadyen (2011) describes examples of poor corehole restoration and makes recommendations on how it should be done.

10.5 Landform design

'From place to place the environment is different in fundamental ways; unless these differences are made part of the information base for decision making, we will continue to build mis-sized and unsustainable infrastructures and land use systems' (Marsh, 1997, p. 5). He cites the case of Franconia Notch, a narrow valley in the White Mountains of New Hampshire, United States, where federal and state transportation planners had designed a standard four-lane interstate highway squeezed into the centre of the valley with huge entry and exit ramps. Fortunately the local people objected and after a decade of argument, including court action, the highway planners agreed to modify their standard approach and adopted a smaller design better suited to the topography, drainage and scale of the landscape. This has resonances of McHarg's (1995) famous plea to *Design with Nature.*

But it is not just the design of infrastructure and buildings that needs to be considered. New landforms are also being created by excavation and construction and examples are given in Table 10.5. This section will examine some of the impacts in more detail, paying particular attention to conservation of sensitive physical landscapes and design of authentic landforms. Several of these issues are discussed in more detail in Jarman (1994), Marsh (1997), Gray (1997a, 2002) and Gray and Jarman (2003) while Gregory (2000) has discussed the role of geomorphologists in environmental design as part of a 'cultural physical geography'. Haigh (2002) asks 'what greater challenges and what greater vindication can there be for a discipline than to create a new landscape or to recycle land that has been sacrificed to human well-being?'

Many parts of western Europe have been densely populated for centuries and have lost most of their natural vegetation. It is often said that natural landscapes no longer exist. But while this may be true of the natural vegetation, the same is not true of the geological and geomorphological landscape elements. Although there has been much engineering work done to alter the form of the European land surface since the Industrial Revolution (Hoskins, 1955), not to mention the smoothing out of irregularities achieved by centuries of ploughing, the shape of the land surface over large parts of rural Europe has remained substantially unaltered since the major changes of the Pleistocene and modification by Holocene and modern processes. In contrast to the natural vegetation, the natural topography or land form generally remains intact. Note the use of the term 'land form' rather than 'landforms' here, since in the wider countryside the topographic form of the land may not have formal landform names, yet may still be natural. Similar views are expressed by Kiernan (1997a, p. 9) in arguing that 'Landforms are defined by their contours. Hence any unnatural changes to the contours of a landform by definition damages the natural geomorphology.... The geoconservation significance of the

Table 10.5 Some anthropogenic landforms (modified after Haigh, 1978; Goudie, 2013).

Excavational	
Digging	e.g. drainage ditches
Cutting	e.g. road and rail cuttings
Quarrying	e.g. pits and quarries
Cratering	e.g. bomb craters
Constructional	
Tipping	e.g. tailings heaps, landraising, coastal reclamation
Infilling	e.g. infilling of hollows
Mounding and bunding	e.g. for visual and noise screening
Embanking	e.g. river and coastal flood defence, road embankments
Excavation and construction	
Terracing	e.g. vineyards and rice terraces on slopes
Ridge and furrow	e.g. agriculture
Water features	e.g. agriculture reservoirs, canals, ponds, moats
Remodelling	e.g. golf courses, regrading of slopes
Other landform effects	
River engineering	e.g. channelisation, straightening, dredging, damming
Coastal engineering	e.g. sea walls, armouring
Subsidence	e.g. due to extraction of fluids and minerals underground
Slope failure	e.g. due to loading, lubrication, undercutting

damage is what is important. . . .'. It is the aim of this section to argue that society ought to pay greater attention to respecting, conserving and designing with this natural geomorphological heritage, which is a very important element in landscape character and local distinctiveness (see earlier).

First, the loss of natural landform character often accompanies development. Jarman (1994, p. 42) noted that 'Human agency has long been ironing out the surface irregularities of the land. Ploughing smooths over breaks of slope; farmers fill wet hollows; hill paths are surfaced with fill from glacial hummocks. . . . Wherever development takes place, there is an inevitable tendency to eliminate rather than incorporate local landform character'. Secondly, the design of new landforms may be even more damaging, particularly where engineering standards or traditions dictate regularity. Jarman (1994, p. 42) decries the 'evenly-graded road embankments, rectilinear cut-and-fill shapes, level earth dams, improbably convex screen bunds'.

This is not to say that there is no scope for artificial landscaping, including large-scale landscape art projects. In *The Hitchhiker's Guide to the Galaxy* (1979) by Douglas Adams, Slartibartfast worked as planet designer and was particularly proud of his award for designing Norway's 'lovely crinkly edges' on planet Earth. In real life, we have examples of the landform design work of landscape architects like the stone structures of Andy Goldsworthy or the artistic land remodelling of Charles Jencks. For example, Jencks has designed a restoration scheme using excavated soil at a mining site near Cramlington in Northumberland, England, that features a giant reclining female figure. Named Northumberlandia (or 'Slag Alice' by the local residents) the figure is 400 m long with breasts 30 m high.

It is hoped that it will become a tourist attraction as is another of Jencks' schemes at St Ninian's opencast coal mine in Fife, Scotland. This one features as its centrepiece a loch in the shape of Scotland surrounded by differently shaped mounds representing the continents settled by migrating Scots.

10.5.1 Golf courses

There was a time when golf course design utilised existing landforms. An example is the British Open links courses at St Andrews, Troon and Turnberry where raised beach sands have been blown into dunes, stabilised by vegetation growth and turned into golf courses with a minimum of remodelling. Price (1989, 2002) was even able to classify Scotland's golf courses by their geomorphological characteristics. However, the construction of modern golf courses often involves huge re-engineering works to remodel the topography (Figure 10.16). The creation of raised tees, bunkers, water features, mounds and slopes can radically alter the topographic landscape. This can result in a land form that is out-of-keeping with the local landscape. An example occurs at Dunston Hall near Norwich, England, where the natural medium-scale, rolling, agricultural landscape has been carved into a much finer scale landscape of hummocks, fairway channels and ponds (see Figure 10.17). Apart from the land form and hydrological impacts, the topographic changes decrease the reversibility of the area back to agricultural land (Jones, 1996).

Yet, in the United Kingdom at least, there have been signs of the need for regulation of golf course design through planning and policy guidance. A Countryside Commission publication on *Golf Courses in the Countryside* (Countryside Commission, 1993b) has a chapter on topographical change. This concluded that while all new golf courses will require some earth movement

Figure 10.16 Remodelling work underway to create a new golf course in Algarve, Portugal.

Figure 10.17 Topographic change created by golf course construction, Dunston Hall, Norfolk, England. On the right is the natural topography of the landscape with long, low-angle slopes. On the left, this landform has been remodelled to create fairways, mounds, etc.

during their construction, 'large-scale remodelling is not essential to the quality of a golf course and can be highly inappropriate. In particular, topographical changes should reflect the local topographical character so that the final landform is indistinguishable from the surrounding landform' (Countryside Commission, 1993b, p. 25). It argued that of all the features of a new golf course, 'mounding is often the most alien to the landscape setting'. It notes that 'a flat landscape can accept very little by way of grading and almost certainly no mounding', whereas on steeply sloping or hilly sites, significant earthworks can be justified as long as care is taken to avoid 'slopes that are unacceptably severe because of the risk of appearing wholly unnatural; they may also be liable to erosion' (Countryside Commission, 1993b, p. 25).

An example of the latter problem occurs at Fraser Hill in Malaysia, where massive cut-and-fill slopes have been created by construction of a golf course in a tropical mountain rainforest. In spite of immediate hydroseeding aimed at stabilising the slopes and surfaces, not surprisingly in this environment some slope failures have occurred (Bayfield, 2001).

A guidance booklet for Scottish golf course development (Scottish Natural Heritage and Scottish Golf Course Wildlife Group, 2000) has very little to say on topographical change. However, the organisations did achieve revisions to the planned extension to Dunkeld Golf Course, including the elimination of bunkers on some holes, because of the impact on one of Scotland's National Scenic Areas.

10.5.2 Landraising

Another example of incongruous landform creation is in the above-ground dumping of waste, a practice known as landraising. The hills of waste are created to allow for settlement of waste as it decomposes and to provide land drainage rather than water infiltration, but a further reason is simply to get more waste in, particularly where there is a shortage of landfill void space (Gray, 1998b, 2002).

Some spectacular landforms have been created in this way. For example, the Packington site near Birmingham, England rises 50 m above the surrounding landscape and intrudes 12 m into Birmingham Airport's airspace! However, others have been refused permission partly because of the incongruous landforms proposed. This is particularly true of hills of waste in areas of flat topography such as till plains, river terraces or coastal flats. For example, at Rivenhall Airfield in Essex, England, a planning application was made in 1993 to extract 10 million tonnes of sand and gravel from a 100-ha area and overfill the void with household and commercial waste to create two waste hills rising to 18 m and 12 m above the surrounding till plain and separated by a small valley (Figure 10.18). According to the applicants, the object was to replace the current 'flat and featureless' airfield site with a more undulating and interesting topography. Fortunately, the scheme was refused permission, partly because 'the proposed landform would not relate well visually to any existing feature and when completed would not appear to be an authentic part of the local scene' (Gray, 1998b, p. 187).

But even when attempts have been made to design landraised sites in keeping with the local topography, the results have not always been very successful. For example, at Fleetwood in Lancashire, England, a waste hill was designed to resemble the local drumlins around the Wyre estuary but an analysis of the hill proposed demonstrated that it falls short of an authentic local drumlin in terms of size, shape and orientation (Gray and Jarman, 2003). The applicants claimed that this was a 'carefully designed landform (that) will greatly assist in integrating the landfill into the surrounding landscape' but Gray and Jarman (2003) concluded that this has been a rather weak attempt to design a drumlin in keeping with the local morphology. One suspects that what we have here is a *post hoc* justification for a landform that was designed with little reference to the natural landforms of the area. Yet a simple landform redesign could have achieved an authentic drumlin morphology whilst retaining a similar waste capacity.

10.5.3 Bunding

One method of screening unsightly developments such as landfill sites, gravel pits or new roads is by constructing soil mounds or ridges to obscure them from view and reduce noise. As the UK's Landscape Institute (1995, p. 59) states 'major works in themselves may create adverse landscape and visual impact, and care should be taken to ensure that a new landform looks natural and appears as an integral part of the landscape'. This is particularly good advice in low relief areas where steep-sided, rectilinear ridges and even low mounds can appear completely out of place in their landscape context (Figure 10.19). Approved planting schemes in the countryside now generally encourage the use of native species, and the same approach needs to be taken to geomorphological landscaping. In many areas the linear bund can be regarded as the geomorphological equivalent of the *leylandii* hedge.

A detailed study of bunding was undertaken in South Norfolk, England, by Gray (1997a). The area has a predominantly flat topography as part of the

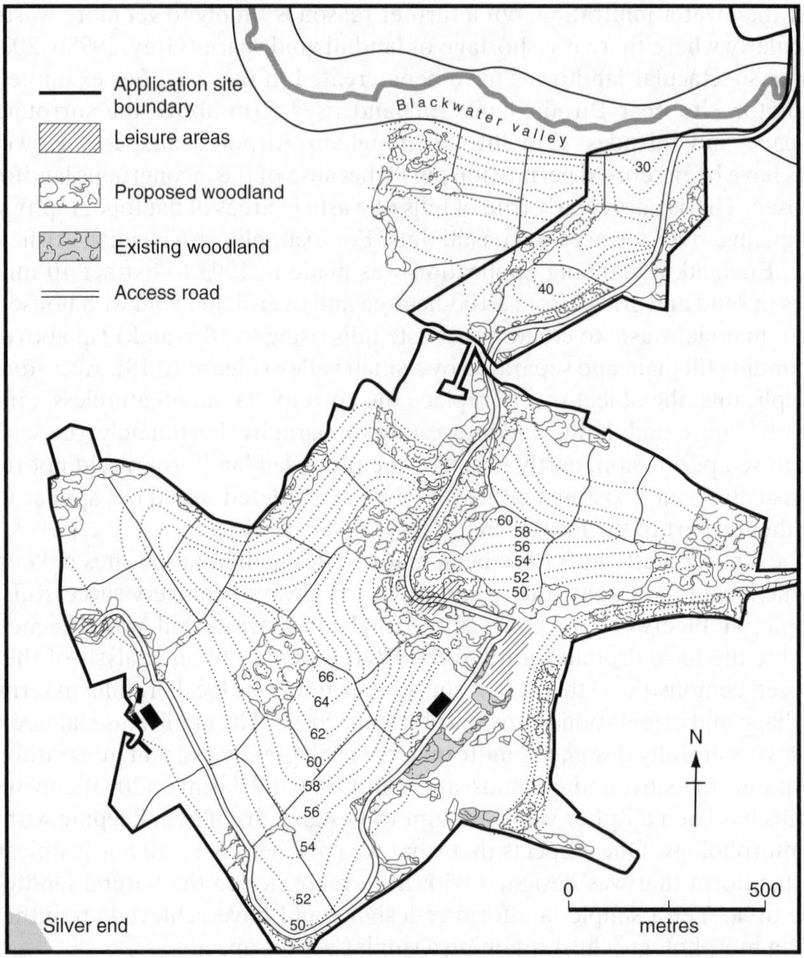

Figure 10.18 The landraising scheme proposed at Rivenhall Airfield, Essex, England in the 1990s.

East Anglian till plain, and is therefore very sensitive to topographical change such as bunding. The study demonstrated a growing and worrying tendency to include bunding and mounding as part of landscaping schemes. Part of the reason is the cost of removing subsoil and topsoil from the sites, including payment of Landfill Tax (see later). It is usually cheaper and easier to simply mound this material on site

10.5.4 Pond and reservoir formation

The digging of ponds and agricultural reservoirs raises at least three geodiversity issues. Firstly, pond shape may not be in keeping with the character of local natural ponds. Secondly, the spoil may be heaped into incongruous bunds and

Figure 10.19 Bunding around an agricultural barn, Wattisfield, Suffolk, England. Note the contrast between the natural topgraphy of low-angle slopes and the steep-sides, straight-edges, and sharp breaks of slope of the bund.

mounds. And thirdly, topsoil may be buried by the spoil rather than being stripped, stored and replaced over the mounds.

Two examples of all these issues may be cited from the South Norfolk study cited earlier (Gray, 1997a). Figure 10.20 shows the design of a garden pond and agricultural reservoir in this area. The natural shape of ponds in Norfolk is circular or oval as in the Breckland meres or Diss Mere. The garden pond however is highly intricate in shape. On the other hand, the agricultural reservoir is rectilinear. Neither has an authentic shape, and this is compounded by the nature of the associated bunding, the burial of topsoil and the absence of a topsoil cover on the bunds. Agricultural reservoirs allowing winter storage of water are being encouraged in the east of England where climate change may lead to wetter winters and drier summers, but their design needs to be more carefully considered.

10.5.5 Urban landform

As well as design of rural land form as described earlier, towns also have a geomorphology that can be eroded or buried under new developments. As Turner (1998, p. 37) states that 'Modern man (sic) has tended to conceal or destroy the landform of cities ... Landscape planners should seek out landform, just as Michelangelo looked at a block of marble and saw a statue concealed within'. He suggests (p. 38) that 'every city needs a landform plan'. Birmingham, England is an example of a city where the landform has been concealed by buildings, whereas Robert Adam's designs for Georgian Edinburgh, Scotland utilised the landform variation of the New Town site.

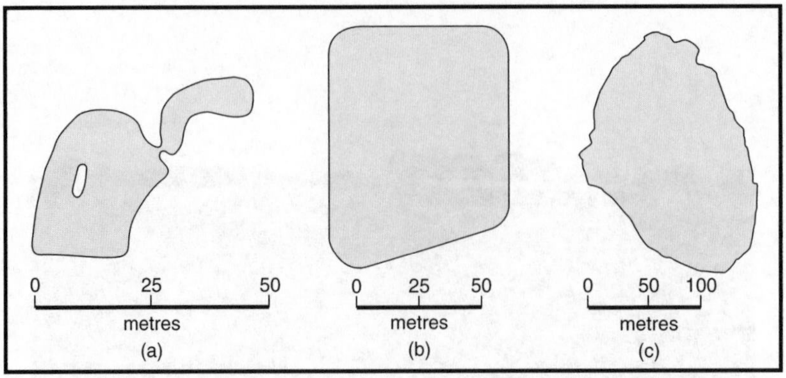

Figure 10.20 Examples of pond design in South Norfolk, England: (a) new garden pond at Fundenhall, (b) new agricultural reservoir at Rushall, and (c) the more natural shape of South Norfolk pond, Diss Mere.

10.6 Conclusions

In this chapter, it has been argued that as well as protected sites/areas we need a 'wider landscape' approach in geoconservation. This is because much of the natural physical landscape layer remains intact and makes a significant contribution to landscape character. International initiatives such as the European Landscape Convention are having an increasing role in this regard and in Section 10.3 I outlined a number of 'wider landscape' initiatives relevant to geoconservation. The chapter also considered georestoration and the human alteration of the form of the land, which has a long and not very distinguished history. Land form may be incongruous in its local area due to its size, height, gradient, shape or detailed variation. This should not be taken to imply that all geomorphological change is unacceptable. There are situations, such as construction of flood defence levées, where landform changes may be essential or desirable. It is also true that the natural landscape contains many incongruous features and that some man-made features eventually become an accepted and even valued part of the landscape, for example archaeological earthworks. But what is needed is an intelligent and aware approach to what we are changing and why we are changing it. Geomorphological character and authenticity ought to be important elements in landscape character assessment. Key characteristics of landform and process character need to be identified and included in character guidelines, geomorphological sensitivity to development should be recognised, and opportunities for restoration or enhancement should be taken.

11

Geoconservation and Land-use Planning

Local planning authorities should maintain the character of the undeveloped coast, protecting and enhancing its distinctive landscapes ...

Para 114, National Planning Policy Framework, England (2012)

11.1 Land-use planning systems

This chapter, and the following one, follow on from the Chapter 10 because some of the most important ways in which geoconservation can be delivered in the wider landscape are through land-use planning systems and policy initiatives.

Most developed countries have some type of land-use planning system established in recognition that unhindered development is likely to be detrimental to the environment. Laws and regulations have therefore been introduced to control the siting, design and impacts of development. In principle, most of these systems could be used, to a greater or lesser extent, to safeguard geodiversity, though their full potential at an international level has yet to be recognised. Protected areas should be notified to local planning authorities so that their status can be taken into account when planning applications within or adjacent to protected areas are submitted and decisions made. Similarly, planning policies can be used to give general recognition and protection to types of protected areas. An example of a notified area being saved from a housing development was described in the Netherlands by Gonggrijp (1993). An area of abandoned meanders and river terraces, close to the town of Roermond and included as a site in the Netherlands' *Nature Policy Plan*, was threatened by urban expansion. However, the listing in the plan saved the site from development at that time thanks to the intervention of a government minister.

Geodiversity: Valuing and Conserving Abiotic Nature, Second Edition. Murray Gray.

Despite this success, a number of issues remain. First, many countries do not have effective land-use planning systems or do not enforce the systems they have in place. This means that buildings are often constructed without consent with the result that they may destroy or damage important geoheritage sites. For example, Wimbledon (2012) has described the bulldozing of important sections on the south coast Crimea, Ukraine commenting that '... normal town and country planning practices and controls are not working here, and there is no geosite protection. Cliff-top development on a more or less virgin coast is normally impossible where there are proper conservation and planning systems in place' (Wimbledon, 2012, p. 5). In the absence of strong planning controls and enforcement, buildings may also be constructed in localities subject to natural hazards and there are many examples of loss of life and/or property in areas where unauthorised building has occurred.

Secondly, in many countries there is little national influence on planning decisions, which are mostly made locally. Concern is being expressed about the cumulative impact of local decision-making in the US Rockies and the 'tyranny of small decisions'. As a result there are calls for better planning policies, legal instruments and development incentives to manage future growth (e.g. Baron, Theobald and Fagre, 2000). In those countries with decentralised planning systems the challenge is to motivate local decision-makers to take geoscience issues fully into account since they make lack any knowledge of them.

But there are examples of sound planning systems. For example, in New Zealand, the *Resource Management Act* requires all local authorities to produce a district scheme (plan) updated every 10 years after public consultation. It is through these local plans that important geological sites can be protected. The Act states that 'the protection of outstanding natural features and landscapes from inappropriate subdivision, use and development' is a matter of national importance (Hayward, 2009). Outstanding natural features may include geological sites, and the Geological Society of New Zealand has had considerable success in having hundreds of geological and geomorphological features included in many district scheme schedules of outstanding natural features (Hayward, 2009). Planning consent is unlikely to be granted to any development that will adversely affect one of these features.

Many authorities have introduced planning zones that attempt to delimit areas of land for certain land uses, leaving other areas for conservation. An early example of this type of approach was Alberta's 'Eastern Slopes' (of the Canadian Rockies) study (Government of Alberta, 1977), which introduced three categories and eight zones. The Prime Protection Zone, for example, comprised the mountain summits and high plateaux above about 2000m thus protecting the mountain scenery and sensitive environments of this zone. Developments that would not be permitted in this zone included mineral exploration and development, petroleum and natural gas exploration and development, commercial timber operations, domestic grazing, cultivation, industrial development, residential development and off-highway vehicle use. Box 11.1 gives another example from Canada of how planning protection was enacted in the Niagara Escarpment area of Ontario in response to threats from aggregate quarrying and urban expansion.

Box 11.1 Planning in the Niagara Escarpment, Canada

The Niagara escarpment in Ontario, Canada is a spectacular large-scale landform stretching 750 km from Niagara Falls to Tobermory on Lake Huron. Its cliffs contain the most extensive Silurian stratigraphy in North America (Davidson *et al.*, 2001). It was recognised in the 1970s that the whole of the escarpment was at risk from aggregate production and uncontrolled housing development. As a result, the Niagara Escarpment Planning and Development Act (1973) was passed, delineating the whole escarpment as a special planning area and establishing the Niagara Escarpment Commission (NEC) to oversee planning and development control within the plan area. Subsequently, the Niagara Escarpment Plan (NEP) was produced in 2005 as Canada's first large-scale environmental land-use plan. It contains provisions to designate *Mineral Resource Extraction Areas* aimed at controlling the impact of mineral extraction operations on the escarpment environment and ensure appropriate restoration and after-uses that may include provision for quarry faces to be left unrestored for aesthetic or educational purposes.

Among the other work that has been undertaken is an inventory of the most significant features, including geological sites and landforms, which has been used to designate nature reserves and conservation areas (e.g. Hockley Valley, Mono Cliffs, Cabot Head, Bruce's Cave and Wodehouse Karst), parks and special planning zones along the length of the Escarpment. 'As a result, the overall character of the escarpment, its geological features, and the lands in its vicinity have been afforded lasting protection' (Davidson *et al.*, 2001, p. 230).

Several local authorities have their own regulations to protect physical features within their boundaries. For example the City of Portland, Oregon, USA has regulations to protect streams, springs and seeps by requiring them to be placed in separate tracts of land at least 15 feet (circa 3.5 m) from the nearest development.

In the UK, the planning system traditionally has been policy led and heavily centrally controlled, but there are differences between the different parts of the UK. In England, until 2012 national planning policy was contained in 25 Planning Policy Statements and Guidelines (PPSs and PPGs) and 10 similar guidelines on mineral planning (MPGs and MPSs). These included PPS9 on *Biodiversity and Geological Conservation* published in 2005, one of the aims of which was to give greater recognition to the need for geoconservation. It was supported by a Good Practice Guide giving detailed guidance on the implementation of national policy at the local level including on geodiversity and geoconservation. However, in 2012 the series was replaced by a single National Planning Policy Framework (NPPF; see Box 11.2). This removed much of the previous guidance on geoconservation, but, under the heading 'Minimise impacts on biodiversity and geodiversity', it still contains the statement that planning policies should 'aim to prevent harm to geological conservation interests'. While this and other statements are useful, there is no encouragement (as there is for biodiversity) to use development in a more creative way, for example retain, maintain or create geological exposures, or to integrate biodiversity and geodiversity in planning decision-making. Again, until recently, Government guidance has then been carried through into Regional Planning Statements (RSSs) and Local Development Frameworks (LDFs). However, the Localism Act (2011) cancelled the former and streamlined the latter.

Box 11.2 Examples of planning policies related to geodiversity in England and Wales

National Planning Policy Framework (2012), states that:

- 'The planning system should contribute and enhance the natural and local environment by protecting and enhancing valued landscapes, geological conservation interests and soils; [and] recognising the wider benefits of ecosystem services ' (para. 109);
- 'Local planning authorities should set criteria-based policies against which proposals for any development on or affecting protected wildlife or geodiversity sites or landscape areas will be judged' (para. 113);
- 'Local planning authorities should maintain the character of the undeveloped coast, protecting and enhancing its distinctive landscapes, particularly in areas defined as Heritage Coast, and improve public access to and enjoyment of the coast' (para. 114);
- 'Proposed development on land within or outside a Site of Special Scientific Interest (SSSI) likely to have an adverse effect on a SSSI interest (either individually or in combination with other developments) should not normally be permitted' (para. 118);

Pembrokeshire Coast National Park Local Plan (2011) Policy 47: Local Sites of Nature Conservation or Geological Interest: 'Development that would be liable to significantly harm ... the main features of interest within a Regionally Important Geodiversity Site, will only be permitted if the importance of the development outweighs the local value of the site and mitigation, minimisation or off-setting has been investigated'. The National Park also has a Supplementary Planning Document (SPD) on *Regionally Important Geodiversity Sites.*

The Lake District National Park Core Strategy (2010) states that: 'We want to lead by example, with policies that conserve and enhance biodiversity and geodiversity both within and outside designated areas. We recognise that these assets are essential landscape components, contributing to local distinctiveness' (Para 4.55.1). Policy CS26 states that 'We will protect the important geodiversity of the Lake District National Park ... '

Suffolk Coastal District Council Draft Development Management Policy DM27 on Biodiversity and Geodiversity (November, 2010) states that 'Development will not be permitted where there is an unacceptable impact on biodiversity and geodiversity ... '

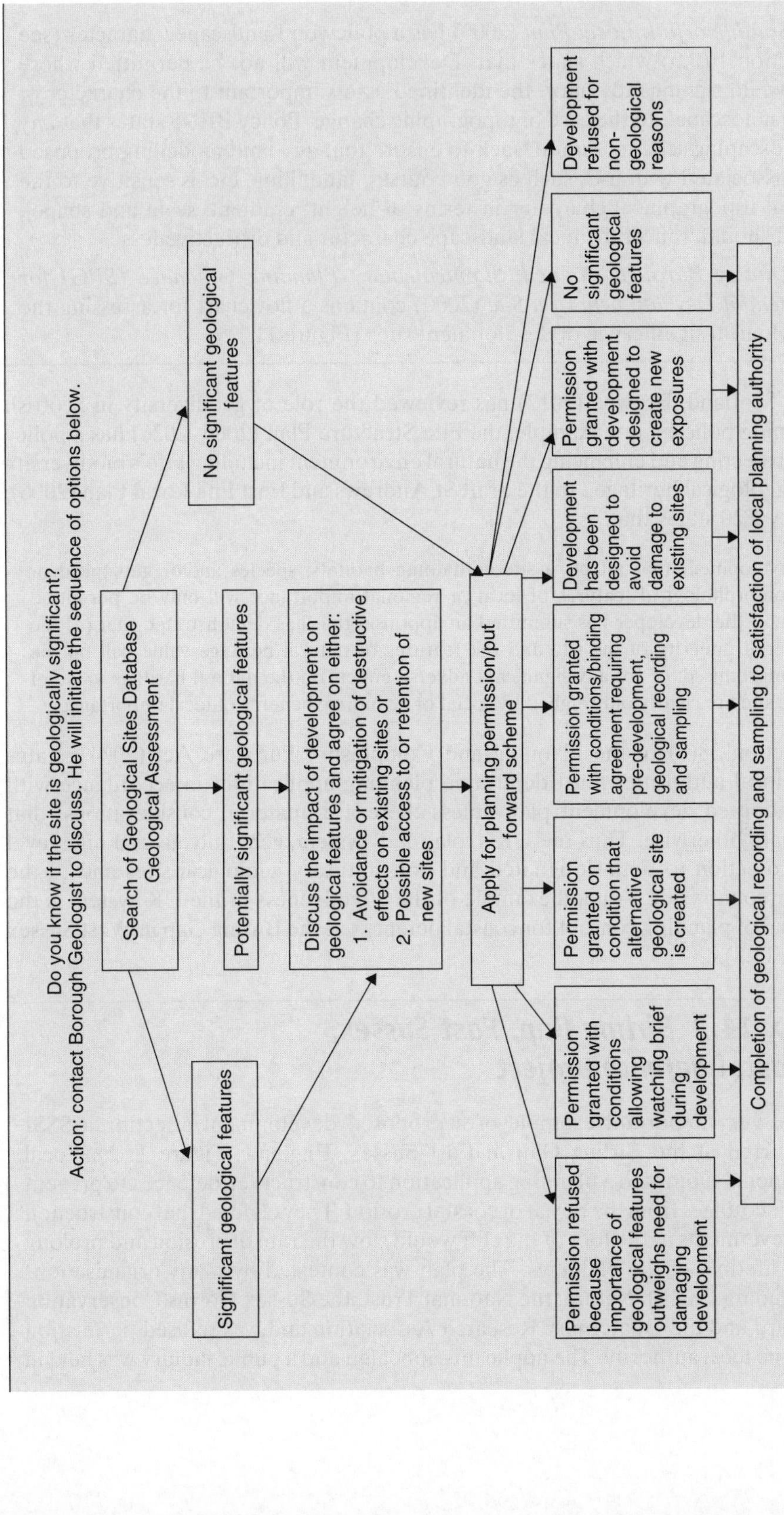

Figure 11.1 Key to decision-making regarding geological impacts, Dudley Borough Council, England, by permission of Dudley Borough Council.

South Norfolk Local Plan (2003) has a policy on Landscape Character (see Section 10.3.3) which states that 'Development will not be permitted where it would significantly harm the identified assets important to the character of the landscape'. In the case of topographic change, Policy BEN5 states that any landscaping scheme should 'seek to ensure that any land modelling proposed as associated with uses such as golf courses, landfilling, etc. is sensitive to the local topographical character in terms of height, gradient, scale and shape', and should 'reflect the local landscape character and distinctiveness'.

Dudley Borough Council, Supplementary Planning Guidance (SPG) for Potential Development of a Site (2000) contains a flowchart for assessing the geological significance of development sites (Figure 11.1).

In Scotland, Browne (2012) has reviewed the role of geodiversity in Scottish planning policies. For example, the Fife Structure Plan (2006–2026) has a policy of 'Protecting and enhancing the natural environment including Fife's biodiversity and geological heritage'. In the draft St Andrews and East Fife Local Plan (2006), Policy E20 states that:

> Development that affects a site containing habitats, species and/or geological or geomorphological features of local or regional importance will only be permitted where the developer has submitted an appraisal that has demonstrated that (a) the overall integrity of the site and the features of natural heritage value will not be compromised; or (b) any significant adverse effects on the natural heritage value of the site are clearly outweighed by social or economic benefits of local importance.

Section 38(6) of the Planning and Compulsory Purchase Act (2004), states that local authorities must determine planning applications in accordance with the adopted development plan unless there are 'material considerations' that indicate otherwise. Thus the UK's 'plan-led' system, generally gives a high level of protection to both designated and non-statutory geological sites and to the wider countryside. A good example of the effectiveness of the UK system is the refusal of planning consent for coastal defences at the Birling Gap in West Sussex (Box 11.3).

Box 11.3 Birling Gap, East Sussex, coastal defence project

One very important example of a proposed development affecting a SSSI occurred at the Birling Gap in East Sussex, England (Figure 11.2). Local residents submitted a planning application to construct sea defences to prevent their cottages from the threat of coastal erosion. They claimed that construction of revetments at the foot of the cliff would slow the rate of erosion and prolong the lifetime of the buildings. The plan was contested by many organisations including English Nature, the National Trust, the Sussex Downs Conservation Board and the Quaternary Research Association and was refused permission by the local authority. The applicants appealed and a public inquiry was held in

2000. Supporting the local authority's case for refusal, English Nature argued that the continued erosion of the cliffs was important to maintain the local beach material supply and thus protect the rest of the cliffline, retain the local cliff exposures of the truncated chalkland dry valley with periglaciated chalk surface and fill of Late Quaternary sediments, and conserve the local spectacular coastal scenery. The objectors argued that the revetments would be intrusive and would themselves eventually be undermined by erosion on their flanks.

Figure 11.2 The cottages under threat and the Chalk cliffs at Birling Gap, East Sussex, England. See plate section for colour version.

The Inquiry Inspector agreed with the objectors, believing that the revetments would harm the beauty of the area and have an adverse impact upon the local nature conservation interests. The Inspector confirmed the importance of GCR and SSSI status and made the point that it is not just scientists who put a value on designated sites since they operate within existing legislation authorised by society at large (Prosser, 2001a).

In England the government funded a 'Pathfinder Progamme' to the tune of £11 million to 15 coastal sites to provide, for example, partial compensation for loss of cliff-top houses. Compensation has been offered to residents amounting to 40–50% of their house value had they not been in an at-risk location (*Eastern Daily Press*, 3 February 2011). Planning policies are also being relaxed to allow individuals affected to rebuild inland on sites that would otherwise not be supported. Such approaches bring benefits of naturally evolving rather than protected coastlines.

A less successful example of the planning system protecting a SSSI is given in Box 11.4.

Box 11.4 Menie Estate, Aberdeenshire, golf course and resort complex

In 2006, Trump International Golf Links Scotland applied for outline planning consent for two 18-hole golf courses together with related development at Menie House in an area of coastal sand dunes, about 14 km north of Aberdeen city centre. One of the courses was to be a championship links course that partly intruded into the Foveran Links SSSI. The application was in outline only but indicative layouts and designs were included in the application that also comprised:

- a golf clubhouse, golf academy, driving range, practice area and ancillary buildings;
- a resort hotel of 450 rooms on 8 floors, conference centre and spa;
- 950 holiday apartments in four blocks;
- 36 golf villas;
- 500 houses for sale;
- accommodation for 400 staff;
- a new access, gatehouse, roads and parking areas.

After consideration at two meetings of Aberdeenshire Council in 2007, the application was ‘called-in’ by the Scottish Government and a four-week public inquiry was held in 2008. Objectors included Scottish Natural Heritage, which was concerned that the active sand sheets and dune system in general would be affected by excavations and stabilisation measures. Most of the back nine holes of the championship course would take up the southern third of the area of the SSSI.

Donald Trump himself gave evidence at the public inquiry and stated that if the championship course was moved away from the southern part of the SSSI and built elsewhere, as objectors had suggested, it would no longer be the truly great course that he planned since it would not include the spectacular high dune part of the dune system. If he was refused permission to develop the course in this area he would withdraw from this project as it would no longer fulfil his vision of building an outstanding course.

At the end of their over 200 page report, the inquiry inspectors concluded that ‘much, though not all, of the geomorphological interest in that affected area of the SSSI would be compromised, as would its overall integrity … The loss of this dynamism cannot be mitigated against’. However, their overall conclusion was that these adverse effects were outweighed by the social and economic benefits that were of national significance’. Scottish Government Ministers agreed with this conclusion and the project was given outline consent. Jonathan Hughes, Scottish Wildlife Trust’s Head of Policy argued that the decision ‘sends dangerous messages that SSSIs could be up for grabs if you write a big enough cheque’.

Other ways in which the UK planning system can assist in geoconservation include the use of restoration conditions in the granting of consents for mineral extraction and landfill sites, including contours on the final landform and proposals for the handling and storage of soil. However, Bridgland (1994) makes the point that such restoration conditions make it difficult to preserve geologically interesting quarry faces because the restoration programme is agreed before quarrying even begins. If important faces become exposed during quarrying 'it is thus necessary to persuade the various interested parties to modify existing plans—a procedure that might prove expensive.' (Bridgland, 1994, p. 88). He therefore calls for geologists to have a greater influence at the planning stage to maximise the research and teaching potential of exposures created by future quarrying. Specifically, he points out that planning authorities in the UK now routinely require developer-funding of archaeological rescue digs prior to mineral excavation (through the use of legal agreements), and makes the case for the same type of arrangement to apply to geology. 'It seems anachronistic that our geological heritage doesn't enjoy the same treatment. Extraction frequently destroys potentially important sediments just as it destroys archaeological remains, yet there is no systematic mechanism for ensuring that the geological evidence is recorded' (Bridgland, 1994, p. 90).

11.2 Environmental Impact Assessment

This is an approach, often linked to land-use planning systems, which has the potential to assess the impact of major development proposals on geodiversity, and also the impact of geomorphological processes on the development. It is therefore an ad hoc rather than a strategic approach to the identification of geodiversity threats and the need for prevention or mitigation, but in areas where detailed information is limited, it may be a useful tool in geodiversity conservation that deserves to be better known and used. For full details see, for example, Glasson, Therivel and Chadwick (2005) and Carroll and Turpin (2009).

Environmental Impact Assessment (EIA) originated in the United States under the National Environmental Policy Act (1969) and has gradually been adopted by many countries around the world, albeit is different forms. EIA is a systematic and integrative process for assessing the possible impacts of major developments prior to a decision being taken as to whether the proposal should be given permission to proceed. If carried out effectively, it should prevent environmentally unacceptable projects from being implemented and mitigate the environmental effects of proposals that are approved. The information assembled during the assessment is published as an Environmental Impact Statement (EIS) that accompanies a planning application and allows consultees and the public to understand the impacts of what is being proposed and the measures taken to reduce these impacts. In essence, EIA is intended to ensure that decisions on all major developments can only be taken in the foreknowledge of their likely environmental consequences.

In the USA, it was quickly assimilated into state and local statutes and since then a host of other industrialised countries have adopted similar procedures, including Canada (1973), Australia and New Zealand (1974), France and Germany (1976), the Netherlands (1981) and Japan (1984). The European Community formally established a Directive on Environmental Impact Assessment in July 1985, and the procedure was adopted by the UK in 1988.

Most development projects can be broken down into six major phases (Rivas *et al.*, 1995a, 1995b):

- *preliminary reconnaissance*—site selection involving preliminary EIA;
- *site investigation*—verification of the technical suitability of the selected site, with detailed EIA studies;
- *detailed planning and design*—architectural and engineering design of the project, with feedback to EIA and development of mitigation measures;
- *construction*—during project construction monitoring of impacts occurs with evaluation and, if necessary, redesign of mitigation measures;
- *operation*—during the operation of the project, monitoring of impacts, evaluation of mitigation measures and redesign as necessary is continued;
- *decommissioning*—after the lifetime of the project, there may be provision or scope for site restoration or further mitigation.

Hodson, Stapleton and Emberton (2001, p. 170) commented that 'Avoiding significant development impacts on soil ultimately protects the whole of the ecosystem from degradation ... (but) relatively few types of development have significant impacts on geology'. While one can thoroughly agree with the importance of soil protection, it is difficult to support the comment on geology (or geomorphology), given the threats outlined in Chapter 5. Development can have significant effects on many aspects of the geological and geomorphological environment, and in turn, the geological environment can impact on developments, particularly through the operation of hazardous processes that may seriously affect the development. Thus before development proceeds, the geological environment needs to be assessed in the broadest sense, including both direct and indirect impacts (Rivas *et al.*, 1995a). It is important that this work takes place in the early phases of the project (site selection, site investigation and design) since retrospective work is usually expensive and rarely provides the best solutions (Erikstad and Stabbetorp, 2001). Earth scientists can also be involved in the later phases (construction, operation and decommissioning) to monitor impacts, assess the effectiveness of mitigation and redesign elements of the scheme as appropriate.

The main steps in the EIA process are:

- *Screening* is the process of deciding whether an EIA is required. Only major projects currently require environmental assessments, and these are specified in national legislation/ regulations. These list the type of projects where an environmental assessment is mandatory, for example oil refineries, power stations and special waste incinerators. In the United Kingdom, these are detailed in Schedule 1, while Schedule 2 lists those where one may be necessary depending on the environmental impacts as perceived by the local authority, for example holiday village near to SSSI or large poultry farm.

- *Scoping* is the process of deciding what are the main environmental issues that need to be investigated, and what issues do not need to be examined in as much detail.
- *Baseline Studies* This is the process whereby the applicant/consultant collects information on the elements of the existing environment relevant to the impacts identified during scoping. The aim is to establish current environmental conditions and consider the character, extent, importance and vulnerability of the various components of the environment. In the geosciences this may involve desk and field studies of geology, topography, processes and soils.
- *Policies and Plans* are also examined at this stage to determine the international, national, regional and local policies relevant to the project. This search will also allow identification of protected sites, designated areas, etc.
- *Impact Predictions and Assessments* This is where detailed work is carried out to identify and assess the impacts of the project on the existing environment. For impacts of, and on, the geological/geomorphological environment, appropriate specialists should be employed. The stage requires the characteristics of the project to be known (e.g. use of natural resources, soil stripping and storage, land remodelling, waste generation and management, etc.) and the impact on receptors to be assessed. Impacts may be direct or indirect; short, medium or long term; reversible or irreversible; permanent or temporary; beneficial or adverse; singular or cumulative. The magnitude or physical extent of the impacts should be quantified where possible.
- *Mitigation* is where measures are introduced to avoid, reduce or recompense some or all of the identified adverse impacts. The most satisfactory method is avoidance, for example by redesign. This emphasises the point that EIA should be an iterative process of refinement of the proposals to reduce the environmental impacts. Reduction may be achieved also by redesign but also by screen planting. Recompense might involve acceptance of the impacts but attempting to compensate by other measures, e.g. providing habitats elsewhere.
- *Alternatives* Proposals involving EIAs may be required to explain what alternatives (e.g. sites, designs, etc.) have been considered and reasons for rejection.

Among the earth science aspects that will need to be assessed are:

- natural hazards that may impact on a project—could be recorded on a hazards/process map;
- impact of the project on soils and geological materials—as Rivas *et al.* (1995a) state: 'Any human activity which consumes, sterilises or degrades these resources, would represent a negative impact' and should be reported as such, even though sustainable use of these resources may be regarded as acceptable. Sterilisation refers to development that permanently or temporarily prevents the usage of a resource. For example, building houses over a mineral-rich vein would sterilise it in that mining the vein is prevented, at least in the short to medium term. Degradation normally refers to pollution of the resource. Geological and soil resources can be recorded on geology/soils maps;

- impact of project on other georesources, including designated sites and areas, topography, landscape aesthetics, processes, hydrology, etc.

These can be brought together in a Geological Impact Assessment (GIA) that is subsequently integrated into the EIA (Rivas *et al.*, 1995a). Assessments can be carried out in various ways. Rivas *et al.* (1995b) give a case study of a motorway in northern Spain, Bergonzoni *et al.* (1995) describe its use in regional park planning in central Italy, Conacher (2002) questions the appropriateness of EIA to problems of land salinisation and degradation in western Australia and Holt-Wilson (2012) describes the use of GIA for a boatyard development in England. Box 11.5 is an example of a planned military area in the Netherlands and Box 11.6 explains the development of an Oil and Gas Management Plan/EIS for National Park Units in Texas, USA. Impacts have traditionally been assessed using tools such as the Leopold Matrix (Leopold *et al.* 1971) prepared by the USGS, but nowadays more sophisticated GIS systems are used. Erikstad *et al.* (2008) discuss the issue of how value assessments are made in EIA. They propose a new scale from international to local to ease communication between different disciplines and management systems.

A further development is the use of Strategic Environmental Assessment (SEA), which is the environmental impact assessment of public policies, plans and programmes (PPPs). According to Therivel and Thompson (1996), this can be an effective way of helping to meet the twin goals of delivering sustainable development and monitoring nature conservation. 'Issues such as significant impacts on our nature conservation resource and inappropriate operations which are likely to exceed the environmental carrying capacity would be assessed at a stage where change can be accommodated without unnecessary cost' (Therivel and Thompson, 1996, p. 5). The European Union adopted the Strategic Environmental Assessment Directive (2001/42/EC) in June 2001 and Member States were required to incorporate it in their national legislation by July 2004 (Holstein, 2002).

Box 11.5 EIA of military area at Ede, the Netherlands

The Dutch Ministry of Defence proposed to use part of the Ginkelse Heide area, near Ede in the central Netherlands as a military training area (Asch and van Dijck, 1995). The area comprises four sub-areas shown on a schematic geomorphological map in Figure 11.3.

- Sijsselt—a forested hill area comprising an ice-push moraine, with podsol soils;
- Ginkelse Heide—an area extending from the toe of the moraine in the west to a glaciofluvial fan in the east where the sand has been blown into low aeolian covers and ridges;
- Ginkelse Zand—an area of aeolian sand dunes;
- Planken Wambuis—a wooded continuation of the glaciofluvial fan.

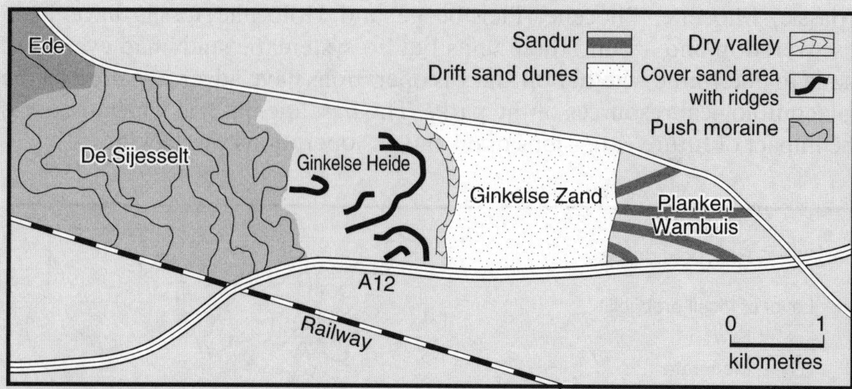

Figure 11.3 Schematic geomorphological map of the proposed military area at Ede, the Netherlands (after Asch, T.W.J. and van Dijck (1995) The role of geomorphology in environmental impact assessment in the Netherlands. In Marchetti, M., Panizza, M. Soldati, M. and Barani, D. (eds) 1995 *Geomorphology and Environmental Impact Assessment*. CNR-CS Geodinamica Alpina e Quaternaria, 63–70, by permission of Professor M. Panizza).

The plan required the following areas:

- an open area of at least 225 ha for movement of vehicles and men;
- a forested area for woodland manoeuvres and bivouacing, with tracks;
- forest edges for digging of shallow defensive holes, again with tracks;
- a 10 m wide perimeter track connecting all training areas;
- a parking place for loading, unloading and cleaning of vehicles.

The wooded areas of Sijsselt and Planken Wambius were excluded as open areas because tree felling was prohibited. These areas were assessed as suitable for woodland manoeuvres, etc. Geomorphologists quickly realised that the other two areas were extremely sensitive to the impacts of soil and sediment disturbance if intensively used.

Among the direct and indirect impacts predicted from intensive use of the two areas were destruction of vegetation cover, destruction of soil profiles, wind erosion of soil and sand and dust hazard; destruction of stabilised dunes and ridges, soil compaction and disturbance of soil water balance, water erosion.

Box 11.6 Oil and Gas Management Plan/EIS for the Lake Meredith and Alibates Flint Quarries, Texas, USA

Lake Meredith National Recreation Area and Alibates Flint Quarries National Monument are two units of the US National Parks System located between two major structural basins in the Texas Panhandle (Figure 11.4). Important

Triassic, Miocene, Pliocene, Pleistocene and Holocene fossils have been discovered in and around these units but no systematic study had ever been made to determine whether oil and gas operations have adversely affected the palaeontological resources of the parks. The baseline data on which to assess the impact of future extensions of oil and gas operations was lacking.

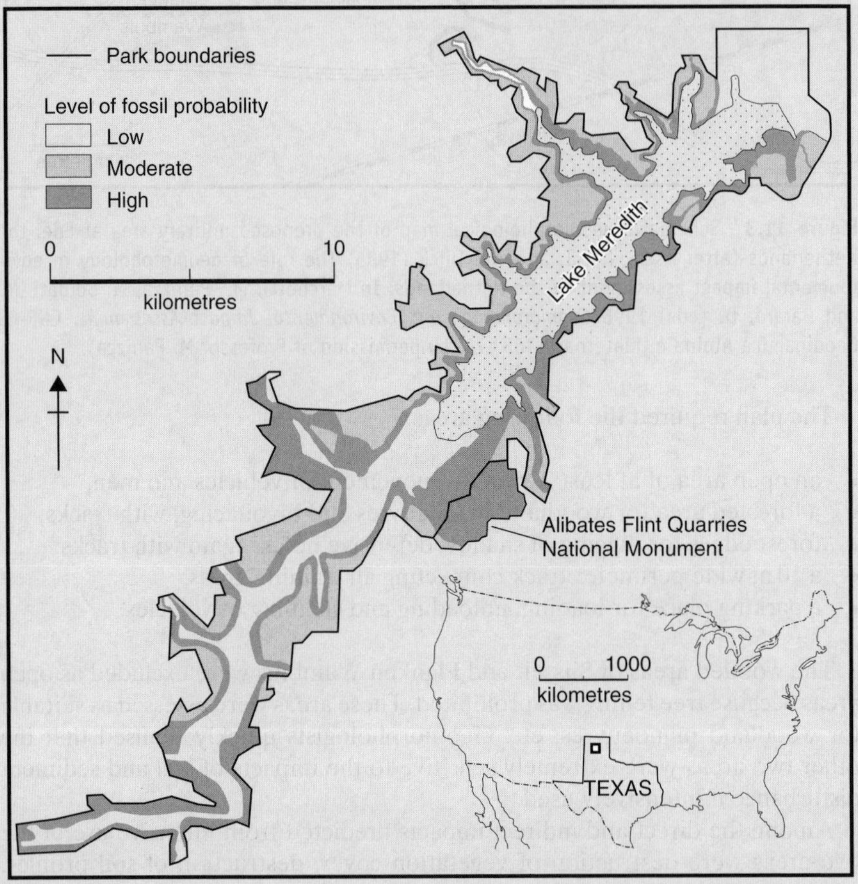

Figure 11.4 Palaeontology Resource Sensitivity Map of the Lake Meredith and Alibates Areas, Texas, United States (after Santucci, V.L., Hunt, A.P. and Norby, L. (2001a) Oil and gas management planning and the protection of palaeontological resources. *Park Science*, 21, 36–38. Reproduced with permission of NPS).

Having identified the need for better baseline information, NPS palaeontologists undertook a comprehensive inventory of the palaeontological resources of the protected areas through literature reviews, museum searches and field surveys. Significant resources identified included Upper Triassic amphibians, reptiles and petrified wood, Miocene—Pliocene root casts, silicified grasses, insect burrows, mammal bone beds and a mastodon tooth, and Pleistocene mammals including a complete skull of the giant bison *Bison latifrons*. A palaeontology resource sensitivity map has been produced (Figure 11.4) and the important sites have been recorded in a GIS database and are to be monitored periodically.

The final stage has been the development of an oil and gas management plan/ Environmental Impact Statement identifying circumstances in which a palaeontological survey will be necessary and the procedures for carrying out such a survey. For example, in high-probability fossil areas where ground disturbance is planned, a full palaeontological survey and proposals to minimise fossil disturbance would be required. Guidance is also given on procedures when unanticipated fossil discoveries occur during approved operations or fossils are damaged within previously identified sites (Santucci, Hunt and Norby, 2001).

11.3 Conclusions

Land-use planning, including Environmental Impact Assessment procedures, can be a very important tool in conserving geodiversity because they provide a means by which impacts on the environment can be assessed prior to development starting. This can mean that development is refused or modified, or that conditions can be imposed to mitigate the impacts. Unfortunately, few countries have totally effective planning systems or fail to enforce them, and no planning system known to me is capable of giving full regard to impacts on geodiversity.

12 Geoconservation and Policy

Principles and guidance to facilitate the management of geodiversity in a changing climate are now required

Colin Prosser *et al.* (2010)

The role of policy has already been introduced in places in previous chapters, for example in relation to National Park or Geopark Management Plans, but this chapter will deal with policy not just in protected or specific areas but for the wider landscape.

12.1 Sustainable mining and mineral policies

Because of the poor reputation of mining and quarrying in environmental management and other areas, the Global Mining Initiative (GMI) was launched in 1998 by nine international mining companies in order to redefine the role of the global mining industry in relation to sustainable development. The initiative has involved three elements. First, a Mining Minerals and Sustainable Development Project (MMSD) was carried out to analyse the challenges facing the industry and ways in which the issues might be resolved. The final report (MMSD, 2002) recognises that 'simply meeting market demand for mineral commodities falls far short of meeting society's expectations of industry'. Instead it proposes an 'agenda for change' and a 'vision for the mineral sector' to create a picture of what it would look like if it were to maximise its contribution to sustainable development.

Secondly, the International Council on Mining and Metals (ICMM) was founded in 2001 (replacing its predecessor, the International Council on Metals and the Environment), with the aim of being 'the clear and authoritative global voice of the world's mining and metals industries, developing and articulating their sustainable development case, discovering and promoting best practice on

Geodiversity: Valuing and Conserving Abiotic Nature, Second Edition. Murray Gray.

sustainable development issues within the industries and acting as the principal point of engagement with the industries for stakeholders at the global level'. Among the issues being confronted in the ICMM Charter (www.icmm.com) are mineral economics, environmental stewardship, social and cultural responsibility, human health and safety, product stewardship, stakeholder engagement and innovation in technology (Box 12.1).

Thirdly, a series of international conferences have been held, starting in Toronto, Canada in 2002, to debate the ideas emerging from the MMSD Project.

Box 12.1 ICMM's 'Fundamental Principles of Sustainable Development'

In order to meet society's requirements for minerals and metals, while contributing to sustainable development and enhancing shareholder value, ICMM members will:

- Implement and maintain ethical business practices and sound systems of corporate governance;
- Integrate sustainable development considerations within the corporate decision-making process;
- Uphold fundamental human rights and respect cultures, customs and values in dealings with employees and others who are affected by our activities;
- Implement risk management strategies based on valid data and sound science;
- Seek continual improvement of our health and safety performance;
- Seek continual improvement of our environmental performance;
- Contribute to conservation of biodiversity and integrated approaches to land use planning;
- Facilitate and encourage responsible product design, use, re-use, recycling and disposal of our products;
- Contribute to the social, economic and institutional development of the communities in which we operate;
- Implement effective and transparent engagement, communication and independently verified reporting arrangements with our stakeholders.

One of the reasons for the interest in sustainable mining is the looming shortage of some minerals, in some cases related to potential political instability. Many earth materials 'are so much a part of everyday living that … they are generally taken for granted and little heed is taken of the impact of their use on the longer-term reserves' (Kelk, 1992, p. 34). But as Prentice (1990, p. x) remarks, as raw materials become scarcer, 'The proper use of our mineral resources can only be achieved if we use all our geological skills to ensure that they are used to the best advantage'. One of the aims of the International Year of Planet Earth (IYPE) in 2009 was to promote the sustainable use of geomaterials.

Bulk minerals are unlikely to be exhausted for a very long time, but there is the possibility that local shortages will occur because of transport costs involved in importation. In south-east England, traditional river terrace and valley floor aggregate resources are largely exhausted and instead most aggregate is obtained by offshore dredging in the southern North Sea or English Channel or as crushed rock from more distant hard rock quarries in Scotland and Scandinavia, thus increasing transport costs.

Fossil fuel shortages, particularly oil, are also looming. Hubbert (1956) predicted that American oil production would peak in 1969. Although castigated at the time, American oil production in fact peaked in 1970 and has declined since then. Using similar methods several authors predicted when global 'peak oil' production would occur (e.g. Deffeyes, 2001: Rifkin, 2002; Roberts, 2004; Sorrell *et al.*, 2009; Gorelick, 2010). Some think that 'peak oil' has already passed (2013). Most easily obtained reserves are in the Middle East, and political instability could easily produce oil shortages and higher prices within a few years. WikiLeaks releases indicate that Saudi Arabia, previously thought to have the world's largest oil reserves, may actually have over-estimated these by 300bn barrels or nearly 40% (*The Guardian*, 8 February 2011, p. 23). Apart from biofuels, most alternative energy sources are abiotic in nature (see Section 4.5.3) and their diversity certainly provides society with considerable flexibility. Kerr (1998, p. 1128) believed that history is likely to judge the Age of Oil as 'a two-century binge of profligate energy use'. At a meeting on Peak Oil at the Geological Society of London in 2008, the petroleum geologist Colin Campbell made the point that 'the Stone Age didn't end for the lack of stones' and hoped that new energy sources might emerge. But instead, new unconventioanal oil sources are being exploited and new oil technologies are being developed, including 'fracking', the fracturing of shale to release the 'tight oil' that does not easily flow out of the rock. The technique is already exploiting over half a million barrels of oil per day from the Bakken Shale in North Dakota alone and there are over 20 other suitable formations known to exist in the USA, including the Eagle Ford Shale in Texas (Maugeri, 2012).

Fiscal Instruments are economic measures designed to manage resource use. It has been noted that when the price of a mineral rises so does the economic attraction of recycling, and fiscal instruments can be introduced to stimulate this process. They are designed to discourage the extraction of virgin materials and encourage the use of recycled or secondary materials and new recycling processes. Examples from the UK include the Landfill Tax, introduced in 1996 to discourage landfilling of waste, including inert waste, and the Aggregates Levy introduced in 2002 at £1.60/tonne, which is designed to reduce the use of virgin aggregate. Part of the income from the levy, amounting to several million pounds each year, was used to establish an Aggregates Levy Sustainability Fund (ALSF) in England that funded research and conservation related to aggregate extraction. Unfortunately this source of funding has been terminated.

Another strategy is to introduce 'producer responsibility' codes or Pigouvian taxes, named after the 1920s British economist Arthur Pigou. These make manufacturers responsible for some or all of the costs of recycling or waste disposal. For example, the EU introduced a new Directive in 2002 dealing with Waste Electronic and Electrical Equipment (WEEE). Under this, producers are required to take responsibility for their products at the end of their useful life. Similarly 12 million cars end up on European refuse dumps each year (Jones and Hollier, 1997) and the EU's End-of-Life Vehicle Directive now requires car manufacturers to take back their scrap cars. As the costs of doing so have to be passed on to the customer, there is an incentive for manufacturers to use recyclable parts and materials, reduce packaging, etc. (Barrow, 1999). BMW and Mercedes-Benz cars have led the way in using recycled and recyclable materials in car manufacture. In turn, recycling of cars is being assisted by new technology such as high-powered fragmentizers that can quickly convert a pre-compacted scrap car into accurately sorted bundles of ferrous, non-ferrous and other materials.

At the local level, Prosser (2001b) argues that a number of pressures have impacted on the minerals industry and made it more receptive to geoconservation issues. This is leading to increased dialogue, agreements and partnerships with the minerals industry, including *Statements of Intent* and *Memoranda of Understanding* between the minerals industry and nature conservation agencies (Prosser, 2001b). English Nature (2003) published guidance on how the minerals industry can contribute to geodiversity conservation. This includes:

- Careful planning of new quarries by identifying as early as possible any features of geoheritage value on the site and taking appropriate steps (recording, protecting, etc.).
- Operating quarries so that new discoveries are brought to geologists' attention, allowing research and recovery to take place, and facilitating long-term management.
- Restoring quarries to retain features of interest and provide safe access to these (see earlier).
- Taking a corporate perspective by developing Company Geodiversity Action Plans (see later) and company targets, and reporting on successes.

Scottish Natural Heritage has undertaken a GIS-based approach to evaluating the sensitivity of different components of the natural environment to mineral extraction in the Midland Valley of Scotland where development pressures are greatest (Scottish Natural Heritage, 2000b).

Another policy being promoted is the reintroduction of stone rather than concrete in building construction. A Scottish Stone Liaison Group (SSLG) was formed in 2000 followed by a Welsh Stone Forum (Fforwm Cerrig Cymru) in 2003 and an English Stone Forum in 2006. The aim of these organisations is to promote understanding, and encourage the use of natural stone as a sustainable building material. Unfortunately, due to funding problems, the SSLG was dissolved as an independent operation in 2010 but continues some work under the auspices of Historic Scotland.

In rural areas the importance of using appropriate designs and materials for buildings, walls, paths and car parks is being increasingly recognised, through the production of design guides, etc. The Countryside Commission published technical guidance on *Design in the countryside* in England (Countryside Commission 1993c, p. 12) that argued forcefully 'for the retention of regional diversity, local distinctiveness and harmony between buildings, their settlements and the landscape' and urged all planning authorities 'to endorse the principle that new development in the countryside must reflect and respect the diversity and distinctiveness of the local landscape character'. Part of this involves the use of local geological materials and many design guides now encourage the use or reuse of local building stone. Scottish Natural Heritage has produced a design guide for *Car Parks in the Countryside* that promotes the use of local materials unmetalled parking areas, access tracks and walls. 'Where such materials are available locally they can ensure that the hard surfaces are sympathetic to the local soils and geology' (Scottish Natural Heritage, 2000c, p. 53). The guidance argues that it is essential 'that the local construction method/style is employed. This demands that traditional walling skills are utilised to ensure the new walls complement rather than detract from earlier workmanship'.

12.2 Agricultural Policy

Since large areas of land are in agricultural use, agricultural policy and practice can have a huge impact on the physical landscape (see Section 5.8). Over the past 30 years or so, agricultural policy in western Europe, through the Common Agricultural Policy (CAP), has subsidised intensive farming systems, overproduction of food (hence butter mountains and wine lakes) and small farms too unprofitable to survive in a free market. All this has had a detrimental effect on the environment. Reform of the CAP has proved difficult, not least because of vested farming interests and the huge social and economic impact that subsidy withdrawal causes. Even with subsidies, many farmers have struggled to operate economically and have left agriculture so that farm amalgamation has occurred. At the same time, subsidies for food production are being reduced while payments for sustainable land management are being increased. These agri-environment schemes encourage more traditional land management practices and environmental protection and although mainly aimed at enhancing biodiversity, they also have the potential to support geodiversity conservation and management through, for example, reducing diffuse pollution, retaining landscape distinctiveness and protecting soils (DEFRA, 2002b). An example is the Tir Gofal scheme in Wales (Box 12.2).

Box 12.2 The Tir Gofal agri-environment scheme in Wales

Tir Gofal is an agri-environment scheme managed by the Countryside Council for Wales (CCW) in partnership with many other agricultural, forestry and land-management organisations and is co-funded by the EU. It is aimed at 'helping to maintain the fabric of the countryside' (Countryside Council for Wales, 1999, p. 3). It is a whole farm scheme, offering a 10-year agreement with a 5-year break clause. It comprises four elements:

1. *Land management*—mandatory compliance involving management of key habitats. Among the geoconservation measures included are:
   - retain all existing traditional field boundaries (hedges, walls, banks, slate fences (see Figure 12.1);
   - safeguard rock features and geological sites;

- protect ponds, streams and rivers with a 1 m buffer strip (increased to 10 m for operations involving farmyard manure or slurry)'.

2. *Creating new permissive access*—voluntary options including providing access for educational purposes;
3. *Capital works*—payment for capital schemes which protect and manage habitats or support access provision;
4. *Training for farmers*—including habitat management and skills such as drystone walling.

Figure 12.1 Slate fencing, Nan Ffrancon Valley, Wales.

There is significant scope here for farmers to involve themselves with geoconservation as well as biodiversity. However, Tir Gofal is being phased out and replaced by a new scheme—Glastir—in 2014.

In England and Wales, the Environmental Stewardship Scheme managed by DEFRA brings similar benefits, particularly for biodiversity but also for geoconservation, including incentives for preventing soil erosion, maintaining drystone walls and providing access and educational opportunities. The Environment Agency in England and Wales is encouraging farmers to produce voluntary Whole Farm Plans that include agricultural chemical usage and measures to reduce diffuse pollution, soil protection and management of river bank erosion. Pilot projects have been successful and the work has been expanded. In Australia, Carroll *et al.* (2001) describe a 'Neighbourhood Catchment' approach to land and stream management whereby discussions take place between government scientists and all landowners within a catchment in order to encourage sustainable land-use practices for the good of all.

12.3 Soil Policy

Soil is one of the most vulnerable Earth resources. Moreover, impacts can occur with rapidity in response to land-use changes or new technologies. The implications have now been recognised by many decision-makers and soil management and conservation have therefore become a very active field for policy development.

Brady and Weil (2002) comment on the need for a global perspective on soils. 'Changes in soil productivity in one area affect food security and food prices, as well as biodiversity and water quality, in both nearby and distant places. This growing global perspective is paralleled by the growing acceptance of the ecosystem concept as the prime basis for decisions on natural-resource management' (Brady and Weil, 2002, p. 871). After noting that in the United States, Europe and East Asia soil quality has declined as a result of intensive agricultural practices and from land application of waste materials, they promote the concept of soil health. They define this as 'the capacity of a soil to function within (and sometimes outside) its ecosystem boundaries to sustain biological productivity and diversity, maintain environmental quality, and promote plant and animal health' (Brady and Weil, 2002, p. 873). This will only be achieved by better soil management strategies.

A European Soil Charter was developed in the 1990s (Harcourt, 1990) that recognises soil as a finite and valuable resource that requires protection (Box 12.3).

Box 12.3 European Soil Charter (after Harcourt, 1990)

(a) Soil is one of humanity's most precious assets. It allows plants, animals and man to live on the Earth's surface.
(b) Soil is a limited resource which is easily destroyed.
(c) Industrial society uses land for agriculture as well as for industrial and other purposes. A regional planning policy must be conceived in terms of the properties of the soil and the needs of today's and tomorrow's society.
(d) Farmers and foresters must apply methods that preserve the quality of the soil.
(e) Soil must be protected against erosion.
(f) Soil must be protected against pollution.
(g) Urban development must be planned so that it causes as little damage as possible to adjoining areas.
(h) In civil engineering projects, the effects on adjacent land must be assessed during planning, so that adequate protective measures can be reckoned in the cost.
(i) An inventory of soil resources is indispensible.
(j) Further research and interdisciplinary collaboration are required to ensure wise use and conservation of the soil.
(k) Soil conservation must be taught at all levels and be kept to an ever-increasing extent in the public eye.
(l) Governments and those in authority must purposefully plan and administer soil resources.

This was further developed by the European Commission in 2006 as *a Thematic Strategy on Soil Protection* including a proposal for a framework Directive addressing such issues as erosion, decline of soil organic matter and soil contamination. Unfortunately, a blocking minority of states (UK, France, Germany, Austria and Netherlands) has meant that the Commission has so far been unable to reach agreement to introduce the Directive

In the UK, a *Code of Good Agricultural Practice* was first published in 1993 (and updated in 2009) and includes sections on good practice for soil management. In 1996, the *Royal Commission on Environmental Pollution* recommended that a soil strategy be drawn up, based on the following principles:

- soils must be conserved as an essential part of life support systems;
- soils should be accorded the same priority in environmental protection as air or water;
- integrated environmental management must include soil sustainability as a key element;
- where practicable, contaminated sites should be recovered for beneficial use;
- further contamination of soils from any source should be avoided.

As a result, a *Scottish Soil Framework* was published in 2008 and a *Soil Strategy for England* was produced in 2009 (Box 12.4). Most Scottish soil classes are, fortuitously, included within existing SSSIs, so that a measure of protection is already in place, but Gauld and Bell (1997) believed that greater attention should be given to soils in SSSI management plans. A *Soil Protection Act* (1987, updated 1994) has been introduced in the Netherlands with the goal of ensuring the health and multifunctionality of its soils.

Box 12.4 Safeguarding our Soils: a Soil Strategy for England (DEFRA, 2009)

The Vision of this Strategy is that by 2030 'all England's soils will be managed sustainably and degradation threats tackled successfully. This will improve the quality of England's soils and safeguard their ability to provide essential services for future generations'. The Vision will mean that:

- agricultural soils will be better managed and threats to them addressed;
- soils will play a greater role in the fight against climate change and in helping us to adapt to its impacts;
- soils in urban areas will be sufficiently valued for the ecosystem services they provide and given appropriate weight in the planning system;
- where development occurs, construction practices will ensure that vital functions can be maintained;
- pollution of soils is prevented and our historic legacy of contaminated land is being dealt with.

12.4 Geoconservation and climate change

Climate change has the potential to impact on both static and dynamic elements of geodiversity. In the UK, the latest climate projections (UKCP09) indicate warmer and wetter winters, hotter and drier summers and rising relative sea-levels but with variations across the UK. Prosser *et al.* (2010) have reviewed the issues involved in managing and conserving geodiversity in the UK resulting from climate change. They conclude that all types of geodiversity site will be impacted to some extent by changes in active processes, but that sites located on the coast, near rivers or active slopes are most likely to experience the greatest changes, particularly from sea-level rise, increased erosion or flooding. The likely societal response to these changes is likely to be a public demand for 'hard' defences and slope engineering, but Prosser *et al.* urge the use of a wider range of adaptation strategies that address the needs of geodiversity as well as biodiversity and society. These may include:

- Demonstrating and raising awareness of the dynamic nature of landscapes and of the need to work with changing geomorphological processes. There is a need to recognize the inevitability of natural change, maintain or restore the capacity of natural systems to absorb change and understand landscape sensitivity and thresholds. A conservation rather than a preservation approach is needed (Burek and Prosser, 2008).
- Developing and applying new conservation techniques, for example by adopting methods used in different climatic regimes, adapting existing techniques to address accelerated processes, developing techniques for *in situ* and rescue conservation, and reviewing the practicality of conserving all geodiversity sites in the face of anticipated impacts of climate change.
- Developing strategies to enable important geomorphological features to be identified, safeguarded and managed in situations where they are likely to be increasingly mobile. Existing systems of spatially fixed boundaries and current levels of site management are unlikely to be adequate. Adaptive management will be needed such a 'creating room for rivers' and encouraging coastal realignment (Orford and Pethick, 2006).
- Working with governments at all levels, policy makers, the planning system and affected communities to plan for geoconservation in a changing climate. This will include addressing the potential conflict between geoconservation and social demands, for example by allowing coastal communities the time and means to adapt to increased coastal erosion and threats to cliff top homes (see also Section 11.1).

Brown, Prosser and Stevenson (2012) have also assessed how geodiversity is likely to be affected by climate change and have proposed a number of the key adaptive principles. In Tasmania, Sharples (2011) has investigated in detail the potential climate change impacts on geodiversity in the Tasmanian Wilderness World Heritage Area based on the Climate Futures Project projections.

Box 12.5 summarises some of the higher priority potential impacts and possible management responses. Organic and organo-mineral soils are a key carbon store but are easily degraded by overgrazing and climatic effects. Lilly *et al.* (2009) reviewed the factors that may lead to the erosion of these soils on Scottish and Northern Irish uplands and concluded that climate change was likely to increase erosion through soil desiccation in dry weather and by extreme precipitation events.

Biodiversity is also certain to be affected by climate change, but geodiverse landscapes are likely to provide a more resilient range of habitats for species. In particular, hill and mountain areas allow species to move uphill as the climate warms, whereas no such opportunity exists in flat landscapes. In Section 14.3 there is further discussion on the potential role of geodiversity in enabling biodiversity to adapt to climate change through the concept of protecting the stage, rather than the actors.

Box 12.5 Some higher priority potential impacts on TWWHA geodiversity and possible management responses (after Sharples, 2011)

Changes to geodiversity are likely, albeit of widely varying magnitudes. The main impacts can be summarized as follows:

- Widespread degradation of moorland organic soils due to increased seasonal drying, warming and fire, especially on better-drained slopes;
- Vegetation stress and increasing fire risks leading to increased wind and water erosion;
- Increased aeolian process activity;
- Drying and degradation of swamps, bogs and peat;
- Decreased periglacial processes in the longer term;
- Increased frequency of landslides due to increased frequency of high-intensity rainfall events;
- Increasingly seasonal runoff and streamflow variation;
- Increased catchment and channel erosion and sedimentation;
- Likely increased flooding of low-lying coastal areas in response to sea-level rise;
- Likely more frequent flash flooding and sedimentation in caves;
- Accelerated loss of some relict soft sediment deposits and landforms.

Prevention or mitigation of these impacts is unlikely to be possible or beneficial. Shaples argues that the goal of management should therefore be to manage the consequences of change rather than attempting to stop them happening. Four fundamental management responses are discussed:

1. Do nothing—likely to be the only realistic response in many cases;
2. Recording, sampling and preserving irreplaceable stratigraphic, palaeoenvironmental, cultural or other information likely to be lost through accelerated erosion;

3. Monitoring and researching climate change impacts on geodiversity;
4. Selective, limited intervention to mitigate projected impacts on geodiversity, e.g. changed fire management regimes or protection of specific features or systems that could result in catastrophic change.

12.5 Geodiversity audits and action plans

Geodiversity audits are comprehensive surveys of the geological resources of an area aimed at developing, supporting and promoting active geological conservation. They summarise current knowledge about the geology of the area and aim to make the data more accessible to local authorities, planners, conservation bodies, statutory organisations, educational establishments and the general public. Although they inevitably include issues of site conservation, they can and should include the geology, geomorphology and soils in the wider landscape, threats to the wider georesource and maintenance of geological databases for the area. One of the first to be carried out was the *Peterborough Geological Audit*, UK (Peterborough Environment City Trust, 1999) (see Box 12.6).

Box 12.6 Peterborough Geological Audit, UK

The Peterborough Geology Audit includes the following sections:

- geological description of the area, solid and drift;
- shorter descriptions of landscape, soils, working quarries, important historical geological sites, museum collections, urban geology and archaeology.
- methods of data assembly and site selection;
- threats to the georesource, including mineral extraction, land restoration, loss of soils, landfilling and loss of temporary exposures;
- a description of 20 important sites in the Peterborough area including one SSSI, one NNR, six RIGS, and the remainder as locally valuable sites. For each of the RIGS the current status of the site is given and actions are listed to maintain or improve site conditions. The other sites each have a brief statement of objectives;
- a set of targets under five headings: raising awareness of geological conservation, site surveying and documentation, longer tern site protection, site management and enhancement, and geological collections in Peterborough. Each target lists a lead group(s), the funding source (mostly 'to be arranged') and a target date;
- details of how the objectives of the audit relate to the other local processes including the local planning system (see earlier), LocalAgenda 21, UK Biodiversity Action Plan, English Nature's Natural Areas approach (see earlier);
- proposals for site monitoring.

The *Durham Geodiversity Audit* (Lawrence, Vye and Young, 2004) was as a collaborative venture between the British Geological Survey and Durham

County Council with funding from the Aggregates Levy Sustainability Fund (see Section 12.1). It describes the geological evolution of County Durham and its rocks, geological structures, Quaternary deposits, landforms, soils, fossils, minerals as well as geophysics and geochemistry. It also contains chapters on the use and understanding of the geological resource. Geodiversity Audits may be extended into Geodiversity Action Plans and the two are often combined as in the case of the *North Pennines Geodiversity Audit and Action Plan* (2004) and the *Northumberland National Park Geodiversity Audit and Action Plan* (Lawrence *et al.*, 2007).

Other parts of the UK have gone straight to producing a Local Geodiversity Action Plan (LGAP), the equivalent of the successful Local Biodiversity Action Plans (LBAPs). Carson (1996) argued that Biodiversity Action Plans could not ignore the abiotic foundations and the more local the plan the greater the need to consider such details. Consequently some Biodiversity Action Plans contain geodiversity sections or integrate geodiversity with biodiversity, for example in Devon. In Scotland, Scottish Natural Heritage has worked with other organisations to ensure that the important links between biodiversity and geodiversity are recognised in LBAPs. One example is the *Tayside Biodiversity Action Plan* (2009) which states that 'Tayside's complex biodiversity only exists because of its underlying 'geodiversity'.

Burek and Potter (2004, 2006) reviewed the early work on LGAPs, identified common threads and made recommendations on how to progress them successfully. The aim of LGAPs is to provide a structured approach to local geoconservation activities, and according to Burek and Potter (2004):

- Plans should be simple and easy to implement;
- Objectives must be clear and straightforward;
- Targets should be achievable and timed;
- Local audits should be undertaken first to establish the size and shape of the task;
- Monitoring of Action Plans must be straightforward.

More recently, Burek (2012b) has carried out an analysis of seven UK LGAPs and concluded that they are important drivers for geoconservation and geotourism. A more comprehensive review of 41 published or planned LGAPs in England was published by Haffey (2008) in a Natural England Research Report. The main conclusions of this study are summarised in Box 12.7.

Box 12.7 LGAPs in England: conclusions of the review by Haffey (2008)

- LGAPs are widely seen as being very effective for raising the profile of geodiversity with public, private and voluntary sector organisations, especially local authorities and the minerals industry, and in creating a structured approach to the delivery of geoconservation at the local level.

- A broad understanding of the geology of the LGAP area needs to exist, or be acquired, before an LGAP can be developed. There is no evidence to suggest that an audit is required prior to the production of an LGAP.
- If LGAPs are to be effective, they should be pitched at a level that takes account of the sometimes limited geological knowledge of the audience and presented in a way that can capture peoples' imagination and interest.
- Funding has been obtained for the production of LGAPs but there is often a lack of resources for implementation. There is concern at the heavy reliance on the voluntary sector and lack of commitment from local authorities in some areas.
- Some local authorities in Devon have developed an integrated biodiversity and geodiversity action plan on the grounds that this reflects their interdependence. a down-side is that this approach leads to geodiversity issues being subsumed within a more high profile biodiversity agenda and being allocated an inadequate share of resources.

In some LGAPs there is a tendency to focus on geological sites rather than a more inclusive approach which values local landscapes, active processes and soils. Once finalised LGAPs need to be regularly updated and should be cross-referenced and embedded in other regional and local authority plans including community strategies and local plans. There is a real opportunity here for a radical, bottom-up approach to valuing natural physical qualities. A further recent development includes the development of Company Geodiversity Action Plans (cGAPs), and guidance has been produced on their production for aggregate companies (Thompson *et al.*, 2006).

Gray (1997a, p. 323) commented that 'In the long term, perhaps one day we will see a ... Geodiversity Action Plan for the UK to rank alongside its biological counterpart'. And sure enough, in 2006 work was initiated to produce a UK Geodiversity Action Plan that was launched in 2011 (Box 12.8).

Box 12.8 Extracts from the UKGAP—a framework for action (2011)

The UKGAP aims to be 'A framework for enhancing the importance and role of geodiversity' and is divided into six themes:

Theme 1 Furthering our understanding of geodiversity.
Theme 2 Influencing planning policy, legislation and development design.
Theme 3 Gathering and maintaining information on our geodiversity.
Theme 4 Conserving and managing our geodiversity.
Theme 5 Inspiring people to value and care for our geodiversity.
Theme 6 Sustaining resources for our geodiversity.

The UKGAP also contains a total of 14 Objectives, over 50 Targets and 17 Indicators of success. An example for Theme 1 is shown here.

Theme 1: Furthering our understanding of geodiversity

Objective	Target
Objective 1. To foster UK-based pure and applied geoscience research in order to better understand our geodiversity and its role in understanding and managing our natural environment.	Continue to undertake research that enhances our understanding of UK geodiversity. Continue to undertake research that interprets UK geodiversity to support and better understand the Ecosystem Service Approach. Continue to undertake research to better understand landform and surface processes to contribute to our understanding of landscape-scale management and change Use geodiversity evidence to help us better understand past climate and environmental changes and to forecast future change.

Indicator 1. Recognition within research
The number of refereed research papers relating to UK geodiversity

12.6 Strategies, codes and charters

Other national or sub-national strategies, codes and charters can also be noted. For example, Tasmania's *Nature Conservation Strategy 2002–06* in Australia recognised that 'Rocks, soil and landforms are the basis on which all life exists'. In Spain, there is an *Andalusian Strategy for the Preservation of Geodiversity* (2003) that provides for geoheritage to be protected in rural areas of Andalusia.

Another way to conserve sites, particularly fossil and mineral ones, is by promoting collecting codes. There is a general one for the United Kingdom promoted by the Geologists' Association, and also a number of local codes, for example for West Dorset (Edmonds, 2001). More recently, A Scottish Fossil Code was published in 2008 (Box 12.9) and reviewed in 2012.

Box 12.9 The essentials of the Scottish Fossil Code (2008)

- Seek permission—you are acting within the law if you obtain permission to extract, collect and retain fossils;
- Access responsibly—Consult the Scottish Outdoor Access Code prior to accessing land. Be aware that there are restrictions on access and collecting at some locations protected by statute;
- Collect responsibly—Exercise restraint in the amount collected and the equipment used. Be careful not to damage fossils and the fossil resource.

Record details of both the location and the rocks from which the fossils are collected;
- Seek advice—if you find an exceptional or unusual fossil do not try to extract it; but seek advice from an expert. Also seek help to identify fossils or dispose of an old collection;
- Label and look after—collected specimens should be labelled and taken good care of;
- Donate—if you are considering donating a fossil or collection choose an accredited museum, or one local to the collection area.

The most impressive Charter incorporating geodiversity is the *Australian Natural Heritage Charter* (Australian Heritage Commission, 1996; revised 2002). This ground-breaking document is based on a number of fundamental principles:

- intergenerational equity—the present generation should ensure that the health, diversity and productivity of the environment is maintained or enhanced for the benefit of future generations;
- existence value—living organisms, earth processes and ecosystems may have value beyond the social, economic or cultural values held by humans;
- uncertainty—accepts that our knowledge of natural heritage and the processes affecting it is incomplete, and that the full potential significance or value of natural heritage remains unknown;
- precautionary principle—where there are threats or potential threats of serious or irreversible environmental damage, lack of full scientific certainty should not be used as a reason for postponing measures to prevent environmental degradation.

The Charter applies both inside and outside protected areas. It has a section on definitions (Article 1 of the Charter) which includes definitions of 'geodiversity' and 'earth processes' and these terms are then threaded through many of the subsequent 43 Articles of the revised edition. Box 12.10 gives some examples of these Articles.

Box 12.10 Examples of Articles of the Australian Natural Heritage Charter (2002) relevant to geodiverity conservation

Article 2—The basis for conservation is the assessment of the natural significance of a place, usually presented as a statement of significance.
Article 3—The aim of conservation is to retain, restore or reinstate the natural significance of a place.
Article 5—Conservation is based on respect for biodiversity and geodiversity. It should involve the least possible physical intervention to ecological processes, evolutionary processes and earth processes.

Article 11—Elements of geodiversity and biodiversity that contribute to the natural significance of a place should not be removed from the place unless this is the sole means of ensuring their survival, security or preservation and is consistent with the conservation policy.

Article 12—The destruction of elements of habitat or geodiversity that form part of the natural significance of a place is unacceptable unless it is the sole means of ensuring the security of the wider ecosystem or the long-term conservation of the natural significance.

Article 20—Reinstatement is appropriate only if there is evidence that the species or habitat elements or features of geodiversity that are to be reintroduced have existed there naturally at a previous time, returning them to the place contributes to retaining the natural significance of that place, and processes that may threaten their existence at that place have been discontinued.

Article 21—Enhancement is appropriate only if there is evidence that the introduction of additional habitat elements, elements of geodiversity or individuals of an organism which exist at that place are necessary for, or contribute to, the retention of the natural significance of the place.

Article 22—Where organisms or elements of geodiversity are introduced to a place for the purpose of enhancement, the individuals introduced to the place should not alter the natural species diversity, genetic diversity or geodiversity of the place if that would reduce its natural significance.

Article 23—Enhancement in existing natural systems should be limited to a minor part of biodiversity or geodiversity of a place and should not change ecosystem processes nor constitute a majority of the habitats or features of geodiversity of the place.

Article 31—Work on a place should be preceded by research and by review of the available physical, oral, documentary and other evidence about the existing biodiversity and geodiversity. . . .

Article 33—Evidence of the existing biological diversity, geodiversity, and any other significant features of the place ... should be recorded before any disturbance of the place.

The inclusion of 'geodiversity' in several Articles and the overall approach of the Charter represent a model that other countries could usefully follow. Since the Charter was prepared by the Australian Heritage Commission (the government conservation body), it demonstrates that there is clear government support for the principles of geodiversity conservation.

The Charter is supported by the *Natural Heritage Places Handbook* (Australian Heritage Commission, 1999) which clarifies the 10-step process for natural heritage conservation in Australia, ranging from identifying sites and obtaining information about them through determining the natural significance of the place, developing a conservation policy and plan, to implementing the plan and monitoring the results. Box 12.11 is an example of a conservation policy for the Red Rock Craters and Lakes, near Alvie in Victoria.

Box 12.11 Conservation policy for Red Rock Craters and Lakes, Victoria, Australia

Information about the place: Red Rock is a complex of well-preserved maar craters and scoria cones covering 550 ha. The maars are broad craters which extend below the general ground level. They are surrounded by low rims with steep inner walls and gentle slopes away from the craters. The maars have scalloped edges suggesting that they were formed by multiple eruptions. The deposits associated with them consist of volcanic ash with particles the size of sand and gravel with abundant large volcanic bombs and blocks. The scoria cones formed after the maars. They are small, steep-sided volcanoes, that have been progressively superimposed on each other. They consist of scoria fragments and minor layers of ash and fused lava spatter.

Statement of significance: The Red Rock area is a well-preserved example of a complex volcanic landform, of which only a few are known to exist in Australia. It is a significant site for both geological and limnological scientific and educational purposes. Red Rock has an exceptional range of eruptive phenomena. It is a notable multi-orifice volcano with an exceptional range of pyroclastic materials and structures. Red Rock has a complex relationship to the water table and has a wide range of lakes which have been extensively studied.

Condition and other management information: The Red Rock volcanoes are well preserved, having suffered only negligible erosion. Current land use includes seven major quarries in the area, but most are not operational. Tenure is mainly freehold.

Compatible uses: Most uses are compatible if they do not cause erosion or changes in the natural drainage patterns of the place. Compatible land uses would be tourism, grazing, catchment protection activities and limited quarrying (if this is restricted to places that do not demonstrate unique features essential to understanding the significance of the place). Where there are exposures of the geology, educational use could be made of the area.

Desirable future condition: The place should remain in a condition in which the geomorphology of the landform remains intact and the natural processes of change are not accelerated, directly or indirectly, by human activity. Extractive industry should avoid any places that demonstrate unique features not well represented elsewhere in the place. Tenure is not a critical factor for management, and existing land uses are appropriate to maintain the natural heritage significance identified.

Proposed conservation procedure: A conservation plan can be developed and incorporated in the local planning scheme to guide decisions on management and on future developments and activities proposed for the area.

In Scotland, a proposal to introduce a 'geodiversity duty', analogous to the biodiversity duty introduced by the Natural Environment and Rural Communities Act (2006) in England and Wales, was not accepted by the Scottish Government (Browne, 2012). But the Scottish Government, together with Scottish Natural

Heritage, the British Geological Survey and GeoConservationUK is supporting the *Scottish Geodiversity Forum* and its publication of the *Scottish Geodiversity Charter* (www.scottishgeodiversityforum.org). Launching the Charter on 6 June 2012, Stewart Stevenson, Minister for Environment and Climate Change, said: 'I welcome Scotland's Geodiversity Charter which not only encourages understanding and appreciation of our geodiversity but also promotes awareness and more integrated management of something so fundamental to all our lives'. By October 2012 the Charter had already been signed by 30 organisations. In addition, Gordon and Barron (2011) have proposed a *Geodiversity Framework for Scotland* that should identify strategic priorities and provide a framework for action in a similar way to the *Scottish Soil Framework*. The proposed vision is that 'Geodiversity is recognised as an integral and vital part of our environment, economy and heritage to be safeguarded for existing and future generations in Scotland'. Following the Scottish lead, Geodiversity Charters are being developed for England and Wales.

12.7 Conclusions

In Chapters 10, 11 and 12, I have tried to describe some of the ways in which geoconservation is being extended beyond protected areas, spatially into the wider landscape and thematically into natural resource management. It has been argued by some that the protected area approach described in Chapters 6, 7, 8 and 9 works well for densely populated lowland areas and for particular themes such as igneous rocks or palaeontology, whereas the wider approaches outlined in this part are more relevant for less populous, upland areas and landscapes. For example, Daly, Erikstad and Stevens (1994) argued that in Britain and Ireland the preponderance of low relief areas, poor rock exposure and high level of threat led to a strongly protected area approach, whereas in Norway with its high relief, good rock exposure and generally low level of threat there has been an emphasis on protecting landscapes and landform systems. However, they point out that in Norway, Quaternary deposits and geomorphological features are often locally threatened particularly in the more populous southern coastal zone of the country, and in Scotland there are analogies with the situation in most of Norway. They therefore conclude that the differences in current approach appear more to do with pragmatism and tradition rather than fundamental differences in the types of features or landscapes. It is easy to agree with this conclusion and to emphasise that geoconservation needs both site-based and wider approaches if it is to maximise its effectiveness. As Chapters 10, 11 and 12 have shown, both these strands are highly complex, but it is important that geoscientists become less insular and utilise the many opportunities that arise in public life to promote the interests of geoconservation in the both protected areas and the wider landscape.

Part V

Putting It All Together

13 Geodiversity and Geoconservation: an Overview

Climb the mountains and get their good tidings ... nature's sources never fail

John Muir (1901)

Having described international and national approaches and programmes for geoconservation in Chapter 6, Chapter 7, Chapter 8 and Chapter 9 and for geoconservation in the wider landscape in Chapter 10, Chapter 11 and Chapter 12, this chapter aims to summarise how geodiversity is a key basis for geoconservation before explaining how the aims and methods of geoconservation vary depending on which element of geodiversity is being considered.

13.1 Geodiversity as a basis for geoconservation

As explained in Chapter 1, the word 'geodiversity' was first used only in the 1990s, but the principles behind its application to nature conservation have a much longer history. For example, in the UK, the Report of the *Wild Life Conservation Special Committee* (Huxley, 1947) that led two years later to the establishment of the Nature Conservancy and Sites of Special Scientific Interest (SSSIs), contains the following quote:

> Great Britain presents in a small area an extremely wide range of geological phenomena ... the supply of a steady flow of trained geologists for industrial work at home and overseas, requires that there shall be available in this country a sufficient number of representative areas for geological study' (Huxley, 1947, para 64).

For 'range of geological phenomena' in this quote we could easily substitute 'geodiversity', and 'representative areas' must logically mean areas representative

Geodiversity: Valuing and Conserving Abiotic Nature, Second Edition. Murray Gray.

of the country's geodiversity. So as early as 1947 the principle of using geodiversity as the basis for geoconservation was being promoted.

The Geological Conservation Review (GCR) that undertook a major site selection programme in Britain between 1977 and 1990 was intended to 'reflect the range and diversity of Great Britain's Earth heritage' (Ellis *et al.* 1996, p. 45). Site selection was achieved on three main criteria, one of which was 'sites that are nationally important because they are **representative** of an Earth Science feature, event or process which is fundamental to Britain's Earth history' (Ellis *et al.* 1996, p. 45). Note the use of the words 'range', 'diversity' and 'representative' in this quote.

However, Kiernan (1997b, p. 21) made the important point that 'Identifying representative examples of phenomena obviously requires the existence of some form of classification framework that answers a basic question: examples of what?' He also noted (p. 22) that 'Unlike in the life sciences, no generally agreed comprehensive classification of landforms has ever evolved among geomorphologists, nor does one seem likely to arise', and the same is true of physical processes and landscapes. However, we do have some internationally accepted classification systems for rocks, minerals, fossils and soils, so that for these at least we can begin to base geoconservation on the principle of representativeness.

We can also detect the use of geodiversity principles in nature conservation site selection in many other countries. The United States has two main conservation programmes. The National Parks network is world famous and new units can be added if they meet certain criteria, one of which is that they must not represent a feature already adequately represented in the system. Similarly, to be included on the National Natural Landmarks network, units must be 'one of the best examples of a type of biotic community or geologic feature'. In other words, in the USA there is an attempt to conserve different types of geologic features, i.e. geodiversity.

Similarly, in Section 9.5, we saw that the Irish Geological Heritage programme has recently identified 16 geological themes, e.g. Precambrian, coastal geomorphology. 'Each theme is intended to provide a national network of Natural Heritage Area sites and will include all components of the theme's scientific interest' (Parkes and Morris 2001, p. 82), i.e. it is clear that the system is intended to establish a representative selection of Ireland's geodiversity.

In Section 9.8 we noted that Kenny and Hayward (1993) believed that 'The overriding objective of earth science conservation in New Zealand should be to ensure the survival of the best representative examples of the broad diversity of geological features, landforms, soil sites and active physical processes'. In effect, this is a call to have a site network that recognises and protects the geodiversity of New Zealand.

In Spain the Natural Heritage and Biodiversity Act (2007) represented a breakthrough in that:

> ... for the first time since the beginnings of environmental legislation in Spain, a law is inspired by the principle of conservation of geodiversity (Art. 2), defines concepts like geodiversity, geoparks and geological heritage (Art. 3) and establishes (Art. 9) that the Ministry of Environment ... will elaborate and maintain an updated

Inventory of Sites of Geological Interest representative of ... the 20 Geological Frameworks identified in the Spanish Global Geosites project, as well as seven additional geological units representative of Spanish geodiversity' (Garcia-Cortés, Gallego and Carcavilla, 2012, p. 336).

Thus, Spain is another example of a country where geoconservation is being based on representing the country's geodiversity.

In Chapter 7 we saw that until recently World Heritage Sites (WHS) have been proposed by countries and accepted by UNESCO if they met the criterion of 'outstanding universal heritage', i.e. UNESCO adopted a reactive role. In the past few years, the IUCN has been working on behalf of UNESCO to become more analytical and proactive, and this includes the geological component of the World Heritage List. We saw that Dingwall, Weighell and Badman (2005) have examined the list to determine whether the geological column is fully represented, and the report indicates a significant gap at the Silurian with no sites of this age represented. In also proposing the adoption of 13 geothemes (e.g. fossil sites, coastal systems, see Table 7.3) to help in assessing future WHS applications and identifying possible gaps in representation, there is confirmation that IUCN/UNESCO are trying to ensure that the world's geodiversity is represented in the World Heritage List. In Chapter 7 we also noted the detailed thematic case studies recently carried out for the IUCN of caves and karst by Williams (2008) and of volcanoes by Wood (2009). In reviewing the global distribution of World Heritage Sites containing these features, they concluded that there are gaps in the geographical and/or thematic representation that could be filled by further nominations, i.e. that the World Heritage List could be more representative of the global geodiversity of caves, karst and volcanic features.

In conclusion, the concept of representativeness related to geodiversity has been used by many countries in selecting national geoconservation sites and is increasingly used to guide the nomination of geological World Heritage Sites.

13.2 Geoconservation Management Aims and Methods

It is widely accepted that geoconservation sites must be managed in different ways depending on the type of site involved. For example, we saw in Section 9.4.3 that Natural England recognizes three types of geoconservation site, each with its own approach to site management (Prosser, Murphy and Larwood, 2006):

- Exposure sites are those where a deposit is extensive underground so that the principle management aim is to maintain exposure of the strata, whether by quarrying, periodic clearing and cleaning, or by coastal or fluvial erosion.
- Finite sites occur where geological features are of limited extent, so that removal may cause depletion of the resource. Management generally controls removal of material.
- Integrity sites are geomorphological sites where the dynamics of active processes or the integrity of landform contouring need to be retained.

These ideas can be further refined by identifying geoconservation management aims for different elements of geodiversity. An example of this is shown in Table 13.1. It is clear from this that geoconservation management is complex since it must acknowledge the very different aims that must be applied, i.e. the conservation of rare fossils must involve very different strategies to the conservation of soils which in turn is very different to the conservation of natural physical processes.

And what holds for geoconservation management aims, also holds for geoconservation methods. Table 13.2 is an attempt to classify the methods available and map their usage onto a grid showing the elements of geodiversity. Some methods can be applied to all elements of geodiversity but others are much more specialized. In many cases several methods are applied at the same site/area. An outline of the available methods is given below.

13.2.1 Site monitoring and management

Site monitoring and management include a range of sub-methods. Monitoring of *in situ* fossil sites is described by Santucci, Kenworthy and Mims (2009) who describe a number of methods including repeat photography, erosion stakes and digital mapping. Other chapters in Young and Norby's (2009) book include monitoring of cave and karst, coastal, fluvial, glacial and periglacial, geothermal and volcanic sites. The Palaeontological Resources Preservation Act (2009) in the USA requires federal agencies to protect information about the nature and

Table 13.1 Geoconservation management aims for different elements of geodiversity.

Element of geodiversity	Rare or common	Management aims
Rocks and minerals	Rare	Maintain integrity of outcrop and subcrop. Remove samples for curation.
	Common	Maintain exposure and encourage responsible collecting. Encourage sustainable use. Value historical and modern uses of geomaterials.
Fossils	Rare	Wherever possible, preserve *in situ*. Otherwise remove for curation.
	Common	Encourage responsible collecting and curation.
Topography/Land form		Maintain integrity of topography/land form. Encourage authentic contouring in restoration work and new landscaping schemes.
Landscape		Maintain contribution of natural topography, rock outcrops and active processes to landscape. Encourage geomorphologically authentic design in restoration work and new landscaping schemes.
Processes		Maintain dynamics and integrity of operation. Encourage restoration of process and form using geomorphologically authentic design principles.
Soil		Maintain soil quality, quantity and function.

Table 13.2 A classification of geoconservation methods (xx = major method, x = significant method, blank = rarely used method).

Geoconservation method	Sub-method	Element of geodiversity						
		Rocks	Minerals	Fossils	Landforms	Landscapes	Processes	Soils
Site Management	Secrecy	x	x	xx	x			
	Signage	x	x	x	x		x	
	Physical barriers	x	x	xx	xx		xx	
	Reburial			xx				
	Site clearance	xx	x	xx				
Curation	Specimens		x	xx				
	Casts			x				
	Documents, photos					N/A		
Permits, licences		x	x	xx	x			
Supervision	Static/mobile rangers			x	x			
	Remote				x			
	Local residents			x				
Benevolent Ownership		xx	xx	xx	xx	xx	xx	xx
Restoration					xx	xx	xx	
Legislation	Nature conservation (statutory sites)	xx	xx	xx	xx	xx	xx	xx
	Planning	xx	xx	xx	xx	xx	xx	xx
	Environmental				x	xx	xx	xx
Policy	Non-statutory sites (e.g. WHS, geoparks, local sites)	xx	xx	xx	xx	xx	xx	x
	Management plans	xx	xx	xx	xx	xx	xx	xx
	Policies/ position statements	xx	xx	xx	xx	xx	xx	xx
	Codes of conduct	x	xx	xx	x			
Education	Visitor centres & museums	xx	xx	xx	xx	xx	xx	x
	Publications	xx	xx	xx	xx	xx	xx	xx
	Panels	xx	xx	xx	xx	xx	xx	x
	Activities and clubs	xx	xx	xx	x			
	TV programmes	xx	x	xx	x		xx	
	Training and CPD	xx	xx	xx	x		x	
	Other							

specific location of fossils where this is warranted to protect them. Secrecy is in fact an important site management method used principally at fossil and mineral sites, where discovery is not immediately advertised until research work is completed and even then the whereabouts of the site may be not be made publicly available. Examples of this applying to bedrock exposures are some of the rare Ediacaran fossil sites in South Australia. This method also applies to some landforms, e.g. the location of rare forms of cave speleothem may be kept secret.

Signage involves the use of signs and notices to dissuade potential transgressors to desist from intruding into and/or damaging sites. The signage may involve words or diagrams or both. An example of the latter from Craters of the Moon National Monument is shown in Figure 13.1. At Petrified Forest National Park there is a sign that encourages visitors to buy petrified wood from the many suppliers outside the park boundaries rather than taking samples from within the park where minimum fines of $275 apply.

The use of physical barriers is an important method, particularly to prevent public access to very sensitive geological locations or hazardous sites. The permeability of the barrier may vary. In the case of caves containing fragile

Figure 13.1 Signage and words to discourage rock collecting at Craters of the Moon National Monument, Idaho, United States.

speleothems, the entrances can be gated and locked, thus preventing public access and providing a very impermeable barrier. An example is the entrance to Shooting Star Cave in Tasmania where a lockable metal grid prevents access to all except *bona fide* cave research workers. In some cases, visitor centres/museums are constructed over important geosites, thus restricting access to opening times when the sites are supervised (see below). An example is the Fossil Quarry site at Dinosaur National Monument in Utah, United States where the quarry exposure is covered by a visitor centre/research facility (Figure 13.2), though this is partly also to protect the site from weathering effects. Slightly less secure are sites that are surrounded by high fencing. Examples here include the petrified tree near Mammoth in Yellowstone National Park (Figure 13.3) and the fossiliferous Silurian ripple beds at Wren's Nest National Nature Reserve at Dudley in the English Midlands (Figure 9.18). At Birk Knowes, Scotland, because of fossil thefts, a restricted fossil outcrop has been fenced, collecting permits have been stopped and notice boards with anti-collecting messages in four languages have been erected (MacFadyen, 2001a). More permeable physical restraints may involve lower fencing together with signage requesting visitors not to cross this. In the case of physical processes, fencing is often used for health and safety reasons to avoid injury to visitors as well as preventing disturbance to the operation of natural processes. Examples occur at several hot spring sites in Yellowstone National Park and in Iceland. Another example of the separation of visitors from geosystems are the boardwalks across sand dune areas constructed to protect the stabilising vegetation.

Reburial is a rather specialised and rare method of geoconservation that can be applied to fossil sites in particular to prevent access by covering sites with soil following exposure. This method allows future study of fossils *in situ*. An example

Figure 13.2 Visitor Centre at Dinosaur National Monument, Utah, United States.

Figure 13.3 Fencing around a petrified tree in Yellowstone National Park, Wyoming, United States to protect it from souvenir hunters who have denuded an adjacent one.

occurs in Sheffield, England where Boon (2004) described some Westphalian age fossil tree stumps, originally contained within wooden sheds to protect them from the elements and souvenir hunters, but recently reburied by soil.

Site clearance is used on sites that have become degraded and/or vegetated in order to re-expose a geological section or other feature. The Face Lift project run by English Nature in the 2000s is one organised example of this type of approach, with over 250 sites being cleared (Murphy, 2002).

13.2.2 Curation and licensing

Curation is a commonly accepted method of geoconservation, particularly for vulnerable fossils which are removed from their *in situ* location, cleaned, mounted and documented and either put on public display or placed in store. An example is the West Runton elephant skeleton from West Runton, Norfolk, England, which was exposed by coastal erosion and subsequently removed to the safety of a museum. Also in this category we can include the processes of taking moulds of fossils and then making casts of them for use in education where the originals could not be used or replicated. Moulds of whole outcrops including trackways are also made and the casts put on display in museums or visitor centres (e.g. Edwards and Williams, 2011). Curation of documents, photographs, sketches, etc., is also important in this category.

Permitting/Licensing is used in many places to control access and/or collecting by visitors and research workers. For example, it is against the law in Nova Scotia, Canada to dig fossils without a permit and this is also the case in many other countries/provinces. Another example is access to 'The Wave', a series

Figure 13.4 'The Wave' on the Utah/Arizona border, United States. See plate section for colour version.

of smooth water cut channels through the finely dune-bedded, red and white, Jurassic Navajo Sandstone (Figure 13.4) mid-way between Page and Kanab on the Utah–Arizona border, USA. Access here is restricted by the Bureau of Land Management by issuing permits to a maximum of 20 persons per day.

13.2.3 Supervision

This is rarely used in geology because of cost, but a number of examples can be given of different supervision strategies used for both facilitating enjoyment as well as protecting sites. Occasionally rangers in the US National Parks will be positioned at important sites to provide visitor information and guard against deliberate damage. This is the case, for example, at Mesa Verde National Park in Colorado, USA where archeological remains in some cliff alcoves are managed in this way. Mobile rangers are more common and rely on occasional ranger patrols of park trails or roads by foot or vehicle. Ranger-led tours can be used to ensure that visitors touring a site are supervised throughout. Examples include some tours through sensitive caves at Carlsbad Caverns National Park, New Mexico, USA where rangers at both front and rear of tour parties can supervise groups and answer questions. At Wadi El-Hitan World Heritage Site in Egypt, access is restricted to organized tours with official guides. At Petrified Forest National Park, Arizona, USA, visitors' cars may be searched on the exit roads from the park to try to ensure that large quantities of fossil wood are not removed from within the park boundaries. The Countryside Council for Wales employs some 'wildlife crime' officers whose jobs include investigating geodiversity infringements.

In some cases cameras are installed to supervise visitors to sites. An example of this remote surveillance occurs at the Stump Cross Caves in the Yorkshire Dales National Park, England where a number of CCTV cameras oversee key speleothem locations (also protected by metal grills). This can be a viable option at vulnerable and sensitive sites.

Public surveillance can also take place where sites lie close to residential areas. Local residents, aware of the importance of nearby geological sites, may voluntarily supervise the access points. An example occurs at Valentia Island, Ireland (Box 9.12) where access to the world's oldest *in situ* tetrapod trackway is overlooked by residents (Parkes, 2001). Another well-known example of community stewardship occurs at Joggins Fossil Cliffs (Section 7.7.2) where local resident and former coal miner Don Reid is known as 'Keeper of the cliffs'.

13.2.4 Benevolent ownership

This applies to sites and areas that are owned by organizations or individuals with a clear commitment to conservation. It is therefore likely that all elements of geodiversity within the owned land will be managed in a way that protects the geoheritage interests. An example is the National Trust in England, Wales and Northern Ireland which owns over 1000 km of coastline. The John Muir Trust, which owns land in Scotland including Ben Nevis, aims to protect wild landscapes from development threats. The Museum of the Rockies in Montana, USA now owns Egg Mountain, an important area for the discovery of Maiasaur fossils (Horner and Dobb, 1997). A further example is the Valentia Island tetrapod trackway site in Ireland, which has been bought by the Irish government.

13.2.5 Restoration

This was considered in some detail in Section 10.4.

13.2.6 Legislation

Nature conservation legislation is widely used to give formal protection to geological and geomorphological sites. An example is *the* Nature Conservation (Scotland) Act (2004) that allows Scottish Natural Heritage to designate Sites of Special Scientific Interest, including geological and geomorphological sites. Section 52 of the same Act also requires Scottish Natural Heritage to prepare and issue a Scottish Fossil Code. Many other countries also have legislation to protect geosites. Finland, for example, has a Nature Conservation Act (1996) that allows geological Natural Monuments such as erratic boulders to be designated. Nova Scotia, Canada has a Special Places Protection Act (1980) that makes it illegal to dig fossils or artifacts without a Heritage Research Permit. This Act is one of the many pieces of provincial and local legislation protecting the Joggins Fossil Cliffs World Heritage Site as shown in Figure 13.5.

Other types of legislation are also useful in giving protection to sites. For example, the planning legislation in Britain requires Local Planning Authorities

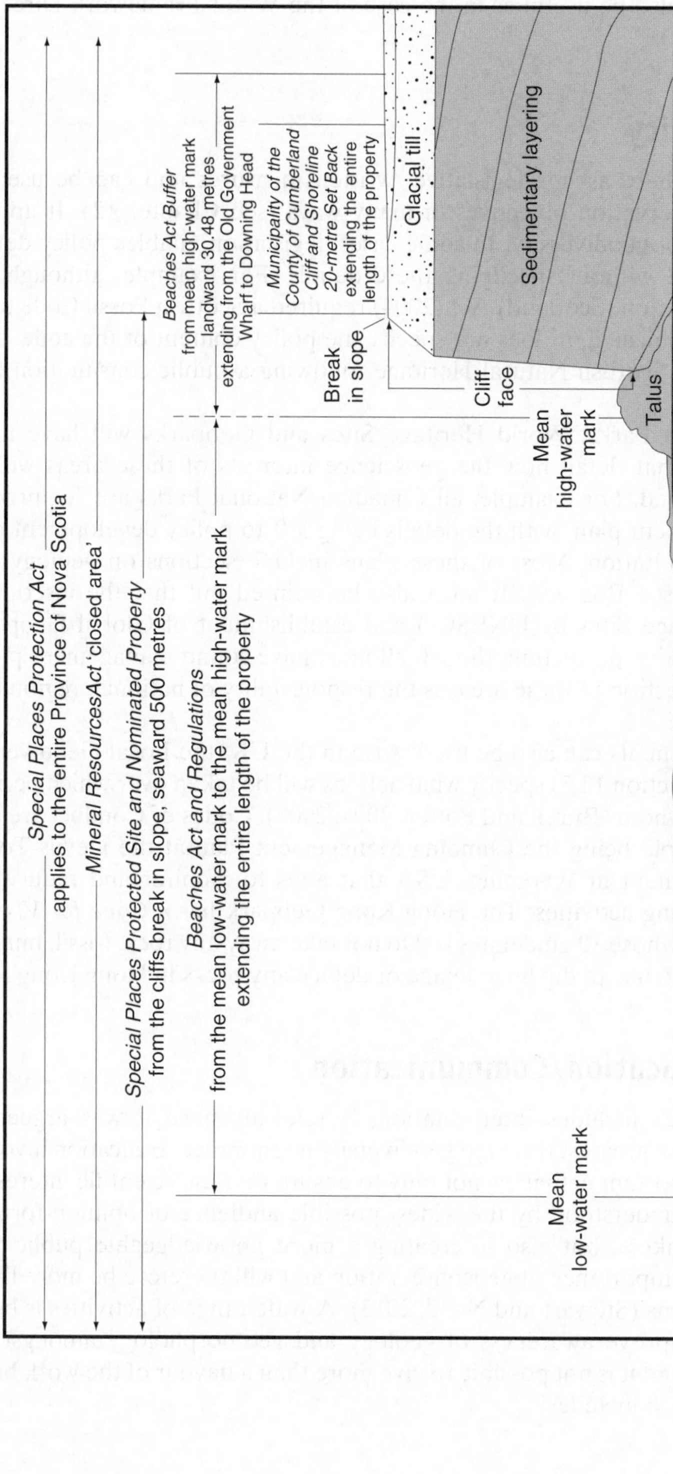

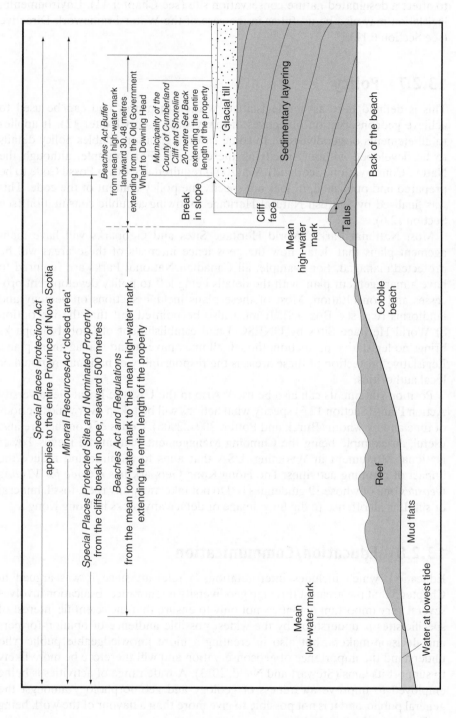

Figure 13.5 The legislation protecting the Joggins World Heritage Site. (Reproduced by permission of the Joggins Fossil Institute.)

to consult the nature conservation agencies on any planning applications likely to affect a designated nature conservation site (see Chapter 11). Environmental legislation may also be useful as in the case of the Water Framework Directive (see Section 6.11).

13.2.7 Policy

This is defined here as non-legislative, written measures and can be used to achieve geoconservation objectives in many ways (see Chapter 12). It applies to all elements of geodiversity. In some cases, legislation enables policy details to be developed without specifying these details. For example, although the Nature Conservation (Scotland) Act (2004) required a Scottish Fossil Code to be prepared and published, it does not specify the policy content of the code. This was finalised by Scottish Natural Heritage following a public consultation (see Section 12.6).

Most National Parks, World Heritage Sites and Geoparks will have management plans that detail how the geoscience interests of these areas will be protected/managed. For example, all Canadian National Parks are required to have a management plan, with the details being left to policy development processes and consultation. Most of these plans include sections on geology and landforms (e.g. see Box 9.8). It must also be pointed out that the inscription of World Heritage Sites by UNESCO and establishment of Global Geoparks brings no legislative protection, though all must have sound management plans. Legislative protection of these areas is the responsibility of national, regional or local authorities.

Position statements can also be used. Also in the UK, the Local Geodiversity Action Plans (Section 12.5) specify what actions will be taken over which periods of time and by whom (Burek and Potter, 2004, 2006). Codes of Conduct are also useful, an example being the Climbing Management Plan at the Devils Tower National Monument in Wyoming, USA that aims to monitor and reduce the impact of climbing activities. The Hong Kong Geopark has a *Code for Visiting Geosites* one of whose 12 guidelines is 'Do not take away any rock, fossil, mineral or silt. It is an offence to dig up, damage or deface any rocks in Hong Kong'.

13.2.8 Education/Communication

Education, which includes interpretation, is relevant since it was argued in Chapter 5 that the greatest threat to geodiversity is ignorance. Education involves several very important activities, not only to ensure that the scientific interest of specific sites is understood by the widest possible audience of opinion-formers and decision-makers, but also in creating a more knowledgeable public who understand the importance of geoconservation and will therefore be more likely to support its aims (Stewart and Nield, 2013). A wide range of activities is being employed to improve awareness of geology and geomorphology amongst the general public, and it is not possible to give more than a flavour of the work being undertaken, which includes:

- museums—for example the Royal Tyrell Museum, Drumheller, Alberta, Canada, and the Natural History Museum, London, United Kingdom;
- visitor centres—for example the Dynamic Earth visitor centre in Edinburgh, Scotland (Monro and Davison, 2001) and the George C. Page Museum at the La Brea Tar Pits, Los Angeles, United States;
- theme parks—for example the Sand World theme park at Travemünde, Germany, the west coast fossil park in South Africa (Roberts, 2002) and the Gletschergarten in Lucerne, Switzerland;
- disused mines—for example the Wieliczka salt mine in Poland (Hallett, 2002), the Sardinia Mining Park, Italy and the Great Orme Copper Mine in North Wales;
- site interpretation panels—for example Figure 13.6 (see also Anderson and Brown, 2010; Mansur and da Silva, 2011; Moreira, 2012);
- books, interpretation leaflets, maps, postcards, etc.—for example the *Landscape Fashioned by Geology* series of booklets from Scottish Natural Heritage;
- displays of fossil and mineral specimens;
- field guides—for example the *Landscapes from Stone* project in the north of Ireland (McKeever and Gallagher, 2001) as self-guided car and cycle routes;
- geological pedestrian trails (geotrails), either self-guided or guide-led (Figure 13.7)—for example Mount Unzen Geopark, Japan has several suggested 'Geo Saraku' walking routes; the Sharp Island 'Geo Trail' in Hong Kong Geopark demonstrates 'the great diversity of coastal landforms'; a Soil Trail has been developed on Anglesey, Wales (Conway, 2010) and 10 walks in the National Forest in the English Midlands are described by Ambrose *et al.* (2012); a 300 km long hiking trail has been established in Germany (Wrede *and* Mügge-Bartolovic, 2012);

Figure 13.6 Example of an interpretation panel: Hateg Country Dinosaur Geopark, Romania.

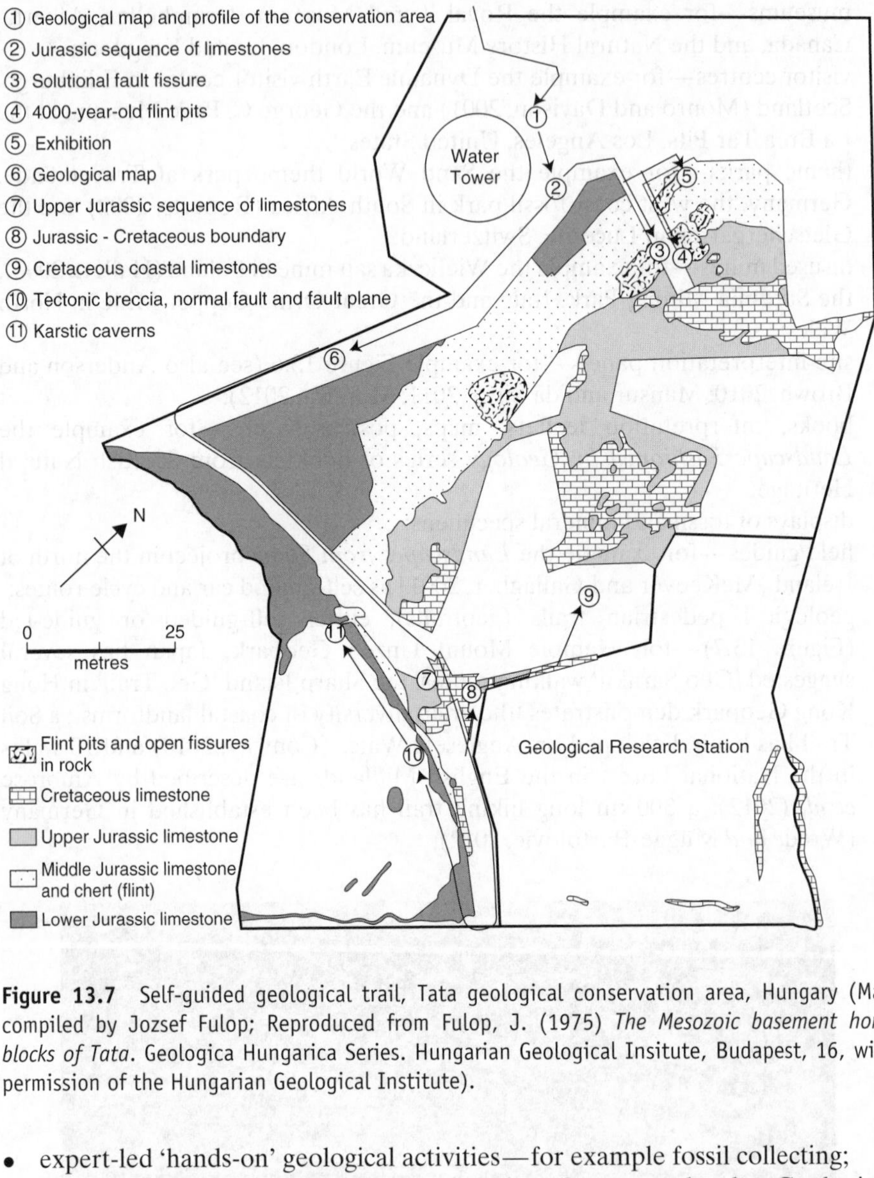

Figure 13.7 Self-guided geological trail, Tata geological conservation area, Hungary (Map compiled by Jozsef Fulop; Reproduced from Fulop, J. (1975) *The Mesozoic basement horst blocks of Tata*. Geologica Hungarica Series. Hungarian Geological Insitute, Budapest, 16, with permission of the Hungarian Geological Institute).

- expert-led 'hands-on' geological activities—for example fossil collecting;
- geological highway guides or road trails—for example the Geological Highway Maps of the Canadian provinces, including Nova Scotia and Saskatchewan, and a set of 11 Geological Highway Maps for groups of US states produced by the American Association of Petroleum Geologists (see also Figure 13.8);
- geological train trails—for example the Llangollen Steam train trail, 'Explore the Severn Valley Railway' and the Eskdale railway trail in Cumbria, all in the United Kingdom;
- audio-visual presentations;

Figure 13.8 Example of an information sign on the geological highway over the Bighorn Mountains, Wyoming, United States.

- television programmes—for example *Walking with Dinosaurs, Earth Story, Planet Dinosaur* and *Early Life*;
- web sites—for example the interactive, web-based maps promoting geotourism in the Titel loess plateau of Vojvodina, Serbia (Vasiljevic *et al.*, 2009), and the Canadian Geovistas website for Jasper National Park (www.earthsciencecanada.com/geovistas/);
- iPhone/iPad apps—for example the UK Natural History Museum's Evolution app for the iPad;
- festivals and events—for example the annual *Scottish Geology Festival* and Finland's annual *Day of Geology*;
- children's clubs and educational initiatives—for example the UK's *Rockwatch* geology club, school-quarry industry links in Canada (Milross and Lipkewich, 2001) and initiatives by Earth Science Teachers' associations;
- links with industrial archaeology, landscape history, the arts and other fields;
- professional training and continuing professional development—for example for US National Park staff or the educational programmes organised by Lesvos Petrified Forest Geopark, Greece;
- student training—for example Masters course in Geoheritage and Geoconservation at University of Minho, Portugal;
- community awareness programmes—for example the Bulimba Creek Catchment Co-ordinating Committee in Australia (Cameron, Conover and Micic, 2001) and the Community Earth Heritage Champions initiative in Hereford and Worcester, England (Miles, 2009);

- state of the environment reports, etc.—for example Stace and Larwood's (2006) on geodiversity in England;
- urban geology guides—for example urban geology trails and walks in London (e.g. Robinson, 1984, 1985), Edinburgh, Chester, Bristol, Exeter, Gloucester, Bath and Malvern in the United Kingdom; there is even a trail associated with a single building, viz. Worcester Cathedral.

13.3 'Point' and 'diffuse' threats and their management

In Chapter 5 we saw that in pollution management, the terms 'point pollution' and 'diffuse pollution' are often used to distinguish between those activities where the pollutants are discharged from a readily identifiable source such as a pipe or drain, and those activities that occur over a wide area associated with particular land uses such as agriculture or urban runoff. It is important to recognize that different management approaches are used in each case. Point pollution is tackled by using powers and regulatory methods to control the specific sources of pollution and by prosecution of the polluters if appropriate. Diffuse pollution, on the other hand, is often tackled through policy development and implementation, education and partnership working.

This analogy can be usefully applied to geoconservation to distinguish between activities that may destroy or damage specific sites or features ('point threats') from those that have an effect on the wider landscape ('diffuse threats'). It is suggested that the former impacts are best managed through the protected area approach by using legislation to designate important scientific and other sites and then by site management and enforcement. This approach will apply most to certain elements of geodiversity, particularly rock, mineral, fossil and landform sites. On the other hand, different approaches, similar to those used to tackle diffuse pollution are more appropriate to protect other elements of geodiversity including topography/land form, physical processes and soils in the wider landscape. Indeed, diffuse pollution is one of the diffuse threats to the wider landscape.

13.4 Conclusions

Despite the diversity of geoconservation systems and networks currently in use, many countries and organisations now agree that the aim of geoconservation should be to maintain the range of earth science features within their borders. Although few yet use the terminology, what this means is that they aim to conserve a representative selection of the geodiversity of their country/province. They can do so using a range of methods, the choice of which will depend on the element of geodiversity being considered, the nature of the area and the objectives of the work.

14 Comparing and integrating geodiversity and biodiversity

All the world's a stage

As You Like It (1599) by William Shakespeare

This chapter revisits some important aspects of geodiversity and biodiversity. First, it discusses some criticisms that have been levelled at the use of the term 'geodiversity'. Secondly, it discusses how geodiversity might be measured and monitored. Thirdly it assesses whether geological extinction is a threat in the same way as in biology. Fourthly, having established geodiversity as an important and independent field of nature conservation, it goes on to explore the paradox that it needs to be better integrated with biodiversity. But it then goes further to argue for this integrated nature conservation to be part of a truly holistic approach to sustainable land management more generally. Finally, it examines the possibility of conflicts between geodiversity and biodiversity conservation and discusses ways of resolving these conflicts.

14.1 Criticisms of 'geodiversity'

I began this book with a discussion of how the term 'geodiversity' had emerged in response to the spectacular success of 'biodiversity' in regalvanizing efforts in wildlife conservation. Although this can be seen as 'jumping on the biodiversity bandwagon', it is also the case that geoconservationists have come to realise that 'diversity' is a basic guiding principle underlying all nature conservation, not just bioconservation. As noted in Section 13.1, the concept of representative sites has been long recognised in geoconservation but the obvious point that what is

Geodiversity: Valuing and Conserving Abiotic Nature, Second Edition. Murray Gray.

needed is a representative sample of geodiversity had not dawned until the focus on biodiversity made it obvious.

But as we noted in Chapter 1, there have been criticisms of the use of the term. These criticisms may be summarised as follows, with some responses in italics:

1. We 'may be attempting to draw too strong a parallel between sites, landscape features and processes in biology and geology', particularly given 'the amounts of time commonly involved' (Joyce, 1997, p. 39).
   This comment on timescales is unclear since it should be noted that processes in biology and geology can both occur on very short or very long timescales.
2. There is 'a contrast between an active ecosystem involving living organisms which may come to an end completely, e.g. by the death of the organisms, and the non-living processes of the landscape or earth's subsurface, where processes may stop, restart, and change their nature with no parallel to the death or extinction of living organisms' (Joyce, 1997, p. 39).
   This is partly accepted but as we shall see below, there is a geological equivalent of extinction if particular types of minerals, rocks, landforms, etc. are lost by over-exploitation or other human or natural processes.
3. The heritage significance of a geological site, landform or region may in some cases 'lie not in its diversity but in its uniformity. Examples might be a gibber plain, an active dune field, a thick and uniform sedimentary sequence, a mass extinction fossil site' (Joyce, 1997, p. 39). Similarly, Stock (1997, p. 42) believes that we should 'avoid the notion that 'geo-diversity' has some objective quality to earth scientists as 'biodiversity' may have to biologists and ecologists. Diversity of landscapes (say) is not necessarily an over-riding priority in heritage'.
   This point is understood (Figure 14.1) but is matched in biology. The spectacle of a dune field stretching as far as the eye can see is matched in biology by

Figure 14.1 Repetition of one aspect of geodiversity can be impressive. Tertiary basalt columns cut across by marine erosion, Giant's Causeway, Northern Ireland.

swarms of Monarch butterflies, flocks of flamingos on African lakes or herds of caribou in arctic North America. As Sharples (2002a) and Houshold and Sharples (2008) have pointed out, what is important is the overall geodiversity of the planet and not of particular systems.

4. There appears to be 'no scientific advantage in trying to force geodiversity into being of the same conceptual type as biodiversity, which is based on biological and ecological theory involving species, genera, families, etc. Indeed, there could be considerable intellectual disadvantages in such usage' (Stock, 1997, p. 42).
   It is accepted that the nature of biodiversity and geodiversity are not identical. Kiernan (1996, p. 12) for example, notes that 'there are important differences between the way landforms come into being and are related to one another and the evolutionary relationships that exist between biological organisms'. But it is not accepted that this precludes the use of the term 'geodiversity'. The important issue is whether it fulfils a particular role or leads to new insights, and it is the contention of this book that exploring and valuing the planet in terms of its geodiversity achieves precisely these objectives. It is also important to recognise that there are important similarities between biodiversity and geodiversity. For example, the concept of soil 'orders' is established in soil classification as in biological classification (Kiernan, 1997a, p. 4). The terms 'species' and 'varieties' have long been used for minerals and are an essential part of palaeontology. We can also recognise landform assemblages as the equivalent of plant or animal assemblages. As Kiernan (1996, p. 12) points out 'There are different types and assemblages or systems of landforms just as there are species and communities of plants and animals, and, like plants and animals, there are some landforms that are common and others that are rare, some that are robust and some that are fragile ... The case for conserving geodiversity is just as compelling as the case for conserving biodiversity, whether undertaken for intrinsic or for utilitarian reasons, whether judged on scientific, moral or economic terms'.
5. Since an ecosystem includes the abiotic components of habitat, there is little need for separate recognition of geoconservation or geodiversity.
   This is a very biocentric view of the world. As demonstrated in Chapter 4, geodiversity has its own independent set of values, only one of which is the foundation for biodiversity. Geodiversity has its own independent conservation strategies, while at the same time recognising that the biotic and abiotic elements of nature conservation need to be better integrated (see later).
6. We know that geology is diverse so isn't 'geodiversity' just another name for geology? (Brocx and Semeniuk, 2007; Brocx, 2008).
   Biologists certainly have not taken this view, using 'biodiversity' at both global and local levels even though 'biology' is known to be diverse. It is true, however, that in recent years the word 'geodiversity' has often been used very loosely in the literature, often as a synonym for geology. But it has a very specific definition (see Chapter 1) focusing on the diversity of abiotic nature and the need to conserve this diversity. It is this focus on the value of geological/geomorphological/ pedological/hydrological diversity that distinguishes the geodiversity approach because geological diversity has traditionally never been fully valued and celebrated in geology research, teaching or public communication.

7. Geodiversity is inappropriate to encapsulate the principles implied by geoheritage (Brocx and Semeniuk, 2007).
   Geodiversity and geoheritage have different definitions (see Chapter 1). Geodiversity has resulted from billions of years of Earth history (see Chapter 2) and therefore clearly does encapsulate the principles of Earth heritage.

14.2 Measuring geodiversity

A huge amount of work has been carried out world-wide in measuring biodiversity, drawing up lists of threatened species and implementing habitat action plans and species recovery programmes. In comparison to this, the measurement of geodiversity has barely begun.

Biologists refer to at least six measures of biodiversity (Table 14.1). They have also combined measures in order to develop indices of diversity. For example the Shannon index, (H′) which takes into account both the number of species and the evenness of occurrence of the species:

$$H' = \sum n_i/N \, \mathrm{Ln} \, n_i/N$$

where:

Ln = Napierian logarithm
n_i = area covered by the ith category
N = total area studied.

This section discusses some ideas, initiatives and issues related to the measurement of geodiversity.

14.2.1 Spatial diversity

Various authors have tried to measure spatial geodiversity, but it is unclear how this could be done for all elements of geodiversity given the differences between biodiversity and geodiversity. For some aspects it would be relatively easy. In palaeontology, for example, we could simply catalogue the fossil species discovered and establish rarity or abundance at international, national or more local levels. It is likely that we would discover many threatened and vulnerable fossil species. Similarly, international stratotypes have been identified (see Section 6.9)

Table 14.1 The main measures of biodiversity.

α-diversity	Species numbers in each habitat
ß-diversity	Proportion of habitats in which a given species is present
γ-diversity	Total number of species in a region
Habitat diversity	Number of habitats in an area
Age diversity	Frequency distribution of ages of a species in an area or habitat
Genetic diversity	A measure of clonal variation in a species

and their loss would be fairly obvious. In the case of minerals, the task would be less straightforward because of the many mineral species and secondary species, but nonetheless it could be done in much the same way. In the case of rocks, landforms, processes and soils, we would have to rely on internationally agreed classification systems. As we saw in Chapter 13, these exist for rocks and soils, but not yet for landforms and processes though there are many commonly used landform and process names. Hardest of all would be landscapes since these are ill-defined and continuous, but there are similarities here with habitat definition and measurement. However, in both this situation and elsewhere, we start to confront issues of scale. For example, if we are trying to measure the geodiversity of a valley, do we measure the diversity of its soils, its slopes, its fluvial processes, its channel gravels or what? Hjort and Luoto (2012) have argued that it is possible to measure geodiversity from space using digital elevation models and remote sensing data, but this is unlikely to be able to assess small-scale geodiversity or dynamic aspects, e.g. diversity of channel gravels or fluvial processes.

An important distinction between geodiversity and biodiversity also becomes evident here. In the geosciences we place great value on interpretations of, for example, rocks and landforms and these may change through time as ideas and theories develop. Measurement of geodiversity must therefore involve more than simply counting the number of different rocks and landforms and must also include the interpretations that we place upon them as well as the interrelationships between them.

Some recent attempts have been made to measure geodiversity. Serrano and Ruiz-Flaño (2007) proposed a Geodiversity Index (Gd) as:

$$Gd = EgR/LnS$$

where:

Eg = number of different physical elements in the unit
R = coefficient of roughness of the unit
S = surface area of the unit (km^2)
Ln = Napierian logarithm.

They show a map of part of central Spain where Gd has 5 categories from very low to very high. This index has also been used by Hjort and Luoto (2010) to map the geodiversity of high-latitude landscapes in northern Finland.

Also in Spain, Carcavilla, Durán and López-Marinez (2008) explored the issue of measuring geodiversity but without giving formulae. They recognised:

- *variety*—the number of geodiversity classes in a region, termed the primary or intrinsic geodiversity;
- *frequency*—the number of times that each class recurs. 'It is important to ascertain whether the classes are equal in area, whether one predominates, whether any are highly fragmented and so on';
- *distribution*—how the classes of geodiversity are spatially arranged.

Figure 14.2 illustrates these concepts and shows some theoretical patterns of geodiversity distribution.

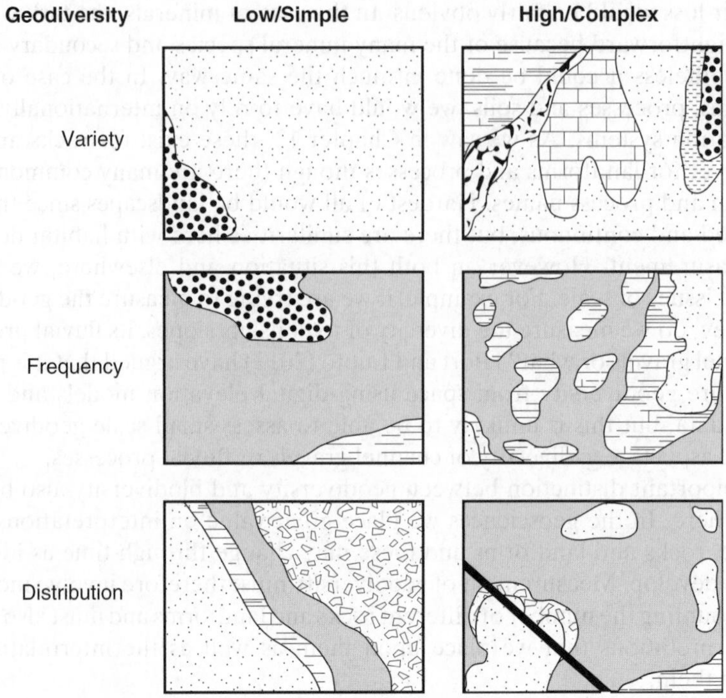

Figure 14.2 Spatial patterns of geodiversity (after Carcavilla *et al.* (2008) Different approaches for the study of geodiversity. Paper presented to the 33rd International Geological Congress, Oslo. Reproduced with permission).

Pereira *et al.* (2013) have developed a methodology for mapping the 'variety' or 'primary' geodiversity of Paraná State in Brazil. Their sources were small-scale maps of geology, geomorphology, palaeontology and pedology, together with information on mineral occurrences. All these maps were overlain with a 25 × 25 km grid to generate partial geodiversity index values (e.g. in the case of geology, the number of different stratigraphical and lithological units in each grid square). These values were then added together to produce a Geodiversity Index isoline map that is intended to focus attention on high geodiversity areas for land-use planning purposes such as conservation, land management and use of natural resources.

Ruban (2010) saw geodiversity as referring to the diversity of geoheritage geosites and developed a number of equations including the following:

- Geodiversity 1 = total quantity of geosite types occurring on a given territory
- Geoabundance 1 = Total quantity of geosites on a given territory
- Georichness 1 = Quantity of geosites, where each type is represented

Ruban also presents equations for geodiversity loss. It should be noted, however, that geosites represent only a limited part of geodiversity. Furthermore, geosites are identified and classified differently in different countries so that comparing one country or even region with another becomes difficult or impossible. Knight (2010) provides further criticisms of Ruban's approach.

Ibáñez *et al.* (1998) used quantitative indices including the Shannon index to measure global pedodiversity of the major soil groups (MSG). They found that the climatic zones closest to the Poles (boreal and cold) have the lowest pedorichness, possibly because the Quaternary glaciations removed the soils and a new pedodiversity has not had time to develop. Benito-Calvo *et al.* (2009) used simple morphometric, morphoclimatic and geological classifications in the Iberian peninsula to calculate geodiversity indices. Not surprisingly, they found that the highest diversity values are associated with Alpine collision zones whereas the lowest diversity values are found in Cenozoic basins and Mesozoic areas with no significant tectonic deformation.

Measurement ought to have a purpose or it becomes little more than a descriptive, academic exercise. The main aim of measuring geodiversity ought to be either (a) so that repeat measurement helps us to understand what geodiversity losses are occurring, or (b) as a tool for sustainable land/georesource management. Let us hope that we can bear this in mind and not engage in years of fruitless research and argument about what geodiversity indices to use.

14.2.2 Geological extinction?

In Chapter 4 we saw that there is economic value in the diversity of geological materials. In Section 12.1 we noted that the material resources of the planet are finite and the principles of sustainable development dictate that we should use these resources wisely for the sake of future generations who might also want to use them. This is very important since most geological resources are non-renewable, or renewable only on very long timescales. According to Milton (2002, p. 123) 'Biodiversity conservation is ... about sustaining the widest possible variety of living things—protecting rare and vulnerable organisms and their habitats, trying to prevent extinction'. The question arises therefore whether there is a geological equivalent of biological extinction?

This will occur where a particular type of rock, sediment, mineral, fossil, landform, process, landscape or soil has been so heavily quarried, utilised, impacted or degraded that it is danger of disappearing completely. Many minerals can be considered extinct because their single known occurrences have been destroyed by quarrying, overcollecting or other human intervention. One example of this is the highest quality, pure white marble from the famous Carrara marble quarries in north Italy that has been utilised by sculptors through the centuries, for example Michaelangelo's *David* in Florence. This resource has become exhausted and lower quality marble is now being used (*The Times*, 26 March 2001). Martinez-Torres, Alonso and Valle (2011) refer to the very limited remaining extent of an amber deposit in northern Spain. And we also know that the 'Blue John' banded fluorspar deposits in Derbyshire, England are almost fully exploited (Ford, 2005). In Section 4.5.4 the shortage of Rare Earth Elements was discussed, although some are rarer than others. In Section 5.11 we saw how overcollecting including the use of hammers and power tools has devastated many important fossil and mineral sites. In some cases this has led to loss of fossils. Swart (1994, p. 321) believes that overcollecting at the world famous Ediacara Fossil Reserve in South Australia has lead to the virtual destruction of not just the site, but also the loss of many of the fossil species. This immediately raises three particular issues:

1. Most of the fossil species from the site have been curated in museums or exist in private collections. Therefore we must distinguish between field extinction of fossils and minerals and extinction more generally.
2. It is difficult to prove field extinction since other examples of the same fossils, minerals or rocks may exist at depth at a site, or in geologically unexplored parts of the world, or even beneath the Greenland or Antarctic ice-sheets, for example.
3. Again we run into issues of scale, since extinction locally and nationally may also be regarded as important, as well as global extinction.

In comparing biological and geological extinction, Pemberton (2001) points out that 'in a lot of instances, rare or threatened species can be propagated or bred in captivity. On the contrary many geo features have formed under conditions, climatic or geological, that are now inactive. They are essentially relict or 'fossil' features which, once disturbed, will never recover or will be removed forever'. In similar vein, Kiernan (1991) commented that since many landforms are relic features, 'They are irreproducible, at least over a human timescale. Conservation of genetic material in laboratories or botanical gardens is not an option in conservation of these landforms, only recognition and appropriate management of the stocks we have now'. The fact is that most georesources are non-renewable and it is simply not possible to recreate many aspects of geodiversity once they are lost. For example, if the international stratotype section of the Ordovician-Silurian boundary at Dobb's Linn in Scotland is lost, it is lost forever. This means that the case for conserving geodiversity is often stronger than that for biodiversity.

14.2.3 Increasing geodiversity

A widely used method for assessing the availability of geological resources is the McKelvey box (Figure 14.3). In this scheme the outer box is the resource base, which is the total amount of a material that exists on Earth (= total resource). However, much has yet to be discovered (= hypothetical resource) and many geomaterials could not be economically or technically exploited (= conditional reserves). Therefore there is an interior box representing the reserves, which is that part of the total resource whose exploitation might conceivably be economic and technically feasible in the foreseeable future. However, estimates of reserves and resources will vary with changes in economic conditions and geological knowledge. Shortages in reserves will raise the price of a commodity and therefore increase the pace of geological exploration that in turn will increase the reserves. Shortage of reserves will also increase price, making it more economic to exploit lower-grade and less-accessible resources.

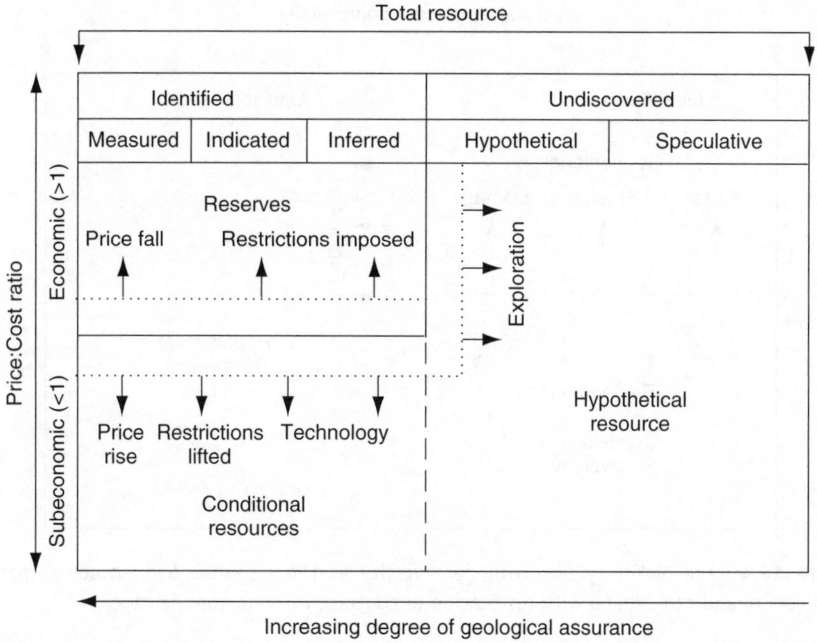

Figure 14.3 The McKelvey Box showing the relationship between resources and reserves. The size of a reserve relative to the resource varies, expanding because of exploration, price rises, improved technology, etc., but contracting with price falls, governmental restrictions, etc. (From Bennett, M.R. & Doyle, P. (1997) Environmental Geology. Reproduced with permission of John Wiley & Sons.)

We can now apply this resource/reserve concept to geodiversity and geoheritage (Figure 14.4). The outer box would then refer to all the geodiversity in the world, at least some of which has yet to be described and mapped, and which has been increasing since the Earth formed (see Section 2.5). But it is not necessary or desirable to conserve all this diversity given the pragmatic problems of doing so and the needs of society to utilise some of the georesource. We therefore have an inner box related to the geoheritage. These are the elements of geodiversity that are judged to be significant and worthy of conservation because of their values (see Section 1.3 and Chapter 4). Geoconservation should aim to avoid the further loss of geoheritage, while georestorartion (Section 10.4) and field research should seek to increase it. It is estimated that there are still millions of wildlife species awaiting discovery, description and classification. And what is true for biodiversity is also true for geodiversity.

New mineral discoveries are still being made. For example, at the Broken Hill silver-lead-zinc ore deposit in western New South Wales, Australia, geoscientists from Museum Victoria have reported over 70 secondary mineral species, three of which are new (mawbyite, segnitite and kintoreite). At the Lake Boga Granite in northern Victoria, amongst the 50 phosphate species are at least two new species (ulchrite and bleasdaleite) and at the nearby Wycheproof Granite, two

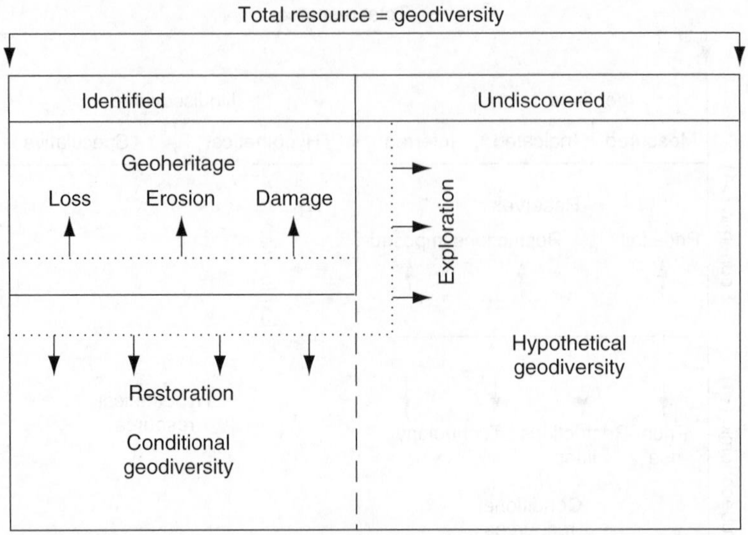

Figure 14.4 The McKelvey Box with geodiversity as the resource. Geoheritage is part of geodiversity and can expand with restoration or contract with erosion, damage, etc.

new zirconium minerals have been discovered (wycheproofite and selwynite). Brooks (2012) describes over 200 rare minerals from two sites in Greenland. These cases illustrate the scope for the recognised geodiversity of minerals to increase significantly in future. Since most of these new minerals are likely to be rare, it follows that they could be easily lost without geoconservation.

Similarly in palaeontology, discoveries of new species are very common. For example, over the past 15 years, Professor Xu Xing from the Chinese Academy of Science together with colleagues and collaborators have described many new species of dinosaur from sites in China (e.g. Xu *et al.* 2003, 2012). There is no doubt that there are probably millions of new fossil species still undiscovered. As noted above, other elements of geodiversity have less clear classification systems and new names, for example, for rocks, soils and landforms are likely to be introduced in future.

14.3 Integrating geodiversity and biodiversity

Having put forward the case for valuing and conserving geodiversity in this book, it is essential to end it with a plea for greater integration of nature conservation objectives, recognising the roles of both geodiversity and biodiversity. This is not to say that there is not a risk in linking geodiversity too closely to biodiversity since the former may come to be seen as simply the foundation of biodiversity rather than having many other values as explained in Chapter 4. This should be avoided.

Several authors have, however, reached the conclusion that integrated nature conservation is important and some (e.g. Huggett, 1995; Brierley and Fryirs, 2005) have begun using terms such as biophysical or geoecological management. Knoll, Canfield and Konhauser (2012) have written a whole book on *Fundamentals of geobiology*. Pemberton (2001) has made the point that 'Treating the entire natural environment as an intricately related system is quite logical from a land management and conservation perspective and quite obvious from the way the natural environment works ... To be serious and consistent about nature conservation there needs to be a return to the approach where the entire natural environment is considered not just the above ground or living part of the environment'. As Carson (1996, p. 8) points out, the Victorian' naturalists' understood the value of integration but somewhere the two were separated'.

At the global scale, recognition of the interdependency of geological and biological systems leads us inevitably back to the Gaia philosophy, pioneered by Lovelock (1979, 1995) and outlined in Section 4.2. As Manning (2001, p. 20–21) puts it, '... we owe to James Lovelock the general recognition that life has not been merely a passenger on the planet; it has affected certain crucial processes throughout most of Earth's history'. This integrated approach to geology and biology has recently found favour with scientists throughout the world through the development of Earth Systems Science, which investigates the linkages through geological time between lithosphere, biosphere, atmosphere and hydrosphere. This new thinking has been stimulated by new observational tools (e.g. remote sensing), new observational platforms (e.g. satellites), new analytical methods (e.g. chemical and biochemical analyses) and new theories and models, aided by increased computational power (Lamb and Singleton, 1998; Boulton, 2001). Parks and Mulligan (2010, p. 2751) consider that geodiversity 'incorporates many of the environmental patterns and processes that are considered drivers of biodiversity'. These drivers include climate, topography, geology and hydrology that together provide the resources of energy, water, space and nutrients. They then investigated the global datasets that would allow links between geodiversity and biodiversity indices to be explored and conclude that 'geodiversity has potential as a conservation planning tool, especially where biological data are not available or sparsely distributed'.

At the regional scale, a number of authors have concluded that conserving biodiversity may be best achieved by conserving geodiversity, and have used the metaphor of conserving the stage rather than the actors (Hunter, Jacobson and Webb, 1988; Anderson and Feree, 2010; Beier and Brost, 2010). This is likely to be particularly relevant to the issue of protecting biodiversity in times of climate change. Anderson and Feree (2010) hypothesised that geological factors may be the primary influence on biodiversity patterns in 14 Northeastern US States and three Canadian Maritime Provinces. The results of linear regression analysis demonstrated that four geophysical factors: the number of rock types, latitude, elevation range and the amount of calcareous bedrock, predicted species diversity with a high degree of certainty ($R^2 = 0.94$). They also tested 18 700 locations that contained 885 rare species and found that 40% of species are restricted to a single lithological class. They concluded that biodiversity can best be protected under current and future climates by protecting geophysical settings. Similarly, Beier and Brost (2010, p. 701) advocated conserving the arenas of biological activity, rather than the temporary occupants of those arenas (Hunter, Jacobson and

Webb, 1988). In other words, they proposed that 'land facets' with uniform topographic and soil attributes should be identified and used to design reserves and linkages that would function through times of climate change. This has been echoed by Bruneau, Gordon and Rees (2011) who believe that 'spatially integrated approaches at the landscape/ecosystem scale are arguably most critical in a changing world'. And Brazier *et al.* (2012) have reviewed how an understanding of geodiversity can assist biodiversity conservation at times of climate change. They conclude (p. 227) that 'conservation management for the benefit of specific habitats or species is compromised where it does not refer to their physical setting and an understanding of the role of geomorphological processes'.

At the more local level, Hopkins (1994) commented that the ecological research community has paid little attention to the geologically-influenced patterns of plant distributions, partly because experimental investigation and interpretation has proved so difficult. However, he suggests that 'if biodiversity is to be developed scientifically, it is now timely for ecologists, geologists and geomorphologists to collaborate more closely on the interactions between species habitats and geological phenomena' (p. 3). In other words, an important first step in achieving integration of biodiversity and geodiversity is to establish more clearly the scientific links between the two.

Box 14.1 Geodiversity/biodiversity links in Hickory Run State Park, Pennsylvania, United States

Through Hickory Run State Park in Pennsylvania runs the outer limit of the last ice-sheet to flow southwards into the USA about 20 000 years ago. As a result, the park displays two very different landscape types that in turn have produced two distinctive sets of wildlife habitat. The undulating nature of the western part of the park reflects the glacial deposition associated with the end moraine of the ice-sheet and the valley erosion associated with glacial meltwater rivers. The eastern part of the park is higher and was not covered by the ice, but was affected by periglacial processes. These included the frost disturbance of rock outcrops, the frost weathering of boulders, and the downslope movement of these boulders to accumulate in the famous Hickory Run Boulder Field, a National Natural Landmark and State Park Natural Area (Figure 9.11). On the glaciated western side of the park, the end moraine is dominated by thin and moist soils, evergreen trees and sphagnum moss bogs. Blackburnian warbler, redbreasted nuthatch and northern water thrush inhabit this area, and in the spring spotted and Jefferson salamanders and wood frogs flock to the bogs to breed. On the other hand, the unglaciated eastern side of the park is dominated by beech and chestnut oak trees inhabited by the American redstart, red-eyed vireo and Louisiana water thrush. Hickory Run State Park therefore illustrates how the geological evolution of a landscape has produced a diversity of landforms and materials that in turn have provided a range of habitats in which biodiversity has evolved (Gray, 2005).

Box 14.1 gives an example of such links, and detailed research investigating the links is beginning to be carried out. Moles and Moles (2004), for example, carried out detailed geochemical work on the soils of part of The Burren National Park, Ireland, and then related the variations to biodiversity (Moles and Moles, 2004). At the landscape scale (km^2), distribution patterns of vegetation and species are linked to bedrock geology and the distribution of glacial sediments. At the landscape element scale (ha), weathering and mass movement processes have created a pattern of alkaline soils in exposed limestone till near drumlin crests, and more acid soils in colluvium on lower slopes. This is illustrated in Figure 14.5. At a patch scale (10 s m^2) geological influences on vegetation include the occurrence of patches of clay-rich soil derived from glacially transported shale-boulders. Finally, at the site scale (1–10 m) boulder strewn surfaces inhibit access by grazing animals (hares, goats, cattle) thus resulting in more successful seed production than in exposed grassland. Their conclusion is that effective conservation of

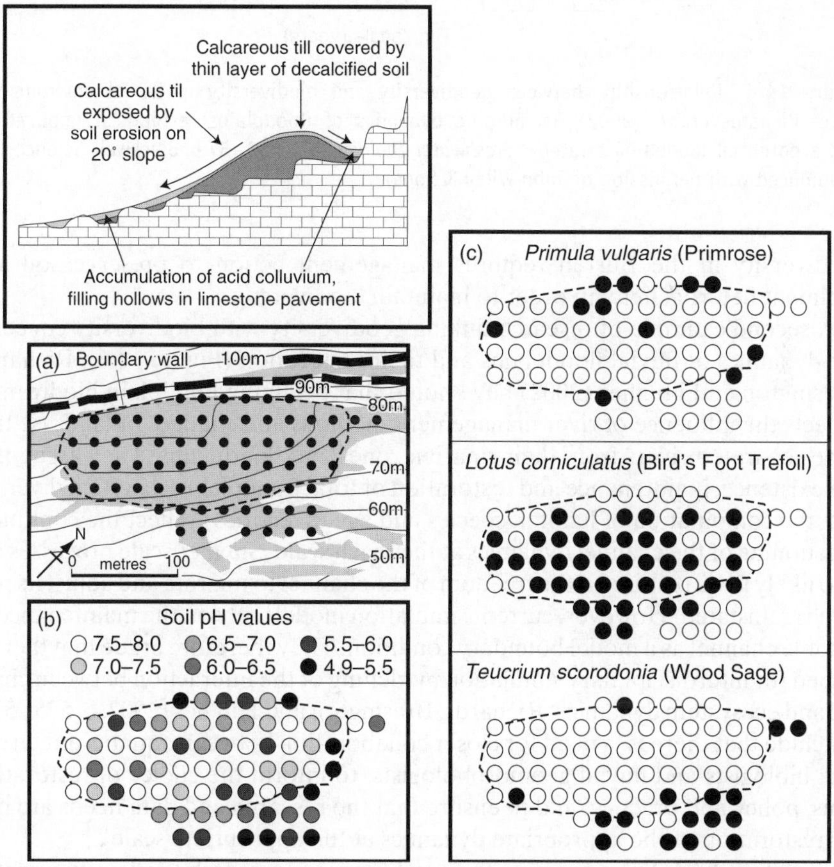

Figure 14.5 Relationship between geodiversity and biodiversity in part of the Burren National Park, Ireland. (Reproduced with permission of Norman Moles.)

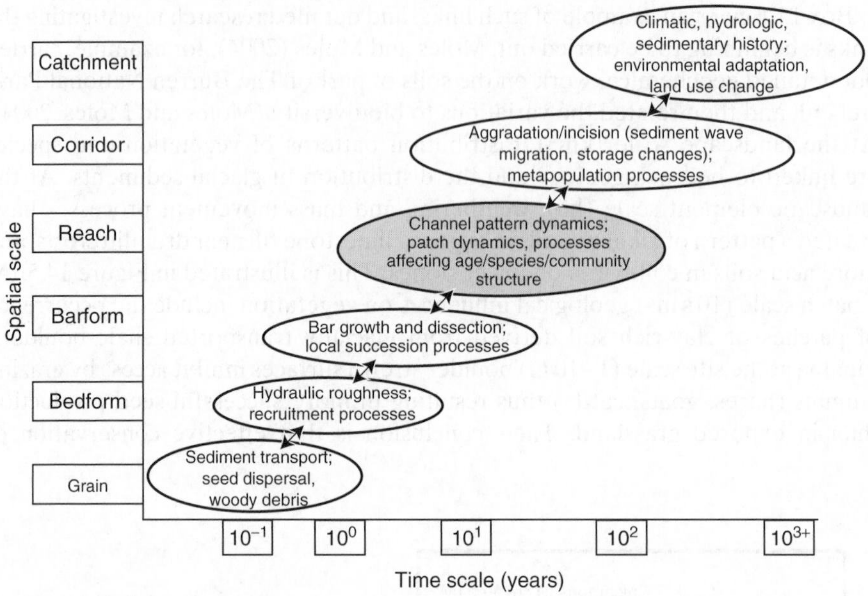

Figure 14.6 Relationships between geodiversity and biodiversity in fluvial environments (after Richards *et al.* (2002) Geomorphic dynamics of floodplains: ecological implications and a potential modelling strategy. *Freshwater Biology*, 47, 559–579, Blackwell Science Ltd. Reproduced with permission of John Wiley & Sons).

biodiversity in the Burren requires management action to preserve soil and sediment heterogeneity in order to maintain biodiversity.

A second example of research linking geodiversity with biodiversity concerns the dynamics of fluvial floodplains and their restoration (Figure 14.6). Richards, Brasington and Hughes (2002) have noted that declining floodplain biodiversity reflects the influence of river management methods and changes of land-use that have separated rivers from their riparian zones and floodplains. They argue that the existence, maintenance and restoration of total floodplain ecosystem diversity (i.e. the combination of habitat, species and age diversities) reflect the continued functioning of the channel dynamics at the reach scale. Smaller scale processes are also likely to be restored if the freedom of the channel to migrate and adjust its pattern is reinstated. However, current simulation models rely on the maintenance of a stable channel as a model boundary condition. They therefore argue that there is a need for interdisciplinary simulation modelling of the interaction between channel and ecosystem dynamics. Richards, Brasington and Hughes (2002, p. 575–576) conclude that there is ‘a need for closer collaboration between aquatic and terrestrial biologists and fluvial geomorphologists, to inform the choice of restoration aims, policy and practice, and to ensure that the research and data needs are met for restoration of the appropriate dynamics at the appropriate scale’.

Sharples (2002a) also foresees the need for more research to identify in detail the role of geodiversity in sustaining ecological systems, and the extent to which changes in abiotic variables affect biodiverity:

> ...if we accept the value of geodiversity in ecological processes then we need to determine the role which particular elements of geodiversity play in ecological processes. Such a determination is a matter of scientific research and monitoring, which allows us to determine whether the disturbance, degradation or destruction of a "geo-phenomenon" will result in an unacceptable degree of change or degradation to the broader natural environment and ecological processes of which it is part; if so then it is a thing of significant geoconservation value which should be managed accordingly to avoid such detrimental effects' (Sharples, 2002a, p. 6).

Gordon *et al.* (2001) stress the need for an integrated 'geo-ecological' approach to the conservation management of fragile upland environments that are sensitive to both natural and human impacts, and the same issues apply to mountains in the Czech Republic and Sweden (Gordon *et al.*, 2002).

14.4 Integrated land management

But it is also necessary to go beyond the integration of geodiverity and biodiversity to see nature conservation as an important part of land management and planning in the widest sense. Usher (2001) uses three spheres 'geosphere', 'biosphere' and 'anthroposphere' to illustrate the integrated nature of the issues that confront nature conservation (Figure 14.7) emphasising the need for integration of natural and cultural landscapes and management objectives. Valuing and conserving geodiversity must be at least an equal partner in this approach: in fact it is difficult to envisage a workable land management approach that does not include geoconservation.

Examples of holistic approaches both in theory and practice are beginning to emerge. The Council of Europe's Pan-European Biological and Landscape

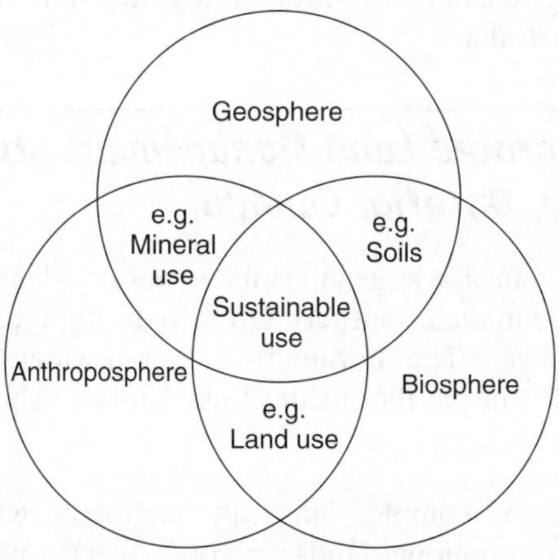

Figure 14.7 The overlapping nature of the geosphere, biosphere and anthroposphere (after Usher, M.B. (2001) Earth Science and the natural heritage: a synthesis. In Gordon, J.E. and Leys, K.F. (eds) *Earth Science and the Natural Heritage*. Stationary Office, Edinburgh, 315–324, by permission of Scottish Natural Heritage).

Figure 14.8 The Integrated Coastal Zone Management Project for beach and dune systems in Donegal, Ireland.

Diversity Strategy (PEBLDS) emphasises the value of landscapes as a mosaic of cultural, ecological and geological features. Integrated Catchment Management and Integrated Coastal Zone Management are now widely applied. An example of the latter is the management project carried out on the beach and dune systems of Donegal, Ireland by Donegal County Council, Limavady Borough Council and the University of Ulster (Figure 14.8). Similarly, English Nature has proposed a new integrated approach to coastal management (Covey and Laffoley, 2002). Boxes 14.2 and 14.3 give examples of fully integrated land-use planning projects from Canada and Australia.

Box 14.2 Integrated Land Management Strategy, Oak Ridges Moraine, Ontario, Canada

The Oak Ridges Moraine is a large interlobate moraine deposited at the margin of the Laurentide ice mass in southern Ontario during the Late Wisconsinian Stage of the Pleistocene. The moraine is 2–20 km wide, 50–250 m high and extends for over 160 km parallel to the Lake Ontario shoreline It performs several functions:

- it has a 'diverse and complex landscape character, which offers an aesthetic contrast to a regional landscape of lacustrine plains dominated by agricultural and urban development' (Frazer and Johnson, 1997, p. 18);
- it forms the headwaters and aquifer recharge area for the region's surface water and groundwater supplies;

- it contains very large sand and gravel deposits close to the developing conurbations of southern Ontario;
- it has significant biodiversity and prime agricultural areas.

The moraine has been increasingly subject to development pressures in the Greater Toronto area where the population is predicted to rise from 4.25 million in 1997 to 6.5 million by 2022. Because of these development pressures on a valuable geological, ecological and landscape resource, the Provincial Government undertook a long-term planning study of the moraine as a landscape unit and involved local residents by forming a Citizens Advisory Committee to work alongside a Technical Working Committee. The aim was to try to gain consensus on the future sustainable planning of the area and to produce an integrated land-use plan. In 2001 the Ontario Provincial Government passed the *Oak Ridges Moraine Protection Act* and developed the *Oak Ridges Moraine Conservation Plan*. Geology and geomorphology are prominent in the plan, one of the objectives being to ensure that the area 'is maintained as a continuous natural landform and environment for the benefit of present and future generations'.

'Any land use changes requires a Landform Conservation Plan which must convince the approval body that the essential landform character will be maintained' (Frazer and Johnson, 1997, p. 19). The Landform Conservation System is accompanied by a Natural Heritage System to sustain biodiversity, and a Water Resources System to sustain surface and groundwater resources. Local planning policies have been amended to require municipalities to have regard to natural heritage values in judging planning applications. This is an example of the development of integrated land-use strategies landscape character assessment, planning policies and partnership working (Davidson *et al.*, 2001).

Box 14.3 Integrated catchment management at Mole Creek karst, Tasmania, Australia

The Mole Creek karst area lies in northern Tasmania (Figure 14.9) and most of the rivers flow underground. The area has been subject to a number of the problems characteristic of karst in agricultural areas, including vegetation removal, soil erosion and cave sedimentation, stock access to streams and reductions in water quality, tipping of chemical drums and stock carcasses in caves and sinkholes, and stream damming and diversion (Eberhard and Houshold, 2001). In addition, a number of potentially damaging operations have been proposed in recent years including logging and intensive pig rearing. Landownership is mixed with private ownership being predominant in the lower more fertile lands and public ownership in the higher areas.

However the situation is more complex with public areas also occurring on lower areas, including the Mole Creek Karst National Park, and these complexities of landownership make land management more difficult where surface operations can damage geomorphological interests both underground and downstream.

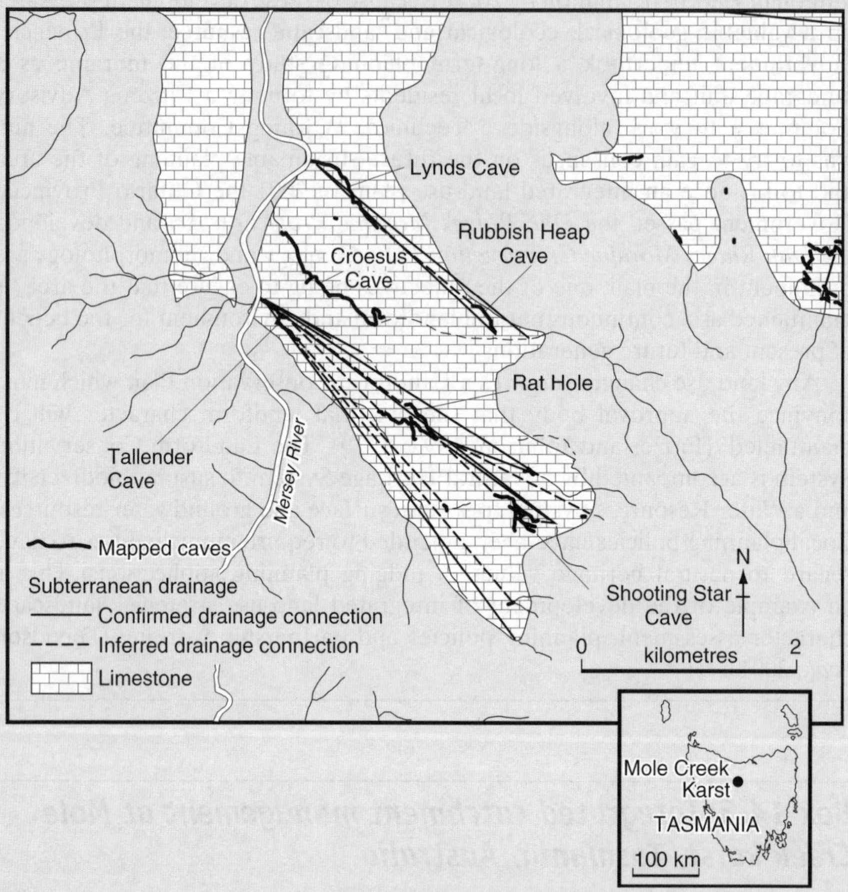

Figure 14.9 Map of the Mole Creek Karst Area, Tasmania, Australia (Map compiled and reproduced with permission of Rolan Eberhard).

As a result of these problems, Tasmania's Department of Primary Industries, Water and Environment (DPIWE) in co-operation with Australia's National Heritage Trust, is developing an integrated catchment management strategy for the Mole Creek karst. The elements of the strategy include:

- *Inventory and zoning*. This involves systematic documentation and GIS-storage of geomorphology and hydrology by mapping of surface landforms and caves and water tracing with dyes. This is followed by sensitivity zoning, a map-based tool for modelling the management implications of spatial variations in geomorphology, hydrology and ecology across the area.

- *Cross tenure protocol.* Agreement is being sought from private owners and public agencies for a protocol to promote sustainable land management approaches and practices in the karst area and in the knowledge of its special value and sensitivity. A project development committee has been established with representatives from state organisations and agencies, local authorities, farmers and other landowners.
- *Education.* Community awareness is being promoted through publications, media publicity and formation of a 'karstcare' group to start restoring cave degradation resulting from past recreational activities inside caves.
- *Ground works.* This includes fencing off karst features and streams, revegetation of cleared land and construction of off-stream water storage to avoid stock access to streams (Eberhard and Houshold, 2001).

14.5 Potential geodiversity/biodiversity conflicts

Having established the case for greater integration of geodiversity and biodiversity, it would be wrong to suggest that this will always be easy and straightforward. There are likely to be situations in which the interests of the two are likely to conflict. Conflict is a common occurrence even within biodiversity, particularly where there have been animal introductions which disturb existing populations (Milton, 2002).

Some biologists have taken the view that since biodiversity is about 'sustaining the widest possible variety of living things' (Milton, 2002, p. 123), the aim of Biodiversity Action Plans should be to maximise biodiversity, perhaps by restructuring landscapes to provide a greater range of habitats. For example, it has been suggested that in river restoration the main aim should be to maximise the types of habitat in order to maximise species richness. However, these measures can conflict with the objective of retaining and restoring authentic landforms and landscapes and is certainly contrary to Articles 21–23 of The Australian Natural Heritage Charter (see Box 12.10). In other words, it needs to be appreciated by biologists and others that the aim of river restoration is not just about enhancing habitats and increasing species richness, but is more fundamentally about restoring the river because of the existence and aesthetic values of a natural river. There are many examples of, for example, pool and riffle sequences being introduced into sluggish rivers where they quickly become silted up. In addition, in attempting to respond to the impacts of climate change on biodiversity, there may be proposals to remould landscapes to create a range of habitats into which species can adapt. There therefore needs to be a greater understanding of geomorphological landforms and processes amongst landscape planners and an integrated approach to landscape management.

There is also a tendency for some biologists to see natural geomorphological processes as threatening to biodiversity, and therefore to argue for preservation of the status quo. A UNEP-WCMC Report (2002) on mountain areas saw South America's Mountains as 'particularly vulnerable to 'destructive earthquakes' with approximately 88% of the mountain land area deemed at risk'. Similarly much has been written on the impact of climate change on wildlife, but climate and sea-level have been changing continuously, albeit rather more slowly, since the world formed and plant and animal life has had to adapt. Indeed, it could be argued that continuous planetary change and the operation of earth surface processes, have been important drivers in creating heterogeneity and new habitats and in delivering modern biodiversity. Habitats may change but change is part of nature.

Finally it should also be noted that there may be conflicts within geoconservation and that these may increase as climate changes. For example, should we favour the operation of a geomorphological process such as erosion over the retention of a finite site such as a fossil-bearing mine dump or a limited glacial deposit?

14.6 Conclusions

Despite objections from some geologists, the term and concept of 'geodiversity' appear to have validity and to reflect much current practice. We have barely begun to measure geodiversity but should only do so if there is a clear objective, for example so that we may know our heritage and whether it is being lost. Perhaps a Geodiversity Red Book can be established to sit alongside its biodiversity equivalent which lists those species that are rare or threatened with extinction and therefore in need of conservation. Geoconservation should be established as an independent field of nature conservation but also needs to be integrated with biodiversity conservation and both need to become part of sustainable land management strategies more generally, resolving any conflicts along the way by careful assessment of values on a case-by-case basis.

15
A Future for Geodiversity?

Innumerable shades of grey,
Innumerable shapes...
these stones have far more differences in colour, shape and size
Than most men to my eyes...

'On a raised beach' (1956) by Hugh MacDiarmid

15.1 Valuing and conserving geodiversity

Several conclusions arise from this book:

- The Earth has not only a huge biodiversity but also an enormous geodiversity and this diversity is of value both intrinsically and in utlitarian ways that are barely appreciated by human societies, and certainly not adequately in the 'ecosystem services' approach as currently applied.
- Not all of the geodiversity of the planet needs to be conserved: only those elements that are seen as being significant for one or more of their values. This proportion of total geodiversity is known as the geoheritage.
- There are significant threats to this geoheritage from a variety of human activities either directly or through influencing the rates of operation of earth surface processes. Different aspects of geodiversity vary in their sensitivity and vulnerability to different types of threat.
- Protection of geoheritage is a complex task with a wide variety of approaches and methods available. Much will depend on the element of geodiversity to be conserved (e.g. fossil site, geomorphological process or soil type). Most countries now agree that a principal aim of geoconservation should be to maintain a representative sample of a country's or area's geodiversity.
- While there are significant efforts at protection and management at the international level and in several nations, sub-nations and regions, some countries have barely begun the processes of geoconservation, particularly in

Geodiversity: Valuing and Conserving Abiotic Nature, Second Edition. Murray Gray.

the developing world. Consequently it is likely that significant elements of geoheritage are being lost.
- There is a huge need for education about geodiversity and geoheritage among the public and decision-makers for as Hillel (1991, p. 9) has put it, 'we cannot protect what we do not understand'.
- Geonconservation needs to be established as an independent subject in its own right but also to be integrated with wildlife conservation in a holistic approach to nature conservation, and beyond that to comprehensive and sustainable land management and natural resource management strategies.

As part of this wider approach, Crofts (2001) discussed six principles for sustaining the Earth's resources:

1. Accept that natural change is inevitable;
2. Work with natural functions and processes;
3. Manage natural systems within the limits of their capacity;
4. Manage natural systems in a spatially integrated manner;
5. Use non-renewable Earth resources wisely and sparingly at a rate that does not restrict future options;
6. Use renewable resources within their regeneration capacity.

Although it is possible to add, subtract or modify these principles, they do present the basis for a future strategy for sustaining geodiversity and ought to be embedded in each country's nature conservation strategy. Furthermore, we can conclude that the justification for geoconservation rests on two main principles (Gordon and Leys, 2001b, p. 7):

- the duty to future generations to preserve our geoheritage so that it may become theirs; and
- the direct benefits for mankind and the natural world from the sustainable use of natural resources.

In other words, everything we have we inherit from the past, and everything we use or lose deprives future generations of it. That, in a nutshell, is the case for valuing and conserving abiotic nature.

15.2 Predictions reassessed

So what is the future for geodiversity, geoheritage and geoconservation? In the first edition of this book, published in 2004, I made some predictions for the year 2025. Some of those predictions have already come true, but it is clear that others will not. In addition, some measures not predicted have been introduced.

- For example, at the international level, The European Landscape Convention is starting to have a significant impact and the IUCN and Council of Europe's Resolutions (Chapter 6) were not anticipated. However, I did predict that major efforts would be made to 'recognise, value designate and conserve a representative selection of the world's geodiversity' (Gray, 2004, p. 368), and sure enough, as we saw in Chapter 7, the IUCN has been assisting UNESCO towards making the World Heritage List more representative of global geodiversity (Dingwall *et al.*, 2005; Williams, 2008; Wood, 2009). Although, it is now unlikely that hundreds more geological and geomorphological sites will be included in the World Heritage List by 2025, it does seem possible that increased development pressures and conflicts over resources will increase the number of World Heritage Sites in Danger. The Global Geosites network, so strongly heralded 10 years ago, has failed to gain continuing support from the IUGS, but as predicted, the Geoparks initiative has expanded rapidly and countries are now being limited on the numbers they can bring forward (Chapter 8). On the other hand, we are still some way from a global Convention on Geodiversity or a European Directive.

- At the national level, there has been some progress to value geodiversity on a par with biodiversity, particularly notable being the new Norwegian and Spanish Laws that specifically include geodiversity (see Section 9.6). Certainly, the terms 'geodiversity', 'geoconservation' and 'geoheritage' are much more widely used and have been included in some national dictionaries. More countries have decided to identify a representative selection of their nation's geodiversity and designate these as protected areas. It is proving more difficult to organise 'the community, political, legislative, technical and financial support necessary to establish protection mechanisms and make them work' (Kiernan, 1991). The 'georegionalisation approach' has become more widespread and has also allowed recognition of national landscape diversity. There has been less recognition of natural topographic character and the need for its conservation, but there has been an increasing interest in river restoration and in allowing the continued operation of natural, dynamic geomorphological processes. My predictions were overoptimistic about the extent to which these initiatives would be supported by national legislation and policies such as codes of conduct, land-use planning policies, environmental impact assessments, land management strategies and 'State of the Environment' reports. Only in the United Kingdom have Geodiversity Action Plans expanded, but the United Kingdom has also launched a national UK Geodiversity Action Plan. Some progress has been made by governments to work in partnership with mineral and quarrying companies to minimise the use of virgin materials, the environmental impact of mining and the production of waste materials. There is also much greater public recognition that oil resources are finite and criticisms of governments and oil companies for not doing more to develop alternative energy sources. More academic research has been undertaken to establish the relationships between geodiversity and biodiversity at all scales from global Earth Systems Science to patch scale dynamics. Some university course modules and Masters degrees in Geoheritage and Geoconservation have been established, the most noteworthy of which is the Masters degree at the University of Minho, Portugal, begun in 2005. My prediction that a *Journal of Geodiversity and Geoconservation* would be launched has proved correct, albeit under the title *Geoheritage*.

And Sir David Attenborough's short series on *Early Life* confirmed my prediction that he would present a television series on fossils. Many other television geology series have also been screened.

- At the local/community/individual level, progress has been most spectacular in some areas, particularly in Geoparks, at geotourism hotspots and in Geodiversity Action Plan areas, all of which are providing employment opportunities often in economically deprived areas. Local and regional authorities are doing more than previously to ensure that quarries, river and coastal restoration schemes are appropriate and authentic, and in particular, that quarry restoration conditions retain the geological interest of the quarry (Section 10.4.1). Geotourism holiday offerings are becoming more numerous and organised, but perhaps less successful than predicted has been Earth science teaching in schools and interest in the geosciences amongst the young, except perhaps in relation to dinosaurs and natural hazards.

15.3 The benefits of geodiversity

In summary, geodiversity gives us:

- A strong basis for valuing the abiotic world. We saw in Chapter 4 that geodiversity creates many goods and services some of which are independent of biodiversity and some of which are shared with the biotic world. For example, geodiversity gives us an enormous range of building materials, industrial minerals and energy sources, it provides a platform for a diverse range of human activities, it forms the basis of all our productive soils and agricultural land, it contributes to our spectacular landscapes, it informs us about past climates and the impact of future climate change, it tells us about the evolution of life on Earth, and it is a training ground for future geologists.
- A main basis for geoconservation. We saw in Chapter 10 that several countries/regions are selecting geoconservation sites as representative examples of the areas' geodiversity.
- An integrating mechanism for the geosciences - geology, geomorphology and soil science are taught in separate departments in many of the world's universities. Geodiversity provides us with an integrating geoscience paradigm.
- A means of promoting the importance of the geosciences through their role in sustaining many elements of both modern society and biodiversity;
- A means of promoting the role of the geosciences in integrated and sustainable land resource management.
- A means of promoting the geosciences among politicians, planners, decision-makers and the public most of whom already understand the concept of biodiversity. Crawford and Black (2012) report the encouraging statistic that 45% of 150 visitors to the Giant's Causeway in Northern Ireland had heard of the terms 'geodiversity' and 'geoconservation'. As Semeniuk (1997, p. 54) argued 'it is necessary to make strong connections between biodiversity and geodiversity if politicians, planners and decision makers are to be convinced that geodiversity has a place in general conservation'.

- Potential resources for the future. Just as biologists often argue the case for conserving plants and animals because of their future potential uses to society, a similar argument can be applied to the abiotic world and provides a further reason for conserving geodiversity.

15.4 The future

Geodiversity and geoconservation have come a long way in recent years (Henriques *et al.* 2011) but still have a long way to go in every country in the world. But there is a particular need for progress to be made in the developing world where geoheritage losses are probably very significant. We certainly need more research about the losses that are occurring so that perhaps a Geodiversity Red Book can be established to sit alongside its biodiversity equivalent which lists those species that are rare or threatened with extinction and therefore in need of conservation. Finally, we need to try to achieve equal status for biodiversity and geodiversity in as many strategies, policies and plans as possible. This will involve continuing efforts to educate and inform those decision-makers responsible for developing these policies and ensuring that they are implemented. The task is a huge one but geodiversity as a paradigm has made significant strides in the last few years for the simple reason that it is too important to ignore any longer.

The Irish Nobel Laureate poet, Seamus Heaney has talked about 'giving glory to things because they are', and for the final words of this book, we quote from the words of the Scottish poet Hugh MacDiarmid. Contemplating the beach shingle above the shoreline in the Shetlands, he wrote these important words:

What happens to us
Is irrelevant to the world's geology
But what happens to the world's geology
Is not irrelevant to us.
We must reconcile ourselves to the stones,
Not the stones to us...
Let men find the faith that builds mountains
Before they seek the faith that moves them.

Hugh MacDiarmid, (1956) *On a raised beach*

References

Abbott, P.L. 2009 *Natural disasters*, 7th edn. McGraw-Hill, Maidenhead.

Abernethy, B. and Wansborough, T.M. 2001 Where does river rehabilitation fit in catchment management? In Rutherfurd, I., Sheldon, F., Brierley, G. and Kenyon, C. (eds) *Third Australian Stream Management Conference*. Co-operative Research Centre for Catchment Hydrology, 1–6.

Abrahams, A.D. and Parsons, A.J. (eds) 1994 *Geomorphology of desert environments*. Chapman & Hall, London.

Achuff, P.L. 1994 *Natural Regions, subregions and natural history themes of Alberta: a classification for protected areas management*. Government of Alberta, Edmonton.

Adams, W.M. 1996 *Future nature: a vision for conservation*. Earthscan, London.

Adams, W.M. 1998 Landforms, authenticity and conservation value. *Area*, 30, 168–169.

Adams, W.M. and Mulligan, M. (eds) 2003 *Decolonizing nature*. Earthscan, London, 108–134.

Alcalá, L. 1999 Spanish steps towards geoconservation. *Earth Heritage*, 11, 14–15.

Alcalá, L. and Morales, J. 1994 Towards a definition of the Spanish palaeontological heritage. In O'Halloran, D., Green, C., Harley, M., *et al.* (eds) *Geological and landscape conservation*. Geological Society, London, 57–61.

Alexandrowicz, Z. 2012 Poland. In Wimbledon and Smith-Meyer (eds) *Geoheritage in Europe and its conservation*. ProGEO, Oslo, 254–263.

Alexandrowicz, Z., Urban, J. and Miskiewicz, K. 2009 Geological values of selected Polish properties of the UNESCO World Heritage List. *Geoheritage*, 1, 43–52.

Alvarez, L.W., Alvarez, W., Asaro, F. and Michel, H.V. 1980 Extraterrestrial cause of the Cretaceous—Tertiary extinction. *Science*, 208, 1095–1108.

Amadio, V., Amadei, M., Bagnaia, R., *et al.* 2002 The role of geomorphology in landscape ecology: the landscape unit map of Italy, scale 1 : 250000 (Carta della Natura Project) In Allison, R.J. (ed) *Applied geomorphology: theory and practice*. John Wiley & Sons, Ltd, Chichester, 265–282.

Ambrose, K., McGrath, A., Weightman, G., *et al.* 2012 *Exploring the landscape of the National Forest*. British Geological Survey, Keyworth.

Amrikazemi, A. 2010 *Atlas of geopark and geotourism resources of Iran*. Geological Survey of Iran, Ministry of Industries and Mines.

Anderson, D. and Brown, E.J. 2010 Perspectives on Quaternary outreach and aspirations for the future. *Proceedings of the Geologists' Association*, 121, 455–467.

Anderson, M.G. and Feree, C.E. 2010 Conserving the stage: climate change and the geophysical underpinnings of species diversity. *PLoS One*, e11554.

Andrasanu, A. and Grigorescu, D. 2012 Romania. In Wimbledon, W.A.P. and Smith-Meyer, S. (eds) *Geoheritage in Europe and its conservation*. ProGEO, Oslo, 274–287.

Geodiversity: Valuing and Conserving Abiotic Nature, Second Edition. Murray Gray.

van den Ancker, H. and Jungerius, P. 2012 The Netherlands. In Wimbledon, W.A.P. and Smith-Meyer, S. (eds) *Geoheritage in Europe and its conservation*. ProGEO, Oslo, 232–245.

Anon 2000 Pebble problems. *Earth Heritage*, 13, 6.

Arendt, A.A., Echelmeyer, K.A., Harrison, W.D., *et al.* 2002 Rapid wastage of Alaska glaciers and their contribution to rising sea levels. *Science*, 297, 382–386.

Arisci, A., De Waele, J., Di Gregorio, F., *et al.* 2004 The human factor in natural and cultural landscapes: the geomining park of Sardinia. In Parkes, M. (ed) *Natural and cultural landscapes: the geological foundation*. Royal Irish Academy, Dublin, 287–290.

Ásbjörnsdóttir, L., Einarsson, S. and Jónasson, K. 2012 Iceland. In Wimbledon, W.A.P. and Smith-Meyer, S. (eds) *Geoheritage in Europe and its conservation*. ProGEO, Oslo, 170–179.

Asch, T.W.J. and van Dijck, S.J.E. 1995 The role of geomorphology in environmental impact assessment in the Netherlands. In Marchetti, M., Panizza, M., Soldati, M. and Barani, D (eds) 1995 *Geomorphology and environmental impact assessment*. CNR–CS Geodinamica Alpina e Quaternaria, Milan, 63–70.

ASH Consulting Group 1994 *Coastal erosion and tourism in Scotland: a review of protection measures to combat coastal erosion related to tourist activities and facilities*. Scottish Natural Heritage Review, 12.

Ashurst, J. and Dimes, F.G. (eds) 1990 *Conservation of building and decorative stone*. Butterworth-Heinemann, Oxford.

Ashman, M. and Puri, G. 2002 *Essential soil science*. Blackwell, Oxford.

Ashworth, J., Stoker, B. and Aish, A. 2010 *Ecological network guidance on Marine Conservation Zones*. Natural England and Joint Nature Conservation Committee.

Asplund, D. 1996 Energy use of peat. In Lappalainen, E. (ed) *Global peat resources*. International Peat Society, Finland, 319–325.

Attenborough, D. 1990 Foreward. In *Earth science conservation in Great Britain: a strategy*. Nature Conservancy Council, Peterborough.

Attfield, R. 1999. *The ethics of the global environment*. Edinburgh University Press, Edinburgh.

Aust, H. and Sustrac, G. 1992 Impact of development on the geological environment. In Lumsden, G.I. (ed) *Geology and the environment in western Europe*. Oxford University Press, Oxford.

Australian Heritage Commission 1996 *Australian Natural Heritage Charter*, 1st edn. Australian Heritage Commission, Canberra.

Australian Heritage Commission 1999 *Natural Heritage Places Handbook*. Australian Heritage Commission, Canberra.

Australian Heritage Commission 2002 *Australian Natural Heritage Charter*, 2nd edn. Australian Heritage Commission, Canberra.

Badman, T. and Bomhard, B. 2008 *World heritage and protected areas: an initial analysis of the contribution of the World Heritage Convention to the global network of protected areas presented to the 32nd session of the World Heritage Committee, Quebec City, Canada, in July 2008*. IUCN, Gland.

Balmford, A., Bruner, A., Cooper, P., *et al.* 2002 Economic reasons for conserving wild nature. *Science*, 297, 950–953.

Barrentino, D., Vallejo, M. and Gallego, E. (eds) 1999 *Towards the balanced management and conservation of the geological heritage in the new millennium*. Sociedad Geológica de Espana, Madrid.

Baron, J.S., Theobald, D.M. and Fagre, D.B. 2000 Management of land use conflicts in the United States Rocky Mountains. *Mountain Research and Development*, 20, 24–27.

Barrow, C.J. 1999 *Environmental management: principles and practice*. Routledge, London.

Barry, J. 1999. *Environment and social theory*. Routledge, London.

Bartley, R. and Rutherfurd, I. 2001 Statistics, snags and sand: measuring the geomorphic recovery of streams disturbed by sediment slugs. In Rutherfurd, I., Sheldon, F., Brierley,

G. and Kenyon, C. (eds) *Third Australian Stream Management Conference*. Co-operative Research Centre for Catchment Hydrology, Monash, 15–22.

Barton, M.E. 1998 Geotechnical problems with the maintenance of geological exposures in clay cliffs subject to reduced erosion rates. In Hooke, J. (ed) *Coastal defence and Earth science conservation*. Geological Society of London, 32–45.

Battarbee, R.W. 1992 Holocene lake sediments, surface water acidification and air pollution. In Gray, J.M. (ed) *Applications of Quaternary Research*. John Wiley & Sons, Ltd, Chichester.

Battarbee, R.W., Flower, R.J., Stevenson, A.C. and Rippey, B. 1985. Lake acidification in Galloway: a palaeoecological test of competing hypotheses. *Nature*, 314, 350–352.

Baucon, A. 2009 *Geology in art: an unorthodox path from visual arts to music*. http://www.tracemaker.com/geology_and_art.html. Milan.

Bayfield, N. 2001 Mountain resources and conservation. In Warren, A. and French, J.R. *Habitat conservation: managing the physical environment*. John Wiley & Sons, Ltd, Chichester, 7–38.

Beavis, S. and Lewis, B. 2001 The impact of farm dams on the management of water resources within the context of total catchment development. In Rutherfurd, I., Sheldon, F., Brierley, G. and Kenyon, C. (eds) *Third Australian Stream Management Conference*, 23–28.

Beckerman, W. and Pasek, J. 2001 *Justice, posterity, and the environment*. Oxford University Press, Oxford.

Beier, P. and Brost, B. 2010 Use of land facets to plan for climate change: conserving the arenas, not the actors. *Conservation Biology*, 24, 701–710.

Bell, F. 1999 *Geological hazards*. Spon Press, London.

Bell, F.G., Hälbich, T.F.J. and Bullock, S.E.T. 2002 The effects of acid mine drainage from an old mine in the Witbank Coalfield, South Africa. *Quarterly Journal of Engineering Geology and Hydrogeology*, 35, 265–278.

Belshaw, C. 2001 *Environmental philosophy: reason, nature and human concern*. Acumen, Chesham.

Belsky, A.J. 1995 Spatial and temporal landscape patterns in arid and semi arid savannas. In Hansson, L., Hahrig, L. and Merriance, G. (eds) *Mosaic landscapes and ecological processes*. Chapman & Hall, London, 31–56.

Benito-Calvo, A., Pérez-Gonzalez, A., Magn, O. and Meza, P. 2009 Assessing regional geodiversity: the Iberian Peninsula. *Earth Surface Processes and Landforms*, 34, 1433–1445.

Benn, D.I and Evans, D.J.A. 2010 *Glaciers and glaciation*, 2nd edn. Hodder Education, London.

Bennett, M.R. 1994 The management of Quaternary SSSI in England, with specific reference to disused pits and quarries. In Stevens, C., Gordon, J.E., Green, C.P. and Macklin, M.G. (eds) *Conserving our landscape*, English Nature, Peterborough, 36–40.

Bennett, M.R. and Doyle, P. 1997 *Environmental geology*. John Wiley & Sons, Ltd, Chichester.

Bennett, M.R. and Glasser, N.F. 2010 *Glacial geology: ice sheets and landforms*, 2nd edn. Wiley-Blackwell, Chichester.

Bennett, N. 2003 Brighton rocked by cliff dilemma. *Earth Heritage*, 19, 7.

Bentley, M. 1996 *Moss of Achnacree and Achnaba landforms*. Earth Science Documentation Series, Scottish Natural Heritage, Edinburgh.

Benton, M. and Harper, D. 2009 *Introduction to palaeobiology and the fossil record*. Wiley-Blackwell, Chichester.

Bergonzoni, M., Vezzani, A., Lugaresaresti, J.I., *et al.* 1995 Environmental Impact Assessment studies in the regional park of Sassi di Roccamalatina (Northern Apennines, Italy). In Marchetti, M., Panizza, M., Soldati, M. and Barani, D (eds) *Geomorphology and environmental impact assessment*. CNR–CS Geodinamica Alpina e Quaternaria, Milan, 139–156.

Best, M.G. 2003 *Igneous and metamorphic petrology*, 2nd edn. Blackwell, Oxford.

Best, M.G. and Christiansen, E.H. 2001 *Igneous petrology*. Blackwell Science, Oxford.

Bird, E. 2000 *Coastal geomorphology*. John Wiley & Sons, Ltd, Chichester.

Bjorlykke, K. 2011 *Petroleum geoscience: from sedimentary environments to rock physics*. Springer, Heidelberg.

Bland, W. and Rolls, D. 1998 *Weathering: an introduction to scientific principles*. Arnold, London.

Bloom, A.L. 1998 *Geomorphology: a systematic analysis of Late Cenozoic landforms*, 3rd edn. Prentice Hall, New Jersey.

Boardman, J. 1988 Public policy and soil erosion in Britain. In Hooke, J.M. (ed) *Geomorphology in environmental planning*. John Wiley & Sons, Ltd, Chichester, 33–50.

Bolner-Takács, K., Cserny, T. and Horváth, G. 2012 Hungary. In Wimbledon, W.A.P. and Smith- Meyer, S. (eds) *Geoheritage in Europe and its conservation*. ProGEO, Oslo, 158–169.

Boon, G. 2004 Buried treasure: Sheffield's lost fossil forest laid to rest (again). *Earth Heritage*, 22, 8–9.

Boulton, G.S. 2001. The Earth system and the challenge of global change. In Gordon, J.E. and Leys, K.F. *Earth science and the natural heritage*. Stationery Office, Edinburgh, 26–54.

Bowen, N.L. (1928) *The evolution of igneous rocks*. Princeton University Press, Princeton.

Bowersox, G.W. and Chamberlin, B.E. 1995 *Gemstones of Afghanistan*. Geoscience Press, Tucson.

Boylan, P.J. 2008 Geological site designation under the 1972 UNESCO World Heritage Convention. In Burek, C.D. and Prosser, C.D. (eds) *The history of geoconservation*. Geological Society of London, Special Publication 300, 279–304.

Brady, N.C. and Weil, R.R. 2002 *The nature and properties of soil*, 13th edn. Prentice Hall, New Jersey.

Brampton, A.H. 1998 Cliff conservation and protection: methods and practices to resolve conflicts. In Hooke, J. (ed) *Coastal defence and Earth science conservation*. Geological Society, London, 46–57.

Brancucci, G., Burlando, M. and Marin, V. 2002 The role of geological heritage in the Natura 2000 Network (Habitats Directive, 92/43/EEC): a local study in northern Italy (Liguria Region). *Abstract for Natural and Cultural Landscapes Conference*. Royal Irish Academy, Dublin.

Brancucci, G., D'Andrea, M., Gisotti, G., Paliága, G., Poli, G., Zarlenga, F. and Giovagnoli, M.C. 2012 Italy. In Wimbledon, W.A.P. and Smith-Meyer, S. (eds) *Geoheritage in Europe and its conservation*. ProGEO, Oslo, 188–199.

Brazier, V., Bruneau, P.M.C., Gordon, J.E. and Rennie, A.F. 2012 Making space for nature in a changing climate: the role of geodiversity in biodiversity conservation. *Scottish Geographical Journal*, 128, 211–233.

Breeze, D. and Munro,G. 1997 *The Stone of Destiny: symbol of nationhood*. Historic Scotland, Edinburgh.

Bridge, J. 2002 *Rivers and floodplains*. Blackwell, Oxford.

Bridges, E.M. 1994 Soil conservation: an international issue. In O'Halloran, D., Green, C., Harley, M., *et al.* (eds) *Geological and landscape conservation*. Geological Society, London, 11–15.

Bridgland, D. 1994 The conservation of Quaternary geology in relation to the sand and gravel extraction industry. In O'Halloran, D., Green, C., Harley, M., *et al.* (eds) *Geological and landscape conservation*. Geological Society, London, 87–91.

Brierley, G.J. and Fryirs, K. 2005 *Geomorphology and river management: applications of the river styles framework*. Blackwell Publishing, Oxford.

Brilha, J. 2002 Geoconservation and protected areas. *Environmental Conservation*, 29, 273–276.

Brilha, J. 2005 *Património, geológico e geoconservacao*. Palimage, Braga.

Brilha, J. 2012 Portugal. In Wimbledon, W.A.P. and Smith-Meyer, S. (eds) *Geoheritage in Europe and its conservation*. ProGEO, Oslo, 265–273.

Bristow, C.M. 1994 Environmental aspects of mineral resource conservation in southwest England. In O'Halloran, D., Green, C., Harley, M., *et al.* (eds) *Geological and landscape conservation*. Geological Society, London, 79–86.

British Geological Survey 2001 *Building stone resources of the United Kingdom*. British Geological Survey, Keyworth.

Brocx, M. 2008 *Geoheritage: from global perspectives to local principles for conservation and planning*. Western Australian Museum, Perth.

Brocx, M. and Semeniuk, V. 2007 Geoheritage and geoconservation: history, definition, scope and scale. *Journal of the Royal Society of Western Australia*, 90, 53–87.

Brocx, M. and Semeniuk, V. 2010 The geoheritage significance of crystals. *Geology Today*, 26, 216–225.

Bromley, R.G. 1996 *Trace fossils: biology, taphonomy and applications*, 2nd edn. Chapman & Hall, London.

Bronowski, J. 1973 *The ascent of Man*. BBC Publications, London.

Brookes, A. 1985 River channelisation. *Progress in Physical Geography*, 9, 44–73.

Brookes, A. 1986 Response of aquatic vegetation to sedimentation downstream from channelisation works in England and Wales. *Biological Conservation*, 38, 351–367.

Brookes, A. 1988 *Channelised rivers: perspectives for environmental management*. John Wiley & Sons, Ltd, Chichester.

Brookes, A., Baker, J. and Redmond, C. 1996 Floodplain restoration and riparian zone management. In Brookes, A. and Shields, F.D. (eds) *River channel restoration*. John Wiley & Sons, Ltd, Chichester 201–229.

Brookes, A. and Shields, F.D. (eds) 1996a *River channel restoration: guiding principles for sustainable projects*. John Wiley & Sons, Ltd, Chichester.

Brookes, A. and Shields, F.D. 1996b Perspectives on river channel restoration. In Brookes, A. and Shields, F.D. (eds) *River channel restoration*. John Wiley & Sons, Ltd, Chichester, 1–19.

Brookes, A. and Shields, F.D. 1996c Towards an approach to sustainable river restoration. In Brookes, A. and Shields, F.D. (eds) *River channel restoration*. John Wiley & Sons, Ltd, Chichester, 385–402.

Brooks, A.J., Roberts, H. and Kenyon, N.H. 2009 *Accessing and developing the required biophysical and data layers for Marine Protected Areas network planning and wider marine spatial planning purposes*. Report No 8, Task 2A Mapping of Geological and Geomorphological Features. ABP Marine Environmental Research Ltd., Southampton.

Brooks, K. 2012 A tale of two intrusions—where familiar rock names no longer suffice. *Geology Today*, 28, 13–19.

Brown, C., Walpole, M., Simpson, L. and Tierney, M. 2011 Chapter 1: Introduction to the UK National Ecosystem Assessment. In: *UK National Ecosystem Assessment Technical Report*. UNEP-WCMC, Cambridge, pp. 1–10.

Brown, E.J., Prosser, C.D. and Stevenson, N.M. 2012 Geodiversity, conservation and climate change: key principles for adaptation. *Scottish Geographical Journal*, 128, 234–239.

Browne, M.A.E. 2012 Geodiversity and the role of the planning system in Scotland. *Scottish Geographical Journal*, 128, 266–277.

Bruneau, P.M.C., Gordon, J.E. and Rees, S. 2011 *Ecosystem sensitivity and responses to change: understanding the links between geodiversity and biodiversity at the landscape scale*. Report No. 450, JNCC, Peterborough.

Brunsden, D. and Thornes, J.B. 1979 Landscape sensitivity and change. *Transactions of the Institute of British Geographers*, NS4, 463–484.

Bruschi, V.M., Cendrero, A. and Albertos, J.A.C. 2011 A statistical approach to the validation and optimisation of geoheritage assessment procedures. *Geoheritage*, 3, 131–149.

Brussard, P.F., Murphy, D.D. and Noss, R.F. 1992 Strategy and tactics for conserving biodiversity in the United States. *Conservation Biology*, 6, 157–159.

Bucher, K. and Grapes, R. 2011 *Petrogenesis of metamorphic rocks*. Springer, Heidelberg.

Buck, S.J. 1998 *The global commons: an introduction*. Earthscan, London.

Buckeridge, J.S. 1994 Geological conservation in New Zealand: options in a rapidly eroding environment. In O'Halloran, D., Green, C., Harley, M., *et al.* (eds) *Geological and landscape conservation*. Geological Society, London, 263–269.

Budil, P., Lorencova, M., Stanzelova, Z., *et al.* 2012 Czech Republic. In Wimbledon, W.A.P. and Smith-Meyer, S. (eds) *Geoheritage in Europe and its conservation*. ProGEO, Oslo, 92–99.

Bull, P.A. and Morgan, R.M. 2006 Sediment fingerprints: a forensic technique using quartz and sand grains. *Science and Justice*, 46, 64–81.

Bull, L. and Kirkby, M. (eds) 2002 *Dryland rivers*. John Wiley & Sons, Ltd, Chichester.

Bullard, J. and Livingstone, D. 2002 Interactions between aeolian and fluvial systems in dryland environments. *Area*, 34, 8–16.

Burek, C.V. 2008a The role of the voluntary sector in the evolving geoconservation movement. In Burek, C.V. and Prosser, C.D. (eds) *The history of geoconservation*. Geological Society of London, Special Publication 300, 61–89.

Burek, C.V. 2008b History of RIGS in Wales: an example of successful cooperation for geoconservation. In Burek, C.V. and Prosser, C.D. (eds) *The history of geoconservation*. Geological Society of London, Special Publication 300, 147–171.

Burek, C. 2012a Rock miles. *Earth Heritage*, 37, 7–9 .

Burek, C. 2012b The role of LGAPs (Local Geodiversity Action Plans) and Welsh RIGS as local drivers for geoconservation within geotourism in Wales. *Geoheritage*, 4, 45–63.

Burek, C., Ellis, N., Evans, D.H., *et al.* 2013 Marine geoconservation in the United Kingdom—a toe in the water? *Proceedings of the Geologists' Association*, in press.

Burek, C. and Legg, C. 1999 Grike-roclimates. *Earth Heritage*, 12, 10.

Burek, C. and Potter, J. 2004 Local Geodiversity Action Plans: sharing good practice workshop. *English Nature Research Report*, 601.

Burek, C. and Potter, J. 2006 Local Geodiversity Action Plans: setting the context for geological conservation. *English Nature Research Report*, 560.

Burek, C.V. and Prosser, C.D. (eds) 2008 *The history of geoconservation*. Geological Society of London, Special Publication 300.

Burnett, M.R., August, P.V., Brown, J.H. and Killingbeck, K.T. 1998 The influence of geomorphological heterogeneity on biodiversity. 1. A patch scale perspective. *Conservation Biology*, 12, 363–370.

Caitcheon, G., Prosser, I., Wallbrink, P., *et al.* 2001 Sediment delivery from Moreton Bay's main tributaries: a multifaceted approach to identifying sediment sources. In Rutherfurd, I., Sheldon, F., Brierley, G. and Kenyon, C. (eds) *Third Australian Stream Management Conference*, 103–108.

Cairns, J. 1991 The status of the theoretical and applied science of restoration ecology. *The Environmental Professional*, 13, 186–194.

Campbell, S. and Bowen, D.Q. 1989 *Quaternary of Wales*. Nature Conservancy Council, Peterborough.

Campbell, S. and Wood, M. 2002 Scientific vandalism? *Earth Heritage*, 17, 3.

Cameron, W., Conover, K. and Micic, S. 2001 Tools and techniques used by the Bulimba Creek Catchment Coordinating Committee to improve the Bulimba Creek. In Rutherfurd, I., Sheldon, F., Brierley, G. and Kenyon, C. (eds) *Third Australian Stream Management Conference*, 109–114.

Canup, R.M. and Asphaug, E. 2001 Origin of the Moon in a giant impact near the end of the Earth's formation. *Nature*, 412, 708–712.

Carcavilla, L., Durán, J.J. and López-Marinez, y J. 2008 *Different approaches for the study of geodiversity*. Paper presented to the 33rd International Geological Congress, Oslo.

Carcavilla, L., Durán, J.J., Garcia-Cortés, Á. and López-Martinez, y J. 2009 Geological heritage and geoconservation in Spain: past, present, and future. *Geoheritage*, 1, 75–91.

Carroll, B. and Turpin, T. 2009 *Environmental Impact Assessment handbook: a practical guide for planners*, 2nd edn. Thomas Telford, London.

Carroll, C., Rohde, K., Millar, G., *et al.* 2001 Neighbourhood catchments: a new approach for achieving ownership and change in catchment and stream management. In

Rutherfurd, I., Sheldon, F., Brierley, G. and Kenyon, C. (eds) *Third Australian Stream Management Conference*, 121–126.

Carson, G. 1996 Biodiversity's underlying influence. *Earth Heritage*, 6, 8–9.

Carson, R. 1962 *Silent spring*. Houghton Mifflin, Boston.

Carvalho, L. and Anderson, N.J. 2001 Lakes. In Warren, A. and French, J.R. *Habitat conservation: managing the physical environment*. John Wiley & Sons, Ltd, Chichester, 123–146.

Catto, N. 2002 Anthropogenic pressures on coastal dunes, southwestern Newfoundland. *Canadian Geographer*, 46, 17–32.

Cernatic-Gregoric, A. and Zega, M. 2010 The impact of human activities on dolines (sinkholes)—typical geomorphological features of karst (Slovenia) and possibilities of their preservation. *Geographica Pannonica*, 14, 109—117.

Charlton, G. Undated *Foreword—understanding the European Landscape Convention*. English National Park Authorities Association, London.

Charlton, R. 2008 *Fundamentals of fluvial geomorphology*. Routledge, Abingdon.

Clarkson, E.N.K. 1993 *Invertebrate palaeontology and evolution*, 3rd edn. Chapman & Hall, London.

Clarkson, E.N.K. 2001 The palaeontological resource of Great Britain—our fossil heritage. In Bassett, M.G., King, A.H., Larwood, J.G. and Parkinson, N.A. (eds) *A future for fossils*. National Museums and Galleries of Wales, Cardiff, 10–17.

Cleal, C. 2012 United Kingdom. In Wimbledon, W.A.P. and Smith-Meyer, S. (eds) *Geoheritage in Europe and its conservation*. ProGEO, Oslo, 392–403.

Cleal, C., Thomas, B., Bevins, R. and Wimbledon, B. 2001 Deciding on a new world order. *Earth Heritage*, 16, 10–13.

Clifford, N.J. 2001 Conservation and the river channel environment. In Warren, A. and French, J.R. *Habitat conservation: managing the physical environment*. John Wiley & Sons, Ltd, Chichester, 67–104.

Clifton-Taylor, A. 1987 *The pattern of English Building*, 4th edn. Faber and Faber, London.

Collerson, K.D. and Kamber, B. 1999. Evolution of the continents and the atmosphere inferred from Th-U-Nb systematics of the depleted mantle. *Science*, 283, 1519–1522.

Collins, C. (ed) 1995 *The care and conservation of palaeontological materials*. Butterworth-Heinemann, Oxford.

Conacher, A.J. 2002 Geomorphology and Environmental Impact Assessment in relation to dryland agriculture. In Allison, R.J. (ed) *Applied geomorphology: theory and practice*. John Wiley & Sons, Ltd, Chichester, 317–333.

Condie, K.C. 2000. Episodic continental growth models: afterthoughts and extensions. *Tectonophysics*, 322, 153–162.

Conway, J. 2010 A soil trail? A case study from Anglesey, Wales, UK. *Geoheritage*, 2, 15–24.

Cooke, R.U. and Doornkamp, J.C. 1990 *Geomorphology in environmental management*. Oxford University Press, Oxford.

Cooke, R.U., Warren, A. and Goudie, A.S. 1993 *Desert geomorphology*. UCL Press, London.

Cooper, R.A., Nowlan, G.S. and Williams, S.H. 2001 Global Stratotype Section and Point for the base of the Ordovician System. *Episodes*, 24(1), 19–28.

Coratza, P, Bruschi, V.M., Piacentini, D., *et al.* 2011. Recognition and assessment of geomorphosites in Malta at the Il-Majjistral Nature and History Park. *Geoheritage*, 3, 175–185.

Costa, M.P., Nunes, J.C., Constancia, J.P., *et al.* 2008 *Azores volcanic caves*. Amigos dos Acores/Os Montanheiros/GESPEA, Azores.

Costanza, R., d'Arge, de Groot, R.S. *et al.* 1997 The value of the world's ecosystem services and natural capital. *Nature*, 387, 253–260.

Countryside Commission 1993a Landscape Assessment guidance *(CCP423)*. Countryside Commission, Cheltenham.

Countryside Commission 1993b Golf courses in the countryside *(CP438)*. Countryside Commission, Cheltenham.

Countryside Commission 1993c Design in the countryside *(CP418)*. Countryside Commission, Cheltenham.
Countryside Council for Wales 1999 *Tir Gofal: agri-environment scheme for Wales*. Countryside Council for Wales, Bangor.
Covey, R. and Laffoley, D. d'A. 2002 *Maritime state of nature report for England: getting onto an even keel*. English Nature, Peterborough.
Cowie, J.W. and Wimbledon, W.A.P. 1994 The World Heritage List and its relevance to geology. In O'Halloran, D., Green, C., Harley, M., *et al.* (eds) *Geological and landscape conservation*. Geological Society, London, 71–73.
Crawford, K.R. and Black, R. 2012 Visitor understanding of the geodiversity and the geoconservation value of the Giant's Causeway World Heritage Site, Northern Ireland. *Geoheritage*, 4, 115–126.
Creaser, P. 1994a The protection of Australia's fossil heritage through the Protection of Movable Cultural Heritage Act 1986. In O'Halloran, D., Green, C., Harley, M., *et al.* (eds) *Geological and landscape conservation*. Geological Society, London, 69–70.
Creaser, P. 1994b An international earth sciences conservation convention framework for the future. *Australian Geologist*, 90, 45–46.
Cribb, S. and Cribb, J. 1998 *Whisky on the rocks*. British Geological Survey, Keyworth.
Crofts, R. 2001 Sustainable use of the Earth's resources. In Gordon, J.E. and Leys, K.F. *Earth science and the natural heritage*. Stationery Office, Edinburgh, 286–295.
Cullingford, J.B. and Nadin, V. 1994 *Town and country planning in Britain*, 11th edn. Routledge, London and New York.
Cuomo, C.J. 1998 *Feminism and ecological communities*. Routledge, London and New York.
Cutter, S.L. and Renwick, W.H. 1999. *Exploitation, conservation, preservation: a geographic perspective on natural resource use*, 3rd edn. John Wiley & Sons, Ltd, Chichester.
Czudek, T. and Demek, J. 1970 Thermokarst in Siberia and its influence on the development of lowland relief. *Quaternary Research*, 1, 103–120.
Dahlgren, S. 2006 *Gea Norvegica Geopark: Application dossier for nomination as a European Geopark*. Gea Norvegica Geopark, Norway.
Dahlkamp, F.J. 2013 *Uranium deposits of the world*. Springer, Heidelberg.
Daily, G.C. (ed) 1997a *Nature's services: societal dependence on natural ecosystems*. Island Press, Washington DC.
Daily, G.C. 1997b Introduction: what are ecosystem services? In Daily, G.C. (ed) *Nature's services: societal dependence on natural ecosystems*. Island Press, Washington DC, 1–10.
Daly, D. 1994 Conservation of peatlands: the role of hydrogeology and the sustainable development principle. In O'Halloran, D., Green, C., Harley, M., *et al.* (eds) *Geological and landscape conservation*. Geological Society, London, 17–21.
Daly, D., Erikstad, L. and Stevens, C. 1994 Fundamentals in earth science conservation. *Mémoires de la Société de Geologique de France*, 165, 209–212.
D'Andrea, M., Coilacchi, S., Gramaccini, G., *et al.* 2004 The data base of Italian geosites inventory. In Parkes, M. (ed) *Natural and cultural landscapes: the geological foundation*. Royal Irish Academy, Dublin, 103–106.
Dasmann, R.F. 1984 *Environmental conservation*, 5th edn. John Wiley & Sons, Ltd, Chichester.
Davidson, R.J., Kor, P.S.G., Fraser, J.Z. and Cordiner, G.S. 2001. Geological conservation and integrated management in Ontario, Canada. In Gordon, J.E. and Leys, K.F. *Earth science and the natural heritage*. Stationery Office, Edinburgh, 228–233.
Davies, J. and Pearce, N. 1993 Motorways can seriously improve your exposure. *Earth Science Conservation*, 32, 22–23.
Davies, J.L. 1980 *Geographical variation in coastal development*, 2nd edn. Longman, Harlow.
Davis, G.H. and Reynolds, S.J. 1996 *Structural geology of rocks and regions*, 2nd edn. John Wiley & Sons, Ltd, Chichester.
Davis, J.R. and Fitzgerald, D. 2002 *Beaches and coasts*. Blackwell, Oxford.

Decker, R.W. and Decker, B.B. 1998 *Volcanoes*. Freeman, New York.
Deer, W.A., Howie, R.A. and Zussman, J. 2001 *Feldspars*, 2nd edn. Geological Society, London.
Deffeyes, K.S. 2001 *Hubbert's peak: the impending world oil shortage*. Princeton University Press, Princeton.
DEFRA 2002a *Working with the grain of nature: England's biodiversity strategy*. Department of the Environment, Food and Rural Affairs, London.
DEFRA 2002b *The strategy for sustainable farming and food: facing the future*. Department of the Environment, Food and Rural Affairs, London.
DEFRA 2006 *Local Sites: guidance on their identification, selection and management*. Department of Environment, Food and Rural Affairs, London.
DEFRA 2007 *Conserving biodiversity—the UK approach*. Department of the Environment, Food and Rural Affairs, London.
DEFRA 2009 *Safeguarding our soils: a strategy for England*. Department of the Environment, Food and Rural Affairs, London.
DEFRA 2011a *Biodiversity 2020: a strategy for England's wildlife and ecosystem services*. Department of Environment, Food and Rural Affairs, London.
DEFRA 2011b *The natural choice: securing the value of nature*. Department of the Environment, Food and Rural Affairs, London.
Demek, J. 1994 Global warming and permafrost in Eurasia: a catastrophic scenario. *Geomorphology*, 10, 317–329.
Diaz-Martinez, E. 2012 The world's leading nature conservation organisation incorporates geoconservation in its agenda. *ProGEO News*, 2012(4), 4–6.
Dingwall, P., Weighell, T. and Badman, T. 2005, *Geological World Heritage: a global framework*. IUCN, Gland.
Dinis, J.L., Oliveira, F.P., Rey, J. and Duarte, I.L. 2010 Finding geological heritage: legal issues on private property and fieldwork: the case of outstanding early angiosperms (Barremian to Albian, Portugal). *Geoheritage*, 2, 77–90.
Dixon, G. 1995 Aspects of geoconservation in Tasmania: a preliminary review of significant Earth features. Report to the Australian Heritage Commission, Occasional Paper 35, Parks and Wildlife Service, Tasmania.
Dixon, G. 1996a *Geoconservation: an international review and strategy for Tasmania*. Parks and Wildlife Service, Tasmania, Hobart, Occasional Paper 35.
Dixon, G. 1996b *A reconnaissance inventory of sites of geoconservation significance on Tasmanian islands*. Report to Parks and Wildlife Service, Tasmania and Australian Heritage Commission.
Dixon, G. and Duhig, N. 1996 *Compilation and assessment of some places of geoconservation significance*. Regional Forest Agreement, Commonwealth of Australia and State of Tasmania.
Dixon, G., Houshold, I., Pemberton, M. and Sharples, C. 1997 Wizards of Oz: geoconservation in Tasmania. *Earth Heritage*, 8, 14–15.
Dolman, P. 2000 Biodiversity and ethics. In O'Riordan, T. (ed) *Environmental science for environmental management*, 2nd edn. Pearson Education, Harlow, 119–148.
Domas, J. 1994 Damage caused by opencast and underground coal mining. In O'Halloran, D., Green, C., Harley, M., *et al.* (eds) *Geological and landscape conservation*. Geological Society, London, 93–97.
Donnelly, L. 2002 Finding the silent witness. *Geoscientist*, 12(5), 16–18.
Donner, J.J. 1995 *The Quaternary history of Scandinavia*. Cambridge University Press.
Donovan, K. and Suharyanto, A. 2011 The creatures will protect us. *Geoscientist*, 21(1), 12–17.
Doughty, P. 2002 Grottos, granite and gold. *Geodiversity Update*, 4, 1–3.
Dowling, R. 2011 Geotourism's global growth. *Geoheritage*, 3, 1–13.
Dowling, R. 2012 Conference report: the third Global Geotourism Conference, 30 Oct–1 Nov 2011, Sultanate of Oman. *Geoheritage*, 4, 221–223.
Dowling, R. and Newsome, D. (eds) 2006 *Geotourism*. Elsevier, Amsterdam.

Dowling, R. and Newsome, D. (eds) 2010 *Global geotourism perspectives*. Goodfellow Publishers, Oxford.

Downs, P.W. and Gregory, K.J. 2008 Approaches to river channel sensitivity. *Professional Geographer*, 47, 168–175.

Doyle, P. 1989 Threat to Bartonian stratotype. *Earth Science Conservation*, 26, 27.

Doyle, P. 1996 *Understanding fossils*. John Wiley & Sons, Ltd, Chichester.

Doyle, P. and Bennett, M.R. 1998 Earth heritage conservation: past, present and future agendas. In Bennett, M.R. and Doyle, P. (eds) *Issues in environmental geology: a British perspective*. Geological Society, London, 41–67.

Doyle, P. and Bennett, M.R. 2002 *Fields of battle: terrain in military history*. Kluwer Academic Publishers, Dordrecht.

Drew, D. 2001 *Classic landforms of the Burren karst*. Geographical Association, Sheffield.

Dudley, N. 2008 Guidelines for applying protected area management categories. IUCN, Gland, Switzerland, 96 pp.

Duff, K. 1994 Natural Areas: an holistic approach to conservation based on geology. In O'Halloran, D., Green, C., Harley, M., *et al.* (eds) *Geological and landscape conservation*. Geological Society, London, 121–126.

Duffin, C.J. and Davidson, J.P. 2011 Geology and the dark side. *Proceedings of the Geologists' Association*, 122, 7–15.

Eberhard, R. 1994 *Inventory and management of the Junee River Karst System, Tasmania*. Report to Forestry Tasmania, Hobart.

Eberhard, R. 1996 *Inventory and management of karst in the Florentine Valley, Tasmania*. Report to Forestry Tasmania, Hobart.

Eberhard, R. (ed) 1997 *Pattern and process: towards a regional approach to national estate assessment of geodiversity*. Australian Heritage Commission, Canberra.

Eberhard, R. and Houshold, I. 2001 River management in karst terrains: issues to be considered with an example from Mole Creek, Tasmania. In Rutherfurd, I., Sheldon, F., Brierley, G. and Kenyon, C. (eds) *Third Australian Stream Management Conference*, 197–204.

Eden, S., Tunstall, S.M. and Tapsell, S.M. 1999 Environmental restoration: environmental management or environmental threat? *Area*, 31, 151–159.

Eder, W. 1999 Geoparks of the future. *Earth Heritage*, 12, 21.

Edmonds, R. 2001 Fossil collecting on the West Dorset coast: a new voluntary Code of Conduct. In Bassett, M.G., King, A.H., Larwood, J.G. and Parkinson, N.A. (eds) *A future for fossils*. National Museums and Galleries of Wales, Cardiff, 46–51.

Edwards, D. and Williams, D. 2011 Rescue palaeontology. *Geology Today*, 27, 65–69.

Ehrlich, P.R. 1988 The loss of diversity: causes and consequences. In Wilson, E.O. (ed) *Biodiversity*. National Academy Press, Washington DC, 21–27.

Ellis, J. (ed) 1993 *Keeping archives*, 2nd edn. Australian Society of Archivists.

Ellis, N. 2008 A history of the Geological Conservation Review. In Burek, C.D. and Prosser, C.D. (eds) *The history of geoconservation*. Geological Society of London, Special Publication 300, 123–135.

Ellis, N. 2011 The Geological Conservation Review (GCR) in Great Britain: rationale and methods. *Proceedings of the Geologists' Association*, 122, 353–362.

Ellis, N.V., Bowen, D.Q., Campbell, S., *et al.* 1996 *An introduction to the Geological Conservation Review* Joint Nature Conservation Committee, Peterborough.

English Nature 1993 *Natural Areas: English Nature's approach to setting nature conservation objectives: a consultation paper*. English Nature, Peterborough.

English Nature 1998 *Natural Areas: nature conservation in context* (CD-Rom). English Nature, Peterborough.

English Nature 2002 *Revealing the value of nature*. English Nature, Peterborough.

English Nature 2003 *Geodiversity and the minerals industry: conserving our geological heritage*. English Nature Peterborough.

Erikstad, L. 1994 Quaternary geology conservation in Norway, inventory program, criteria and results. *Memoires de la Société Géologique de France*, 165, 213–215.

Erikstad, L. 2008 History of geoconservation in Europe. In Burek, C.D. and Prosser, C.D. (eds) *The history of geoconservation*. Geological Society of London, Special Publication 300, 249–256.

Erikstad, L. 2012 Norway. In Wimbledon, W.A.P. and Smith-Meyer, S. (eds) *Geoheritage in Europe and its conservation*. ProGEO, Oslo, 246–253.

Erikstad, L., Lindblom, I., Jerpasen, G., *et al.* 2008 Environmental value assessment in a multidisciplinary EIA setting. *Environmental Impact Assessment Review*, 28, 131–143.

Erikstad, L. and Stabbetorp, O. 2001 Natural areas mapping: a tool for environmental impact assessment. In Gordon, J.E. and Leys, K.F. *Earth science and the natural heritage*. Stationery Office, Edinburgh, 224–227.

Essex County Council 2009 *Tendring Geodiversity Characterisation Report*. Essex County Council, Chelmsford.

Etches, J. 2006 *Wadi El Rayan Protected Area educational/interpretive plan and interpretive product development*. IUCN, Gland.

Evans, A.M. 1997 *An introduction to economic geology and its environmental impact*. Blackwell Science, Oxford.

Falcon-Lang, H.J. 2006 A history of research at the Joggins Fossil Cliffs, the world's finest Pennsylvanian section. *Proceedings of the Geologists' Association*, 117, 377–392.

Fassoulas, C., Mouriki, D., Dimitriou-Nikolakis, P. and Iliopoulos, G. 2012 Quantitative assessment of geotopes as an effective tool for geoheritage management. *Geoheritage*, 4, 177–193.

Fedonkin, M.A., Ivantsov, A.Y., Leonov, M.V., *et al.* 2009 Palaeo-piracy endangers Vendian (edicaran) fossils in the White Sea—Arkhangelsk region of Russia. In Lipps, J.H. (ed) *PaleoParks: our paleontological heritage protected and conserved in the field worldwide*. Carnets de Géologie e-book.

Feick, J. and Draper, D. 2001 Valid threat or "tempest in a teapot"? An historical account of tourism development and the Canadian Rocky Mountain Parks World Heritage Site designation. *Tourism Recreation Research*, 26, 35–46.

Ferron, P. Bélanger, M., Madore, L. and Verpaelst, P. 2010 *Guidelines for managing outstanding geological sites*. Government of Quebec, Quebec.

Fettes, D. and Desmons, J. (eds) 2007 *Metamorphic rocks: a classification and glossary of terms*. Cambridge University Press.

Fiffer, S. 2000 *Tyrannosaurus Sue*. W.H. Freeman and Co., New York.

Fishman, I. and Kazakova, Y. 2012 Kazakhstan. In Wimbledon, W.A.P. and Smith-Meyer. S. (eds) *Geoheritage in Europe and its conservation*. ProGEO, Oslo, 200–207.

Ford, T.D. 2005 *Derbyshire Blue John*. Landmark Publishing Ltd., Derbyshire.

Ford, D. and Williams, P. 2007 *Karst hydrogeology and geomorphology*. John Wiley & Sons, Ltd, Chichester.

Fordyce, F. and Johnson, C. 2002 The rock diet. *Planet Earth*, Autumn 2002, 18–20. Natural Environment Research Council, Swindon.

Forster, M.W.C. 1999 *An overview of fossil collecting with particular reference to Scotland*. Research, Survey and Monitoring Report 115, Scottish Natural Heritage, Edinburgh.

Fortey, R. 1997 *Life: an unauthorised biography*. Flamingo, London.

Fortey, R. 2002 *Fossils: the key to the past*, 3rd edn. Natural History Museum, London.

Foster, J. (ed) 1997 *Valuing nature?* Routledge, London.

Fox, W. 1990 *Towards a transpersonal ecology: developing new foundations for environmentalism*. Shambala Publications, Boston.

Francis, P. 1993 *Volcanoes: a planetary perspective*. Oxford University Press, Oxford.

Francis, P. and Oppenheimer, C. 2004 *Volcanoes*. Oxford University Press, Oxford.

Frazer, J.Z. and Johnson, F.M. 1997 Earth Science Heritage protection for the Oak Ridges Moraine, Ontario, Canada. *Earth Heritage*, 7, 18–19.

Fredericia, J. 1990 Saturated hydraulic conductivity of clayey tills and the role of fractures. *Nordic Hydrology*, 21, 119–132.

French, H.M. 2007 *The periglacial environment*, 3rd edn. John Wiley & Sons, Ltd, Chichester.

French, J.R. and Reed, D.J. 2001 Physical contexts for saltmarsh conservation. In Warren, A. and French, J.R. *Habitat conservation: managing the physical environment*. John Wiley & Sons, Ltd, Chichester, 179–228.

French, J.R. and Spencer, T. 2001 Sea-level rise. In Warren, A. and French, J.R. *Habitat conservation: managing the physical environment*. John Wiley & Sons, Ltd, Chichester, 305–347.

Fukuyama, F. 2001 We remain at the end of history. *The Independent*, 11 October 2001.

Funnell, B.M. 1994 A geological heritage coast: North Norfolk (Hunstanton to Happisburgh). In Stevens, C., Gordon, J.E., Green, C.P. and Macklin, M.G. (eds) *Conserving our landscape*. English Nature, 59–62.

Gagen, P., Gunn, J. and Bailey, D. 1993 Landform replication experiments on quarried limestone rock slopes in the English Peak District. *Zeitschrift fur Geomorphologie, Supplement-Band*, 87, 163–170.

Garavaglia, V. and Pelfini, M. 2011 Glacial geomorphosites and related landforms: a proposal for a dendrogeomorphological approach and educational trails. *Geoheritage*, 3, 15–25.

Garcia-Cortés, A. (ed) 2009 *Spanish geological framework and geosites: an approach to Spanish geological heritage of international significance*. Instituto Geologico y Minero de Espana, Madrid.

Garcia-Cortés, A., Gallego, E. and Carcavilla, L. 2012 Spain. In Wimbledon, W.A.P. and Smith- Meyer, S. (eds) *Geoheritage in Europe and its conservation*. ProGEO, Oslo, 334–343.

Garcia-Guinea, J. and Calafarra, J.M. 2001 Mineral collectors and the geological heritage: protection of a huge geode in Spain. *European Geologist*, 11, 4–7.

Garrels, R.M. and Mackenzie, F.T. 1971 *Evolution of sedimentary rocks*. W.W.Norton, New York.

Gatley, S. and Parkes, M. 2012 In Wimbledon, W.A.P. and Smith-Meyer, S. (eds) *Geoheritage in Europe and its conservation*. ProGEO, Oslo, 180–187.

Gauld, J.H. and Bell, J.S. 1997 *Soils and nature conservation in Scotland*. Scottish Natural Heritage Review, 62.

Gerrard, J. 2000 *Fundamentals of soils*. Routledge, London.

Giardino, J.R., Schroder, J.F. and Vitek, J.D. (eds) 1987 *Rock glaciers*. Allen & Unwin, London.

Gibbons, S. and McDonald, H.G. 2001 Fossil sites as National Natural Landmarks: recognition as an aaproach to protection of an important resource. In Santucci, V.L. and McClelland, L. (eds) *Proceedings of the 6th Fossil Resource Conference*. Technical Report NPS/NRGRD/GRDTR-01/01, 130–136.

Gilbert, O. 1996 *Rooted in stone: the natural flora of urban walls*. English Nature, Peterborough.

Gill, R. 2010 *Igneous rocks and processes: a practical guide*. Wiley-Blackwell, Chichester.

Gillerman, V.S., Wilkins, D., Shellie, K. and Bitner, R. 2006 Geology and wine II: terroir of the Western Snake River Plain, Idaho, USA. *Geoscience Canada*, 33, 37–48.

Gillespie, C. 2008 Scar sands. *Canadian Geographic*, 128(3), 64–78.

Gippel, C.J., Seymour, S., Brizga, S. and Craigie, N.M. 2001 The application of fluvial geomorphology to stream management in the Melbourne Water area. In Rutherfurd, I., Sheldon, F., Brierley, G. and Kenyon, C. (eds) *Third Australian Stream Management Conference*, 239–244.

Gisotti, G. and Burlando, M. 1998 The Italian job. *Earth Heritage*, 9, 11–13.

Glasser, N.F. 2001 Conservation and management of the Earth heritage resource in Great Britain. *Journal of Environmental Planning and management*, 44, 889–906.

Glasson, J., Therivel, R. and Chadwick, A. 2005 *Introduction to environmental impact assessment*, 3rd edn. Routledge, London.

Glypteris, S. 1991 *Countryside recreation*. Longman, Harlow.

Goldie, H.S. 1994 Protection of limestone pavement in the British Isles. In O'Halloran, D., Green, C., Harley, M., *et al.* (eds) *Geological and landscape conservation*. Geological Society, London, 215–220.

Goldstein, A. 2009 Managing fossil resources at the Falls of the Ohio, Indiana and Kentucky, USA: a fossil park in an urban setting. In In Lipps, J.H. and Granier, R.B.C. (eds) *PaleoParks: our paleontological heritage protected and conserved in the field worldwide*. Carnets de Géologie e-book, 97–101.

Gomez, N. 1991 La protection des sites a oeufs de dinosaures de la Sainte-Victoire (Aix en Provence, France) (abstract). *Terra Nova*, 3, 13.

Gonggrijp, G.P. 1993 River meanders or houses: a case study. In Erikstad, L. (ed) *Earth science conservation in Europe. Proceedings from the Third Meeting of the European Working Group of Earth Science conservation*, 35–38.

Goodwin, R. 1992. *Green political theory*. Polity, Cambridge.

Gordon, J.E. 1994 Conservation of geomorphology and Quaternary sites in Great Britain: an overview of site assessment. In Stevens, C., Gordon, J.E., Green, C.P. and Macklin, M.G. (eds) *Conserving our landscape*, English Nature, Peterborough, 41–45.

Gordon, J.E. 2012 Engaging with geodiversity: 'stone voices', creativity and ecosystem cultural services in Scotland. *Scottish Geographical Journal*, 128, 240–265.

Gordon, J.E. and Barron, H.F. 2011 *Scotland's geodiversity: development of the basis for a national framework*. Scottish Natural Heritage Commissioned Report, 417.

Gordon, J.E. and Barron, H.F. 2012 Valuing geodiversity and geoconservation: developing a more strategic ecosystem approach. *Scottish Geographical Journal*, 128, 278–297.

Gordon, J.E., Barron, H.F., Hansom, J.D. and Thomas, M.F. 2012 Engaging with geodiversity—why it matters. *Proceedings of the Geologists' Association*, 123, 1–6.

Gordon, J.E., Brazier, V., Thompson, D.B.A. and Horsfield, D. 2001 Geo-ecology and the conservation management of sensitive upland landscapes in Scotland. *Catena*, 42, 323–332.

Gordon, J.E., Dvorák, I.G., Jonasson, C., Josefsson, M., Kociánová, M. and Thompson, D.B.A. 2002 Geo-ecology and management of sensitive montane landscapes. *Geografiska Annaler*, 84A, 193–203.

Gordon, J.E. and Leys, K.F. (eds) 2001a *Earth science and the natural heritage: interactions and integrated management*. Stationery Office, Edinburgh .

Gordon, J.E. and Leys, K.F. 2001b Earth science and the natural heritage: developing a more holistic approach. In Gordon, J.E. and Leys, K.F. *Earth science and the natural heritage*. Stationery Office, Edinburgh, 5–18.

Gordon, J.E. and MacFadyen, C.C.J. 2001 Earth heritage conservation in Scotland: state, pressures and issues. In Gordon, J.E. and Leys, K.F. *Earth science and the natural heritage*. Stationery Office, Edinburgh, 130–144.

Gore, J.A. 1996 Foreword. In Brookes, A. and Shields, F.D. (eds) *River channel restoration*. John Wiley & Sons, Ltd, Chichester, xiii–xv.

Gorelick, S.M. 2010 *Oil panic and the global crisis*. Wiley-Blackwell, Chichester.

Goudie, A.S. 2013 *The human impact on the natural environment*, 7th edn. Wiley, Chichester.

Goudie, A.S., Livingstone, I. and Stokes, S. 1999. *Aeolian environments, sediments and landforms*. John Wiley & Sons, Ltd, Chichester.

Government of Alberta 1977 *A policy for resource management of the Eastern Slopes*. Government of Alberta, Edmonton.

Government of Alberta 1994a *A framework for Alberta's special places: Natural Regions Report, 1*. Government of Alberta, Edmonton.

Government of Alberta 1994b *Alberta protected areas: system analysis (1994)*. Government of Alberta, Edmonton.

Government of Alberta 1996 *Natural history overview and theme evaluation of the Canadian Shield Natural Region*. Government of Alberta, Edmonton.

Government of Quebec 2002 *Geological heritage of Quebec: geosites in Quebec*. Government of Quebec, Montreal.

Gradstein, F.M., Ogg, J.G., Schmitz, M. and Ogg, G. 2012 *Geologic time scale 2012*. Elsevier, Amsterdam.

Graf, W. 1988 *Fluvial processes in dryland rivers*. Springer-Verlag, Berlin.

Graf, W. 1992 Science, public policy, and western American rivers. *Transactions of the Institute of British Geographers*, NS17, 5–19.

Graf, W. 1996 Geomorphology and policy for restoration of impounded American rivers: what is "natural"? In Rhoades, B.L. and Thorn, C.E. (eds) *The scientific nature of geomorphology: Proceedings of the 27th Binghampton Symposium in Geomorphology*. John Wiley & Sons, Ltd, Chichester.

Grant, A. 1995 Human impacts on terrestrial ecosystems. In O'Riordan, T. (ed) *Environmental science for environmental management*. Longman, Harlow, 66–79.

Gray, J.M. 1975 The Loch Lomond Readvance and contemporaneous sea-levels in Loch Etive and neighbouring areas of western Scotland. *Proceedings of the Geologists' Association*, 86, 227–238.

Gray, J.M. 1993 Quaternary geology and waste disposal in South Norfolk, England. *Quaternary Science Reviews*, 12, 899–912.

Gray, J.M. 1996 Environmental policy and municipal waste management in the UK. *Transactions of the Institute of British Geographers*, NS22, 69–90.

Gray, J.M. 1997a Planning and landform: geomorphological authenticity or incongruity in the countryside. *Area*, 29, 312–324.

Gray, J.M. 1997b The origin of the Blakeney esker, Norfolk. *Proceedings of the Geologists' Association*, 108, 177–182.

Gray, J.M. 1998a Landforms, authenticity and conservation value: a reply. *Area*, 30, 273–274.

Gray, J.M. 1998b Hills of waste: a policy conflict in environmental geology. In Bennett, M.R. and Doyle, P. (eds) *Issues in environmental geology: a British perspective*. Geological Society, London, 173–195.

Gray, M. 2001 Geomorphological conservation and public policy in England: a geomorphological critique of English Nature's 'Natural Areas' approach. *Earth Surface Processes & Landforms*, 26, 1009–1023.

Gray, M. 2002 Landraising of waste in England, 1990–2000: a survey of the geomorphological issues raised by planning applications. *Applied Geography*, 22, 209–234.

Gray, M. 2004 *Geodiversity; valuing and conserving abiotic nature*. John Wiley & Sons, Ltd, Chichester.

Gray, M. 2005 Geodiversity and geoconservation: what, why and how? *The George Wright Forum*, 22, 4–12.

Gray, M. 2008a Geodiversity: the origin and evolution of a paradigm. In Burek, C.D. and Prosser, C.D. (eds) *The history of geoconservation*. Geological Society of London, Special Publication 300, 31–36.

Gray, M. 2008b Geodiversity: a new paradigm for valuing and conserving geoheritage. *Geoscience Canada*, 35, 51–58.

Gray, M. 2009 Landscape: the physical layer. In Clifford, N.J., Holloway, S.L., Rice, S.P. and Valentine, G. (eds) *Key concepts in geography*. Sage, London, 265–285.

Gray, M. 2011 Other nature: geodiversity and geosystem services. *Environmental Conservation*, 38, 271–274.

Gray, M. 2012 Valuing geodiversity in an "ecosystem services" context. *Scottish Geographical Journal*, 128, 177–194.

Gray, M., Gordon, J.E. and Brown E.J. 2013 Geodiversity and the ecosystem approach: the contribution of geoscience in delivering integrated environmental management. *Proceedings of the Geologists' Association*, 124, 659–673.

Gray, M. and Jarman, D. 2003 Creating new "glacial" landforms from waste materials: two case studies from the UK. *Scottish Geographical Journal*, 119, 311–324.

Greensmith, J.T. 1989 *Petrology of the sedimentary rocks*. Unwin Hyman, London.

Gregory, K.J. 2000 *The changing nature of physical geography*. Arnold, London.

Griffiths, J.S. (ed) 2001 *Land surface evaluation for engineering purposes*. Geological Society of London, Engineering Geology Special Publication 18.

Grigorescu, D. 1994 The geo-ecological education and geological site conservation in Romania. In O'Halloran, D., Green, C., Harley, M., *et al.* (eds) *Geological and landscape conservation*. Geological Society, London, 467–472.

Gritsenko, V., Rudenko, K. and Stetsyuk, V. 2012 Ukraine. In Wimbledon, W.A.P. and Smith-Meyer, S. (eds) *Geoheritage in Europe and its conservation*. ProGEO, Oslo, 378–391.

de Groot, R.S. 1992 *Functions of nature*. Wolters-Noordhoff, Gronoingen.

Gross, R. 1998 *Dinosaur country: unearthing Alberta badlands*. Badland Books, Wardlow, Alberta.

Grotzinger, J. and Jordan, T.H. 2010 *Understanding Earth*, 6th edn. W.H. Freeman, New York.

Guerin, B. 1992 Review of scoria and tuff quarrying in Victoria. *Geological Survey of Victoria Report*, 96.

Guiomar, M. and Pages, J.-S. 2012 France. In Wimbledon, W.A.P. and Smith-Meyer, S. (eds) *Geoheritage in Europe and its conservation*. ProGEO, Oslo, 124–131.

Gulliford, A. 2000 *Sacred objects and sacred places: preserving tribal traditions*. University Press of Colorado, Boulder.

Gunn, J. 1993 The geomorphological impacts of limestone quarrying. *Catena Supplement*, 25, 187–197.

Gunn, J. 1995 Environmental change and land management in the Cuilcagh karst, Northern Ireland. In McGregor, D.F.M. and Thompson, D.A. (eds) *Geomorphology and land management in a changing environment*. John Wiley & Sons, Ltd, Chichester, 195–209.

Gutiérrez-Marco, J.C., Sa, A.A., Garcia-Bellido, D.C., *et al.* 2009 Giant trilobite clusters from the Ordovician of Portugal. *Geology*, 37, 443–446.

Haarhoff, P. and Prosser, C. 2007 South Africa sets example for fossil tourism. *Earth Heritage*, 27, 22–23.

Haas, J. and Hamor, G. 2001 Geological garden in the neighbourhood of Budapest, Hungary. *Episodes*, 24, 257–261.

Haffey, D. 2008 *Local Geodiversity Action Plans: a review of progress in England*. Natural England Research Report 027.

Haigh, M.J. 1978 *Evolution of slopes on artificial landforms—Blaenavon*, UK. Research Paper 183, Department of Geography, University of Chicago.

Haigh, M.J. 2002 Land reclamation and Deep Ecology: in search of a more meaningful physical geography. *Area*, 34, 242–252.

Hallett, D. 2002 The Wieliczka salt mine. *Geology Today*, 18, 182–185.

Hanley, N., Mourato, S. and Wright, R.E. 2001 Choice modelling approaches: a superior alternative for environmental valuation? *Journal of Economic Surveys*, 15, 435–462.

Hansom, J., Crick, M. and John, S. 2000 *The potential application of Shoreline Management Planning to Scotland*. Scottish Natural Heritage Review, 121.

Hansom. J. and Gordon, J.E. 1998 *Antarctic environments and resources: a geographical perspective*. Longman, Harlow.

Harcourt, H. 1990 Quality as well as quantity. *Naturopa*, 65, 4–6.

Hardie, R. and Lucas, R. 2001 Geomorphic categorisation of streams in the Hawkesbury-Nepean Catchment: an application and review of the River Styles Framework. In Rutherfurd, I., Sheldon, F., Brierley, G. and Kenyon, C. (eds) *Third Australian Stream Management Conference*, 271–276.

Hardwick, P. and Gunn, J. 1994 The conservation of cave systems in mixed lithology catchments: a case study of the Castleton Karst, England. In Stevens, C., Gordon, J.E., Green, C.P. and Macklin, M.G. (eds) *Conserving our landscape*, English Nature, Peterborough, 198–202.

Harrison, S.J. and Kirkpatrick, A.H. 2001 Climatic change and its potential implications for environments in Scotland. In Gordon, J.E. and Leys, K.F. (eds) *Earth science and the natural heritage: interactions and integrated management*. Stationery Office, Edinburgh, 296–305.

Hart, D.D. and Poff, N.L. 2002 How dams vary and why it matters for the emerging science of dam removal. *Bioscience*, 52, 659–668.

Hart, M. 2012 Geodiversity, palaeodiversity or biodiversity: where is the place of palaeobiology and an understanding of taphonomy? *Proceedings of the Geologists' Association*, 123, 551–555.

Haslett, S.K. 2008 *Coastal systems*, 2nd edn. Routledge, Abingdon.

Haynes, S.J. 1999 Geology and wine I: concept and terroir and the role of geology. *Geoscience Canada*, 26, 190–194.

Haynes, V.M., Grieve, I.C., Price-Thomas, P. and Salt, K. 1998 *The geomorphological sensitivity of the Cairngorm high plateaux*. Research, Survey and Monitoring Report, 66.

Hayward, B.W. 1989 Earth science conservation in New Zealand. *Earth Science Conservation*, 26, 4–6.

Hayward, B.W. 2009 Protecting fossil sites in New Zealand. In Lipps, J.H. and Granier, R.B.C. (eds) *PaleoParks: our paleontological heritage protected and conserved in the field worldwide*. Carnets de Géologie e-book, 49–64.

Heffernan, K. and O'Brien, J. 2010 *Earth materials*. Wiley-Blackwell, Chichester.

Heldal, T., Kjolle, I., Meyer, G.B. and Dahlgren, S. 2008 National treasure of global significance. Dimension stone deposits in larvikite, Oslo igneous province, Norway. *Geological Survey of Norway Special Publication* 11, 5–18.

Henderson, P., Gluyas, J., Gunn, G., *et al.* 2011. *Rare Earth elements briefing note*. Geological Society of London, London.

Henriques, M.H. 2004 Jurassic heritage of Portugal: state of the art and open problems. *Revista Italiana di Palaeontologia e Stratigrafia*, 110, 389–392.

Henriques, M.H., dos Reis, R.P., Brilha, J. and Mota, T. 2011 Geoconservation as an emerging geoscience. *Geoheritage*, 3, 117–127.

Heritage Council 2002 *Integrating policies for Ireland's landscape*. The Heritage Council, Kilkenny.

Heritage Council 2006 *Landscape Character Assessment in Ireland: baseline audit and evaluation*. Julie Martin Associates, Richmond.

Hildreth-Walker, V. and Werker, J.C. 2006 *Cave conservation and restoration*. National Speleological Society, Huntsville, Alabama.

Hilgen, F.J., Iaccarino, S., Krijgsman, W., *et al.* 2000 The Global Boundary Stratotype Section and Point (GSSP) of the Messinian Stage (uppermost Miocene). *Episodes*, 23(3), 172–178.

Hillel, D. 1991 *Out of the Earth: civilization and the life of the soil*. University of California Press, Berkeley.

Hilson, G. 2002 The environmental impact of small-scale gold mining in Ghana: identifying problems and possible solutions. *Geographical Journal*, 168, 57–72.

Hjort, J. and Luoto, M. 2010 Geodiversity of high-latitude landscapes in northern Finland. *Geomorphology*, 115, 109–116.

Hjort, J. and Luoto, M. 2012 Can geodiversity be predicted from space? *Geomorphology*, 153–154, 74–80.

Hlad, B. 2012 Slovenia. In Wimbledon, W.A.P. and Smith-Meyer, S. (eds) *Geoheritage in Europe and its conservation*. ProGEO, Oslo, 322–333.

HMSO 1994 *Biodiversity: a UK Action Plan*. Her Majesty's Stationery Office, London.

Hodson, M.J., Stapleton, C. and Emberton, R. 2001 Soils, geology and geomorphology. In Morris, P. and Therivel, R. (eds) *Methods of environmental impact assessment*, 2nd edn. Spon Press, London, 170–196.

Hofmann, T. and Schönlaub, H.P. 2012 Austria. In Wimbledon, W.A.P. and Smith-Meyer, S. (eds) *Geoheritage in Europe and its conservation*. ProGEO, Oslo, 30–39.

Holden, J. (ed) 2012 *An introduction to physical geography and the environment*, 3rd edn. Pearson Education, Harlow.

Holm, L. 2012 Denmark. In Wimbledon, W.A.P. and Smith-Meyer, S. (eds) *Geoheritage in Europe and its conservation*. ProGEO, Oslo, 100–105.

Holstein, T. 2002 Moving to a SEA Green Paper. *Town and Country Planning*, 71, 218–220.

Holt-Wilson, T. 2012 EIA for geodiversity. *Earth Heritage*, 37, 4.

Hooke, J. 1994a Conservation: the nature and value of active river sites. In Stevens, C., Gordon, J.E., Green, C.P. and Macklin, M.G. (eds) *Conserving our landscape*, English Nature, Peterborough, 110–116.

Hooke, J. 1998 Issues and strategies in relation to geological and geomorphological conservation and defence of the coast. In Hooke, J. (ed) *Coastal defence and Earth science conservation*. Geological Society, London, 1–9.

Hooke, J.M. 1999 Decades of change: contributions of geomorphology to fluvial and coastal engineering and management. *Geomorphology*, 31, 373–389.

Hooke, R.L. 1994b On the efficacy of humans as geomorphic agents. *GSA Today*, 4, 217–225.

Hootsmans, M. and Kampf, H. 2004 *Ecological networks; Experiences in the Netherlands*. Ministry of Agriculture, Nature and Food Quality, The Hague.

Hopkins, J. 1994 Geology, geomorphology and biodiversity. *Earth Heritage*, 2, 3–6.

Horner, J.R. and Dobb, E. 1997. *Dinosaur lives: unearthing an evolutionary saga*. Harcourt Brace and Co, San Diego.

Hose, T. 2000 European geotourism—geological interpretation and geoconservation promotion for tourists. In Barrentino, D., Wimbledon, W.P. and Gallego, E. (eds) *Geological heritage: its conservation and management*. Instituto Tecnologico Geominero de Espana, Madrid, 127–146.

Hose, T. 2008 Towards a history of geotourism: definitions, antecedents and the future. In Burek, C.D. and Prosser, C.D. (eds) *The history of geoconservation*. Geological Society of London, Special Publication 300, 37–60.

Hose, T.A. 2012a Editorial: geotourism and geoconservation. *Geoheritage*, 4, 1–5.

Hose, T.A. 2012b 3G's for modern geotourism. *Geoheritage*, 4, 7–24.

Hose, T.A. and Vasiljevic, D.A. 2012 Defining the nature and purpose of modern geotourism with particular reference to the United Kingdom and south-east Europe. *Geoheritage*, 4, 25–43.

Hoskins, W.G. 1955 *The making of the English landscape*. Hodder and Stoughton, London.

Houshold, I. and Sharples, C. 2008 Geodiversity in the wilderness: a brief history of geoconservation in Tasmania. In Burek, C.V. and Prosser, C.D. (eds) *The history of geoconservation*. Geological Society of London, Special Publication 300, 257–272.

Houshold, I., Sharples, C., Dixon, G. and Duhig, N. 1997 Georegionalisation—a more systematic approach for the identification of places of geoconservation significance. In Eberhard, R. (ed) *Pattern and process: towards a regional approach to national estate assessment of geodiversity*. Australian Heritage Commission, Canberra, 65–89.

Howard, A.D. 1967 Drainage analysis in geologic interpretation: a summation. *American Association of Petroleum Geologists Bulletin*, 51, 2246–2259.

Hubbert, M.K. 1956 Nuclear energy and the fossil fuels. In *Drilling and Production Practice*. American Petroleum Institute, Washington DC, 7–25.

Huggett, R.J. 1995 *Geoecology: an evolutionary approach*. Routledge, London.

Huggett, R.J. 2003 *Fundamentals of geomorphology*. Routledge, London.

Hughes, F.M.R. and Rood, S.B. 2001 Floodplains. In Warren, A. and French, J.R. *Habitat conservation: managing the physical environment*. John Wiley & Sons, Ltd, Chichester, 105–121.

Hulbert, C.A.V., Wharton, G. and Copas, R. 2009 *Integrated Post-Project Appraisal of an Urban River Restoration Scheme: The River Quaggy, Sutcliffe Park, South East London*. Environment Agency RandD Report, 80 pp.

Hunt, E.L.R. 1988 *Managing growth's impact on the mid-south's historic and cultural resources*. National Trust for Historic Preservation, Washington DC.

Hunter, M.L., Jacobson, G.L. and Webb, T. 1988 Palaeoecology and the coarse filter approach to maintaining biological diversity. *Conservation biology*, 2, 375–385.

Husain, S. 2003 Salt in wounds. *Guardian Society*, 15 January 2003, 8–9.

Hutton, J. 1795 *Theory of the Earth*. William Creech, Edinburgh.

Hurlstone, P. and Long, M. 2000 Striving for a balance. *Earth Heritage*, 14, 24–25.

Huxley, J. 1947 *Report of the Wild Life Conservation Special Committee*. HMSO, London.

Hyndman, D.W. and Hyndman, D.W. 2010 *Natural hazards and disasters*, 3rd edn. Brooks Cole, Pacific Grove.

Ibáñez, J.J., De-Alba, S. and Boixadera, J. 1995 The pedodiversity concept and its measurement: application to soil information systems. In King, D., Jones, R.J.A. and Thomasson, A.J. (eds) *European Land Information System for agro-environmental monitoring*. EU, Brussels, 181–195.

Ibáñez, J.J., De-Alba, S., Bermudez, F.F. and Garcia-Alvarez, A. 1995 Pedodiversity concepts and tools. *Catena*, 24, 215–232.

Ibáñez, J.J., De-Alba, S., Lobo, A. and Zucarello, V. 1998 Pedodiversity and global soil patterns at coarse scales (with Discussion). *Geoderma*, 83, 171–192.

IPCC 2007 *Fourth Assessment Report: Climate change 2007*. Intergovernmental Panel on Climate Change, Geneva, and Cambridge University Press.

Iversen, T.M., Kronvang, B., Madsen, B.L., Markmann, P. and Nielsen, M.B. 1993 Re-establishment of Danish streams: restoration and maintenance measures. *Aquatic Conservation: Marine and Freshwater Ecosystems*, 3, 1–20.

Jackli, H. 1979 Opening address. In Schlüchter, C. (ed) *Moraines and varves*. Balkema, Rotterdam, 5–7.

Jackson, P.W. 2010 *Introducing palaeontology: a guide to ancient life*. Dunedin Academic Press.

Jacobs, P. 2012 Belgium. In Wimbledon and Smith-Meyer (eds) *Geoheritage in Europe and its conservation*. ProGEO, Oslo, 52–61.

Jarman, D. 1994 Geomorphological authenticity: the planning contribution. In Stevens, C., Gordon, J.E., Green, C.P. and Macklin, M.G. (eds) *Conserving our landscape*, English Nature, Peterborough, 41–45.

Jarzembowski, E.A. 1989 Writhlington Geological Nature Reserve. *Proceedings of the Geologists' Association*, 100, 219–234.

Jennings, J.N. 1985 *Karst geomorphology*. Blackwell, Oxford.

Jenny, H. 1941 *Factors of soil formation*. McGraw-Hill, New York.

Jerie, K., Houshold, I. and Peters, D. 2001 Stream diversity and conservation in Tasmania: yet another new approach. In Rutherfurd, I., Sheldon, F., Brierley, G. and Kenyon, C. (eds) *Third Australian Stream Management Conference*, 329–335.

Jiang, P. 1994 Conservation of national geological sites in China. In O'Halloran, D., Green, C., Harley, M., *et al.* (eds) *Geological and landscape conservation*. Geological Society, London, 243–245.

JNCC 2006 *Common Standards Monitoring for designated sites: first six year report*. Joint Nature Conservation Committee, Peterborough.

Joggins Fossil Institute 2007 *Nomination of the Joggins Fossil Cliffs for Inscription on the World Heritage List*. Joggins Fossil Institute, Nova Scotia.

Johansson, C.E. (ed) 2000 *Geodiversitet i Nordisk Naturvård*. Nordisk Ministerråad, Copenhagen.

Johns, C. 2006 Hallowed ground: nothing is ever safe. *National Geographic*, October 2006, 42–43.

Johnson, B.R. 1997 Monitoring peatland rehabilitation. In Parkyn, L., Stoneman, R.E. and Ingram, H.A.P. *Conserving peatlands*. CAB International, Oxford, 323–331.

Jones, G. 1996 Planning lead needed when golf boom is back on track. *Planning for the natural and built environment*, 1175, 26–27.

Jones, G. and Hollier, G. 1997 *Resources, society and environmental management*. Paul Chapman Publishing, London.

Jongman, R.H.G. and Pungetti, G. (eds) 2004 *Ecological networks and greenways: concept, design, implementation*. Cambridge University Press, Cambridge.

Joyce, E.B. 1995 *Assessing the significance of geological heritage*. A methodology study for the Australian Heritage Commission.

Joyce, E.B. 1997 Assessing geological heritage. In Eberhard, R. (ed) *Pattern and process: towards a regional approach to national estate assessment of geodiversity*. Australian Heritage Commission, Canberra, 35–40.

Joyce, E.B. 1999 Different thinking. *Earth Heritage*, 12, 11–13.

Joyce, E.B. 2010 Australia's geoheritage: history of study, a new inventory of geosites and application to geotourism and geoparks. *Geoheritage*, 2, 39–56.

Jungerius, P.D., Matundura, J. and van den Ancker, J.A.M. 2002 Road construction and gulley erosion in west Pokot, Kenya. *Earth Surface Processes and Landforms*, 27, 1237–1247.

Kananoja, T., Suominen, V. and Nenonen, K. 2012 Finland. In Wimbledon and Smith-Meyer (eds) *Geoheritage in Europe and its conservation*. ProGEO, Oslo,114–123.

Karis, L. 2002 The Russian perspective. *ProGEO News*, 2002(1), 8.

Katz, E. 1992 The ethical significance of human intervention in nature. *Restoration and Management Notes*, 9, 90–96.

Kavcic, M.G. 2012 Geological heritage in the Idrija Region (Slovenia). *European Geologist*, 34, 39–43.

Kelk, B. 1992. Natural resources in the geological environment. In Lumsden, G.I. (ed) *Geology and the environment in western Europe*. Clarendon Press, Oxford, 34–138.

Kelley, K.B. and Francis, H. 1994 *Navajo sacred places*. Indiana University Press.

Kemp, T.S. 1999 *Fossils and evolution*. Oxford University Press, Oxford.

Kendall, R. 2012 Wales first. *Earth Heritage*, 38, 33–35.

Kenny, J.A. and Hayward, B.W. (eds) 1993 *Inventory of important geological sites and landforms in the Northland region*. Geological Society of New Zealand, Miscellaneous Publication, 67.

Kerr, J.W. 1990 *Frank slide*. Baker Publishing Ltd., Calgary.

Kerr, R.A. 1998. The next oil crisis looms large—and perhaps close. *Science*, 281, 1128–1131.

Khalid, F. and O'Brien, J. (eds) 1992 *Islam and ecology*. Cassell, London.

Kibblewhite, M.G., Ritz, K. and Swift, M.J. 2008 Soil health in agricultural systems. *Philosophical Transactions of the Royal Society, Series B*, 363, 685–701.

Kiernan, K. 1991 Landform conservation and protectiom. Fifth regional seminar on national parks and wildlife management, Tasmania 1991, resource document. Tasmanian Parks, Wildlife and Heritage Department, Hobart, 112–129.

Kiernan, K. 1994 *The geoconservation significance of Lake Pedder and its contribution to geodiversity*. Unpublished Report to the Lake Pedder Study Group.

Kiernan, K. 1995 *An atlas of Tasmanian karst*, 2 vols. Tasmanian Forest Research Council, Hobart, Report 10.

Kiernan, K. 1996 *The conservation of glacial landforms*. Forest Practices Unit, Hobart.

Kiernan, K. 1997a *The conservation of landforms of coastal origin*. Forest Practices Board, Hobart.

Kiernan, K. 1997b Landform classification for geoconservation. In Eberhard, R. (ed) *Pattern and process: towards a regional approach to national estate assessment of geodiversity*. Australian Heritage Commission, Canberra, 21–34.

Kiernan, K. 2010 Human impacts on geodiversity and associated natural values of bedrock hills in the Mekong Delta. *Geoheritage*, 2, 101–122.

Kiernan, K. 2012 Impacts of war on geodiversity and geoheritage: case studies of karst caves from northern Laos. *Geoheritage*, 4, 225–247.

King, A.H. and Larwood, J.G. 2001 Conserving our most 'fragile' fossil sites in England: the use of OLD25. In Bassett, M.G., King, A.H., Larwood, J.G. and Parkinson, N.A. (eds) *A future for fossils*. National Museums and Galleries of Wales, Cardiff, 24–31.

King, R.B. 1987 Review of geomorphic description and classification in land resource surveys. In Gardiner, V. (ed) *International geomorphology 1986 Part II*. John Wiley & Sons, Ltd, Chichester, 384–403.

Kirkbride, M.P., Duck, R.W., Dunlop, A., Drummond, J., Mason, M., Rowan, J.S. and Taylor, D. 2001 *Development of a geomorphological database and geographical information system for the North West Seaboard: pilot study*. Scottish Natural Heritage Commissioned Report BAT/98/99/137. Scottish Natural Heritage, Edinburgh.

Klein, C. 2002 *Mineral science*, 22nd edn. John Wiley & Sons, Ltd, Chichester.

Klein, C. and Philpotts, T. 2012 *Earth materials: introduction to mineralogy and petrology*. Cambridge University Press, Cambridge.

Klincharov, S. and Petkovska, J. 2012 F.Y.R. of Macedonia. In Wimbledon, W.A.P. and Smith- Meyer, S. (eds) *Geoheritage in Europe and its conservation*. ProGEO, Oslo, 224–231.

Knight, J. 2010 Evaluating geological heritage: correspondence on Ruban, D.A. 'Quantification of geodiversity and its loss' (PGA, 2010, 121(3): 326–333). *Proceedings of the Geologists' Association*, 122, 508–510.

Knight, J. and Harrison, S. (eds) 2009 *Periglacial and paraglacial processes and environments*. Geological Society of London, Special Publications, 320.

Knighton,, A.D. 1998 *Fluvial forms and processes*. Arnold, London.

Knill, J. 1994 An international earth heritage convention—convenience or confusion. *Earth Heritage*, 1, 7–8.

Knoll, A.H., Canfield, D.E. and Konhauser, K.O. 2012 *Fundamentals of geobiology*. John Wiley & Sons, Ltd-Blackwell, Chichester.

Knutz, A. 1984 The production of road construction material from till. *Striae*, 20, 99–100.

Kondolf, G.M. and Downs, P.W. 1996 Catchment approach to planning channel restoration. In Brookes, A. and Shields, F.D. (eds) *River channel restoration*. John Wiley & Sons, Ltd, Chichester, 129–148.

Koster, E.A. 2009 The "European Aeolian Sand Belt" geoconservation of drift sand landscapes. *Geoheritage*, 1, 93–110.

Kozlowski, S. 2004 Geodiversity: the concept and scope of geodiversity. *Przglad Geologiczny*, 52, 833–837.

Kukla, G. 1975 Loess stratigraphy of Central Europe. In Butzer, K.W. and Isaac, L.I. (eds) *After Australopithecines*. Mouton Publishers, The Hague, 99–187.

Kukla, G. 1977 Pleistocene land-sea correlations. *Earth Science Reviews*, 13, 307–374.

Kusky, T.M., Zhai, M.G., and Xiao, W.J., 2010 *The Evolving Continents: Understanding processes of continental growth*. Geological Society of London Special Publication, 338, 424 pp.

Laity, J. 2008 *Deserts and desert environments*. Wiley-Blackwell, Chichester.

Lake, P.S. 2001 Restoring streams: re-building and re-connecting. In Rutherfurd, I., Sheldon, F., Brierley, G. and Kenyon, C. (eds) *Third Australian Stream Management Conference*, 369–372.

Lamb, S. and Singleton, D. 1998 *Earth story: the shaping of our world*. BBC Books, London.

Lancaster, N. 1995 *Geomorphology of desert dunes*. Routledge, London.

Landscape Institute 1995 *Guidelines for landscape and visual impact assessment*. E. and F.N. Spon, London.

Langford, N. 1870 In Haines, A.L. (ed) 1972 *The discovery of Yellowstone Park, Journal of the Washburn Expedition to the Yellowstone and Firehole Rivers in the year 1870*. University of Nebraska Press, Lincoln.

Lapo, A. and Vdovets, M. 2012 Russia. In Wimbledon, W.A.P. and Smith-Meyer, S. (eds) *Geoheritage in Europe and its conservation*. ProGEO, Oslo, 288–299.

Lappalainen, E. 1996 General review on world peatland and peat resources. In Lappalainen, E. (ed) *Global peat resources*. International Peat Society, Finland, 53–56.

Larwood, J. 2003 The camera never lies. *Earth Heritage*, 19, 10–11.

Larwood, J. and Prosser, C. 1998 Geotourism, conservation and society. *Geologica Balcanica*, 28, 97–100.

Lawrence, D.J.D., Arkley, S.L.B., Everest, J.D., *et al.* 2007 *Northumberland National Park: geodiversity audit and action plan*. British Geological Survey, Nottingham.

Lawrence, D.J.D., Vye, C.L. and Young, B. 2004 *Durham geodiversity audit*. Durham County Council, Durham.

Lawton, J. 2010 *Making space for nature: a review of England's wildlife sites and ecological network*. Department of Environment, Food and Rural Affairs, London.

Lean, G. 2003 Huge dust cloud threatens Asia. *Independent on Sunday*, 26 January 2003, 20.

Lees, G. 1997 *Coastal erosion and defence: II Coastal erosion and coastal cells*. Information and Advisory Note, 72. Scottish Natural Heritage, Edinburgh.

Leman, M.S., Ghani, K.A., Komoo, I. and Ahmad, N. 2007 *Langkawi Geopark*. Institute for Environment and Development, Malaysia.

Leopold, S.A. 1959 *Wildlife in Mexico*. University of California Press, Berkeley.

Leopold, L.B., Clark, F.E., Hanshaw, B.B. and Blasley, J.R. 1971 *A procedure for evaluating environmental impact*. US Geological Survey Circular, 645. Department of the Interior, Washington DC.

Levy, A and Scott-Clark, C. 2001 *Stone of Heaven*. Weidenfeld and Nicholson, London.

Lillie, R.J. 2005 *Parks and plates: the geology of our National Parks, Monuments and Seashores*. W.W. Norton and Co., New York.

Lilly, A., Grieve, I.C., Jordan, C., *et al.* 2009 *Climate change, land management and erosion in the organic and organo-mineral soils in Scotland and Northern Ireland*. Scottish Natural Heritage Commissioned Research, 325.

de Lima, F.F., Brilha, J.B. and Salamuni, E. 2010 Inventorying geological heritage in large territories: a methodological proposal applied to Brazil. *Geoheritage*, 2, 91–99.

Lipps, J.H. 2009 PaleoParks: Our paleontological heritage protected and conserved in the field worldwide. In Lipps, J.H. and Granier B.R.C. (eds), *PaleoParks: the protection and conservation of fossil sites worldwide*. Carnets de Géologie e-book.

Lipps, J.H. and Granier, R.B.C. (eds) 2009 *PaleoParks: our paleontological heritage protected and conserved in the field worldwide*. Carnets de Géologie e-book.

Liscak, P. 2012 Slovakia. In Wimbledon, W.A.P. and Smith-Meyer, S. (eds) *Geoheritage in Europe and its conservation*. ProGEO, Oslo, 310–321.

Liscak, P. and Nagy, A. 2012 Information system on important geosites in the Slovak Republic. *European Geologist*, 34, 35–38.

Livingstone, I. and Warren, A. 1996 *Aeolian geomorphology*. Longman, Harlow.

Lloyd, J.W. 1986 Hydrogeology of beer. *Proceedings of the Geologists' Association*, 97, 213–219.

Lockwood, J.P. and Hazlitt, R.W. 2010 *Volcanoes: global perspectives*. Wiley-Blackwell, Chichester.

Lockwood, M., Worboys, G.L. and Kothari, A. (eds) 2006 *Managing protected areas: a global guide*. Earthscan, London.

Lovejoy, T.E. 1997 Biodiversity: what is it? In Reaka-Kudla, L., Wilson, D.E. and Wilson, E.O. (eds) *Biodiversity II: understanding and protecting our biological resources*. Joseph Henry Press, Washington DC, 7–14.

Lovelock, J. 1979 *Gaia*. Oxford University Press, Oxford.

Lovelock, J. 1995 *The Ages of Gaia*, 2nd edn. Oxford University Press, Oxford.

Lundqvist, S., Fredén, C., Johansson, E., *et al.* 2012 Sweden. In Wimbledon, W.A.P. and Smith-Meyer, S. (eds) *Geoheritage in Europe and its conservation*. ProGEO, Oslo, 344–357.

Lugeri, F.R., Farabollini, P., Graziano, G.V. and Amadio, V. 2012 Geoheritage: nature and culture in a landscape approach. *European Geologist*, 34, 23–28.

Lynch, F. 1990 Mynydd Rhiw axe factory. In Addison, K., Edge, M.J. and Watkins, R. (eds) *North Wales field guide*. Quaternary Research Association, Coventry, 48–51.

MacArthur, R.H. and Wilson, E.O. 1967 *The theory of island biogeography*. Princeton University Press, Princeton.

MacDonald, G.A. 1972 *Volcanoes*. Prentice-Hall, Englewood Cliffs.

MacEwen, A. and MacEwen, M. 1987 *Greenprints for the countryside: the story of Britain's national parks*. Allen & Unwin, London.

MacFadyen, C. 1999 *Fossil collecting in Scotland*. Information and Advisory Note, 110. Scottish Natural Heritage, Edinburgh.

MacFadyen, C. 2001a Getting to grips with asset strippers. *Earth Heritage*, 15, 10.

MacFadyen, C. 2001b Plain-speaking blueprints. *Earth Heritage*, 15, 11.

MacFadyen, C. 2011. Irresponsible coring. *Earth Heritage*, 36, 11–12.

Mackay, A.L. 1991 *A dictionary of scientific quotations*. Adam Hilger, Bristol.
Mackenzie, A.F.D. 1998. *Land, ecology and resistance in Kenya, 1880–1952*. Routledge, London.
Madsen, B.L. 1995 *Danish watercourses: ten years with the new Watercourse Act: collected examples of maintenance and restoration*. Danish Environmental Protection Agency, Denmark.
MAFF 1995 *Shoreline Management Plans: a guide for coastal defence authorities*. Ministry of Agriculture, Fisheries and Food, London.
Magome, H. and Murombedzi, J. 2003 Sharing South Africa's National Parks: community land and conservation in a democratic South Africa. In Adams, W.M. and Mulligan, M. (eds) *Decolonizing nature*. Earthscan, London.
Maltby, E. and Proctor, M.C.F. 1996. Peatlands: their nature and role in the biosphere. In Lappalainen, E. (ed) *Global peat resources*. International Peat Society, Finland, 11–19.
Maltman, A. 2003 Wine, beer and whisky: the role of geology. *Geology Today*, 19(1), 22–29.
Mankell, H. 2011 *The Troubled Man*. Vintage Books, London.
Manning, A. 2001 Towards a new synthesis of the biological and Earth sciences. In Gordon, J.E. and Leys, K.F. *Earth science and the natural heritage*. Stationery Office, Edinburgh, 19–25.
Mansur, K.L. and da Silva, A.S. 2011 Society's response: assessment of the performance of the "Caminhos Geologicos" ("Geological paths") project, State of Rio de Janeiro, Brazil. *Geoheritage*, 3, 27–39.
Maran, A. 2012 Geoconservation in Serbia—state of play and future perspectives. *European Geologist*, 34, 29–34.
Marchetti, M., Panizza, M. Soldati, M. & Barani, D. (eds) 1995 *Geomorphology and Environmental Impact Assessment*. CNR-CS Geodinamica Alpina e Quaternaria, 63–70
Marjanac, L. 2012 Croatia. In Wimbledon, W.A.P. and Smith-Meyer, S. (eds) *Geoheritage in Europe and its conservation*. ProGEO, Oslo, 80–91.
Marsh, G.P. 1864 *Man and nature*. C. Scribner, New York.
Marsh, W.M. 1997 *Landscape planning: environmental applications*, 3rd edn. John Wiley & Sons, Ltd, New York.
Marshak, S. 2012 *Earth: portrait of a planet*. 4th edn. W.W. Norton and Co., New York.
Martill, D.M. 2011. Protect - and die. *Geoscientist*, 21(10), 12–17.
Martinez-Torres, L.M., Alonso, J. and Valle, J.M. 2011 The Upper Aptian-Lower Albian amber deposit of the Penacerrada II Geosite (Basque-Cantabrian Basin, Northern Spain): geological context and protection. *Geoheritage*, 3, 55–61.
Martini, G. 1994 Actes du Premier Symposium International sur la Protection du Patrimoine *Géologique. Mémoires de la Société Géologique de France*, NS 165.
Martini, I.P., Brookfield, M.E. and Sadura, S. 2001 *Principles of glacial geomorphology and geology*. Prentice Hall, New Jersey.
Masalu, D.C.P. 2002 Coastal erosion and its social and environmental aspects in Tanzania: a case study in illegal sand mining. *Coastal Management*, 30, 347–359.
Masselink, G., Hughes, M.G. and Knight, J. 2011 *Introduction to coastal processes and geomorphology*. Hodder Education, London.
Mather, A.S. and Chapman, K. 1995 Environmental resources *Longman*, Harlow.
Maugeri, L. 2012 *Oil: the next revolution*. Belfer Center for Science and International Affairs, Harvard Kennedy School, Discussion Paper 2012–10.
Mayor, A. 2005 *Fossil legends of the first Americans*. Princeton University Press, Princeton.
McBratney, A.B.1995 Pedodiversity. *Pedometron*, 3, 1–3.
McCoy, E.D. and Bell, S.S. 1991 Habitat structure: the evolution and diversification of a complex topic. In Bell, S.S., McCoy, E.D. and Mushinsky, H.R. (eds) *Habitat structure: the physical arrangement of objects in space*. Chapman & Hall, London.
McHarg, I. 1995 *Design with nature*. Natural History Press, New York.
McIntosh, A. 2001 *Soil and soul: people versus corporate power*. Aurum Press, London.
McKee, E.D. (ed) 1979 *A study of global sand seas*. US Geological Survey Professional Paper, 1052.

McKeever, P. 2010 *Communicating geoheritage: an essential tool to build a strong geopark brand*. Paper presented to the 4th Global Geoparks Conference, Langkawi, Malaysia.

McKeever, P.J. and Gallagher, E. 2001 Landscapes from stone. In Gordon, J.E. and Leys, K.F. *Earth science and the natural heritage*. Stationery Office, Edinburgh, 262–270.

McKenzie, D.I. 1994 Earth science assessments for heritage waterways: the French River and others in Canada. In O'Halloran, D., Green, C., Harley, M., *et al.* (eds) *Geological and landscape conservation*. Geological Society, London, 127–131.

McKinney, M.L. and Schoch, R.M. 1998 *Environmental science: systems and solutions*. Jones and Bartlett, Sudbury, Massachusetts.

McKirdy, A. 1990 Resolving the Naze conflict: the crenulate bay model. *Earth Science Conservation*, 27, 16–17.

McKirdy, A. 1993 Coastal superquarries: a blot on the Scottish landscape. *Earth Science Conservation*, 33, 18–19.

McKirdy, A.P., Threadgold, R. and Finlay, J. 2001 Geotourism: an emerging rural development opportunity. In Gordon, J.E. and Leys, K.F. *Earth science and the natural heritage*. Stationery Office, Edinburgh, 255–261.

McMillan, A.A., Gillanders, R.J. and Fairhurst, J.A. 1999 *The building stones of Edinburgh*, 2nd edn. Edinburgh Geological Society, Edinburgh.

McNeely, J. 1988 Protected areas. In Pitt, D.C. (ed) *The future of the environment: the social dimensions of conservation and ecological alternatives*. Routledge, London, 126–144.

McNeely, J. 1989 Protected areas and human ecology: how national parks can contribute to sustaining societies of the twenty-first century. In Western, D. and Pearl, M. (eds) *Conservation for the twenty-first century*. Oxford University Press, Oxford, 150–157.

Melendez, G. and Soria, M. 1994 The legal framework and scientific procedure for the protection of palaeontological sites in Spain: recovery of some special sites affected by human activity in Aragón. In O'Halloran, D., Green, C., Harley, M., *et al.* (eds) *Geological and landscape conservation*. Geological Society, London, 329–334.

Mellor, D. 2001 *American rock*. Countryman Press.

Metsähallitus 2000 *The principles of protected area management in Finland*. Metsähallitus [Forest and Park Service], Vantaa.

Midgley, M. 2001 Does the Earth concern us? *Gaia Circular*, Winter 2001/Spring 2002, 4–9.

Mijovic, D. 2012 Serbia. In Wimbledon, W.A.P. and Smith-Meyer, s. (eds) *Geoheritage in Europe and its conservation*. ProGEO, Oslo, 300–309.

Miles, E. 2009 Champion idea. *Earth Heritage*, 31, 8–9.

Millennium Ecosystem Assessment 2005 *Ecosystems and human well-being: a framework for assessment*. Island Press, Washington DC.

Miller, A.J. and Gupta, A. (eds) 1999 *Varieties of fluvial forms*. John Wiley & Sons, Ltd, Chichester.

Miller, G.T. 1998 *Living in the environment*, 10th edn. Wadsworth, Belmont, California.

Milne Home, D. 1872a Scheme for the conservation of remarkable boulders in Scotland, and for the indication of their position on maps. *Proceedings of the Royal Society of Edinburgh*, 7, 475–488.

Milne Home, D. 1872b First report by the Committee on Boulders appointed by the Society. *Proceedings of the Royal Society of Edinburgh*, 7, 703–775.

Milross, J. and Lipkewich, M. 2001 Millenium connections: fostering healthy Earth Science education and industry links. In Gordon, J.E. and Leys, K.F. *Earth science and the natural heritage*. Stationery Office, Edinburgh, 250–254.

Milton, K. 2002 *Loving nature: towards an ecology of emotion*. Routledge, London.

Mitchell, C. 2001 Natural Heritage Zones: planning the sustainable use of Scotland's natural diversity. In Gordon, J.E. and Leys, K.F. (eds) *Earth science and the natural heritage: interactions and integrated management* Stationery Office, Edinburgh, 234–238.

MMSD 2002 *Breaking new ground: the report of the Mining, Minerals and Sustainable Development Project*. Earthscan, London.

Moles, N.R. and Moles, R.T. 2004 Geodiversity as an explanation for biodiversity in The Burren National Park. In Parkes, M. (ed) *Natural and cultural landscapes: the geological foundation* . Royal Irish Academy, Dublin, 61–64.
Monro, S.K. and Davison, D. 2001 Development of integrated educational resources: the Dynamic Earth experience. In Gordon, J.E. and Leys, K.F. *Earth science and the natural heritage*. Stationery Office, Edinburgh, 242–249.
Moreira, J.C. 2012 Interpretative panels about the geological heritage—a case study at the Iguassu Falls National Park (Brazil). *Geoheritage*, 4, 127–137.
Morgan, R.M. and Bull, P.A. 2007 The philosophy, nature and practice of forensic sediment analysis. *Progress in Physical Geography*, 31, 43–58.
Morgan, R.M., Little, M., Gibson, A., *et al.* 2008 The preservation of quartz grain textures following vehicle fire and their use in forensic enquiry. *Science and Justice*, 48, 133–140.
Muir, J. 1901 *Our National Parks*. Houghton Mifflin, Boston.
Murphy, D.D., Freas, K.E. and Weiss, S.B. 1988 An environment–metapopulation approach to population viability analysis for a threatened invertebrate. *Conservation Biology*, 4, 41–51.
Murphy, M. 2001 Minerals in the hands of the collectors. *Earth Heritage*, 15, 14–15.
Murphy, M. 2002 Face lift reclaims geological SSSIs. *Earth Heritage*, 18, 8–9.
Murray, R.C. and Tedrow, J.C.F. 1975 *Forensic geology: Earth sciences and criminal investigation*. Rutgers University Press, New York.
Murray, R.C. and Tedrow, J.C.F. 1992 *Forensic geology*. Prentice Hall, Englewood Cliffs.
Myers, N. 2002 Biodiversity and biodepletion: a paradigm shift. In O'Riordan, T. and Stoll-Kleemann, S. *Biodiversity, sustainability and human communities: protecting beyond the protected*. Cambridge University Press, Cambridge, 46–60.
Nagler, T.F. and Kramers, J.D. 1998. Nd isotopic evolution of the upper mantle during the Precambrian: models, data and the uncertainty of both. *Precambrian Research*, 91, 233–252.
do Nascimento, A.L., Ruchkys, U.A. and Mantesso-Neto, V. 2008 *Geodiversidade, geoconservacao e geotourismo*. Sociedade Brasileira de Geologia.
Nash, R.F. 1990 *The Rights of Nature: a history of environmental ethics*. Primavera Press, New South Wales.
National Research Council 2001 Disposition of high-level waste and spent nuclear fuel: the continuing societal and technical challenges. National Academy Press, Washington DC.
Natural England 2009 *European Landscape Convention*: *Natural England's 2009/2010 Action Plan*. Natural England, Peterborough.
Nature Conservancy Council 1984 *Nature conservation in Great Britain*. Nature Conservancy Council, Peterborough.
Nature Conservancy Council 1990 *Earth science conservation in Great Britain: a strategy* Nature Conservancy Council, Peterborough.
NPCA (1994) *This land is your land*. National Parks Conservation Association, Washington DC.
Newsome, D. and Dowling, R. (eds) 2010 *Geotourism: the tourism of geology and landscape*. Goodfellow Publishers, Oxford.
Nichols, R.F. and Leatherman, S.P. 1995 Sea-level rise and coastal management. In McGregor, D.F.M. and Thompson, D.A. (eds) *1995 Geomorphology and land management in a changing environment*. John Wiley & Sons, Ltd, Chichester, 229–244.
Nichols, W.F., Killingbeck, K.T. and August, P.V. 1998 The influence of geomorphological heterogeneity on biodiversity. II. A landscape perspective. *Conservation Biology*, 12, 371–379.
Nikiforuk, A. 2008 *Tar sands: dirty oil and the future of a continent*. Greystone Books, Vancouver.
Nordic Council of Ministers 2003 *Diversity in nature*. Nordic Council of Ministers, Copenhagen.

Norman, D.B. 1994 Fossil collecting: international issues, perspectives and a prospectus. In O'Halloran, D., Green, C., Harley, M., *et al.* (eds) *Geological and landscape conservation*. Geological Society, London, 63–68.

Norton, B. 1988 Commodity, amenity, and morality: the limits of quantification in valuing biodiversity. In Wilson, E.O. (ed) *Biodiversity*. National Academy Press, Washington DC, 200–205.

Nusipov, E., Fishman, I.L. and Kazakowa, Y.I. 2001 *Geosites of Kazakhstan*. Kazakh Scientific Institute of Mineral Resources.

O'Connor, N.A. 1991 The effect of habitat complexity on the macrovertebrates colonizing wood substrates in a lowland stream. *Oecologia*, 85, 504–512.

O'Donoghue, M 1988 *Gemstones*. Chapman & Hall, London.

O'Halloran, D. 1990 Caves and agriculture: an impact study. *Earth Science Conservation*, 28, 21–23.

O'Halloran, D., Green, C., Harley, M., *et al.* 1994 *Geological and landscape conservation*. Geological Society, London.

Orford, J.D. and Pethick, J.S. 2006 Challenging assumptions of coastal habitat formation in the 21st century. *Earth Surface Processes and Landforms*, 31, 1625–1642.

Oliver, P. and Allbutt, M. 2000 RIGS groups join forces for a face lift. *Earth Heritage*, 14, 22–23.

Outhet, D., Brooks, A., Hader, W., *et al.* 2001 Gravel bed river riffle restoration in New South Wales. In Rutherfurd, I., Sheldon, F., Brierley, G. and Kenyon, C. (eds) *Third Australian Stream Management Conference*, 483–488.

Owens, S. and Cowell, R. 1996 Rocks and hard places: mineral resources planning and Sustainability. *Council for the Protection of Rural* England, London.

Page, K.N. 1998 England's Earth heritage resources: an asset for everyone. In Hooke, J. (ed) *Coastal defence and Earth science conservation*. Geological Society, London, 196–209.

Page, K.N. 2004 The protection of Jurassic sites and fossils: challenges for global Jurassic science (including a proposed statement on the conservation of palaeontological heritage and statotypes). *Revista Italiana di Palaeontologia e Stratigrafia*, 110, 373–379.

Page, K.N. 2005 Radnorshire LANDMAP—Geological Landscapes Aspect Layer: Technical Report. Countryside Council for Wales (for web-based publication).

Page, K.N. (ed) 2008 Geological Landscapes. LANDMAP Methodology: Guidance for Wales. Countryside Council for Wales (http://landmap.ccw.gov.uk/).

Page, K.N. 2011 Consultation on fossil collecting within the 'Jurassic Coast' World Heritage Site, Dorset and East Deven UK: response by ProGEO and the International Subcommission on Jurassic Stratigraphy. *ProGEO News*, 2011(4), 1–7.

Page, K.N., Meléndez, G., Henriques, M.-H. 2008 *Jurassic Global Stratotype Section and Points (GSSPs)—a potential serial World Heritage Site?* Volumina Jurassica, VI, 155–162.

Panizza, M. 2009 The geomorphodiversity of the Dolomites (Italy): a key of geoheritage assessment. *Geoheritage*, 1, 33–42.

Parkes, M.A. 2001 Valencia Island tetrapod trackway—a case study. In Bassett, M.G., King, A.H., Larwood, J.G. and Parkinson, N.A. (eds) *A future for fossils*. National Museums and Galleries of Wales, Cardiff, 71–73.

Parkes, M.A. 2004 Graves and grave stones: geological influence on culture. In Parkes, M. (ed) *Natural and cultural landscapes: the geological foundation*. Royal Irish Academy, Dublin, 151–152.

Parkes, M.A. 2008 A history of geoconservation in the Republic of Ireland. In Burek, C.V. and Prosser, C.D. (eds) *The history of geoconservation*. Geological Society of London, Special Publication 300, 237–248.

Parkes, M.A. and Morris, J.H. 2001 Earth science conservation in Ireland: the Irish Geological Heritage Programme. *Irish Journal of Earth Sciences*, 19, 79–90.

Parks, K.E. and Mulligan, M. 2010 On the relationship between a resource based measure of geodiversity and broad scale biodiversity patterns. *Biodiversity Conservation*, 19, 2751–2766.

Parsons, A.J. 1988 *Hillslope form*. Routledge, London.

Parsons, D.J. 2002 Editorial: understanding and managing impacts of recreation use in mountain environments. *Arctic and Alpine Research*, 34, 363–364.

Passmore, J. 1980 *Man's responsibility for nature*, 2nd edn. Duckworth, London.

Patzak, M. and Eder, W. 1998 "UNESCO Geopark". A new programme—a new UNESCO label. *Geologica Balcanica*, 28, 33–35.

Patzak, M. and Missotten, R. 2010 *UNESCO Geoparks activities and new aspiring geoparks*. Paper presented to the 4th Global Geoparks Conference, Langkawi, Malaysia.

Pavils, G., Nulle, U., Ozola, D. and Markots, A. 2012 Latvia. In Wimbledon, W.A.P. and Smith- Meyer, S. (eds) *Geoheritage in Europe and its conservation*. ProGEO, Oslo, 208–215.

Pearce, D.W. and Turner, R.K. 1990 *Economics of natural resources and the environment*. Harvester Wheatsheaf, Hemel Hempstead.

Pemberton, M. 2001a *Conserving geodiversity, the importance of valuing our geological heritage*. Paper presented to the Geological Society of Australia National Conference, 2001.

Pemberton, M. 2001b Macquarie Island—geodiversity and geoconservation values. Unpublished paper.

Peng, H. and Eder, F.W. 2012 *Danxiashan Global Geopark*. Springer, Heidelberg.

Pereira, D., Pereira, P., Brilha, J., and Santos, L. 2013 Geodiversity assessment of Paraná State (Brazil): an innovative approach. *Environmental Management* (in press).

Pereira, P., Pereira, D. and Caetano-Alves, M.I. 2007 Geomorphosite assessment in Montesinho Natural Park (Portugal). *Geographica Helvetica*, 62(3), 159–168.

Perkins, D. 2002 *Mineralogy*, 2nd edn. Pearson Education, New Jersey.

Perry, D.J. 1962 General report on lands of the Alice Springs area 1956–7. *Land Research Series*, 6, CSIRO, Australia.

Peterborough Environment City Trust 1999 *The Peterborough geology audit*. Peterborough Environment City Trust, Peterborough.

Peterson del Mar, D. 2012 *Environmentalism*. Pearson Education, Harlow.

Pfeffer, K. 2003 Integrating spatio-temporal environmental models for planning ski runs. *Netherlands Geographical Studies*, 311, University of Utrecht.

Phillips, R.H. 2001 Where the limestone meets the sea. Retrieved from Lyric Poetry Website 29 March 2013. http://web.northnet.org/minstrel/limestone.htm.

Phillips, M. and Mighall, T. 2000 *Society and exploitation through nature*. Pearson Education, Harlow.

Piacente, S. and Coratza, P. 2005 Geomorphological sites and geodiversity. *Il Quaternario*, 18(1).

Piccardi, L. and Masse, W.B. (eds) 2007 *Myth and geology*. Geological Society of London, Special Publication 273.

Pigram, J.J. 2000 Options for rehabilitation of Australia's Snowy River: an economic perspective. *Regulated Rivers: Research and Management*, 16, 363–373.

Pimental, D. 1993 *World soil erosion and conservation*. Cambridge University Press, Cambridge.

Pirrie, D., Power, M.R., Rollinson, G.K., *et al*. 2009. Automated SEM-EDS (QEMSCAN) mineral analysis in forensic soil investigations: testing instrumental variability. In Ritz, K., Dawson, I. and Miller, D. (eds) *Criminal and environmental soil forensics*. Springer,Berlin, 411–430.

Pirrie, D. and Rollinson, G.K. 2011. Unlocking the applications of automated mineral analysis. *Geology Today*, 27, 226–235.

Pizzuto, J. 2002 Effects of dam removal on river form and process. *Bioscience*, 52, 683–692.

Plumb, K.A. 1991 New Precambrian time scale. *Episodes*, 14, 139–140.

Pohl, W.L. 2011 *Economic geology: principles and practice*. Wiley-Blackwell, Chichester.
Powell, K. 2002 Open the floodgates. *Nature*, 420, 356–358.
Prasad, K.N. 1994 Geological and landscape conservation in India. In O'Halloran, D., Green, C., Harley, M., *et al.* (eds) *Geological and landscape conservation*. Geological Society, London, 255–258.
Prentice, J.E. 1990 *Geology of construction materials*. Chapman and Hall, London.
Press, F. and Siever, R. 2000 *Understanding Earth*, 3rd edn. W.H. Freeman, New York.
Price, M. 2002a Sacred peak. *Geographical*, 74(5), 9.
Price, R.J. 1989 *Scotland's golf courses*. Aberdeen University Press, Aberdeen.
Price, R.J. 2002b *Scotland's golf courses*, 2nd edn. Mercat Press, Edinburgh.
Pritchard, J.A. 1999 *Preserving Yellowstone's natural conditions: science and the perception of nature*. University of Nebraska Press, Lincoln.
Prosser, C.D. 2001a Spectacular coastline saved. *Earth Heritage*, 16, 4–5.
Prosser, C. 2001b *Geological conservation and the minerals industry—breaking new ground*. In Addison, K. and Reynolds, J.R. (eds) Proceedings of the fourth UK RIGS Conference. UKRIGS, 1–12.
Prosser, C. 2002a Terms of endearment. *Earth Heritage*, 17, 12–13.
Prosser, C. 2002b Terminology: speaking the same language. *Earth Heritage*, 18, 24–25.
Prosser, C. 2003 Going, going, gone ... but not forgotten: Webster's Clay Pit SSSI. *Earth Heritage*, 19, 12.
Prosser, C.D. 2008 The history of geoconservation in England: legislative and policy milestones. In Burek, C.V. and Prosser, C.D. (eds) *The history of geoconservation*. Geological Society of London, Special Publication 300, 113–122.
Prosser, C. 2009 The Piltdown Skull Site: the rise and fall of Britain's first geological National Nature Reserve and its place in the history of nature conservation. *Proceedings of the Geologists' Association*, 120, 79–88.
Prosser, C.D. 2011 Principles and practice of geoconservation: lessons and case law arising from a legal challenge to site based conservation on an eroding coast in eastern England, UK. *Geoheritage*, 3, 277–287.
Prosser, C.D. 2013a Planning for geoconservation in the 1940s: an exploration of the aspirations that shaped the first national geoconservation legislation. *Proceedings of the geologists' Association*, in press.
Prosser, C.D. 2013b Our rich and varied geoconservation portfolio: the foundation for the future. *Proceedings on the Geologists Association, in press.*
Prosser, C.D., Bridgland, D.R., Brown, E.J. and Larwood, J.G. 2011 Geoconservation for science and society: challenges and opportunities. *Proceedings of the Geologists' Association*, 122, 337–342.
Prosser, C.D., Burek, C.V., Evans, D.H., Gordon, J.E., Kirkbride, V.B. Rennie, A.F. and Walmsley, C.A. 2010 Conserving geodiversity sites in a changing climate: management challenges and responses. *Geoheritage*, 2, 123–136.
Prosser, C. and Hughes, I. 2001 CROW flies for geological conservation. *Earth Heritage*, 16, 19.
Prosser, C. and King, A.H. 1999 The conservation of historically important geological and geomorphological sites in England. *The Geological Curator*, 7, 27–33.
Prosser, C., Murphy, M. and Larwood, J. 2006 *Geological conservation: a guide to good practice*. English Nature, Peterborough.
Prosser, I.P., Hughes, A., Rustomji, P., *et al.* 2001 Predictions of the sediment regime of Australian rivers. In Rutherfurd, I., Sheldon, F., Brierley, G. and Kenyon, C. (eds) *Third Australian Stream Management Conference*, 529–536.
Prothero, D.R. and Schwab, F. 1996 *Sedimentary geology*. W.H. Freeman, New York.
Puri, G., Willson, T. and Woolgar, P. 2001 The sustainable use of soil. In Gordon, J.E. and Leys, K.F. *Earth science and the natural heritage*. Stationery Office, Edinburgh, 190–196.
Putnis, A. 1992 *Introduction to mineral sciences*. Cambridge University Press, Cambridge.

Pye, K. and Croft, D. (eds) 2004. *Forensic geoscience: principles, techniques and applications.*Geological Society, London. Special Publication 232.

Ramsay, T., Weber, J., Kollmann, H. and Zouros, N. 2010a Education in European Geoparks. *European Geoparks Magazine*, 7, 14–17.

Ramsay, T., Weber, J., Kollmann, H. and Zouros, N. 2010b Regional development in European Geoparks. *European Geoparks Magazine*, 7, 18–21.

Ramsay, T., Weber, J., Kollmann, H. and Zouros, N. 2010c Research in European Geoparks. *European Geoparks Magazine*, 7, 22–25.

Ramsay, T., Weber, J., Kollmann, H. and Zouros, N. 2010d European Geoparks: destinations for tourism and geotourism. *European Geoparks Magazine*, 7, 26–29.

Raudsep, R. 1994 Geological protected areas and features in Estonia. In O'Halloran, D., Green, C., Harley, M., *et al.* (eds) *Geological and landscape conservation* Geological Society, London, 237–241.

Raudsep, R. 2012 Estonia. In Wimbledon, W.A.P. and Smith-Meyer, S. (eds) *Geoheritage in Europe and its conservation.* ProGEO, Oslo, 106–113.

Raup, D.M. *et al.* 1987 *Palaeontological collecting.* National Academy Press, Washington DC.

Raven, P.J., Fox, P., Everard, M., Holmes, N.T.H. and Dawson, F.H. 1997 River Habitat Survey: a new system for classifying rivers according to their habitat quality. In: Boon, P.J. and Howell, D.L. (eds) *Freshwater quality: defining the indefinable?* Stationery Office, Edinburgh, 215–234.

Read, P.G. 1999 *Gemmology*, 2nd edn. Blackwell, Oxford.

Reading, H.G. 1996 *Sedimentary environments*, 3rd edn. Blackwell Science, Oxford.

Reed, M.A. 1998 Making good use of re-use. *Planning*, 1253, 12–13.

Remane, J., Bassett, M.G., Cowie, J.W., *et al.* 1996 Revised guidelines for the establishment of global chronostratigraphic standards by the International Commission on Stratigraphy. *Episodes*, 19, 77–81.

Reynard, E. 2012 Geoheritage protection and promotion in Switzerland. *European Geologist*, 34, 44–47.

Reynard, E. and Coratza, P. (eds) 2007 Geomorphosites and geodiversity:a new domain of research. *Geographica Helvetica*, 62(3).

Reynard, E. and Coratza, P. (eds) 2011 Geomororphosites and geotourism. *Geoheritage*, 3(3).

Reynard, E., Coratza, P. and Regolini-Bissig, G. (eds) 2009 *Geomorphosites.* Verlag Dr Friedrich Pfeil, Munich.

Rice, C.M., Trewin, N.H. and Anderson, L.I. 2002 Geological setting of the Early Devonian Rhynie cherts, Aberdeenshire, Scotland: an early terrestrial hot spring system. *Journal of the Geological of London*, 159, 203–214.

Richards, K.L., Brasington, J. and Hughes, F. 2002 Geomorphic dynamics of floodplains: ecological implications and a potential modelling strategy. *Freshwater Biology*, 47, 559–579.

Richards, L. 1996 Reclamation—making the best of derelict land. *Earth Heritage*, 6, 16–18.

Richardson, R.D. 1986 *Henry Thoreau: a life of the mind.* University of California Press, Berkeley

Rifkin, J. 2002 *The hydrogen economy.* Polity Press, Cambridge.

Righter, R.W. 2005 *The battle over Hetch Hetchy: America's most controversial dam and the birth of modern environmentalism.* Oxford University Press, Oxford.

Ritchie, D. and Gates, A.E. 2001 *Encyclopedia of earthquakes and volcanoes*, 2nd edn. Checkmark Books, New York.

Rivas, V., Rix, K., Francés, E., Cendrero, A. and Collison, A. 1995a Geomorphology and environmental impact assessment in Spain and Great Britain: a brief review of legislation and practice. In Marchetti, M., Panizza, M., Soldati, M. and Barani, D. (eds)

Geomorphology and environmental impact assessment. CNR–CS Geodinamica Alpina e Quaternaria, Milan, 83–97.

Rivas, V., Rix, K., Francés, E., Cendrero, A. and Brunsden, D. 1995b The use of indicators for the assessment of environmental impacts on geomorphological features. In Marchetti, M., Panizza, M., Soldati, M. and Barani, D. (eds) *Geomorphology and environmental impact assessment*. CNR–CS Geodinamica Alpina e Quaternaria, Milan, 157–180.

Robert, A. 2003 *River processes: an introduction to fluvial dynamics*. Arnold, London.

Roberts, D. 2002 The west coast fossil park: serving science and the community. *Geoclips*, 2, 1–2.

Roberts, P. 2004 *The end of oil: on the edge of a perilous new world*. Houghton Mifflin, Boston.

Robertson, R.M. 1995 *Selected Highland folk tales*. House of Lochar, Colonsay.

Robinson, E. 1984 *London—illustrated geological walks*. Scottish Academic Press, Edinburgh.

Robinson, E. 1985 London—illustrated geological walks, Book 2. Scottish Academic Press, Edinburgh.

Robinson, E. 1987 A geology of the Albert Memorial and vicinity. *Proceedings of the Geologists' Association*, 98, 19–37.

Robinson, E. 1996 'The paths of glory ...' In Bennett, M.R., Doyle, P., Larwood, C.D. and Prosser, C.D. (eds) *Geology on your doorstep*. Geological Society, London, 39–58.

Rochefort, L. and Campeau, S. 1997 Rehabilitation work on post-harvested bogs in South Eastern Canada. In Parkyn, L., Stoneman, R.E. and Ingram, H.A.P. *Conserving peatlands*. CAB International, Oxford, 287–294.

Rodes, B.K. and Odell, R. (1997) *Dictionary of environmental quotations*. Johns Hopkins University Press ISBN 0801857384

Rodwell, J.S. 1991 *British plant communities, vol.* 1: woodlands and scrub. Cambridge University Press, Cambridge.

Röhling, H-G., Schmidt-Thomé, M. and Goth, K. 2012 Germany. In Wimbledon, W.A.P. and Smith- Meyer, S. (eds) *Geoheritage in Europe and its conservation*. ProGEO, Oslo, 132–143.

Rollinson, H. 2007. *Early Earth Systems: a geochemical approach*. Blackwell Publishing, Oxford.

Rose, J. and Letzer, J.M. 1977 Superimposed drumlins. *Journal of glaciology*, 18, 471–480.

Rosengren, N.J. 1994 The Newer Volcanic Province of Victoria, Australia: the use of an inventory of scientific significance in the management of scoria and tuff quarrying. In O'Halloran, D., Green, C., Harley, M., *et al.* (eds) *Geological and landscape conservation*. Geological Society, London, 105–110.

Rosgen, D. 1996 *Applied river morphology*. Wildland Hydrology, Pagosa Springs.

RRC 2002 *Manual of river restoration techniques*. River Restoration Centre, Cranfield.

Ruban, D.A. 2010 Quantification of geodiversity and its loss. *Proceedings of the Geologists' Association*, 121, 326–333.

Ruffell, A. and McKinley, J. 2008 *Geoforensics*. Wiley-Blackwell, Chichester.

Rutherfurd, I., Sheldon, F., Brierley, G. and Kenyon, C. (eds) 2001 *Third Australian stream management conference*. Co-operatice Research Centre for Catchment Hydrology, Monash.

Sacks, J. 2002 *The dignity of difference*. Continuum Books, London.

Saiu, E.M. and McManus, J. 1998 Impacts of coal mining on coastal stability in Fife. In Hooke, J. (ed) *Coastal defence and Earth science conservation*. Geological Society, London, 58–66.

Salminen, R., Kousa, A., Ttesen, R.T., *et al.* 2008 Environmental Geology. *Episodes*, 31, 155–162.

Salo, J., Kalliola, R., Häkkinen, *et al.* 1986 River dynamics and diversity of Amazon lowland forest. *Nature*, 322, 254–258.

Santos, J.J. 2000 *La Palma: history, landscapes and customs*. J.J. Santos, La Palma.
Santucci, V.L. 1999 *Palaeontological resource protection survey report*. National Park Service, Ranger Activities Division and Geolkogic Resources Division.
Santucci, V.L. (ed) 2005 Geodiversity and geoconservation. *George Wright Society*. Forum, 22.
Santucci, V.L. and Hughes, M. 1998 *Fossil Cycad National Monument: a case of palaeontological resource mismanagement*. Technical Report NPS/NRGRD/GRDTR-98/1.
Santucci, V.L., Hunt, A.P. and Norby, L. 2001 Oil and gas management planning and the protection of palaeontological resources. *Park Science*, 21, 36–38.
Santucci, V.L., Kenworthy, J. and Kerbo, R. 2001 *An inventory of palaeontological resources associated with National Park Service caves*. Technical Report NPS/NRGRD/GRDTR-01/02.
Santucci, V.L., Kenworthy, J.P. and Mims, A.L. 2009 Monitoring *in situ* paleontological resources. In Young, R. and Norby, L. (eds) *Geological monitoring*. Geological Society of America, Boulder, 189–204.
Satkunas, J., Lincius, A. and Mikulenus, V. 2012 Lithuania. In Wimbledon, W.A.P. and Smith- Meyer, S. (eds) *Geoheritage in Europe and its conservation*. ProGEO, Oslo, 216–223.
Sax, J.L. 2001 Implementing the public trust in palaeontological resources. In Santucci, V.L. and McClelland, L. (eds) *Proceedings of the 6th Fossil Resource Conference*. Technical Report NPS/NRGRD/GRDTR-01/01, 173–177.
Schullery, P. 1999 *Searching for Yellowstone: ecology and wonder in the last wilderness*. Mariner Books, Boston.
Schumm, S.A. 1979 Geomorphic thresholds: the concept and its applications. *Transactions of the Institute of British Geographers*, NS4, 485–515.
Schumm, S.A. 2005 *River variability and complexity*. Cambridge University Press, Cambridge.
Schumm, S.A., Dumont, J.F. and Holbrook, J.M. 2000 *Active tectonics and alluvial rivers*. Cambridge University Press, Cambridge.
Scottish Natural Heritage 2000a *A guide to managing coastal erosion in beach/dune systems*. Scottish Natural Heritage, Edinburgh.
Scottish Natural Heritage 2000b *Minerals and the Natural Heritage in Scotland's Midland Valley*. Scottish Natural Heritage, Edinburgh.
Scottish Natural Heritage 2000c *Car parks in the countryside: a practical guide to planning, design and construction*. Scottish Natural Heritage, Edinburgh.
Scottish Natural Heritage 2002 Natural Heritage Futures *(21 Prospectuses, 6 Themes and CD ROM)*. Scottish Natural Heritage, Edinburgh.
Scottish Natural Heritage and Scottish Golf Course Wildlife Group 2000 *Nature conservation and golf course development: best practice advice*. Scottish Golf Course Wildlife Group, Dalkeith.
Scottish Wildlife Trust 2002 *Geodiversity: policy summary*. Scottish Wildlife Trust, Edinburgh.
Sear, D.J. 1994 River restoration and geomorphology. *Aquatic Conservation: Marine and Freshwater Ecosystems*, 4, 169–177.
Sear, D.J. and Darby, S. 2008 *River restoration: managing the uncertainty in restoring physical habitat*. John Wiley & Sons, Ltd, Chichester.
Seijmonsbergen, A.C., Sevink, L.H., Cammeraat, L.H. and Recharte, J. 2010 A potential geoconservation map of the Las Lagunas area, Northern Peru, using GIS and remote sensing techniques. *Environmental Conservation*, 37, 107–115.
Seilacher, D. 2008. *Fossil art*. CBM-Publishing, Laasby.
Selby, M.J. 1993 *Hillslope materials and processes*, 2nd edn. Oxford University Press, Oxford.
Selinus, O., Finkelman, R.B. and Centeno, J.A. 2010 *"Medical geology" a regional synthesis*. Springer, Berlin.

Sellars, R.W. 1997 *Preserving nature in the national parks: a history*. Yale University Press, New Haven.

Selley, R.C. 1996 *Ancient sedimentary environments*, 4th edn. Chapman & Hall, London.

Selley, R.C. 2002 Geological and climatic controls on English wines. *GA*, 1(2), 6–7.

Semeniuk, V. 1997 The linkage between biodiversity and geodiversity. In Eberhard, R. (ed) *Pattern and process: towards a regional approach to national estate assessment of geodiversity*. Australian Heritage Commission, Canberra, 51–58.

Serjani, A. 2012 Albania. In Wimbledon, W.A.P. and Smith-Meyer, s. (eds) *Geoheritage in Europe and its conservation*. ProGEO, Oslo, 20–29.

Serrano, E. and Ruiz-Flaño, P. 2007 Geodiversity: a theoretical and applied concept. *Geographica Helvetica*, 62, 140–147.

Sharples, C. 1993 *A methodology for the identification of significant landforms and geological sites for geoconservation purposes*. Forestry Commission Tasmania.

Sharples, C. 1995 Geoconservation in forest management: principles and procedures. *Tasforest*, 7, 37–50.

Sharples, C. 1997 *A reconnaissance of landforms and geological sites of geoconservation significance in the western Derwent Forest District*. Forestry Tasmania.

Sharples, C. 2002a Concepts and principles of geoconservation. pdf document, website, Parks and Wildlife Service, Tasmania.

Sharples, C. 2002b *Some basic principles of geoconservation*. Unpublished paper.

Sharples, C. 2011 *Overview of the report: Potential climate change impacts on geodiversity in the Tasmanian Wilderness World Heritage Area*. Department of Primary Industries, Parks, Water and Environment, Hobart.

Siegesmund, S., Vollbrecht, A. and Weiss, T. (eds) 2001 *Natural stone, weathering phenomena, conservation strategies and case studies*. Geological Society of London, Special Publication, 205.

Sijaric, G. 2012. Bosnia and Herzegovena. In Wimbledon, W.A.P. and Smith-Meyer, s. (eds) *Geoheritage in Europe and its conservation*. ProGEO, Oslo, 63–67.

Silvestru, E. 1994 Karst and environment: a Romanian approach. In O'Halloran, D., Green, C., Harley, M., *et al.* (eds) *Geological and landscape conservation* Geological Society, London, 221–225.

Simpson, J.W. 2005 *Dam!: water, power, politics and preservation in Hetch Hetchy and Yosemite National Park*. Pantheon Books, New York.

Smith, B. 2002 Eating soil. *Planet Earth*, Autumn 2002, 21–22. Natural Environment Research Council, Swindon.

Smith, K. 2000 *Environmental hazards: Assessing risk and reducing disaster*. Routledge, London.

Smith, R.B. 2000 Windows into Yellowstone. *Yellowstone Science*, 8(4), 2–13.

Smith, R.B. and Siegel, L.J. 2000 *Windows into the Earth: the geologic story of Yellowstone and Grand Teton National Parks*. Oxford University Press, Oxford.

Smith-Meyer, S. (1994) Protection of Norwegian river systems. *Memoires de la Societe Geologique de France*, 165, 217–219.

Solarska, A. and Jary, Z. 2010 Geoheritage and geotourism potential of the Strzelin Hills (Sudetic Foreland, SW Poland). *Geographica Pannonica*, 14, 118–125.

Sommers, B.J. 2008 *The geography of wine*. Plume (Penguin), New York.

Sorrell, S., Spiers, J., Bentley, R., Brandt, A. and Miller, R. 2009 *Global oil depletion: an assessment of the evidence for a near-term peak in global oil production*. UK Energy Research Centre, London.

Soulsby, C. and Boon, P.J. 2001 Freshwater environments: an Earth science perspective on the natural heritage of Scotland's rivers. In Gordon, J.E. and Leys, K.F. *Earth science and the natural heritage*. Stationery Office, Edinburgh, 82–104.

Spate, O.H.K. and Learmonth, A.T.A. 1967 *India and Pakistan*. Methuen, London.

Speight, M.C.D. 1973 *Outdoor recreation and its ecological effects*. Discussion Papers in Conservation, 4, University College London.

Sprinkel, D.A., Chidsey, T.C. and Anderson, P.B. 2000 *Geology of Utah's parks and monuments*. Utah Geological Association, Utah.

Stace, H. and Larwood, J. G. 2006 *Natural foundations: geodiversity for people, places and nature*. English Nature, Peterborough.

Stanley, M. 2001 Editorial. *Geodiversity Update*, 1, 1.

Stanley, M. 2004 Geodiversity—linking people, landscapes and their culture. In Parkes, M. (ed) *Natural and cultural landscapes: the geological foundation*. Royal Irish Academy, Dublin, 47–56.

Steers, J.A. 1946 Coastal preservation and planning. *Geographical Journal*, 107, 57–60.

Stephens Stephenson (1998) Case study: slate quarries. *Landscape Design*, 269, 15.

Stephenson, M. and Penn, I. 2003 Rebuilding a survey. *Geoscientist*, 13(3), 14–16.

Stevens, C., Gordon, J.E., Green, C.P. and Macklin, M.G. (eds) 1994 *Conserving our landscape*, Peterborough, 41–45.

Stewart, G.A. and Perry, R.A. 1953 Survey of Townsville-Bowen Region (1950). *Land Research Series*, 2 (CSIRO 1120, Australia).

Stewart, I. and Lynch, J. 2007 *Earth: the power of the planet*. BBC Books, London.

Stewart, I. and Nield, T. 2013 Earth stories: context and narrative in the communication of popular geoscience. *Proceedings of the Geologists' Association*, in press.

Stock, E. 1997 Geo-processes as heritage. In Eberhard, R. (ed) *Pattern and process: towards a regional approach to national estate assessment of geodiversity*. Australian Heritage Commission, Canberra, 41–50.

Stout, G. 2002 *Newgrange and the Bend of the Boyne*. Cork University Press, Cork.

Stürm, B. 1994 The geotope concept: geological nature conservation by town and country planning. In O'Halloran, D., Green, C., Harley, M., *et al.* (eds) *Geological and landscape conservation*. Geological Society, London, 27–31.

Stürm, B. 2012 Switzerland. In Wimbledon, W.A.P. and Smith-Meyer, S. (eds) *Geoheritage in Europe and its conservation*. ProGEO, Oslo, 358–365.

Sugden, D.E. 1982 *Arctic and Antarctic: a modern geographical synthesis*. Blackwell, Oxford.

Sugden, D.E. and John, B.S. 1976 *Glaciers and landscape: a geomorphological approach*. Edward Arnold, London.

Summerfield, M.A. 1991 *Global geomorphology*. Longman, Harlow.

Summerfield, M.A. 2000. *Geomorphology and global tectonics*. John Wiley & Sons, Ltd, Chichester.

Sutherland, D.G. 1984 Geomorphology and mineral exploration: some examples from exploration of diamondiferous placer deposits. *Zeitschrift fur Geomorphology, Supplementary Band*, 51, 95–108.

Swanwick, C. and Land Use Consultants 1999 *Interim landscape character assessment guidance*. Countryside Agency and Scottish Natural Heritage.

Swanwick, C. and Land Use Consultants 2002 *Landscape character assessment: guidance for Scotland and England*. Countryside Agency and Scottish Natural Heritage.

Swart, R. 1994 Conservation and management of geological monuments in South Australia. In O'Halloran, D., Green, C., Harley, M., *et al.* (eds) *Geological and landscape conservation*. Geological Society, London, 319–322.

Taylor, A.G. 1997 *Soils and land use planning*. Information and Advisory Note, 85. Scottish Natural Heritage, Edinburgh.

Taylor, G. and Eggleton, R.A. 2001 *Regolith geology and geomorphology*. John Wiley & Sons, Ltd, Chichester.

Taylor, S.R. and McLennan, S.M. 1985. *The continental crust: its composition and evolution*. Blackwell Scientific Publications, Oxford.

Theodosiou, I. 2012 Greece. In Wimbledon, W.A.P. and Smith-Meyer, S. (eds) *Geoheritage in Europe and its conservation*. ProGEO, Oslo, 144–157.

Therivel, R. and Thompson, S. 1996 *Strategic environmental assessment and nature conservation*. English Nature, Peterborough.

Thomas, A. 2008 *Gemstones: properties, identification and use*. New Holland Publishers, London.

Thomas, B.A. and Warren, L.M. 2008 Geological conservation in the nineteenth and early twentieth centuries. In Burek, C.D. and Prosser, C.D. (eds) *The history of geoconservation*. Geological Society of London, Special Publication 300, 17–30.

Thomas, D.S.G. 2011 *Arid zone geomorphology: process, form and change in drylands*. Wiley-Blackwell, Chichester.

Thomas, L. 2002 *Coal geology*. John Wiley & Sons, Ltd, Chichester.

Thomas, L. and Middleton, J. 2003 Guidelines for management planning of protected areas. World Commission on Protected Areas, Best Practice Protected Area Guidelines Series, No. 10. IUCN and Cardiff University.

Thompson, A., Poole, J., Carroll, L., Foweraker, M., Harris, K. and Cox, P. 2006 *Geodiversity Action Plans for aggregate companies: a guide to good practice*. Capita Symonds Ltd, East Grinstead.

Thomson, J.R., Taylor, M.P., Fryirs, K.A. and Brierley, G.J. 2001. A geomorphological framework for river characterization and habitat assessment. *Aquatic Conservation: Marine and Freshwater Ecosystems*, 11, 373–389.

Thorne, C.R. 1998 *Stream reconnaissance handbook: geomorphological investigation and analysis of river channels*. John Wiley & Sons, Ltd, Chichester.

Thorvardardottir, G. and Thoroddsson, T.F. 1994 Protected volcanoes in Iceland: conservation and threats. In O'Halloran, D., Green, C., Harley, M., *et al.* (eds) *Geological and landscape conservation*. Geological Society, London, 227–230.

Todorov, T. and Nakov, R. 2012. Bulgaria. In Wimbledon, W.A.P. and Smith-Meyer, S. (eds) *Geoheritage in Europe and its conservation*. ProGEO, Oslo, 68–79.

Toghill, P. 1972 Geological conservation in Shropshire. *Journal of the Geological Society*, 128, 513–515.

Toth, L.A. 1996 Restoring the hydrogeomorphology of the channelized Kissimmee River. In Brookes, A. and Shields, F.D. (eds) 1996a *River channel restoration*. John Wiley & Sons, Ltd, Chichester, 369–383.

Trewin, N.H. 2001 Scotland's foundations: our geological inheritance. In Gordon, J.E. and Leys, K.F. *Earth science and the natural heritage*. Stationery Office, Edinburgh, 59–67.

Trimmel, H. 1994 Sixty-five years of legislative cave conservation in Austria: experience and results. In O'Halloran, D., Green, C., Harley, M., *et al.* (eds) *Geological and landscape conservation*. Geological Society, London, 213–214.

Trudgill, S. 2001 *The terrestrial biosphere: environmental change, ecosystem science, attitudes and value*. Pearson Education, Harlow.

Tucker, M.E. 1996 *Sedimentary rocks in the field*, 2nd edn. John Wiley & Sons, Ltd, Chichester.

Tucker, M.E. 2001 *Sedimentary petrology*, 3rd edn. Blackwell Science, Oxford.

Turner, B.L., Clark, W.C., Kates, R.W., Richards, J.F., Mathews, J.T. and Meyer, W.B. (eds) 1990 *The Earth as transformed by human action*. Cambridge University Press, Cambridge.

Turner, T. 1998 *Landscape planning and environmental impact design*, 2nd edn. UCL Press, London.

UK National Ecosystem Assessment 2011 *National Ecosystem Assessment: synthesis of key findings. Department of Environment Food and Rural* Affairs, London.

UNEP-WCMC 2002 *Mountain watch*. United Nations Environment Programme/World Conservation Monitoring Centre, Cambridge.

US Department of the Interior 1971 *Impacts of strip mining. US Department of the Interior*, Washington, DC .

Usher, M.B. 2001 Earth science and the natural heritage: a synthesis. In Gordon, J.E. and Leys, K.F. *Earth science and the natural heritage*. Stationery Office, Edinburgh, 315–324.

Valiūnas, J. 1994 Environmental geology maps for national parks and geomorphological reserves in Lithuania. In O'Halloran, D., Green, C., Harley, M., *et al.* (eds) *Geological and landscape conservation*. Geological Society, London, 273–277.

Varnes, D.J. 1978 Slope movement and types and processes. In Schuster, R.L. and Krizek (eds) *Landslides: analysis and control. Transportation Research Board, Special Report, 176*, 11–33, National Academy of Sciences, Washington DC.

Vasileva, D. 1997 Devil's town: truth and legend. *Earth Heritage*, 8, 17.

Vasiljevic, D., Markovic, S.B., Hose, T.A., *et al.* 2009 The use of web-based dynamic maps in the promotion of the Titel Loess Plateau (Vojvodinas, Serbia), a potential geotourism destination. *Geographica Pannonica*, 13, 78–84.

Vasiljevic, D.A., Markovic, S.B., Hose, T.A., *et al.* 2010 The introduction to geoconservation of loess-palaeosol sequences in the Vojvodina region: significant geoheritage of Serbia. *Quaternary International*, 240(1–2), 108–116.

Vernon, R.H. and Clarke, G.L. 2008 *Principles of metamorphic petrology*. Cambridge University Press, Cambridge.

Viney, C. 2001 *Macquarie Island*. Department of Primary Industries, Water and Environment, Tasmania.

Vinokurov, V. and Goldenkov, A. 2012 Belarus. In Wimbledon, W.A.P. and Smith-Meyer, S. (eds) *Geoheritage in Europe and its conservation*. ProGEO, Oslo, 40–51.

Vitaliano, D.B. 1973 *Legends of the Earth: their geological origins*. Indiana University Press, Bloomington.

Vitaliano, D.B. 2007 Geomythology: geological origins of myths and legends. In Piccardi, L. and Messe, W.B. (eds) *Myth and geology*. Geological Society of London, *Special Publication*, 273, 1–7.

de Waal, L.C., Large, A.R.G. and Wade, P.M. (eds) *Rehabilitation of rivers: principles and implementation*. John Wiley & Sons, Ltd, Chichester.

Walliser, O.H., Bultynck, P., Weddige, K., *et al.* 1995, Definition of the Eifelian-Givetian Stage boundary. *Episodes*, 18, 107–115.

Walsh, S.L., Gradstein, F.M. and Ogg, J.G. 2004 History, philosophy, and application of the Global Stratotype Section and Point (GSSP). *Lethaia*, 37, 201–218.

Wang, S. Lee, K.C. and King, A. 1999 Eastern promise. *Earth Heritage*, 12, 12–13.

Wang, S., Sheu, L.-Y. and Tang, H.Y. 1994 Conservation of geomorphological landscapes in Taiwan. In O'Halloran, D., Green, C., Harley, M., *et al.* (eds) *Geological and landscape conservation*. Geological Society, London, 113–115.

Ward, S.D. and Evans, D.F. 1976 Conservation assessment of British Limestone pavements based on floristic criteria. *Biological Conservation*, 9, 217–233.

Ward, P.D. and Brownlee, D. 2000 *Rare Earth: why complex life is uncommon in the universe*. Copernicus Books, New York.

Warnock, S. and Brown, N. 1998 A vision for the countryside. *Landscape Design*, 269, 22–26.

Warren, A. 2001 Valley-side slopes. In Warren, A. and French, J.R. *Habtiat conservation: managing the physical environment*. John Wiley & Sons, Ltd, Chichester, 39–66.

Warren, A. and French, J.R. 2001 Relations between nature conservation and the physical environment. In Warren, A. and French, J.R. *Habitat conservation: managing the physical environment*. John Wiley & Sons, Ltd, Chichester, 1–5.

Warren, C. 2002 Of superquarries and mountain railways: recurring themes in Scottish environmental conflict. *Scottish Geographical Journal*, 118, 101–127.

Washburn, A.L. 1979 *Geocryology: a survey of periglacial processes and environments*. Edward Arnold, London.

Waterton, C., Ellis, R. and Wynne, B. 2012 *Barcoding nature: shifting taxonomic practices in an age of biodiversity loss*. Routledge, London.

Watson, A. 1984 A survey of vehicular tracks in North-east Scotland for land use planning. *Journal of Environmental Management*, 18, 345–353.

Webber, M. 2001 The sustainability of a threatened fossil resource: Lower Jurassic *Caloceras Beds of Doniford Bay, Somerset*. In Bassett, M.G., King, A.H., Larwood, J.G. and Parkinson, N.A. (eds) *A future for fossils*. National Museums and Galleries of Wales, Cardiff, 108–112.

Webber, M., Christie, M. and Glasser, N. 2006 The social and economic value of the UK's geodiversity. *English Nature Research Report* 709.

Wells, R.T. 1996 *Earth's geological history: a conceptual framework for assessment of World Heritage fossil site nominations*. IUCN, Gland.

Welsh Government, 2012. *Sustaining a living Wales: a green paper on a new approach to natural resource management*. Welsh Government, Cardiff.

Wenk, H.-R. and Bulakh, A. 2004 *Minerals: their constitution and origin*. Cambridge University Press.

Werritty, A. and Brazier, V. 1991 *The geomorphology, conservation and management of the River Feshie SSSI*. Nature Conservancy Council Report.

Werritty, A. and Brazier, V. 1994 Geomorphic sensitivity and the conservation of fluvial geomorphology SSSIs. In Stevens, C., Gordon, J.E., Green, C.P. and Macklin, M.G. (eds) *Conserving our landscape*, English Nature, Peterborough, 100–109.

Werritty, A., Duck, R.W. and Kirkbride, M.P. 1998 Development of a conceptual and methodological framework for monitoring site condition in geomorphological systems. Research, Survey and Monitoring Report 105, Scottish Natural Heritage, Edinburgh.

Werritty, A. and Leys, K.F. 2001 The sensitivity of Scottish rivers and upland valley floors to recent environmental change. *Catena*, 42, 251–274.

Westing, A. and Pfeiffer, E.W. 1972 The cratering of Indo China. *Scientific American*, 226(5), 21–29.

Wharton, G. and Gilvear, D. 2007 River restoration in the UK: meeting the needs of both the EU Water Framework Directive and flood defence? *International Journal of River Basin Management*, 5, 143–154.

White, B. 1997 *Principles and practice of soil science*, 3rd edn. Blackwell, Oxford.

White, L. 1967 The historical roots of our ecological crisis. *Science*, 155, 1203–1207.

Whiteley, M.J. and Browne, M.A.E 2013 Local geoconservation groups—past achievements and future challenges. *Proceedings of the Geologists' Association*, in press.

Whitney, G.G. 1994 *From coastal wilderness to fruited plain: a history of environmental change in temperate North America from 1500 to the present*. Cambridge University Press, Cambridge.

Wiedenbein, F.W. 1993 Ein Geotopschutzkonzept für Deutschland. In Quasten, H. (ed) *Geotopschutz, probleme der methodik und der praktischen umsetzung*. 1. Jahrestagung der AG Geotopschutz, Otzenhausen/Saarland, 17. University de Saarlandes, Saarbrucken.

Wiedenbein, F.W. 1994. Origin and use of the term 'geotope' in German-speaking coutries. In O'Halloran, D., Green, C., Harley, M., *et al.* (eds) Geological and landscape conservation *Geological Society*, London, 117–120.

Wilde, S., Valley, J., Peck, W. and Graham, C. 2001 Evidence from detrital zircons for the existence of continental crust and oceans on the Earth at 4.4 Gyr ago. *Nature*, 409, 175–178.

Wilkinson, C. (ed) 2008 *The Observer book of the Earth*. Observer Books, London.

Wilkinson, S.N., Keller, R.J. and Rutherfurd, I.D. 2001 Rehabilitating pool and riffle sequences: the role of alternate bar dynamics and flow constriction. In Rutherfurd, I., Sheldon, F., Brierley, G. and Kenyon, C. (eds) *Third Australian Stream Management Conference*, 681–688.

Williams, P. 2008 *World Heritage caves and karst: a thematic study*. IUCN, Gland.

Wilson, C. (ed) 1994 *Earth heritage conservation*. Geological Society, London and Open University, Milton Keynes.

Wilson, E.O. (ed) 1988 *Biodiversity*. National Academy Press, Washington DC.

Wilson, E.O. 1997 Introduction. In Reaka-Kudla, L., Wilson, D.E. and Wilson, E.O. (eds) *Biodiversity II: understanding and protecting our biological resources*. Joseph Henry Press, Washington DC, 1–3.

Wilson, J.E. 1998 *Terroir: the role of geology, climate and culture in the making of French wines*. Mitchell Beazley, London.

Wimbledon, W.A.P. 2012 International Geoheritage under threat. *ProGEO News*, 2012(4), 4–5.

Wimbledon, W.A.P. and Smith-Meyer, S. (eds) 2012 *Geoheritage in Europe and its conservation*. ProGEO, Oslo,.

Winchester, S. 2002 *The map that changed the World*. Penguin, London.

Windley, B.F. 1995 *The evolving continents*, 3rd edn. John Wiley & Sons Ltd, Chichester.

Withington, C.F. 1998. Building stones of our nation's capital. *US Department of the Interior and US Geological* Survey, Washington DC.

Whittow, J. 1992 *Geology and scenery in Britain*. Chapman & Hall, London.

Wood, C. 2009 *World Heritage volcanoes: a thematic study*. IUCN, Gland.

Woodcock, N. 1994 *Geology and the environment in Britain and Ireland*. UCL Press, London.

Worboys, G.L. and Winkler, C. 2006 Natural heritage. In Lockwood, M., Worboys, G.L. and Kothari, A. (eds) *2006 Managing protected areas: a global guide*. Earthscan, London, 3–40.

Wordsworth, W. 1952 *A guide through the district of the lakes*. Indiana University Press, Bloomington.

Workman, D.R. 1994 Geological and landscape conservation in Hong Kong. In O'Halloran, D., Green, C., Harley, M., *et al.* (eds) *Geological and landscape conservation*. Geological Society, London, 291–296.

Wrede, V. and Mügge-Bartolovic, V. 2012 GeoRoute Ruhr—a network of geotrails in the Ruhr Area National GeoPark, Germany. *Geoheritage*, 4, 109–114.

Wright, J.R. and Price, G. 1994 Cave conservation plans: an integrated approach to cave conservation. In Stevens, C., Gordon, J.E., Green, C.P. and Macklin, M.G. (eds) *Conserving our landscape*, English Nature, Peterborough, 195–197.

Xu, X., Wang, K., Zhang, K., *et al.* 2012 A gigantic feathered dinosaur from the Lower Cretaceous of China. *Nature*, 484,92–95.

Xu, X., Zhou, Z., Wang, X., *et al.* 2003 Four-winged dinosaurs from China. *Nature*, 421, 335–340.

Xun, Z. and Milly, W. 2002 National geoparks initiated in China: putting geoscience in the service of society. *Episodes*, 25, 33–37.

Young, R. and Norby, L. (eds) 2009 *Geological monitoring*. Geological Society of America, Boulder.

Yuan, X., Chen, Z., Xiao, S., *et al.* 2011. An early Ediacaran assemblage of macroscopic and morphologically differentiated eukaryotes, *Nature*, 470, 390–393.

Zhao, Z.-Z., Long, C.-X., Yuan, X.-H and Zheng, Y.-Y. 2012 Global geoparks in China. In Så, A.A., Rocha, D., Paz, A. and Correia, V (eds) *Proceedings of the 11th European Geoparks Conference*, Arouca Geopark Association (AGA), Arouca.

Zouros, N. (ed) 2008 *European Geoparks*. Natural History Museum of Lesvos Petrified Forest, Greece.

Zouros, N. 2012 Measuring progress in European Geoparks: a contribution for a smart, sustainable and inclusive growth of Europe. In Så, A.A., Rocha, D., Paz, A. and Correia, V (eds) *Proceedings of the 11th European Geoparks Conference*, Arouca Geopark Association (AGA), Arouca.

Zouros, N. and McKeever, P. 2008 European Geoparks: tools for Earth Heritage protection and sustainable local development. In Zouros, N. (ed) *European Geoparks*. Natural History Museum of Lesvos Petrified Forest, Greece.

Index

Geodiversity: Valuing and Conserving Abiotic Nature, Second Edition. Murray Gray.